Therapeutic Potentials of Curcumin for Alzheimer Disease

Akhlaq A. Farooqui

Therapeutic Potentials of Curcumin for Alzheimer Disease

 Springer

Akhlaq A. Farooqui
Department of Molecular and Cellular Biochemistry
The Ohio State University
Columbus, OH, USA

ISBN 978-3-319-15888-4   ISBN 978-3-319-15889-1   (eBook)
DOI 10.1007/978-3-319-15889-1

Library of Congress Control Number: 2016931435

Springer Cham Heidelberg New York Dordrecht London

Printed on acid-free paper

Springer International Publishing AG Switzerland is part of Springer Science+Business Media (www.springer.com)

*This monograph is dedicated to my beloved
father "the late Sharafyab Ahmed Saheb"
whose guidance and influence continue
to inspire and support me.*

Preface

Curcumin, a hydrophobic polyphenol, is the yellow pigment in the Indian spice turmeric (curry powder). It is derived from the rhizome of the herb *Curcuma longa*, which belongs to the family Zingiberaceae. Curcumin is sensitive to degradation by visible and ultraviolet light, as well as high pH and oxygen. It has a half-life of around 8 hours in the human blood. Curcumin possesses antioxidant, antiinflammatory, antimicrobial, anticarcinogenic, antihypertensive, antihyperlipidemic, antidiabetic, antipsoriasis, antithrombotic, antihepatotoxic, and neuroprotective properties. Curcumin has been used extensively in Ayurvedic medicine (Indian system of medicine) and Chinese traditional medicine for centuries as an antiinflammatory, antinociceptive, and antishock agent to relieve pain and inflammation not only in the skin and muscles but also in the treatment of numerous pathological conditions, including rheumatism, digestive and inflammatory disorders, intermittent fevers, urinary discharges, and leukoderma as part of traditional medicine. Curcumin acts not only by inhibiting oxidative stress and neuroinflammation, reversing the amyloid pathology, but also by modulating synaptic plasticity and inducing neurogenesis in the hippocampus of transgenic mouse model of Alzheimer disease (AD), a neurodegenerative disorder, which is characterized by the deposition of senile plaques and neurofibrillary tangles, the loss of neurons, and synapses, along with memory impairment and cognitive deficit. Direct injections of curcumin into the brains of the mice with AD not only delay further development of plaque but also reduce the levels of senile plaque. Curcumin not only retards Aβ-mediated oxidative stress but also inhibits the activation of NF-κB and prevents Aβ-induced cell death in a human neuroblastoma cell line supporting the view that regular consumption of curcumin from childhood to adulthood may reduce the risk of developing AD, and thus, curcumin can be used as a potential therapeutic agent for the treatment of AD.

Information on the treatment potentials of curcumin for AD is scattered throughout the literature in the form of original papers, reviews, and few edited books, which are focused on the effect of curcumin on anticarcinogenic activities of curcumin in visceral tissues. At present there are no books on the effects of curcumin on the brain and potential use of curcumin for the treatment of AD. The overarching objective of this monograph is to provide readers with a comprehensive and cutting-edge information on the effects of curcumin on the brain in a manner that is not only useful to students and teachers but also to researchers, nutritionists, and physicians. This monograph has ten chapters. The first chapter describes information on neurochemical changes in AD. Chapter 2 provides information on potential animal models of AD and their importance in investigating the pathogenesis of AD. Chapter 3 deals with metabolism, bioavailability, and biochemical effects of curcumin in visceral organs and the brain. Chapter 4 focuses on cutting-edge information on the effects of curcumin on transcription factors and enzyme activities in visceral organs and the brain. Chapter 5 describes the effect of curcumin on growth factors and their receptors, ion channels, and transporters in the visceral organs and the brain. Chapter 6 narrates the effect of curcumin on oxidative stress in animal models of AD and in AD patients. Chapter 7 describes the cutting-edge information on the effect of curcumin on neuroinflammation in animal models of AD and in AD patients. Chapter 8 provides readers with information on the therapeutic importance of curcumin in neurological disorders other than AD. Chapter 9 deals with cutting-edge information on treatment of AD with phytochemicals other than curcumin. Finally, Chap. 10 deals with perspective and direction for future research on the potential use of curcumin for the treatment of human AD.

My presentation and demonstrated ability to present complicated information on signal transduction processes associated with effects of curcumin on the brain will make this book particularly accessible to neuroscience graduate students, teachers, and fellow researchers. It can be used as a supplemental text for neuroscience, nutrition, and neurochemistry courses. Clinicians, neuroscientists, neurologists, neurochemists, and nutritionists will find this book useful for understanding the molecular aspects of the potential use of curcumin for AD treatment. To the best of my knowledge, this monograph will be the first to provide a comprehensive description of signal transduction processes associated with beneficial effects of curcumin in the brain.

The choices of topics presented in this monograph are personal. They are not only based on my interest on the effects of curcumin on the brain but also in areas where major progress has been made. The key objective of this monograph is to critically evaluate the effect of curcumin on metabolic processes in the brain. Each chapter of this monograph contains an extensive list of references, which are arranged alphabetically to works that are cited in the text. I have tried to ensure uniformity and mode of presentation as well as a logical progression of subjects from one topic to another and have provided an extensive bibliography. For the sake of simplicity and uniformity, a large number of figures with chemical structures of curcumin and its analogs along with line diagrams of colored signal transduction pathways are also included. I hope that my attempt to integrate and consolidate the knowledge on the effect of curcumin in the brain will initiate more studies on molecular mechanisms associated with beneficial effects of curcumin on human health in general and the brain in particular. This knowledge will be useful for the optimal health of young, boomer, and pre-boomer American generations.

Columbus, OH, USA   Akhlaq A. Farooqui

Acknowledgments

I thank my wife, Tahira, for the critical reading of this monograph, offering valuable advice, useful discussion, and evaluation of the subject matter. Without her help and participation, this monograph neither could nor would have been completed. I would also like to express my gratitude to Gregory Baer of Springer, New York, for his quick responses to my queries and professional manuscript handling.

Contents

About the Author

Dr. Akhlaq A. Farooqui is a leader in the field of signal transduction; brain phospholipases A$_2$; bioactive ether lipid metabolism; polyunsaturated fatty acid metabolism; glycerophospholipid-, sphingolipid-, and cholesterol-derived lipid mediators; glutamate-induced neurotoxicity; and modulation of signal transduction by phytochemicals. Dr. Farooqui has discovered the stimulation of plasmalogen-selective phospholipase A$_2$ (PlsEtn-PLA$_2$) and diacyl- and monoacylglycerol lipases in brains from patients with Alzheimer disease. Stimulation of PlsEtn-PLA$_2$ produces plasmalogen deficiency and increases levels of eicosanoids that may be related to the loss of synapses in brains of patients with Alzheimer disease. Dr. Farooqui has published cutting-edge research on the generation and identification of glycerophospholipid-, sphingolipid-, and cholesterol-derived lipid mediators in kainic acid-mediated neurotoxicity by lipidomics. Dr. Farooqui has authored eleven monographs: *Glycerophospholipids in Brain: Phospholipase A$_2$ in Neurological Disorders* (2007); *Neurochemical Aspects of Excitotoxicity* (2008); *Metabolism and Functions of Bioactive Ether Lipids in Brain* (2008); *Hot Topics in Neural Membrane Lipidology* (2009); *Beneficial Effects of Fish Oil in Human Brain* (2009); *Neurochemical Aspects of Neurotraumatic and Neurodegenerative Diseases* (2010); *Lipid Mediators and Their Metabolism in the Brain* (2011); *Phytochemicals, Signal Transduction, and Neurological Disorders* (2012); *Metabolic Syndrome: An Important Risk Factor for Stroke, Alzheimer Disease, and Depression* (2013); *Inflammation and Oxidative Stress in Neurological Disorders* (2014); and *High Calorie Diet and the Human Brain* (2015). All monographs are published by Springer, New York.

In addition, Dr. Akhlaq A. Farooqui has edited 8 books (*Biogenic Amines: Pharmacological, Neurochemical and Molecular Aspects in the CNS* (2010), Nova Science Publisher, Hauppauge, N.Y.; *Molecular Aspects of Neurodegeneration and Neuroprotection*, Bentham Science Publishers Ltd (2011); *Phytochemicals and Human Health: Molecular and Pharmacological Aspects* (2011), Nova Science Publisher, Hauppauge, N.Y.; *Molecular Aspects of Oxidative Stress on Cell Signaling in Vertebrates and Invertebrates* (2012), Wiley Blackwell Publishing Company, New York; *Beneficial Effects of Propolis on Human Health in Chronic*

Diseases (2012), Vol. 1, Nova Science Publishers, Hauppauge, New York; *Beneficial Effects of Propolis on Human Health in Chronic Diseases* (2012), Vol. 2, Nova Science Publishers, Hauppauge, New York; *Metabolic Syndrome and Neurological Disorders* (2013), Wiley Blackwell Publishing Company, New York; and *Diet and Exercise in Cognitive Function and Neurological Diseases* (2015), Wiley Blackwell Publishing Company, New York).

List of Abbreviations

AD	Alzheimer disease
AGE	Advanced glycation end products
ALS	Amyotrophic lateral sclerosis
APP	Amyloid precursor protein
ARA	Arachidonic acid
$A\beta$	Beta-amyloid
BBB	Blood-brain barrier
CAT	Catalase
COX	Cyclooxygenase
CR	Calorie restriction
GPx	Glutathione peroxidase
HD	Huntington disease
IL-1β	Interleukin-1beta
Ins-1,4,5-P_3	Inositol-1,4,5-trisphosphate
LOX	Lipoxygenase
LTP	Long-term potentiation
LTs	Leukotrienes
NO	Nitric oxide
NOS	Nitric oxide synthase
PD	Parkinson disease
PG	Prostaglandins
PKC	Protein kinase C
PLA_2	Phospholipase A_2
PtdCho	Phosphatidylcholine
PtdEtn	Phosphatidylethanolamine
PtdIns	Phosphatidylinositol
PtdIns(4,5)P_2	Phosphatidylinositol 4,5-bisphosphate
PtdIns4P	Phosphatidylinositol 4-phosphate
SOD	Superoxide dismutase
TNF-α	Tumor necrosis factor-alpha
TXs	Thromboxanes

Chapter 1
Neurochemical Aspects of Alzheimer Disease

1.1 Introduction

Alzheimer disease (AD), a major progressive neurodegenerative disorder, encountered by more than 37 million people worldwide with numbers expected to increase substantially over the next several decades. It is 6th leading cause of death in the United States victimizing about 5.5 million in the year 2012 (Alzheimer's Association 2012). By 2050, this number is expected to jump to 16 million, and in the next 20 years it is projected that Alzheimer's will affect one in four Americans, rivaling the current prevalence of obesity and diabetes (Brookmeyer et al. 1998). In 2010, the annual US health care cost for the treatment of AD reached about US$144 billion (Alzheimer's Association 2010). It is predicted that by 2050, the number of people with AD and other forms of dementia in the U.S. will be tripled. Data from Alzheimer's Disease International forecast that the worldwide prevalence of AD will double every 20 years to 65.7 million by 2030 and 115.4 million by 2050 with higher proportions in the developed versus undeveloped countries (Kalaria et al. 2008). This increase in the number of AD patients is not only attributed to increase in longevity, long term consumption of processed food (western diet), but also to the lack of exercise, and environmental exposure to neurotoxins (Alzheimer's Association 2010, 2012, 2013; Farooqui 2010a, 2015). Aging is an important risk factor for the late-onset cases of AD. Late onset AD has an inevitable manifestation of senescence in that the disease, with its neuropathological signatures, is considered a normal phenomenon of aging (Reser 2009). However, there are differences between the pattern of neuronal loss in normal aging and AD, supporting the view that the latter is not an inevitable consequence of the aging process (West et al. 1994). Recent studies have also indicated that much of late-life cognitive decline (60 %) may not be due to common neurodegenerative pathologies such as plaques and NFTs, suggesting that other determinants (epigenetic changes in the brain) of AD are still to be identified (Brewer 2010; Boyle et al. 2013). Studies on comparison of adult neurons from young and old rodent have indicated that adult neurons

© Springer International Publishing Switzerland 2016

A.A. Farooqui, *Therapeutic Potentials of Curcumin for Alzheimer Disease*,

DOI 10.1007/978-3-319-15889-1_1

from young and old rodent brains have similar viability, resting glucose uptake, and resting respiration in culture, but show increased sensitivity to stressors such as Aβ or glutamate despite the uniform culture environment (Brewer 1998), supporting the view that epigenetic changes may be closely associated with aging process. Age-mediated perturbing changes, which influence neuron glial network activity include: (a) increased oxidative damage to cellular lipids, proteins, and nucleic acids (Mattson 2004); (b) reduction in neurotrophic factor signaling (Zuccato and Cattaneo 2009); and (c) dysregulation of neuronal calcium homeostasis due to alterations in ion homeostasis (Bezprozvanny and Mattson 2008). Oxidative stress has been reported to influence neuronal excitability by inhibiting ion-motive ATPases, modifying ligand- and voltage-gated ion channels and altering neurotransmitter signaling pathways (Gleichmann and Mattson 2011). Other risk factors for AD are hypertension, stroke, heart diseases, impairment of cerebral blood flow, depression, type II diabetes, and smoking (Ballard et al. 2011; Reitz and Mayeux 2014). Many theories have been proposed to explain the degeneration of neurons in AD including: (a) selective vulnerability of cholinergic neurons in the basal forebrain, (b) a viral agent, (c) aluminum deposits, (d) lack of trophic factors, (e) protein misfolding and aggregation along with mitochondrial dysfunction, and (f) induction of insulin resistance (Katzman and Saitoh 1991; Farooqui 2013; de la Monte and Tong 2014).

Neuropathologically AD is characterized by the presence of hyperphosphorylated Tau protein containing neurofibrillary tangles (NFTs) and accumulation of beta amyloid (Aβ) containing plaques, which consist of degenerating and frequently swollen axons, neurites and glia cells (Alzheimer's Association 2010, 2012, 2013). Senile plaques are extracellular deposits of the Aβ peptides, and neurofibrillary tangles accumulate intracellularly. Among various regions, hippocampus and entorhinal cortex show the accumulation of NFTs and Aβ containing plaques. The hippocampus, a brain area critical for learning and memory, is especially vulnerable to damage at early stages in AD. It is shown that 25–70 % of neurons are lost in the hippocampus and entorhinal cortex of AD patients (Jahn 2013). This neuronal loss correlates well with the cognitive memory impairment in AD patients. The memory loss is one of the earliest symptoms in AD patients due to the destruction of entorhinal cortex projections, the perforant pathways to the hippocampal formation leading to apathy, changes in behavior and personality including difficulty in reasoning, disorientation, language problems, and depression (Jahn 2013). Additional changes in AD patients include amyloid angiopathy, age-related brain atrophy, synaptic pathology, white matter rarefaction, and granulovacuolar degeneration. Amyloid angiopathy is the most prevalent disturbance, affecting about 70 % of AD patients; depression ranks second, occurring in about 54 % of patients; and agitation ranks third, appearing in about 50 % of patients (Frisoni et al. 1999). Majority of AD cases (>93–95 %) are of sporadic. These patients are older than 65 years. Only 5–7 % cases appear to be primarily genetic involving apolipoprotein E (apoE) gene.

ApoE protein is a polymorphic protein with three common allele variants: ApoE2, ApoE3 and ApoE4. The ApoE4 gene is the strongest and only confirmed genetic risk factor for the development of late onset AD (LOAD), which enhances the risk level by 3 times in heterozygous individuals and by 12 times in homozygous individuals (Bertram 2009). ApoE4 protein regulates Aβ aggregation and clearance and contributes to the pathogenesis of AD in apoE isoforms dependent manner. ApoE4 has been reported to play an important role in cerebral energy metabolism, modulation of chronic inflammation, neurovascular function, neurogenesis, and synaptic plasticity (Kim et al. 2009, 2014). Pathogenesis of sporadic AD is very complex. It may involve aging, dietary, and environmental factors including long term consumption of high calorie western diet, sedentary lifestyle, interaction of multiple genetic and environmental risk factors, together with the disruption of epigenetic mechanisms controlling gene dynamics and expression (Fig. 1.1) (Kanoski and Davidson 2011; Seneff et al. 2011; Hu et al. 2013; Farooqui 2013, 2014, 2015; Swaminathan and Jicha 2014; Grant et al. 2002). In addition, long term consumption of high calorie Western diet contributes not only to insulin resistance and cerebrovascular dysfunction, but also to oxidative damage and mitochondrial dysfunction along with chronic inflammation (Seneff et al. 2011; de la Monte 2013; Farooqui 2013; Farooqui 2014; Farooqui 2015). Advancing age and long term consumption of high calorie diet may be the most prevalent risk factors for the sporadic AD. In AD patients, above factors along with the decline of cellular protein quality control processes contribute to the oligomerization of Aβ and hyperphosphorylation of Tau protein in the brain (Price et al. 2009; Powers et al. 2009). The molecular mechanisms

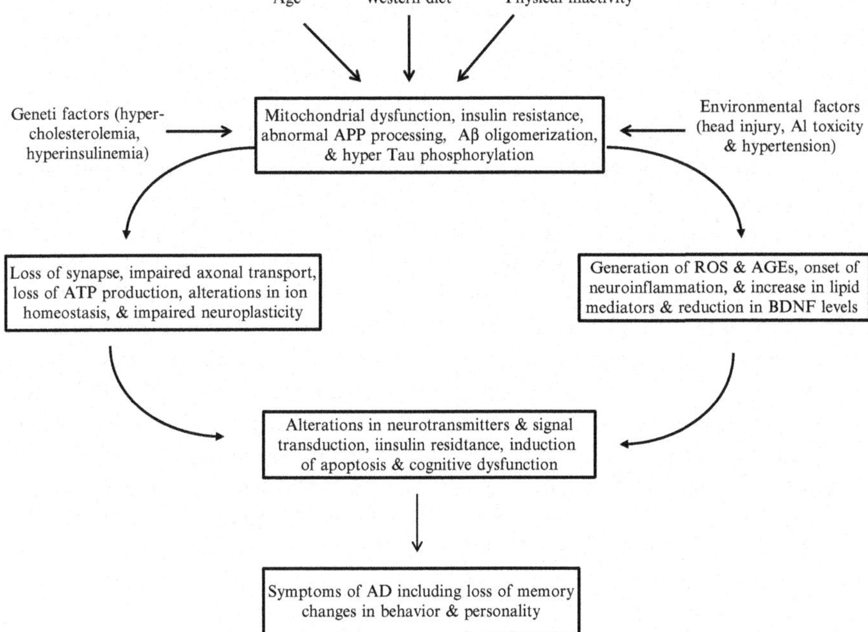

Fig. 1.1 Risk factors and molecular changes associated with the pathogenesis of AD

by which Aβ-mediated oxidative damage, mitochondrial dysfunction and chronic inflammation induce cognitive dysfunction and memory loss are not fully understood. However, it is suggested that these process impair hippocampal plasticity by reducing the expression of brain-derived neurotrophic factor (BDNF), a growth factor, which is abundantly expressed in the hippocampus, hypothalamus, and cerebral cortex (Stranahan et al. 2008; McNay et al. 2010). Post-mortem studies show that BDNF expression is severely decreased in the hippocampus, temporal, and frontal cortex in AD (Siegel and Chauhan 2000). Thus, decreased BDNF in the brain might contribute to advanced aging as well as AD (Ziegenhorn et al. 2007). BDNF contributes to the survival, maintenance, and growth of many types of neurons in the brain. It activates a high-affinity receptor tyrosine kinase called trkB. The activation of trkB receptor is linked with signaling pathways involving PtdIns 3 kinase, Akt kinase and FOXO transcription factors (Stranahan et al. 2008; Bhat 2010; Farooqui 2013, 2015). This pathway is associated with the expression of genes that modulate and enhance synaptic plasticity (glutamate receptor subunits and growth-associated protein 43) and cell survival (due to antioxidant enzyme superoxide dismutase 2, and the anti-apoptotic protein Bcl-2, for example) (Koponen et al. 2004; Bramham and Messaoudi 2005). Furthermore, BDNF also enhances neurogenesis, which may also contribute to increase in cognitive function (Schmidt and Duman 2007). Converging evidence suggests that unhealthy nutrition, lifestyle, and environmental factors may contribute to the pathogenesis of sporadic AD.

Chemical and imaging biomarkers are indicators of specific changes that occur in AD. Many biomarkers have been characterized and developed to monitor sporadic AD in vivo. There are three biomarkers for neuroimaging: hippocampal atrophy is detected by structural magnetic resonance imaging (MRI); decrease in uptake of (18 F)-deoxyglucose in characteristic regions by positron emission tomography (FDG-PET); and increase in amyloid tracer retention by PET (PiB-PET). Similarly, there are three biomarkers for neurochemical changes in the cerebrospinal fluid (CSF) protein levels: low CSF levels of Aβ, elevated total Tau levels in the CSF, and phosphorylated tau (p-tau). Imaging biomarkers are important because they can provide crucial information about topographical changes in the brain (Blennow and Hampel 2003; Klunk et al. 2004; Clark et al. 2011; Ritter and Cummings 2015). Levels of phosphorylated Tau and total Tau in the CSF provide information about the deposition of neurofibrillary tangle and neuroaxonal damage. Decrease in fluoro-deoxyglucose PET in temporo-parietal cortex and (Blennow and Hampel 2003; Herholz 2012), and structural MRI in medial, basal and lateral temporal lobes, and medial parietal isocortex have been used to monitor progression of AD (Fox and Schott 2004; McKhann et al. 1984). Compared with normal aging, levels of Aβ, phosphorylated Tau, and extent of brain atrophy calculated using serial MRI scans are much higher in clinically diagnosed AD subjects and subjects with mild cognitive impairment (MCI) show intermediate rates (Henneman et al. 2009). These markers of AD progression are closely associated with cognitive decline (Evans et al. 2010; Ritter and Cummings 2015).

1.2   Key Molecular Changes Contributing to the Pathogenesis Alzheimer Disease

Neurochemically, AD is characterized by the induction of neuroinflammation, oxidative stress due to increase in levels of phospholipid-, sphingolipid-, and cholesterol-derived lipid mediators along with mitochondrial dysfunction, loss of synapses, neurotrophic factor dysregulation, progressive impairment of memory, decline in cellular protein quality control processes (specifically $A\beta$ precursor and Tau proteins processing), and chronic activation of innate immune responses including those mediated by microglia, the resident CNS macrophages (Price et al. 2009; Farooqui 2010a). Such activation can trigger neurotoxic pathways leading to progressive degeneration. Although the blood-brain barrier (BBB) is able to protect the CNS from immune activation, it becomes more permeable during inflammation, which renders the brain vulnerable to infections. Glutamate is the major excitatory neurotransmitter in the brain. It is rapidly released from nerve terminals in a Ca^{2+}-dependent manner. Released glutamate diffuses across the synaptic cleft to bind with postsynaptic receptors in adjacent neurons (Farooqui et al. 2008).

Phospholipid, sphingolipid, and cholesterol-derived lipid mediators are involved in conveying the message from extracellular signals at the cell surface to the nucleus to induce a biological response at the gene level. Increase in levels of phospholipid-, sphingolipid-, and cholesterol-derived lipid mediators in mitochondrial and synaptic membranes of neurons may contribute to the progression of AD (Farooqui 2010a; Farooqui et al. 2010). Mitochondrial changes include decrease in complex IV of the respiratory chain, damage to complex V, and alterations in voltage-dependent anion channel VDAC, a mitochondrial pore that is involved in redox homeostasis and apoptosis (Ferrer 2009). Mitochondrial function is essential to neuronal metabolism and functions, such as energy production, Ca^{2+} regulation, maintenance of plasma membrane potential, protein folding by chaperones, axonal and dendritic transport and the release and re-uptake of neurotransmitters (Santucci et al. 2008). At the synapse level, neurochemical changes in AD include activation of PlsEtn-PLA$_2$, loss of plasmalogens, and reduction (25 %) in the presynaptic vesicle protein synaptophysin (Farooqui 2010b; Masliah et al. 2001). It is proposed that increase in phospholipid-, sphingolipid-, and cholesterol-derived lipid mediators is not only related to increase in intensity of cross talk among cytoplasmic, mitochondrial, and nuclear compartment, but also increase in the production of misfolded or aggregated synaptic proteins (Masliah et al. 2001). It is well known that aging itself causes synaptic loss in the dentate region of the hippocampus. However, in AD synapses are disproportionately lost relative to neurons, and this loss can be correlated with accumulation of $A\beta$ oligomers, hyperphosphorylation of τ protein and onset of dementia (Terry et al. 1991).

1.2.1 *Neural Membrane Phospholipid Alterations in Alzheimer Disease*

Studies on phospholipid composition of neural membranes have indicated that levels of phospholipids are significantly decreased in different regions of AD patients compared to age-matched control (Söderberg et al. 1991; Wells et al. 1995; Guan et al. 1999; Han et al. 2001; Pettegrew et al. 2001). This decrease in neural membrane phospholipid composition is produced by the stimulation of isoforms of PLA_2 activities (Farooqui et al. 1997, 2003; Stephenson et al. 1999; Rao et al. 2011). Stimulation of PLA_2 isoforms in AD not only produces an elevation in arachidonic acid-derived lipid mediators (prostaglandins, leukotrienes, thromboxanes, isoprostanes, 4-hydroxynonenals, acrolein, and malondialdehyde), but also enhances neural membrane phospholipid metabolism (Fig. 1.2) (Farooqui and Horrocks 2007). Among prostaglandins, J2 prostaglandins contain α,β-unsaturated carbonyl groups and are considered some of the most abundant prostaglandins in the brain (Katura et al. 2010). Its role in AD has been proposed. It is suggested that J2 prostaglandins impair the ubiquitin proteasome pathway (UPP) (Selkoe 2004; Shaw et al. 2007; Figueiredo-Pereira et al. 2015). Defective proteasome activity is not only associated with the early phase of AD (synaptic dysfunction), but also with late AD stages linked and

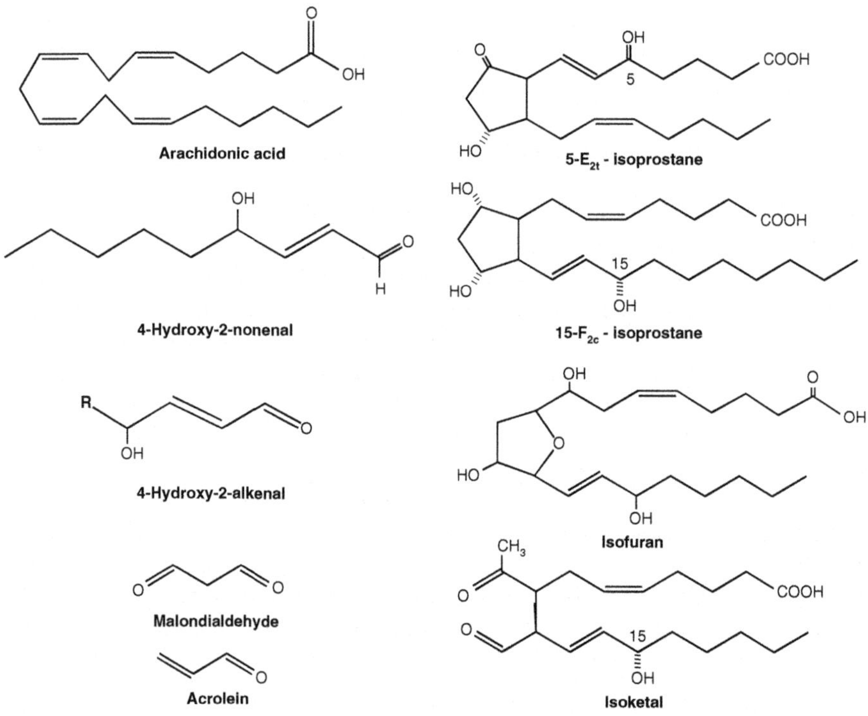

Fig. 1.2 Arachidonic acid-derived nonenzymic lipid mediators associated with the pathogenesis of AD

characterized by the accumulation and aggregation of ubiquitinated (Ub)-proteins in both senile plaques and neurofibrillary tangles (Upadhya and Hegde 2007; Oddo 2008). In addition, J2 prostaglandins also inhibit the de-ubiquitinating enzyme UCH-L1 (Li et al. 2004a; Liu et al. 2011). The activity of this enzyme is down-regulated in AD brains. Down-regulation of UCH-L1 is inversely proportional to the number of neurofibrillary tangles (Choi et al. 2004; Figueiredo-Pereira et al. 2015). Moreover, it is also reported that PGJ2 promotes the accumulation of Ub-proteins, caspase-activation, Tau cleavage at Asp421, and neuritic dystrophy in rat primary cerebral cortical cultures (Arnaud et al. 2009; Metcalfe et al. 2012). Tau cleavage at Asp421 has identified as an early event in AD tangle pathology (Gamblin et al. 2003; Rissman et al. 2004; de Calignon et al. 2010). Collective evidence suggests that J2 prostaglandins promote and support many pathological processes involved in the pathogenesis of AD, including the accumulation and aggregation of ubiquitinated (Ub)-proteins in senile plaques as well as neurofibrillary tangles, activation of caspases, cleavage of Tau, and neuritic dystrophy.

Physicochemical and pathological consequences of enhanced phospholipid degradation in neural membranes not only produce changes in membrane fluidity and permeability and alterations in ion homeostasis, but also triggering changes in activities of membrane-bound enzymes, receptors, and ion channels leading to oxidative stress and chronic inflammation. Among arachidonic acid-derived lipid mediators many eicosanoids produce proinflammatory effects. The proinflammatory effects of eicosanoids are accompanied by the activation of astrocytes and microglia, which synthesize and release proinflammatory cytokines (TNF-α, IL-1β, and IL-6). These cytokines further stimulate isoforms of PLA_2 and cyclooxygenases (COXs) leading to the generation of more proinflammatory eicosanoids and intensification of inflammation (Farooqui and Horrocks 2007; Farooqui 2010a). The cause of increased activities of PLA_2 isoforms in AD brain is not fully understood. However, many studies have indicated that Aβ, which accumulates in AD, may activate $cPLA_2$ activity (Desbène et al. 2012; Sun et al. 2012; Fonteh et al. 2013). In addition, recent studies have indicated that in neurons, N-methyl-D-aspartate receptor (NMDA) and Aβ induce ROS production through NADPH oxidase leading to the activation of MEK1/2/ERK1/2, and $cPLA_2$ (Shelat et al. 2008). Activation of $cPLA_2$ may also promote the release of lysophospholipids, which not only act as detergents, but also through acylation process may be transformed into platelet activating factor, a lipid mediator, which promotes and support inflammation (Farooqui et al. 2000; Lee et al. 2010). Finally, ceramide, a metabolite of sphingolipid metabolism (see below), which accumulates in AD brain (Han et al. 2002) has been reported to stimulate isoforms of PLA_2 (Farooqui 2010a, b). It is reported that in AD, PlsEtn-PLA_2 may be the first PLA_2 isoform that initiates neural injury by altering neural membrane permeability due to the loss of plasmalogens, allowing slow Ca^{2+} influx. Low levels of Ca^{2+} in the presence of ceramide may facilitate translocation of $cPLA_2$ from cytosol to neural membranes and its activation resulting in the hydrolysis of neural membrane PtdCho. As concentration of Ca^{2+} reaches in millimolar level, the $sPLA_2$ may be activated promoting neural cell injury and death. Thus during injury process sequence, PlsEtn-PLA_2 is situated at the proximal end, $cPLA_2$ in the middle, and $sPLA_2$ at the distal end (Farooqui 2010b).

1.2.2 Neural Membrane Sphingolipid Alterations in Alzheimer Disease

Changes in sphingolipid metabolism have also been observed in AD. Thus, levels of ceramide are elevated and sulfatide levels are reduced in brain from AD patients (Fig. 1.3) (Han et al. 2002; Cutler et al. 2004). Increase in ceramide may be due to increase in activities of acid and neutral sphingomyelinases (ASM) and (NSM) in AD (He et al. 2010; Huang et al. 2004; Puglielli et al. 2003). It is well known that increase in ceramide levels may cause changes in multiple enzymes and cell signaling components. Thus, early inhibition of the neuronal survival pathway regulated by phosphatidylinositol-3-kinase/protein kinase B or Akt mediated by ceramide may be a relevant early event in the decision of neuronal survival/death (Arboleda et al. 2009). C6-ceramide not only stabilizes BACE1 (β-secretase) and increases the biogenesis of Aβ by affecting β-but not γ-cleavage of the amyloid precursor protein (APP). Fumonisin B1 inhibits the biosynthesis of endogenous ceramide and reduces the production of Aβ. Furthermore, addition of exogenous C6-ceramide restores not only intracellular ceramide levels, but also induces Aβ production in fumonisin B1-treated cells. These events are specific for APP and are not associated with apoptotic cell death (Puglielli et al. 2003). In addition, ceramide perturbs several molecular and metabolic functions. In particular it may contribute to decrease in glycolysis through rapid modulation of hexokinase activity. This may in turn generate limited amounts of mitochondrial substrates leading to mitochondrial dysfunction and neuronal apoptosis. Subtle and early metabolic changes mediated by the ceramide mediated inhibition of the PtdIns3K/Akt pathway may be potentially associated with genes contributing to the neurodegeneration in AD (Arboleda et al. 2009). Ceramide and ceramide analogs decrease the levels of Tau in PC12 cells (Xie and Johnson 1997). In contrast, the addition of the GM1 ganglioside (Fig. 1.3) increases levels of tau and stabilized the microtubule network in neuroblastoma cells (Wang et al. 1998a). These suggestions are also supported by microarray studies on AD brain. It is shown that there is an upregulation of gene expression of the enzymes involved in the de novo synthesis of ceramide and the downregulation of the enzymes involved in glycosphingolipid synthesis in early AD progression (Katsel et al. 2007).

It is well known that apoE is associated with sulfatide transport and contribute to sulfatide trafficking and homeostasis in the brain through lipoprotein metabolism pathways (Han 2007). Recent studies have indicated that the depletion of sulfatides is not only tightly associated with Aβ pathology in AD (Han 2010), but also linked with the white matter abnormality in AD (Xie et al. 2006; Zhang et al. 2009). The molecular mechanism of sulfatide-mediated changes in Aβ pathology in AD is not fully understood. However, lipidation status of apoE is known to influence the metabolism of Aβ peptides that accumulate as amyloid deposits in the neural parenchyma and cerebrovasculature of AD patients. ApoE not only inhibits the transport of Aβ across the blood–brain barrier (BBB) but also facilitates the proteolytic degradation of Aβ by neprilysin and insulin degrading enzyme (IDE), which are enhanced when apoE is lapidated (Martins et al. 2009). Collective evidence suggests

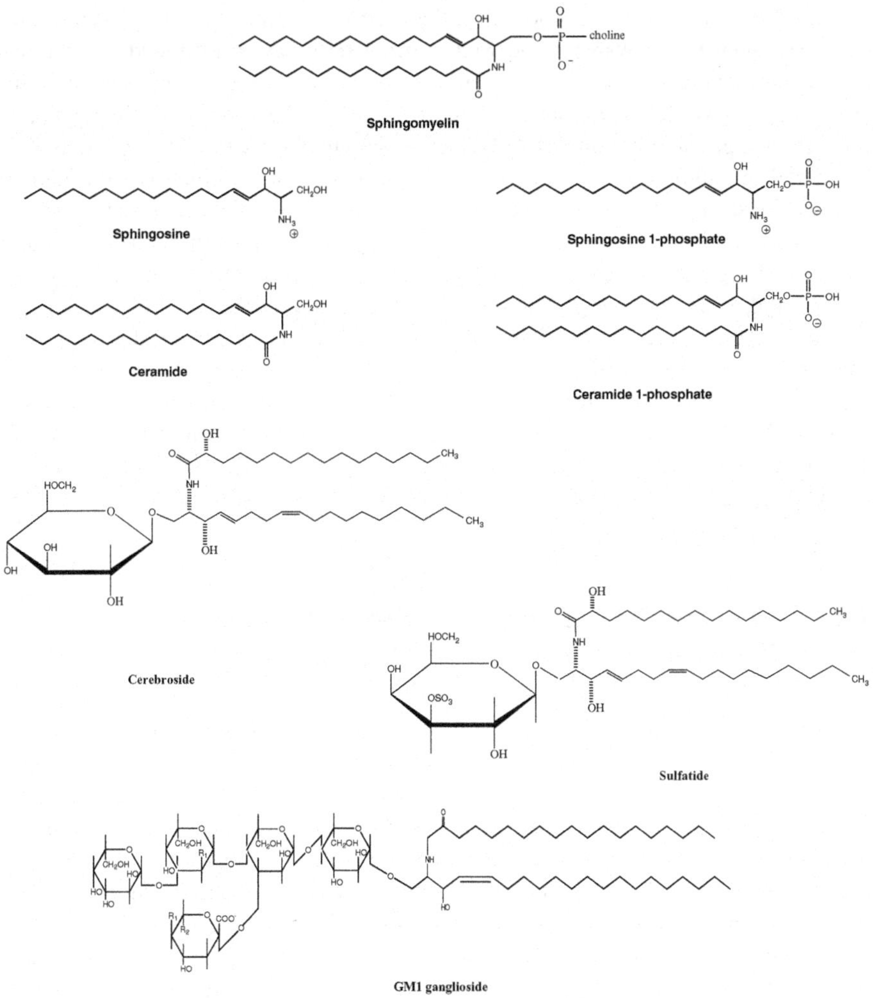

Fig. 1.3 Sphingolipid-derived lipid mediators associated with the pathogenesis of AD

that reduction in sphingosine-1-phosphate levels in the AD brain, together with elevated ceramide and decrease in sulfatide along with changes in Apo E metabolism may contribute to the pathogenesis of AD. This suggestion is also supported by results on accumulation of ceramide in the cortex of APPSL mice, but not in PS1Ki mice compared to age-matched wild-type mice (Barrier et al. 2008).

Like ceramide and sulfatide, the patterns and levels of ganglioside are also altered in AD. Specifically, levels of gangliosides are reduced to 58–70 % of that of control brains in gray matter, and to 81 % in frontal white matter, in early-onset or familial AD cases. Furthermore, levels of gangliosides are also significantly reduced only in the temporal cortex, hippocampus, and frontal white matter in late-onset cases (Svennerholm and Gottfries 1994; Kalanj et al. 1991). Alterations in ganglioside composition have also been reported in the brains of adults with Down syndrome (DS) and in AD (Brooksbank and McGovern 1989). DS is attributed to trisomy (triplication) of chromosome 21, the same chromosome on which APP is located. In vitro studies have indicated that GM1 interacts with Aβ and promotes its aggregation. The ganglioside-bound Aβ (GAβ) possessed unique characteristics, including its altered immunoreactivity, which suggests its distinct conformation from native Aβ, and its strong potency to accelerate Aβ assembly into fibrils. Based on these characteristics, it is hypothesized that GM1 may act by inducing conformational changes during Aβ aggregation and may potentiate the formation of an endogenous seed (ganglioside-enriched (clustered) or raft-like microdomains) for Alzheimer amyloid peptide in the brain (Matsuzaki et al. 2010). It is speculated that neuronal gangliosides may be involved in the accumulation of circulating Aβ to form complexes that are expressed and accumulated in AD brain. Because GM1 possesses neurotrophic properties, administration of GM1 may exert beneficial effects in AD. Additionally, GM1 infusions may be useful for the sequestration of excess Aβ in AD patients to alleviate their toxic effects (Ariga et al. 2008).

1.2.3 Changes in Cholesterol-Derived Metabolites in Alzheimer Disease

Brain is the richest source of cholesterol. In the brain, cholesterol is presents in two pools. One pool, which contains ~70 % of the total cholesterol, is metabolically stable. This pool is associated with myelin membranes of white matter (Davison 1965). The second pool, which represent ~30 % of total cholesterol, is found in the plasma and subcellular membranes of neurons and glial cells of gray matter. This is metabolically active. Due to its unique structure, consisting of a fused rigid ring system, a polar hydroxyl group, and a hydrocarbon tail, cholesterol is essential for the function and organization of lipid bilayer. In plasma membranes, cholesterol accounts for 25–50 mol%. Levels of cholesterol in the endoplasmic reticulum and nuclear membranes are in the range of 1–10 mol%, where as in Golgi cholesterol is present at 10–25 mol% (van Meer et al. 2008; Radhakrishnan et al. 2008; Andreyev et al. 2010). Cholesterol facilitates the formation of "lipid rafts" (microdomains), which are found in the plasma membrane, the trans-Golgi, and endosomal membranes. Lipid rafts are enriched in sphingomyelin, sphingolipids, phospholipid, and cholesterol. They are involved in cellular processes, protein sorting, signaling complex formation, and the initiation of signal transduction pathways. The topology of cholesterol in membrane is well suited for its integration into lipid bilayers, where it aligns itself with phospholipids and sphingolipids so that its isooctyl tail is near the middle of the bilayer and its 3β-OH group is at the water-membrane interface. As stated above, in brain, cholesterol is associated with myelin. However, small amount of cholesterol is present in the nucleus, which contains activities of cholesterol-metabolizing enzymes (Pfrieger 2003). Brain cholesterol levels are five to ten times higher than those in other organs. Due to the limited transport of cholesterol over the BBB, the brain cholesterol levels are largely independent of the serum cholesterol concentrations. In neural membranes, cholesterol not only contribute to physicochemical properties (fluidity and permeability) and endocytosis, but is also involved in antigen expression, exocytosis, synaptic plasticity and transmission, modulation of activities of membrane-bound enzymes, receptors, and ion channels (Simons and Ikonen 2000; Farooqui et al. 2010). Therefore, the maintenance of proper cholesterol synthesis, transport, and intracellular sorting is tightly regulated. Cholesterol also serves as a precursor for the production of steroid hormones, vitamin D, and oxysterols. Although, cholesterol synthesis occurs both in neuronal and glial cells, but the vast majority of brain cholesterol is synthesized by glial cell through de novo synthesis. The transport of cholesterol between glia cells and neurons is mostly performed by clusterin/apolipoprotein J (ApoJ) and apolipoprotein E (ApoE) containing lipoproteins (Grimm et al. 2013). It is proposed that ApoE isoforms may differentially contribute to the pathobiology of AD on a cellular level.

In brain, cholesterol is oxidized by cytochrome P450-dependent oxygenases (CYP46A1), cholesterol oxidases, and acyl-CoA: cholesterol acyltransferase (Fig. 1.4). These enzymes transform cholesterol into hydroxycholesterols or oxysterols (24-hydroxycholesterol, 25-hydroxycholesterol, and 27-hydroxy-cholesterol, keto, hydroperoxy, epoxy, cholesterol oxides, and cholesterol esters) (Mast et al. 2003; Olkkonen et al. 2012). In general, hydroxycholesterols have a markedly shorter biologic half-life than cholesterol, therefore they are considered as products

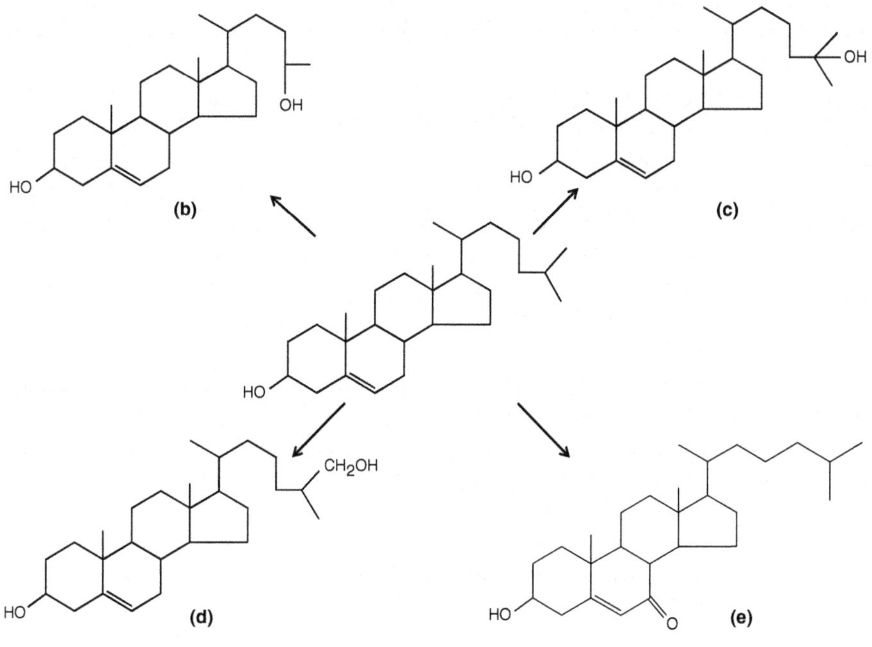

Fig. 1.4 Cholesterol-derived lipid mediators associated with the pathogenesis of AD

of the cholesterol catabolism. Cholesterol-metabolizing enzymes are expressed almost exclusively in neurons in the normal brain, including hippocampal and cortical neurons that are important for learning and memory formation (Russell et al. 2009). 24S-hydroxycholesterol is the major brain cholesterol metabolite, which is not only associated with maintenance cholesterol homeostasis, but also with the removal of excess cholesterol from the brain. 24S-Hydroxycholesterol can cross the blood brain barrier and is transported in the blood stream to the liver where it is metabolized to bile acids (Björkhem et al. 1998, 2001). Increase in CYP27A1 activity in liver and peripheral tissues downregulates cholesterol synthesis through the sterol regulatory element-binding protein (SREBP) pathway as well as enhances the efflux and elimination of cholesterol via liver X receptor (LXR) (Fu et al. 2001).

The human enzyme, CYP46A1, is expressed in neurons in normal human brain and also in glial cells from AD patients (Bogdanovic et al. 2001). Levels of 24S-hydroxycholesterol in normal individual's plasma vary from 50 nM to 350 nM. AD patient's plasma and brain have elevated levels of 24S-hydroxycholesterol (Lutjohann et al. 2000; Leoni 2009). However, associations between plasma 24S-hydroxycholesterol levels and AD have been inconsistent in cross-sectional analyses and may depend on disease stage (Leoni 2009). Patients treated with statins (3-hydroxy-3methylglutaryl (HMG)-CoA reductase inhibitors) for dyslipidemia have been shown to have lower incidence of AD (Jick et al. 2000; Wolozin et al. 2000) and statins have been demonstrated to lower 24S-hydroxycholesterol in plasma levels of AD patients (Vega et al. 2003). Earlier studies on the CYPA1−/− mouse indicate that brain cholesterol homeostasis is maintained via reduced biosynthesis rather than via up-regulation of an alternative metabolizing enzyme (Lund et al. 2003). The CYPA1−/− mouse shows impaired learning and memory due to reduced flow of metabolites through the mevalonate pathway (Russell et al. 2009; Kotti et al. 2008). It is well known that cholesterol contents regulate compartmentation of the APP molecules within the neural cell membrane bilayer. The APP molecules are found inside or outside the rafts. Processes altering the compartmentation of the APP molecules by transferring it to the neural membrane rafts favor its cleavage by secretases and are closely associated with amyloidogenic processing (see below). It is proposed that 24S hydroxycholesterol exerts a unique modulatory effect on APP processing and increases the α-secretase activity as well as the α/β-secretase activity ratio (Famer et al. 2007). The release of 24S-hydroxycholesterol from neurons is coupled with the secretion of ApoE by astrocytes. It is proposed that both processes are related to the intensity of the neurodegenerative process and neuronal injury (Leoni et al. 2010). ApoE also scavenges Tau protein from neurons. The direct correlations between ApoE, 24(S)-hydroxycholesterol, and Tau suggest that cholesterol metabolism may be involved in the production of Aβ (Leoni et al. 2010).

Minor cholesterol metabolites are 22-hydroxycholesterol and 27-hydroxycholesterol. Significant net uptake of 27-hydroxycholesterol occurs from the circulation to the brain tissue. Patients with AD have elevated levels of 27-hydroxycholesterol. It is proposed that 27-hydroxycholesterol may affect the production of β-amyloid in the brain (Ong et al. 2010). The mechanism by which 27-hydroxycholesterol modulates the production of Aβ and Tau protein phosphorylation is not fully understood.

However, recent studies have indicated that leptin, a satiety and energy balance hormone, negatively regulates Aβ production and Tau phosphorylation in neuronal cell lines and primary neurons (Fewlass et al. 2004; Greco et al. 2009). It is also reported that leptin inhibits Aβ production via downregulation of transcription of the γ-secretase components (Niedowicz et al. 2013). Collectively, these studies suggest that leptin may play an important role in accumulation of Aβ and hyperphosphorylation of Tau protein. Feeding rabbits with a 2 % cholesterol-enriched diet for 12 weeks decreases the levels of leptin by approximately 80 %. Incubation of hippocampal organotypic slices with 27-hydroxycholesterol not only reduces leptin levels by approximately 30 %, but also increases Aβ(1–40) and Aβ(1–42) levels to 1.5-fold and 3-fold, respectively along with elevation in phosphorylated Tau protein levels. Treatment of hippocampal organotypic slices with leptin retards the 27-hydroxycholesterol-mediated increase in Aβ and phosphorylated Tau protein by decreasing the levels of β-secretase (BACE1; also called Asp2 and memapsin2) and GSK-3β (glycogen synthase kinase-3β), respectively (Marwarha et al. 2010a, b; Dasari et al. 2010; Mateos et al. 2009). It is also shown that 27-hydroxycholesterol increases BACE1 and Aβ levels in human neuroblastoma SH-SY5Y cells (Marwarha et al. 2013). This increase in BACE1 involves a crosstalk between the two transcription factors NF-κB and the endoplasmic reticulum stress marker, the growth arrest and DNA damage induced gene-153 (gadd153, also called CHOP). Treatment with 27-hydroxycholesterol stimulates the binding of NF-κB with the BACE1 promoter and subsequent increases in BACE1 transcription and Aβ production. The NF-κB inhibitor, sc514, significantly blocks the 27-hydroxycholesterol coupled increase in NF-κB-mediated BACE1 expression and Aβ genesis (Marwarha et al. 2013). Similarly, 27-hydroxycholesterol also increases the expression of gadd153. Silencing gadd153 expression with siRNA retards the 27-hydroxycholesterol-mediated increase in NF-κB activation. Based on detailed investigation, it is proposed that gadd153 and NF-κB work in concert to regulate BACE1 expression and production of Aβ (Marwarha et al. 2013).

22-Hydroxycholesterol has been reported to regulate cell division, ventral midbrain (VM) neurogenesis, and dopaminergic (DAergic) neuronal development (Sacchetti et al. 2009). The retinoic acid receptor-related orphan receptors (RORs) are nuclear receptors, which are closely associated with the transcriptional control of lipid metabolism. RORα plays an essential role in development of the cerebellum and in regulation of the circadian rhythm (Jetten 2009), while RORγ is best known for its role in regulation of T cell development (Winoto and Littman 2002). Both RORα and RORγ are expressed in the liver and play roles in the regulation of glucose and lipid metabolism (Kang et al. 2007). Recently studies have indicated that RORα and RORγ also bind 24S-hydroxycholesterol with high affinity, and 24S-hydroxycholesterol functions as a RORα/γ inverse agonist suppressing the constitutive transcriptional activity of these receptors in cotransfection assays (Wang et al. 2010). These findings support the view that RORα and RORγ serve as novel sensors for oxysterols and display an overlapping ligand preference and functional cross-talk with the LXR. Collective evidence suggests that hydroxycholesterols are major lipid mediators of cholesterol metabolism. They modulate cholesterol trafficking, gene transcription, PtdCho synthesis, apoptosis, and cognitive function. In addition to having a direct physical effect on membranes and modulation of Ca^{2+} signals, hydroxycholesterols also inhibit the phosphorylation of endothelial nitric oxide synthase and isoforms of PLA_2 (Bansal and Shalini 2005; Trousson et al. 2009).

1.3   Protein Metabolism Alterations in Alzheimer Disease

It is well known that protein homeostasis or proteostasis is controlled and regulated by a complex network of cellular mechanisms that monitors the concentration, folding, cellular localization, and interactions of proteins from their synthesis through their degradation (Powers et al. 2009). Despite the activity of the proteostasis network, subsets of aggregation-prone proteins fail to fold properly, escape degradation, and form insoluble aggregates that accumulate within the cell. Decline in proteostasis and accumulation of damaged proteins are hallmark of aging and many neurodegenerative diseases including AD (Selkoe 2011). Typically, individuals that carry neurodegeneration-linked mutations develop the disease during their fifth decade, whereas sporadic neurodegenerative maladies appear during the seventh decade or later (Amaducci and Tesco 1994). The cellular and molecular mechanisms underlying AD are still not fully understood. However, considerable genetic, biochemical and molecular biological evidence support the amyloid-cascade hypothesis. According to the amyloid hypothesis, accumulation of $A\beta$ in the brain, resulting from an imbalance between production and clearance, is the primary factor that drives the pathogenesis of AD pathogenesis (Hardy and Selkoe 2002).

APP is a larger type I transmembrane spanning glycoprotein, which plays an important role in neuroprotection, ion transport, synapse formation, and transcriptional signaling (Fig. 1.5). It is also suggested that APP functions as a molecular switch, which controls both neuroplasticity-related processes, and pathogenesis of AD. There is strong evidence that APP may act as a trophic factor relevant to neurite outgrowth and synaptogenesis, as well as growth and cell proliferation (Dawkins and Small 2014; Hughes et al. 2014). However, more studies are required to fully clarify mechanisms of APP action (Dawkins and Small 2014). APP contains a copper binding domain (not localized in its $A\beta$ part) (Barnham et al. 2003) and possesses ferroxidase (oxidizes Fe^{2+} into Fe^{3+}) activity assisting in plasma membrane Fe^{2+}-export by ferroportin, counteracting iron accumulation and oxidative stress (Duce et al. 2010). APP's ferroxidase activity has been shown to take place on the extracellular plasma membrane side, where APP (in interaction with ferroportin) loads Fe^{3+} into blood transferrin (Duce et al. 2010).

The APP gene is located on chromosome 21 in humans with three major isoforms (APP695, APP751 and APP770) arising from alternative splicing (Goate et al. 1991). APP751 and APP770 are expressed in most tissues and contain a 56 amino acid Kunitz Protease Inhibitor (KPI) domain within their extracellular regions. APP695 is predominantly expressed in neurons and lacks the KPI domain (Rohan de Silva et al. 1997; Kang and Muller-Hill 1990). The APP promoter sequence indicates that the APP gene belongs to the class of housekeeping genes. The APP promoter lacks typical TATA and CAAT boxes, but contains consensus

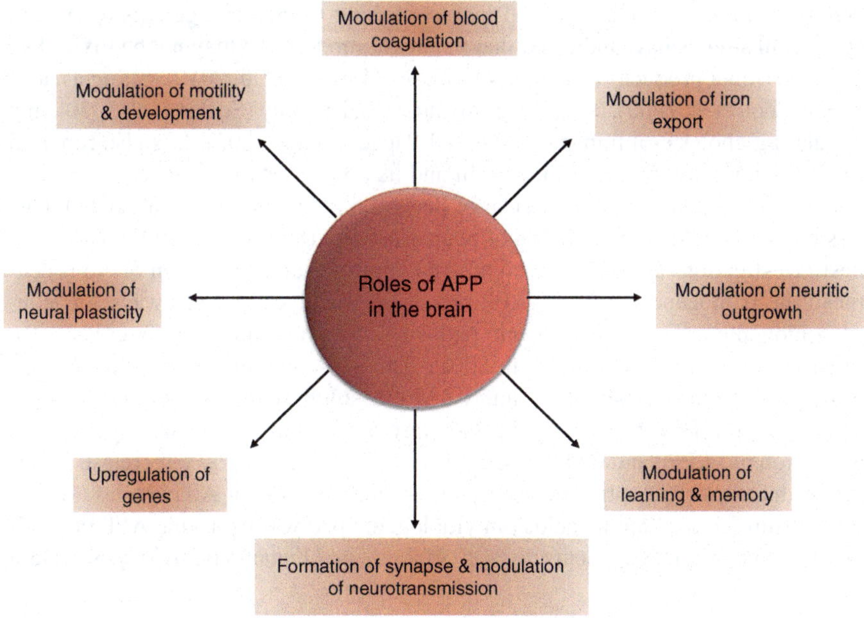

Fig. 1.5 Roles of APP in the brain

sequences for the binding of a number of transcription factors including SP-1, AP-1 and AP-4 sites, a heat shock control element and two Alu-type repetitive sequences (Dawkins and Small 2014). The presence of SP-1, AP-1 and AP-4 sites in the APP promoter suggests that APP may play an important role in cell proliferation, differentiation, neurite outgrowth, cell adhesion and synaptogenesis. The molecular mechanisms associated with APP-mediated cell proliferation, neurite outgrowth and synaptogenesis have not been fully elucidated. However, it is suggested that the structure of APP resembles a cell-surface receptor (Kang et al. 1987), but a receptor function for APP has not been unequivocally established. It is also stated that F-spondin may act as a potential APP ligand (Ho and Sudhof 2004). However, more studies are needed on F-spondin and APP interactions in the brain. AD is accompanied by the abnormal APP processing, accumulation of Aβ aggregates, and formation of neurofibrillary tangles, which are composed of hyperphosphorylated Tau protein. In the brain, Aβ is predominantly produced by neurons. However, astrocytes and other glia, also generate Aβ especially under stress conditions that induce glial activation, as occurs in AD. Aβ accumulation precedes and drives hyperphosphorylation and aggregation of Tau protein (Oddo et al. 2003). Moreover, Aβ-induced degeneration of cultured neurons and cognitive deficits in mice with an AD-like disease requires the presence of endogenous Tau protein (Roberson et al. 2007).

Many old individuals, who have relatively high levels of Aβ deposition, show normal cognitive function (Aizenstein et al. 2008). This observation along with the failure of some anti-Aβ therapies to preserve or rescue cognitive function (Extance 2010) suggests that Aβ may not be universally neurotoxic, but other mechanisms directly or indirectly related to Aβ may contribute to the pathogenesis of AD. In fact, recent studies have indicated that in animal models, Aβ may not be toxic (Lee et al. 2007), but produces activation of kinases (Tabaton et al. 2010), acting as anti-oxidant (Zou et al. 2002), containing anti-microbial activity (Soscia et al. 2010), and modulating cholesterol transport (Yao and Papadopoulos 2002). It is also reported that Aβ is normally present in the brain and its concentration is higher in younger individuals than older in the absence of dementia (van Helmond et al. 2010). One possible role for Aβ is modulation of neuroplasticity (Puzzo et al. 2008). However, most investigations are still focused on Aβ hypothesis and generation of Aβ is central to the pathogenesis of AD. APP processing involves two pathways: (a) non-amyloidogenic or (b) amyloidogenic pathways, which are initiated by either α- or β-secretase cleavage (Schmitz et al. 2002). The non-amyloidogenic pathway leads to the production two peptides namely sAPPα (soluble) and membrane anchored peptice called CTFα. The action of β-secretase in amyloidogenic pathway results in production of sAPPβ and CTFβ, respectively. CTF processing by γ-secretase generates the harmless P3 peptide (non-amyloidogenic pathway) or Aβ peptides ranging in size from 35 to 42 amino acids (amyloidogenic pathway), plus the APP intracellular domain (AICD) fragment (Fig. 1.6) (Takami and Funamoto 2012). γ-Secretase

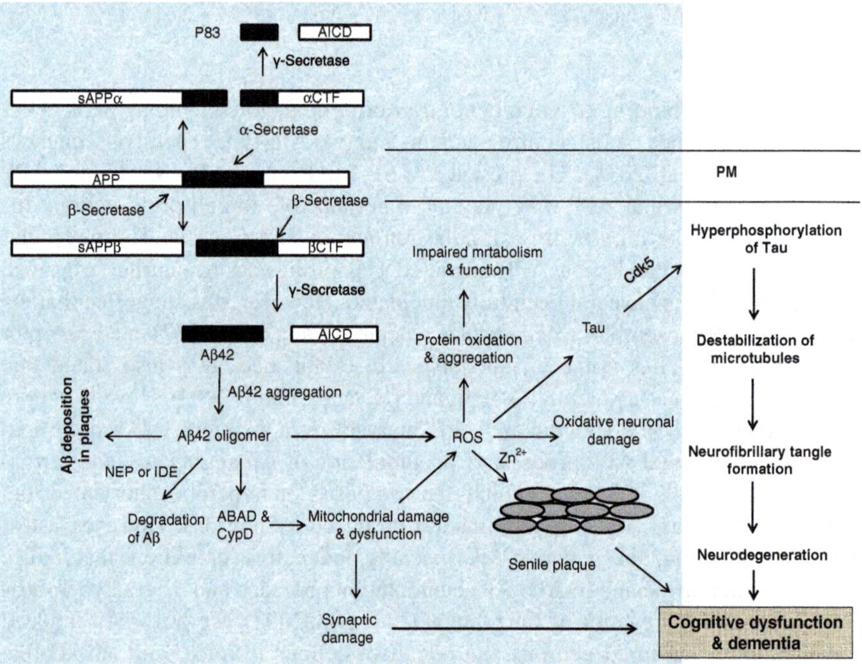

Fig. 1.6 Generation of Aβ and hyperphosphorylation of Tau are associated with cognitive dysfunction and dementia in AD. Plasma membrane (*PM*); amyloid precursor protein (*APP*); presenilin (*PS1*); substrate inhibitory domain (*ASID*); insulin-degrading enzyme (*IDE*); neprilsin (*NEP*); and reactive oxygen species (*ROS*)

is a protein complex consisting of presenilin 1 (PS1)/presenilin 2 (PS2), nicastrin (NCT), anterior pharynx-defective 1 (APH-1), and presenilin enhancer 2 (PEN-2) (Kimberly et al. 2003; Li et al. 2000, 2003; Yu et al. 2000). In addition to APP, it also hydrolyzes Notch (a protein that resides on the surface of signal-receiving cells as a heterodimeric receptor). In the membrane α-secretase is located in phospholipid-rich domains where as both β- and γ-secretases reside in cholesterol-rich lipid rafts of plasma membrane (Cordy et al. 2003), it is proposed that altered levels of cholesterol or/and ratio of cholesterol to phospholipids in cellular membrane may affect secretase activities and determine preferential APP processing pathways (Wolozin 2004; Kaether and Haass 2004). α-Secretase (also known as ADAM10) is involved in the normal turnover of APP. It releases the N-terminal region of APP (called α-APPs) in the extracellular space, whereas the fragment in the membrane left as a residue is called as substrate inhibitory domain (ASID). ASID is cleaved and cleared by γ-secretase to yield the soluble small fragment P3. Thus the cleavage of APP by α-secretase precludes formation of Aβ. In contrast to α-secretase, the release of Aβ involves the action of β- and γ-secretases. Thus, α-Secretase competes with β-secretase for the APP binding and APP hydrolysis (Skovronsky et al. 2000; Postina et al. 2004). In this model, α-secretase plays a dual anti-amyloidogenic role: first, it hydrolyzes APP in the Aβ region precluding Aβ formation, and second, it initiates a feedback loop in which αCTF binds γ-secretase and acts as a γ-secretase modulator which specifically lowers Aβ production (Tian et al. 2010).

1.3.1   Nonamyloidogenic Pathway

In the nonamyloidogenic pathway, the α-secretase (ie, a disintegrin and metalloproteinase domain protein 10 or ADAM10) prevents formation of toxic Aβ peptides from APP and alternatively promote the production of neuroprotective and neurotrophic soluble fragments (sAPPα) and membrane-bound fragment, C83 (Kandalepas and Vassar 2012). C83 is subsequently cleaved by the γ-secretase complex to yield the 3 kDa fragment, P3, and an APP intracellular domain (CTF fragment), which is further processed by γ-secretase to generate Aβ17-40/Aβ17-42 fragments (Fahrenholz 2007; Thinakaran and Koo 2008). sAPPα has been reported to enhance memory when injected into the cerebral ventricles of mice (Meziane et al. 1998) and AICD has been shown to improve memory in transgenic mice (Laird et al. 2005; Ma et al. 2007; Konietzko 2012). The cleavage of sAPP may also have a role in axon pruning during development and trophic factor deprivation stimulates β-secretase-dependent cleavage of an N-terminal fragment of APP, which binds to death receptor-6 and triggers axon degeneration (Nikolaev et al. 2009). CTF also facilitates neurite extension. Collective evidence suggests that the soluble N-terminal sAPPα and CTF have neuroprotective properties (Fig. 1.5). In transgenic mouse models that over-express APP and Aβ peptides, the deposition of Aβ is accompanied by the development of synaptic deficits and dendritic spine loss. The consequences of loss of APP are not fully understood. However, studies in an

APP knock-out (APP−/−) mouse model have indicated that APP plays important roles not only in regulating synaptic structure and function leading to reduction in long-term potentiation (Tyan et al. 2012), but also catalyzes the oxidation of Fe^{2+} through a mechanism, which is analogous to ferritin, a ferroxidase that does not have a multicopper active site (Duce et al. 2010). Ferroxidases are known to prevent oxidative stress caused by Fenton and Haber-Weiss chemistry by oxidizing Fe^{2+} to Fe^{3+}. Losses of ferroxidase activities cause pathological Fe^{2+} accumulation.

α-Secretase is stimulated by that Sirtuin1 (SIRT1), a NAD-dependent deacetylase, which directly activates and transcription of α-secretase. Upregulation of α-secretase activity through the 5-hydroxytryptamine 4 (5-HT4) receptor has been shown to reduce Aβ production, amyloid plaque load and to improve cognitive impairment in transgenic mouse models of AD (Pimenova et al. 2014). The molecular mechanism involved in 5-HT4 receptor-stimulated proteolysis of APP is not fully understood. However, it is recently shown that G protein and Src dependent activation of phospholipase C are required for α-secretase activity (Pimenova et al. 2014). Further elucidation of the signaling pathway indicates that inositol trisphosphate phosphorylation and casein kinase 2 activation are prerequisite for α-secretase activity. In addition, α-secretase also initiates the activation of notch pathway by cleaving the membrane-bound notch receptor thus liberating an intracellular domain that activates nuclear genes for neurogenesis (Costa et al. 2005; Hartmann et al. 2001). The similarity in topology and proteolytic processing between APP and Notch suggest that APP may function as a membrane receptor like Notch.

The presenilins and nicastrin are components of γ-secretase. These proteins have not only been implicated in AD, but also have a role in normal development in the Notch signaling pathway. In response to its ligand, Notch receptor undergoes a series of cleavage events culminating in a final presenilin-mediated intramembranous cleavage that releases the Notch intracellular domain (Kopan and Goate 2000). This fragment is translocated to the nucleus, where it influences transcriptional regulation (Artavanis-Tsakonas et al. 1999). The role of presenilins in processing of the Notch receptor is related to their role in processing APP, the protein that accumulates in AD patients. Presenilin is involved in cleaving the βamyloid precursor protein (βAPP) and Notch in their transmembrane domains. The former process (termed γ-secretase cleavage) generates Aβ, which is involved in the pathogenesis of AD. The latter process (termed S3-site cleavage) generates Notch intracellular domain (NICD), which is involved in intercellular signaling. Nicastrin binds both full-length βAPP and the substrates of γ-secretase (C99- and C83-βAPP fragments), and modulates the activity of γ-secretase (Chen et al. 2001). Expression of Notch-1 is significantly increased in the hippocampus of AD patients compared with normal subjects (Mitani et al. 2014; Wagner et al. 2014). Notch-1 has been reported to modulate neurogenesis and neuronal plasticity in the hippocampus (Albensi and Mattson 2000; Oikawa et al. 2012). It is proposed that the Notch signaling pathway is involved in Aβ-mediated neuronal cell apoptosis, but the molecular mechanisms promoting and supporting this process remain unknown.

1.3.2 Amyloidogenic Pathway

The amyloidogenic pathway leads to the production of Aβ from APP (Fig. 1.5) (Octave 2005; Nathlie and Jean-Noel 2008). The remaining membrane-anchored stubs are cleaved within the membrane plane by the γ-secretase complex, releasing the APP intracellular domain (AICD) into the cytosol and leading to the generation of the multiple forms of Aβ (ranging from 37 to 43 amino acids), which form aggregates and accumulate in different subcellular organelles of neurons in AD patients. The predominantly produced species is Aβ40. A trace amount of Aβ42 is produced in a ratio of approximately 99 to 1. Aβ40 is benign, where as Aβ42 is toxic. The less abundant 42 amino acid variant (Aβ42) aggregates much faster than Aβ40 and may therefore be directly related to the pathogenesis of AD (Haass and Selkoe 2007). Aβ42 is the predominant species of Aβ in senile plaques (Verdile et al. 2004). For the sack of simplicity, I will refer Aβ42 as Aβ. Aβ is expressed normally and ubiquitously throughout life. In vitro studies have indicated that Aβ monomers exist in three major conformation forms: α-helix, β-sheet or random coil (Liu et al. 2006). Since the fibrils mostly consist of β-sheets, while the original hydrophobic part of Aβ is an α-helix, the conformational transition of Aβ from α-helix to β-sheet probably is the very first step in fibril formation, and there is evidence that conformational transitions of Aβ monomers depend on physical and chemical parameters of the environment. Aggregation of Aβ is a multistep process which is not only modulated by conformation of Aβ, pH, electrical charge, hydrophilicity, or hydrophobicity of the environment, but also by interaction with other elements that either promote or inhibit Aβ aggregation (Mclauren Dorrance et al. 2000). Several other factors, such as membrane composition and biophysical properties, phosphorylation, and the presence of metal ions also modulate the aggregation of Aβ (Zhu et al. 2015). Thus, extracellular Aβ undergoes phosphorylation by a cell surface-localized or secreted form of protein kinase A (Kumar and Walter 2011). The phosphorylation of serine residue 8 promotes aggregation by stabilizing the β-sheet conformation of Aβ and promotes the formation of oligomeric Aβ aggregates (dimers or trimers or as large as dodecamers) that initiate fibrillization. Phosphorylated Aβ can be detected not only in the brains of transgenic mice, but also in brains of human AD patients. It is suggested that phosphorylation-mediated aggregation of Aβ may be relevant in the pathogenesis of late onset AD (Kumar and Walter 2011). The behavioral symptoms of AD correlate with the aggregation of Aβ and phosphorylation level of Tau (Bloom 2014). Moreover, Aβ has been shown to be an upstream regulator of Tau in AD pathogenesis that triggers the conversion of Tau from a normal to a toxic status. On the other hand, it was shown that Tau enhances Aβ toxicity via a feedback loop (Ballard et al. 2011). The accumulation of Aβ oligomers (also known as ADDLs) predominantly occurs in brain regions associated with learning and memory, including the hippocampus, and binds to sites that are located at neuronal synapses to cause the disruption to neuronal signaling and ultimately neuronal cell death (Gong et al. 2003; Lacor et al. 2004). Unlike the insoluble fibrils, ADDLs can diffuse and have tendency to interact with synapses in hippocampal and entorhinal cortical areas.

A number of ADDL binding sites have been proposed. These sites include glutamate receptors (both ionotropic and metabotropic), insulin receptors, acetylcholine receptors (both muscarinic and nicotinic), as well as cellular prion protein (PrP^c) which may function as a co-receptor for Aβ (De Felice et al. 2009a, b; Lauren et al. 2009; Kessels et al. 2010; Renner et al. 2010). Although the precise binding site remains controversial, ADDLs presumably act via multiple receptors at synapse, thereby contributing to the range of issues that characterize AD.

Interactions of ADDLs with synaptic protein may destabilize the synapses and cause the synaptic loss in the postmortem AD brains (Reddy et al. 2005; Scheff et al. 2007). The level of synaptic reduction closely parallels the degree of premortem cognitive impairment. These studies suggest that ADDLs elevation and synaptic loss, rather than Aβ plaque load, may represent the best indicators of the severity of dementia or cognitive impairment in AD. Collective evidence suggests that ADDLs may act as pathologic ligands for the pathogenesis of AD (Lacor et al. 2004, 2007). At nanomolar concentrations, ADDLs block hippocampal long-term potentiation, cause dendritic spine retraction from pyramidal cells, and impair rodent spatial memory (Lacor et al. 2004; Walsh et al. 2002; Selkoe 2008). Furthermore, ADDLs have been reported to rapidly disrupt synaptic memory mechanisms at very low concentrations via stress-activated kinases and oxidative/nitrosative stress mediators (Farooqui 2010a). Interactions between synapses and ADDLs produce AD-like pathology not only through neuronal Tau hyperphosphorylation (De Felice et al. 2007) and induction of oxidative stress (De Felice et al. 2007), but also through the synapse deterioration and loss (Roselli et al. 2005). ADDLs also downregulate plasma membrane insulin receptors, via a mechanism involving calcium calmodulin-dependent kinase II and casein kinase II inhibition (De Felice et al. 2009a, b). Most significantly, this loss of surface insulin receptors, and ADDL-induced oxidative stress and synaptic spine deterioration, can be completely blocked by insulin. Accumulating evidence suggests that Aβ oligomer plays a central role in the pathogenesis of AD, and Tau acts downstream of Aβ as a modulator of the disease progression. Aβ produces its neurotoxic effects by enhancing the production of ROS not only directly, but also through indirectly. Thus, Aβ oligomers directly generate H_2O_2 not only through the activation of a copper-dependent superoxide dismutase-like activity (Fang et al. 2010) and activation of NADPH-oxidase in astrocytes, but also through the modulation of mitochondrial ROS generation via regulation of activity of enzymes such as Aβ-binding alcohol dehydrogenase and α-ketoglutarate dehydrogenase (Borger et al. 2013). Mitochondria are the major intracellular targets of ADDLs. High levels of ADDLs produce mitochondrial dysfunction, leading not only to the overproduction of reactive oxygen species (ROS) and inhibition of cellular respiration, but also reduction in ATP production and damage to the mitochondrial structure (Starkov and Beal 2008; Reddy and Beal 2008). ADDL-mediated mitochondrial dysfunction leads to changes in homeostasis and intracellular Ca^{2+} in cultured neurons leading to neuronal apoptosis (Starkov and Beal 2008). In addition, accumulation of calcium in mitochondria results in opening of the mitochondrial permeability transition pore (mPTP), a wide mitochondrial membrane channel that allows the passage of large molecules bidirectionally, resulting in disintegration of mitochondrial structure (Starkov and Beal 2008). In cultured cortical neurons, Aβ oligomers not only induce apoptosis through the JNK–c-Jun–FasL–caspase-dependent extrinsic apoptotic pathway (Li et al. 2009), but also modulates redox factor-1 (Tan et al. 2009). Aβ also binds with nanomolar affinity to the cellular prion protein, inducing synaptic dysfunction through changes in long-term potentiation (Lauren et al. 2009). The finding that Aβ oligomers decrease RACK1 distribution in the membrane fraction of cortical neurons suggesting that the Aβ-induced loss of RACK1 distribution in the cell membrane may underlie the impairment of muscarinic regulation of PKC and GABAergic transmission (Liu et al. 2001, 2011). In addition, Aβ activates microglia cells, which induce and support the inflammatory response by releasing cytokines and activating activities of PLA$_2$, iNOS, COX-2 and 12/15-lipoxygenase in AD (Sun et al. 2007; Pratico et al. 2004; Farooqui 2010a). Collective evidence suggests that Aβ promotes neurodegeneration through the induction of oxidative stress and neuroinflammation, which can also be attributed to the potential immunogenicity of amyloid protein misfolding and aggregation, which is likely to occur 10–15 years before the clinical manifestation of AD (Querfurth and LaFerla 2010). For the misfolding of proteins two features are necessary. One feature is the structural change from α-helix configuration of native status to form a cross-β structure. Second feature is the presence of one or two hydrophobic core fragments. During the formation of misfolded proteins, the hydrophobic interaction is the driving force and the cross-β structure provides the platform for assembling. Compared to the native status of the protein, the misfolded assembly provides a changed microenvironment, which is normally more hydrophobic and is a very important feature for developing sensitive fluorescent probes (Leandro and Gomes 2008). For neuroinflammation it is well known that microglial cells secret proinflammatory cytokines. However, it is not known whether brain inflammation occurs at the early, prodromal, or preclinical stage of AD remains unknown (Farooqui 2014).

1.3.3 Degradation of Aβ in the Brain

In the old age, lifestyle-mediated compromised α-site cleavage and overwhelming β-site cleavage may result in over-production of Aβ and ADDLs (Farooqui 2013, 2015). Aβ is normally removed by both global and local mechanisms, with the former requiring vascular transport across the BBB by binding/transport proteins. Local mechanisms associated with the degradation of Aβ involve several enzymic mechanisms. Thus, Aβ is degraded by amyloid-degrading enzymes (ADEs), which are members of the M13 peptidase family (neprilysin (NEP), NEP2 and the endothelin converting enzymes (ECE-1 and -2)). A distinct metallopeptidase, insulin-degrading enzyme (IDE)\, also contributes to the degradation and removal of Aβ in the brain. The ADE family currently has more than 20 members. Some are membrane-bound and others are soluble. NEPs are mainly located on neuronal cells, especially in the striatonigral pathway (Barnes et al. 1988a, b). However, it is also present in the hippocampus, where it functions to inactivate somatostatin, and in cortical regions (Barnes et al. 1995). NEPs play an important role in brain function terminating neuropeptide signals. Reduction in levels of NEP in specific brain areas during aging or after stroke may contribute to the development of AD pathology (Nalivaeva et al. 2014). The deletion of β-secretase (a novel aspartyl protease) and γ-secretase (a complex of proteins containing presenilin1 (PS1) or presenilin2 (PS2), nicastrin (NCT), Aph1 and Pen2) in APP transgenic mice abolishes neuronal production of Aβ and deposition of amyloid plaques (Luo et al. 2001). Thus, increased activities of β-secretase and γ-secretase may be the primary driver for the neurodegeneration and cognitive dysfunction in sporadic AD (Fig. 1.5). Indeed, increase in β-secretase and γ-secretase protein levels and activity have been observed in brains from AD patients (Li et al. 2004b). Neuronal β-secretase and γ-secretase levels and activity increase not only with age, but also following pathological events such as ischemia and traumatic brain injury along with elevation Aβ levels and an increase in risk of AD (Cole and Vassar 2007). It is interesting to note that activities of α-, β-, and γ-secretases and APP processing are modulated by lipids. Thus, cholesterol and GM1 have been reported to increase the generation of Aβ (Zha et al. 2004; Wolozin 2004), whereas docosahexaenoic acid (DHA) and sphingomyelin decrease amyloidogenic processing of APP (Grimm et al. 2005, 2011a, b, 2012). The production of Aβ also depends on the bioavailability of cholesterol in neurons, since the activity balance of the α- and β-secretases is related to the lipid composition of neurons. High concentrations of cholesterol into the neurons result in an increased in amyloidogenic APP process catalyzed by β-secretase, whereas at lower levels, cholesterol metabolism stimulates the α-secretase-mediated APP processing. The hypothesis that statins, the cholesterol lowering drugs for the treatment of AD is based on above mentioned information (Schmitt et al. 2004).

Studies on DHA-mediated reduction in amyloidogenic pathway indicate that DHA acts by decreasing β- and γ-secretase activity. No changes are observed in expression and protein levels of BACE1 and presenilin1. Furthermore, DHA enhances the protein stability of α-secretase causing an increase in nonamyloidogenic processing. It is also reported that in the presence of DHA, cholesterol shifts from raft to non-raft domains, and this is accompanied by a shift in γ-secretase activity and presenilin1 protein levels. At low plasma membrane cholesterol concentrations, APP is processed through α-secretase-initiated processing, but as concentration of cholesterol rises in the membrane, APP is processed by β- and γ-secretases resulting in Aβ production (Vassar 2004; Cui et al. 2011). These observations support the view that Aβ metabolism and amyloidogenic pathway may be coupled with cholesterol homeostasis (Grimm et al. 2007).

Information on neurotoxic effects of Aβ aggregates remains elusive. Amyloid has been reported to be directly toxic to neuronal cell cultures (Golde et al. 2006). Aβ-mediated oxidative stress and inflammation may play a major role in the pathogenesis of AD (Sultana et al. 2009; Farooqui 2010a). Redox reactions occurring during normal cellular respiration produce ROS, such as superoxide anion, hydroxyl radical and hydrogen peroxide. ROS are also produced by the activation of NADPH oxidase and through the oxidation of arachidonic acid by cyclooxygenase-2 (COX-2) (Farooqui and Horrocks 2007). In the brain, when redox balance is lost, ROS attack and alter cellular components, including nucleic acids, lipids and proteins, causing damage to important cellular structures and impairing membrane integrity their functions leading to neuronal cell death (Farooqui 2010a). In AD, increase in ROS is supported not only by the dysregulation of important biological metals, but also by impairment in mitochondrial function. Increase in APP levels in mitochondrial membranes may be one cause of mitochondrial dysfunction in neurons (Kawahara 2010). Furthermore, Aβ may also disrupt mitochondrial membrane function by inserting into it as oligomers, creating calcium-permeable channels (Reddy 2009; Kawahara 2010; Farooqui 2010a). Aβ oligomers may also incorporate into neuronal membranes, leading to a wider dysregulation of calcium homeostasis in the neuron (Kawahara 2010). Other investigators have proposed that Aβ fibrils form pores in neurons, leading to calcium influx and the neuronal death associated with AD (Demuro et al. 2011). Disruption of calcium homeostasis affects cellular function and can trigger apoptotic pathways of cellular death (Farooqui 2010a). Thus, ROS increases Aβ production, which induces oxidative stress, a vicious circle between ROS and Aβ accumulation that may accelerate the progression of AD. Furthermore, fibrillar Aβ peptides induce neuronal apoptosis through the activation of neutral sphingomyelinase (N-SMase) (Jana and Pahan 2004). Treatment of human primary neurons with fibrillar Aβ promote the formation of ceramide and further activation of N-SMase. Treatment of neuronal cultures with antisense of N-SMase protects neurons from Aβ-induced apoptosis and cell death. These studies support the view that fibrillar Aβ may cause neuronal damage through the stimulation of N-SMase (Jana and Pahan 2004) and the sphingomyelin cycle may play an important role in neurodegeneration in AD brain.

Oxidative stress and neuroinflammation are interrelated processes. In AD, neuroinflammatory changes are caused by the activation of microglia, astrocytes, and macrophages particularly in the amyloid deposits (Heneka and O'Banion 2007). The degeneration of neurons results in the release of large amounts of proinflammatory mediators, including cytokines, free radicals, and nitric oxide (NO), all of which increase the generation of insoluble Aβ (Velez-Pardo et al. 2002). Aβ-mediated respiratory burst in microglia produces ROS and tumor necrosis factor alpha (TNF-α), which aggravates Aβ deposition and further neuronal dysfunction and eventual death (Liu et al. 2002). The potentially significant contribution of inflammatory mechanisms in AD has prompted consideration of anti-inflammatory treatment strategies (Kim et al. 2010). Collective evidence suggests that neurodegeneration in vulnerable regions of the brain in AD may contribute to the release of inflammatory mediators such as cytokines and activated complement components (Agostinho et al. 2010; Farooqui 2010a, b). Induction of neuroinflammation in AD brain is also supported by excitotoxicity that has been reported to occur in AD and accompanied by elevated levels of proinflammatory eicosanoids and platelet activating factor (Farooqui et al. 2007, 2008).

1.3.4   Interactions of Aβ Peptide with Other Proteins

Cellular prion protein (PrP^C) is a transmembrane glycoprotein, which is anchored to the extracellular leaflet of the plasma membrane through a carboxyl (C)-terminal glycosylphosphatidylinositol (GPI) anchor (Biasini et al. 2012). The N-terminal portion of PrP^C is intrinsically unstructured, contains the octarepeat region, and has multiple copper binding sites both within and outside of the octarepeat region (Klewpatinond et al. 2008; Stanyon et al. 2014). PrP^C is converted into an aggregated, β-sheet-rich neurotoxic isoform called PrP^{Sc} in prion diseases. PrP^C serves not only as the substrate for PrP^{Sc} conversion and propagation, but also as a transducer of PrP^{Sc}-associated neuronal death. The N-terminus of PrP^C not only binds with oligomeric Aβ peptides (Lauren et al. 2009; Chen et al. 2010), but also interacts with many different cell surface proteins, including the α7 nicotinic acetylcholine receptor (Beraldo et al. 2010), metabotropic glutamate receptors mGluR1 (Beraldo et al. 2010) and mGluR5 (Beraldo et al. 2010; Um et al. 2013), kainate receptor GluR6/7 (Carulla et al. 2011), and α-amino-3-hydroxy-5-methyl-4-isoxazole propionic acid (AMPA) receptor subunits GluA1 (Watt et al. 2012) and GluA2 (Fig. 1.7) (Kleene et al. 2007; Watt et al. 2012). The high affinity binding of PrP^C with Aβ oligomers results in its insertion in the membrane leading to memory impairment in transgenic AD mice (Lauren et al. 2009). The binding of Aβ oligomers with PrP^C in mice impairs hippocampal long-term potentiation, memory, and

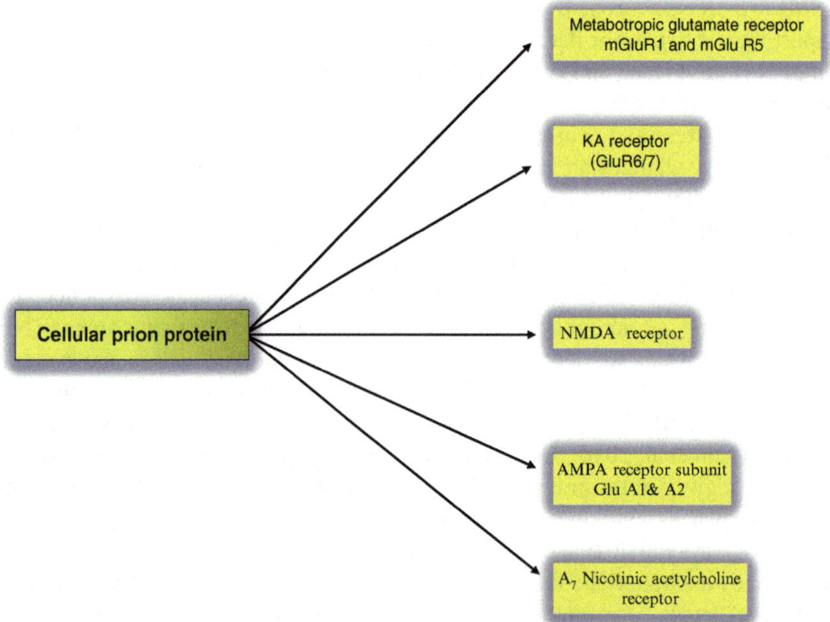

Fig. 1.7 Involvement of prion protein in modulation of various receptors

learning in a manner that involves Fyn, Tau, and glutamate receptors. Furthermore, PrP^C not only catalyzes the formation of certain Aβ oligomeric species in the synapse, but also mediates the toxic effects of other β-sheet rich oligomers as well. Therapeutic approaches utilizing soluble PrP^C ectodomain or monoclonal antibodies targeting PrP^C can at least partially prevent the neurotoxic effects of Aβ oligomers in mice. The binding of Aβ oligomers to PrP^C is coupled with mGluR5-mediated activation of Fyn kinase (Um et al. 2013). This process may contribute to the facilitation of long-term depression (LTD) (Hu et al. 2014). PrP^C also interacts with NMDA receptors and PrP^C-deficient mice display enhanced NMDA receptor-mediated neuronal excitotoxicity (Khosravani et al. 2008) (Fig. 1.8). Tau is required to target Fyn to the dendritic spine where it phosphorylates the NMDA receptor, a prerequisite for recruiting the postsynaptic density (PSD) (95 kD) protein (PSD-95) into a protein complex. This complex then mediates the excitotoxic signaling triggered by Aβ oligomers. The binding of Aβ with PrP^C accounts for only 50 % of the

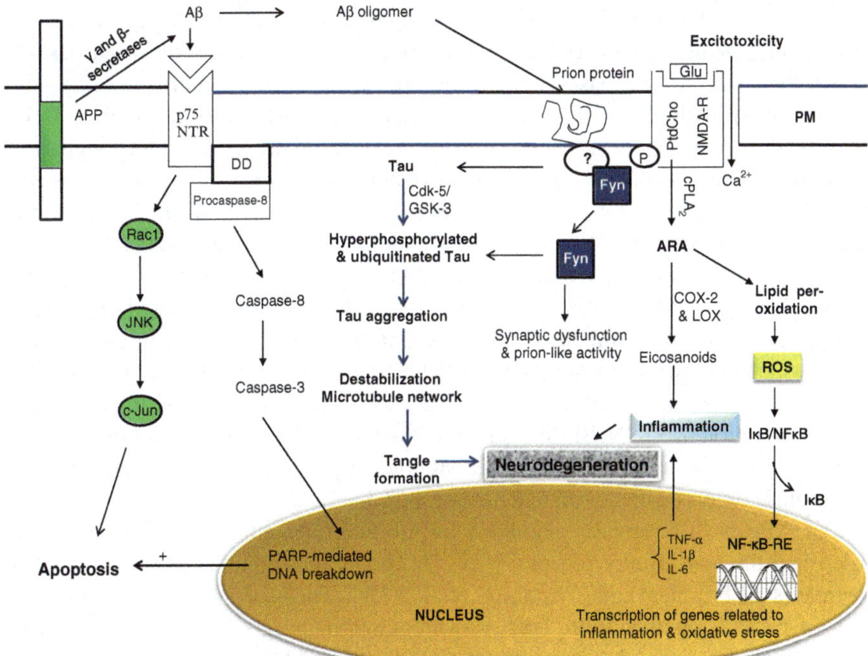

Fig. 1.8 Interactions of Aβ42 oligomer with PrPC protein and Aβ and p75NTR in AD and prion diseases. Amyloid precursor protein (*APP*); β-amyloid (*Aβ*); cellular prion protein (*PrPC*); glutamate (*Glu*); NMDA receptor (*NMDA-R*); phosphatidylcholine (*PtdCho*); Lyso-phosphatidylcholine (*Lyso-PtdCho*); cytosolic phospholipase A$_2$ (*cPLA$_2$*); cyclooxygenase (*COX*); lipoxygenase (*LOX*); arachidonic acid (*ARA*); platelet-activating factor (*PAF*); reactive oxygen species (*ROS*); nuclear factor-κB (*NF-κB*); nuclear factor-κB-response element (*NF-κB-RE*); inhibitory subunit of NF-κB (*I-κB*); tumor necrosis factor-α (*TNF-α*); interleukin-1β (*IL-1β*); interleukin-6 (*IL-6*); inducible nitric oxide synthase (*iNOS*); secretory phospholipase A$_2$ (*sPLA$_2$*); death domain (*DD*); nitric oxide (*NO*); poly(ADP)ribose polymerase (*PARP*)

total neuronal binding of oligomeric Aβ. Additional binding of Aβ is presumably occurs through other receptors, perhaps including low-density lipoprotein receptors, which have been shown to bind several ligands, including Aβ. More studies are required to confirm and elucidate this potential role for PrP^C in Aβ neurotoxicity (Lauren et al. 2009; Chen et al. 2010).

In addition to PrP^C, Aβ also binds with p75^{NTR}, a growth factor that regulates neuron survival, function and structure by acting in concert with a collection of ligands and co-receptors (Chiarini et al. 2006; Longo and Massa 2008). Nerve Growth Factor (NGF) signaling through p75^{NTR} induces death or survival depending on the cellular context, while the pro form of NGF signals through p75^{NTR} and sortilin to produces death (Nykjaer et al. 2004). p75^{NTR} also regulates neurite outgrowth through its interactions with the Nogo and LINGO co-receptors (Mi et al. 2004). Both soluble and aggregated forms of Aβ mediated neuronal apoptosis by binding to p75^{NTR} receptor. These observations strengthen the role of p75^{NTR} in the etiology of AD (Yaar et al. 1997; Yaar et al. 2002; Sotthibundhu et al. 2008; Knowles et al. 2009). Studies with SK-N-BE neuroblastoma cells with and without neurotrophin receptors p75^{NTR} have indicated that p75^{NTR} mediates the Aβ-induced cell death via intracellular death domain (DD). This signaling involves activation of caspase-8, which then activates caspase-3, resulting into apoptotic cell death (Fig. 1.8). The binding of Aβ with p75^{NTR} activates downstream signaling pathways, such as JNK, NF-κB, and PtdIns3 kinase along with the release of proinflammatory signaling involving the release of cytokines. Collective evidence suggests that overexpression of p75^{NTR} in a variety of cell lines confers more sensitivity to Aβ-mediated neurotoxicity (Perini et al. 2002), whereas p75^{NTR}-deficient mouse hippocampal neurons are resistant to Aβ-mediated neurotoxicity (Sotthibundhu et al. 2008). It is proposed that in AD, the activation of the redox-sensitive NF-κB and increased expression and release of pro-inflammatory cytokines amplify neuronal death through astrocytes, which flood neurons with NO and its lethal metabolite, ONOO–. Furthermore, p75^{NTR} and its DD also interact with prion protein fragment PrP106–126 and induce the degeneration of SK-N-BE human neuroblastoma cells. The levels of membrane-associated p75^{NTR} in hippocampus are significantly higher in human postmortem AD brains compared to age-matched controls (Chakravarthy et al. 2012). The cleavage of APP is differentially regulated by the neurotrophin high-affinity receptor TrkA and the low-affinity receptor p75 ^{NTR}. This neurotrophin promotes and TrkA decreases APP β-cleavage (Costantini et al. 2005). Therefore, aberrantly upregulation of p75^{NTR} together with TrkA downregulation in aged brains may result in enhanced generation Aβ (Costantini et al. 2006). In addition, p75^{NTR} may increase Aβ production not only through its ability to stabilize secretases, but also through the activation of sphingomyelinase and consequent ceramide production (Costantini et al. 2006; Puglielli et al. 2003). Collective evidence suggests that neurons expressing p75^{NTR} along with increased expression of proinflammatory cytokine are preferential targets of Aβ and prions toxicity in AD as well as other prion diseases (Chiarini et al. 2006; Bai et al. 2008). Recent studies have also indicated that the binding of Aβ oligomers to PrP^C activates Fyn, a Src family kinase. The activation of Fyn is required for NMDA-R phosphorylation and cell surface distribution, dendritic spine loss, along with the release of LDH (Um et al. 2013) supporting the view that PrP^C may play a central role in AD pathogenesis.

Besides binding with PrP^C, and p75^NTR, Aβ interacts with a variety of cell surface receptors, including scavenger receptors (El Khoury et al. 1996) and NH$_2$-formylpeptide receptor 2 in microglia (Tiffany et al. 2001), advanced glycation end products receptors (RAGE) in neurons and microglia (Yan et al. 1996), serpin-enzyme complex receptor (Boland et al. 1996), α-7-nicotinic acetylcholine receptor (α7NAChR) (Wang et al. 2000), APP (Kuner et al. 1998), β-amyloid binding protein (BBP) containing a G protein-coupling module (Lorenzo et al. 2000); and mitochondrial matrix proteins [Abeta-binding alcohol dehydrogenase (ABAD) and cyclophilin (CypD)] (Lustbader et al. 2004; Du et al. 2008) in neurons (Fig. 1.9). Binding of Aβ with the α7NAChR leads to the internalization of NMDARs and modulation of extrasynaptic NMDAR signaling. Among these protein interactions of Aβ with mitochondrial matrix proteins ABAD and CypD are particular important in the pathogenesis of AD. It is reported that the binding of Aβ with ABAD and CypD induce elevated free radical generation and cause mitochondrial dysfunction in AD neurons (Lustbader et al. 2004; Du et al. 2008). However, the mechanistic link between Aβ and mitochondrial damage is not well understood. It is reported that interactions of ABAD and Aβ interact within the mitochondria, resulting in protein expression changes in the brain. The two best described proteins are peroxiredoxin-2 (an antioxidant protein) and endophilin-1 (a protein involved in synaptic vesicle endocytosis and receptor trafficking). The expression of these proteins is

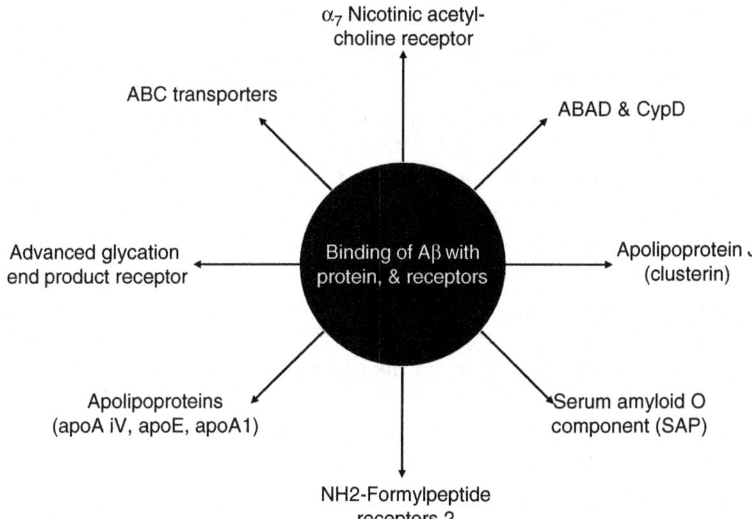

Fig. 1.9 Interactions of β-amyloid with proteins and receptors

linked with synaptic activity (Borger et al. 2013). It is not known that how do the interactions between Aβ and above mentioned proteins are integrated in causing neurodegeneration. Collective evidence suggests that Aβ oligomers bind with numerous neuronal receptors (Yan et al. 1996; Wang et al. 2000; Kuner et al. 1998; Lustbader et al. 2004; Du et al. 2008) and modulate synaptic signaling cascades including MAPK, Akt, Wnt, and Rho pathways. This supports the view that Aβ oligomers may act as a nonspecific pathological receptor ligand/agonist, both at the pre- and postsynaptic membrane. In addition, Aβ oligomers may bind to membranes directly through GM1 ganglioside leading to the induction of structural and functional changes which may impact Ca^{2+} signaling and synaptic plasticity (Hong et al. 2014). A global impact of Aβ oligomers on different signaling pathways and their respective signaling components is consistent with the widely held view that kinase and phosphatase activities are imbalanced early on in the pathogenesis in diseased neurons, resulting in improper hyperphosphorylation of downstream substrates including Tau (Augustinack et al. 2002; Stoothoff and Johnson 2005).

1.3.5 *Contribution of Hyperphosphorylated Tau Protein in the Pathogenesis of AD*

The Tau protein, which is roburstly expressed in neuronal exons, is a disordered protein encoded by a single gene (MAPT) located on chromosome 17q21 in human. It is a member of the heat-stable microtubule-associated proteins (MAP1 and MAP2), which are mainly expressed in mammalian neurons. MAP2 is localized in dendrites, where as Tau binds to microtubules, which are localized in axons. Tau consists of four regions: an N-terminal projection region, a proline-rich domain, a microtubule-binding domain (MBD), and a C-terminal region (Mandelkow et al. 1996). Thus, under normal conditions, Tau proteins stabilize microtubules, which are essential for maintaining cell structure, providing platforms for intracellular transport, forming the spindle during mitosis, and other cellular processes as well. Under abnormal conditions, Tau no longer stabilizes microtubules properly, and this abnormality can result in AD and other dementia such as Lewy body dementia (de Calignon et al. 2010). Tau also plays a role in axonal transport, neurite outgrowth, and adult neurogenesis (Morris et al. 2011). Tau binds to microtubules via the C-terminal microtubule-binding repeats, which consist of three or four imperfect repeats of 31 or 32 amino acids. Tau not only supports the microtubule assembly, but also contributes to the stabilization of microtubules through lateral binding to the surface of microtubules (Matus 1994; Mandelkow et al. 1995). Tau activities are modulated by phosphorylation in the microtubule-binding repeat or the franking region by a number of protein kinases including cyclin-dependant kinase-5 (Cdk-5), glycogen synthase kinase-3 (GSK-3), CaM kinase II, casein kinase II, stress-activated kinase, c-Jun N-terminal kinase (SAPK/JNK) and kinase p38 (Gong and Iqbal 2008; Avila et al. 2010).

The identity and the strict number of tau kinases involved in the pathogenesis of AD remain uncertain. So investigators have focused at specific tau phosphorylation site(s). Tau contains 35 threonine, 45 serine, and 5 tyrosine residues meaning that nearly 20 % of the Tau protein has the potential for phosphorylation. Cdk-5 is a major kinase that phosphorylates Tau. In Tau the chemical nature of phosphorylation site and their number remains controversial. In AD brain, the phosphorylation of Tau mainly occurs at (Ser/Thr)-Pro sequences (Morishima-Kawashima et al. 1995). Among the 16 (Ser/Thr)-Pro sequences, which are present in Tau, cyclin-dependent kinases (Cdk5) phosphorylates 9–13 sites (Chauhan et al. 2005; Hanger et al. 2009). Initial studies using amino acid sequencing have indicated that Cdk5-p25 phosphorylates at Ser202, Thr205, Ser235, and Ser404 (Arioka et al. 1993). Other investigators have reported the phosphorylation of Ser195, Ser202, Thr205, Thr231, Ser235, Ser396, and Ser404 (Paudel et al. 1993) using purified nclk (Cdk5-p25). Similarly, another group of investigators using in vitro studies has demonstrated that Ser202, Thr205, Ser235, and Ser404 to be major sites with Thr153 and Thr212 as minor sites by Cdk5-p25 (Illenberger et al. 1998). Microtubule affinity-regulating kinase (MARK) is another enzyme that phosphorylates Tau in AD. This phosphorylation occurs on KXGS motif in the microtubule binding domains. MARK predominately promotes that phosphorylation of Tau on Ser^{262} in AD. Recent studies have also indicated that Fyn kinase, which contributes to multiple pathways that underlie AD (Haass and Mandelkow 2010), including mediating the toxicity of Aβ oligomers (Chin et al. 2005) and linking Aβ to Tau toxicity (Roberson et al. 2011). Tau also undergoes ubiquitination. The molecular mechanisms by which accumulation of hyperphosphorylated and ubiquitinated Tau aggregates and contributes to the pathogenesis of AD remains unclear. However, there are several possibilities. First, mathematical modeling experiments indicate that bulk accumulation of Tau aggregates in cell bodies may depress neuronal energy metabolism through molecular crowding leading to long term alterations in neuronal physiology (Vazquez 2013). Second, in addition to homotypic aggregation, Tau can coaggregate with other proteins (Giasson et al. 2003), including microtubule associated proteins (Alonso et al. 1997), potentially depressing their levels. Loss of normal Tau protein is well tolerated in animal models, but in AD patients, simultaneous depletion of different classes of microtubule associated proteins may produce severe consequences including the destabilization of microtubule network along with impairment in axonal transport, ultimately leading to NFT formation and neurodegeneration (Martin et al. 2013; Teng et al. 2001). Hyperphosphorylation of Tau makes Tau resistant to calcium activated proteases, calpains and the ubiquitin-proteasome pathway. Finally, some species of hyperphosphorylated and ubiquitinated Tau aggregates and can directly disrupt membrane integrity (Lasagna-Reeves et al. 2014). It is proposed that ubiquitination of hyperphosphorylated Tau along with dysfunction of the ubiquitin-proteasome system may worsen the accumulation of insoluble fibrillar Tau (Fibrillar Tau), which exerts its neurotoxic effects by increasing oxidative stress, neuronal apoptosis, mitochondrial dysfunction, collapse of the microtubule-based cytoskeleton, and subsequent neuronal demise (Mandelkow et al. 2003; Arnaud et al. 2006;

Oddo 2008). These conclusions support the view that Tau oligomers are neurotoxic. The clinical symptoms correlate with Tau pathology and the fact that anomalous Tau hyperphosphorylations constitute a common final pathway for many other dementias. Dephosphorylation of Tau is catalyzed by phosphatases. However, the activity of protein phoshatase-2A (PP2A) is down-regulated in Alzheimer disease brains whereas its inhibitors, I_1^{PP2A} and I_2^{PP2A}, are overexpressed (Iqbal et al. 2009). How Aβ pathology is related to Tau pathology remains unclear? It is becoming increasingly evident that direct or indirect interactions of Aβ with Tau accelerate NFT formation. Aβ and Tau do not act separately and it is proposed that significant crosstalk may occur between these two molecules (Scheff et al. 2007). Tau pathology lies downstream of Aβ deposition in a pathocascade (Blurton-Jones and Laferla 2006). A reduction in endogenous Tau levels in APP transgenic mice not only results in reversal in memory impairment, but also in reduction in susceptibility to experimentally induced excitotoxic seizures, and decrease in early mortality, without altering Aβ levels or plaque load (Roberson et al. 2007). This observation on the role of Tau in synaptotoxicity is supported by the observation that hyperphosphorylated Tau accumulation in dendritic spines of cultured CA3 hippocampal neurons (Zempel et al. 2010). Crossing of human APP transgenic mice with human Tau transgenic mice lead to marked increase in aggregation of Tau with concomitant dendritic spine loss, and acceleration in cognitive impairment (Chabrier et al. 2014). Other processes supporting the crosstalk between Aβ and Tau include: Aβ-mediated neuroinflammation, which may contribute to induction and enhancement of Tau phosphorylation by inflammatory cytokines; Aβ-mediated proteasomal impairment of Tau degradation; and dysregulation of axonal transport with possible bidirectional effects, leading to increasing Aβ as well as Tau (Blurton-Jones and Laferla 2006). Accumulation of Aβ and phosphorylated Tau has been reported to exert an effect on synapses, downregulating the density and function of synapses and consequently leading to abnormal circuiting in neuronal network within the brain (Selkoe 2008; Iqbal et al. 2009). In AD, accumulation of Aβ and the spatial distribution and severity of NFT formation closely correlate with cognitive impairment and brain atrophy observed in AD (Nelson et al. 2012). Aβ42 toxicity can be prevented by Tau reduction or deficiency as evidenced by transgenic mouse studies (Roberson et al. 2007).

1.3.6 Contribution of Insulin and Insulin Resistance in Pathogenesis of AD

Insulin is an anabolic polypeptide hormone consisting of a 21 amino acid α chain and a 30 amino acid β chain. It is produced and secreted by β-cells of the islets of Langerhans in the pancreas. It regulates glucose uptake and utilization by cells, and free fatty acid levels in peripheral blood and therefore plays the central role in regulating energy metabolism in the body. Neural cells also produce small amount of insulin. Peripheral insulin can cross the blood brain barrier and exerts its effects via the insulin receptor (IR), which are found in the olfactory bulb, cerebral cortex, hippocampus, cerebellum and hypothalamus (Unger et al. 1989). The IR is a hetero-tetrameric receptor composed of two extracellular α-subunits that bind insulin, and two transmembrane β-subunits that have intracellular tyrosine kinase activity. In addition to insulin, brain also contains insulin-like growth factors-1 (IGF-1) and -2 (IGF-2) receptors (Zhao et al. 1999). Due to structural and functional homology, insulin and IGF-1 can bind to both IR and IGF-1R (Conejo and Lorenzo 2001). The binding of insulin with the extracellular domain of the IR promotes the autophosphorylation by triggering intrinsic tyrosine kinase activity. The phosphorylation of intracellular substrates leads to the recruitment and activation of multiple proteins and the initiation of several signaling cascades, including PtdIns 3K/Akt and the mitogen-activated protein kinase (MAPK) signaling pathways (Johnston et al. 2003; van der Heide et al. 2006). MAPK pathway is mainly involved in cell differentiation, cell proliferation and cell death, whereas Akt signaling is implicated in cell proliferation, cell growth, and protein synthesis (Brazil and Hemmings 2001; Tremblay and Giguere 2008). In brain, insulin signaling has been reported to regulate neuronal survival, neurotransmission and synaptic activities (Zhao and Alkon 2001). Insulin is also associated with regulation of synaptic plasticity by modulating long-term potentiation (LTP) (Zhao et al. 2010; Nistico et al. 2012), and promoting long-term depression (LTD) (Labouèbe et al. 2013). There are two major mechanisms involved in learning and memory. Collective evidence suggests that in the brain, insulin and IGF regulate neuronal and glial functions such as growth, survival, metabolism, gene expression, protein synthesis, cytoskeletal assembly, synapse formation, neurotransmitter function, and plasticity (Farooqui 2013; de la Monte and Tong 2013, 2014), and therefore play important roles in cognitive function. It has been suggested that AD may represent a metabolic disease of the brain associated with brain insulin and insulin-like growth factor-I (IGF-I) resistance (de la Monte and Tong 2013; Farooqui 2013, 2015), which is defined by the reduction of insulin capacity to activate glucose utilization, either by insulin deficiency or by impairment in its secretion and/or utilization. Insulin resistance is modulated by both genetic and acquired factors. The molecular mechanism associated with insulin resistance is not fully understood. However, consumption of high calorie diet (western diet) and elevations in levels of saturated fatty acids and phospholipid-derived—(diacylglycerol), and sphingolipid-derived (ceramide) lipid mediators in the brain may contribute to insulin resistance (Holland and Summers 2008; Farooqui 2013, 2015). During insulin resistance, one develops reduced sensitivity to insulin resulting in hyperinsulinemia, and this impairment in insulin signaling is closely associated with the pathogenesis of AD. Insulin increases the levels of Aβ in plasma in AD subjects, and these effects may contribute to the risk of developing AD in patients with type II diabetes and MetS. Aβ oligomers contribute to neuronal insulin resistance in the AD brain not only by blocking the insulin network through insulin/Akt pathway (Townsend et al. 2007), but also by dissociating insulin receptors after binding at the dendrites of synaptic sites (Zhao et al. 2008). The fact that insulin and Aβ are hydrolyzed by the similar enzymes has attracted the attention of many investigators on the role of insulin signaling in Aβ clearance. Increases in insulin levels frequently seen in insulin resistance may compete for these enzymes and thus contribute to Aβ accumulation. Indeed, insulin signaling has been shown to regulate expression of metalloproteases such as IDE (Qiu and Folstein 2006), and influence aspects of Aβ metabolism and catabolism (Gasparini et al. 2001). In addition, intracellular Aβ oligomer interferes with insulin receptor signaling by interacting with phosphoinositide-dependent kinase (PDK) and Akt (Lee et al. 2009). Impaired insulin signaling cannot efficiently block GSK3β and therefore, the activated GSK3β triggers APP γ-secretase activity and increases Tau phosphorylation (Hooper et al. 2008) simultaneously aggravating the APP processing and Tau phosphorylation in AD brain. As stated above, Aβ oligomers are small, diffusible aggregates that can specifically attach with synapses in hippocampal and cortical neurons and act as pathogenic ligands for the loss of synapse in AD (De Felice et al. 2009a, b). Furthermore, AD is accompanied by the presence of advanced glycation endproduct (AGE)-modified proteins (Takeuchi and Yamagishi 2008). These modified proteins are ligands for the RAGE, as is Aβ peptide. The interactions between RAGE and its ligands triggers the activation of a key cell signaling pathway, such as p21^{ras} and transcription factor nuclear factor-kappa B (NF-κB) (Takeuchi and Yamagishi 2008). The extent to that this happens in vivo is still unclear, but increased RAGE expression along with increased amounts of AGE-modified proteins is believed to be a major cause of the vascular complications in AD. Blocking interactions between Aβ and RAGE impairs the activation of microglia and reduces the production of proinflammatory mediators (Ramasamy et al. 2009). RAGE also plays an important role not only in the clearance of Aβ, but also contribute to apoE-mediated cellular processing and signaling (Bu 2009). In addition to AGEs and Aβ, RAGE recognizes other ligands including serum amyloid A, S100 protein, and high-mobility group box1. These molecules are often present in the altered tissue environments associated with type 2 diabetes. The binding of Aβ oligomer with RAGE induces AD-like pathology, oxidative stress (Farooqui 2010a), synapse deterioration and loss (De Felice et al. 2009a, b; Lacor et al. 2007), and inhibition of synaptic plasticity (Fig. 1.10) (Lambert et al. 1998).

Insulin dysfunction also contributes to inflammation in visceral tissues and brain. In the visceral tissues, insulin modulates many aspects of inflammatory process. Thus, at low levels, insulin exerts anti-inflammatory effects (Dandona 2002) whereas, during chronic hyperinsulinemia, insulin may exacerbate inflammatory responses by increasing markers of oxidative stress and inflammation (Krogh-Madsen et al. 2004). Similarly, in the brain, several animal studies have indicated that insulin dysfunction may be involved in the neuroinflammation. For example, a significant increase in the number of glial fibrillary acidic protein (GFAP)-reactive astrocytes has been reported to occur in the hippocampi of both nonobese diabetic and stretozotocin-induced mouse models (Saravia et al. 2002). Collective evidence suggests that the desensitization of insulin receptors and insulin resistance reduce the synthesis of several proteins, such as insulin-degrading enzyme (IDE) leading

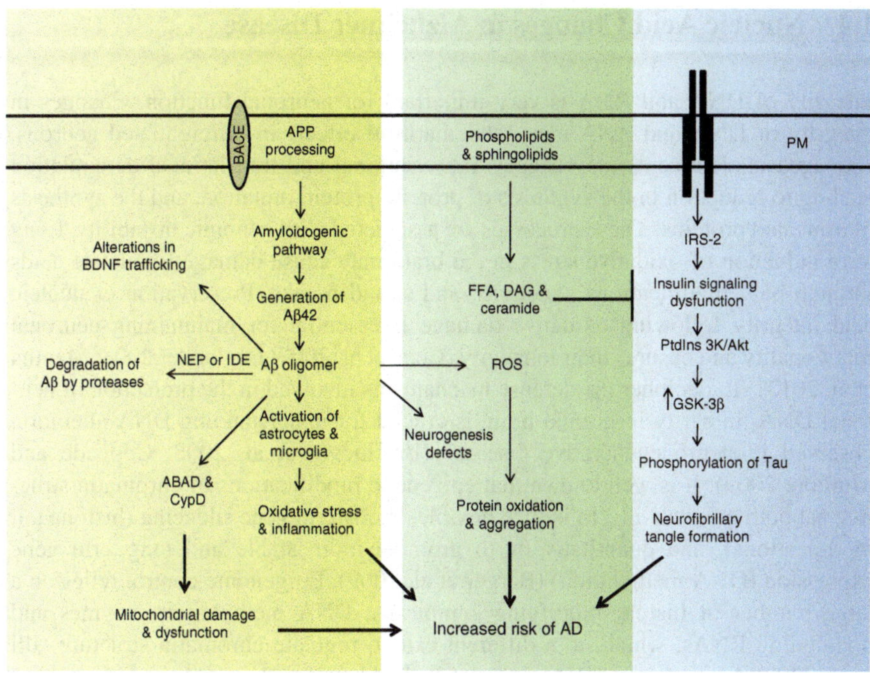

Fig. 1.10 Cross talk among APP processing, phospholipid- and sphingolipid-derived lipid mediators and insulin receptor signaling. Amyloid precursor protein (*APP*); insulin-degrading enzyme (*IDE*); neprilsin (*NEP*); and reactive oxygen species (*ROS*); free fatty acid (*FFA*); diacylglycerol (*DAG*); Aβ-binding alcohol dehydrogenase (*ABAD*); cyclophilin (*CypD*); insulin receptor substrate-2 (*IRS-2*); phosphatidylinositol-3 kinases (*PtdIns 3K*); and glycogen synthase kinase-3β (*GSK3β*)

to greater amyloid deposition. Reduction in insulin signaling produces increase in glycogen synthase kinase-3β activity leading to enhancement in phosphorylation of Tau protein and the formation of NFTs. Stimulation of insulin signaling also protects neurons from Aβ oligomer-induced impairment of LTP (Townsend et al. 2007) and accumulation of hyperphosphorylated Tau (Escribano et al. 2010). Furthermore, in AD defective insulin signaling makes neurons not only energy deficient and vulnerable to oxidative stress and neuroinflammation, but also impairs synaptic plasticity (Bosco et al. 2011). In fact, oxidative stress induced alterations in mitochondrial function are closely associated with loss of synapse in AD (Farooqui 2010a). Hyperinsulinaemia as well as complete lack of insulin causes an increase in Tau phosphorylation resulting in an imbalance in insulin-regulated Tau kinases and phosphatase. Chronic hyperinsulinemia may also exacerbate inflammatory responses and increase markers of oxidative stress (Bosco et al. 2011). These observations are consistent with the concept that high calorie diet-mediated alterations in insulin and insulin receptors may play an important role in the pathogenesis of AD.

1.4 Nucleic Acid Changes in Alzheimer Disease

Integrity of DNA and RNA is very important for neuronal function. Changes in integrity of DNA and RNA may cause harmful effects in differentiated neurons. Damaged nucleic acids may trigger transcriptional and translational deregulation leading to reduction in the synthesis of protein, protein mutation, and the synthesis of truncated proteins. These processes are associated with genomic instability. Long term induction of oxidative stress in the brain may cause damage to nucleic acids through base modifications, deletions, and strand breaks. Preservation of nucleic acid integrity following oxidative damage is essential for maintaining neuronal functionality and ensures their longevity (Chen et al. 2007; Englander 2008; Mantha et al. 2013). To decipher the defense mechanisms involved in the protection of neuronal DNA, integrity in normal brain is crucial for understanding DNA alteration observed in neurodegenerative diseases (Brasnjevic et al. 2008; Coppedè and Migliore 2009). It is well known that epigenetic modifications of chromatin structure act both qualitatively, to induce flexible, short-term gene silencing (histone tail modifications), and quantitatively, to promote more stable and long-term gene expression (DNA methylation) (Berger et al. 2009). Epigenome control relies on a large number of histone-modifying complexes, DNA methylation enzymes and non-coding RNAs, which, to a different extent, regulate chromatin structure (Illi et al. 2009). In the brain, DNA methylation involves covalent addition of a methyl group from the methyl donor (**S**-adenosylmethionine) to a cytosine base within the DNA. This reaction is catalyzed by a family of DNA methyltransferases (Gräff et al. 2011). DNA methylation in the promoter region of a gene is known to alter transcriptional activity (Robertson and Wolffe 2000). CpG islands are extended regions of cytosine and guanine repeats in the promoter region of many mammalian genes. These sites are heavily targeted by DNA methyltransferases and are known to modulate gene expression (Feng et al. 2010). Epigenetic modifications of chromatin structure may be closely associated with the progression of AD (Cencioni et al. 2013). Thus, DNA methylation has been reported to regulate gene transcription through the modulation of promoters. It is not yet entirely clear how changes in DNA methylation are related to neurochemical changes in AD (Farooqui 2010a). However, several studies have indicated that DNA hypomethylation not only correlates with a greater amyloid plaque burden, enhanced APP production, and increased activity of enzymes β-secretase (BACE-1/PS1) along with the onset of oxidative stress, but is also involved in the upregulation of the proinflammatory gene, redox sensitive NF-κB, as well as cyclooxygenase-2 (COX-2), an enzyme, which catalyzes the generation of inflammation inducing prostaglandins (Rao et al. 2012; Deaton and Bird 2011; Chouliaras et al. 2013; Iwata et al. 2014; Gu et al. 2013). Conversely, the hypermethylation of DNA down-regulates promoters for BDNF and c AMP-responsive element (CREB), which interfere with synaptic plasticity. These studies support that view that epigenetic mechanisms involving upregulation of arachidonic acid cascade along with downregulation of BDNF and CREB may contribute to the pathogenesis of AD (Rao et al. 2012). It is not known whether the epigenetic changes observed in AD represent a cause or a consequence of the disease. However, several studies support the view that epigenetic mechanisms may modulate the risk of AD (Chouliaras et al. 2010). Thus, pharmacologically inhibition of DNA methylation in the hippocampus after a learning task has been reported to impair in memory consolidation in mice (Day and Sweatt 2011), and stimulation of histone acetylation improves learning and memory in a mouse model of AD along with increased expression of learning-related gene in aged wild type mice (Fischer et al. 2007; Peleg et al. 2010). These studies support the epigenetic regulation of learning and memory in health and diseases.

Epigenetics is defined as an area of biology, which connects environmental factors with gene expression patterns in cells. It involves sustained and potentially heritable (by meiosis and/or mitosis) alterations in gene expression exerted in the absence of altered DNA sequence (Millan 2013). Epigenetic mechanisms, which are involved in AD range from DNA methylation to altered posttranslational marking of histones to regulatory actions of noncoding RNAs (ncRNAs), with a particularly rich and challenging studies published on microRNAs (miRNAs) (Millan 2013). DNA hypomethylation may lead to an upregulation of the proinflammatory gene, NF-κB, as well as that encoding cyclooxygenase-2 which catalyzes the generation of prostaglandins, leukotrienes and thromoxanes (Rao et al. 2012; Gu et al. 2013) supporting the view that aberrant DNA methylation may contribute to the neuroinflammation. Conversely, the hypermethylation of promoters for BDNF and CREB has been reported to interfere with synaptic plasticity (Rao et al. 2012).

Epigenetic regulation of gene expression includes heritable and reversible DNA modifications, modifications of DNA-binding proteins, such as histones, and generation of miRNAs that interact with messenger RNAs (mRNAs) leading to degradation of the mRNA thereby suppressing gene products (Chuang and Jones 2007). miRNAs are a group of small noncoding RNAs (around 22 nucleotides) that regulate translational repression of target mRNAs. miRNAs act by binding with partial complementarity to messenger RNA (mRNA) sequences, mainly in the 3′ untranslated region (3′UTR). This targeting leads to either degradation or translational repression of the mRNA template(s), causing an overall downregulation in protein output. miRNA play important role in multiple biological processes such as development, proliferation, inflammation, and apoptosis (Xu et al. 2004; Thounaojam et al. 2013). Studies on the effect of oxidative stress on miRNA in hippocampus have indicated that oxidative stress alters the miRNA expression profile of hippocampal neurons, and the deregulated miRNAs might play potential roles in the pathogenesis of AD (Zhang et al. 2014). Aberrant expression and dysfunction of brain-enriched miRNAs has been reported in AD and its animal models. Thus, studies in Drosophila AD models have indicated that levels of microRNAs are consistently dysregulated in adult-onset AD Drosophila brains: eight of which are upregulated (miR- 8, miR-13b, miR-277, miR-279, miR-981, miR-995, miR-998, miR-1017) and nine are downregulated (let-7, miR-1, miR-9a, miR-184, miR-193, miR-263b, miR-276a, miR-285, miR-289) (Kong et al. 2014). Similarly, dysregulation of microRNAs (miRNA-9, miRNA-125b, miRNA-34a, miRNA-155, and miRNA-146a) has also been reported to occur in temporal cortex of AD patient at short postmortem interval (Sethi and Lukiw 2009). The up-regulation of NF-κB-sensitive miRNAs may be involved in the innate immune and inflammatory response and synaptic, neurotrophic, and amyloidogenic functions. In AD a family of above mentioned miRNAs down-regulate the expression key regulatory genes involved interactively in neuroinflammation, synaptogenesis, neurotrophic functions, and amyloidogenesis (Lukiw et al. 2013). Among these miRNAs subfamily, miRNA-125b has been reported to occur in human brain abundantly. Bioinformatics analysis has revealed that an up-regulated miRNA-125b may potentially target the 3′untranslated region (3′-UTR) of the messenger RNA (mRNA) encoding (a) a 15-lipoxygenase (15-LOX) (Zhao et al. 2014), the enzyme that oxidizes and facilitates the conversion of docosahexaneoic acid into neuroprotectin D1 (NPD1), a docosanoid, which is closely associated with neuroprotective effects of docosahexaenoic acid (Farooqui 2009a, 2011). In contrast, elevated miRNA-146a in AD brain is known to specifically target complement factor H (CFH) and the interleukin-1 associated kinase-1 (IRAK-1) mRNAs, and is believed to contribute to altered innate immune responses and neuroinflammation in degenerating human brain cells and tissues in AD (Lukiw et al. 2013). This suggests unless specifically stabilized, certain brain-enriched miRNAs represent a rapidly executed signaling system employing highly transient effectors of CNS gene expression. At present, it is not known whether this upregulation correlates with neuropathological changes in AD brain. Several studies have indicated that Tau-post-translational modification (Tau-PTM) is also controlled by microRNAs in AD (Dickson et al. 2013; Sala Frigerio et al. 2011; Smith et al. 2011). Thus, the synthesis of Tau itself is accelerated by reductions in the levels of miR-27a-3p and 34a, both of which target it directly (Dickson et al. 2013; Sala Frigerio et al. 2011). In addition, it is reported that (b) Tau precursor can be alternatively spliced by the APP-processing enzyme, polypyrimidine tract binding protein (PTPB). Its actions modify the ratio of "4R" to slightly-shorter "3R" isoforms, the former being more prominent in AD. Thus, downregulation of miR-124a and 132, which target PTPB, favors aberrant splicing of Tau into neurotoxic isoforms in the brain (Smith et al. 2011). Finally, downregulation of miR-15a, 103/107 and 27a-3p enhances excess phosphorylation of Tau by extracellular regulated kinase (ERK)1, cyclin-dependent kinase (Cdk)-5 and glycogen synthase kinase (GSK)-P respectively (132; 133). All these studies support the view that both APP processing and synthesis of Tau are modulated by miRNA in AD.

The involvement of epigenetic factors is also supported by studies on mitochondrial DNA (mtDNA) deletions. Thus, significantly higher levels of mitochondrial DNA (mtDNA) deletions have been observed in large vulnerable neurons of the hippocampus and neocortex of AD patients compared to age-matched controls (Hirai et al. 2001). It is proposed that Aβ promotes oxidative stress, which may be responsible for introducing mutations in mtDNA and impair mitochondrial function (Abramov et al. 2004). The main products of ROS-mediated mtDNA base damage are thymine glycol among pyrimidines (Wang et al. 1998b), which has low mutagenicity, and 7,8-dihydro-8-oxo-2′-deoxyguanosine among purines (Bohr 2002), which can cause characteristic $G \rightarrow T$ transversions upon replication (Wang et al. 1998a, b). Although, the molecular mechanisms are not fully understood, oxidative damage-associated single-or double-strand breaks may contribute to the deletions of mtDNA. As a result, these gradually accumulated mtDNA mutations may potentially induce a reduction in the efficiency of the electron transport chain complex, decrease in ATP production and increase in ROS production. In return, the increase in ROS could cause subsequent accumulation of more mtDNA mutations and create a positive feedback loop of increasing mutations and ROS production, followed by eventual cell death (Hsieh et al. 1994). Mutations in mtDNA may cause aging, which is the leading risk factor for AD. In addition, increased levels of 4-HNE in AD inhibit DNA synthesis (Farooqui 2009b). RNA is more susceptible to oxidative damage than DNA because RNA is largely single stranded and its bases are not protected by hydrogen bonding and specific RNA-binding proteins (Nunomura et al. 2009). Also, in cytoplasm, cellular RNA is located close to mitochondria, which are the primary generator of ROS. In AD significant amounts of polyA$^+$ mRNAs are oxidized. The oxidation of RNA oxidation is not random but highly selective. Quantitative analysis in AD brain indicates that some mRNA species are more susceptible to oxidative damage (Shan et al. 2003). Oxidative modification can occur not only in protein-coding RNAs but also in non-coding RNAs. Damage to coding and non-coding RNAs may induce errors in proteins and alter the regulation of gene expression in AD (Nunomura et al. 2009). Significant mRNA oxidation (30–70 %) occurs in the frontal cortex of AD brain (Shan and Lin 2006). In earlier stages of AD, levels of 8-OHG are elevated in the cytoplasm of AD hippocampus, frontal, and occipital neocortex, which is correlated with the β-amyloid load (Lovell and Markesbery 2008; Nunomura et al. 2009).

1.5   Metabolic Syndrome as a Risk Factor for Alzheimer Disease

The metabolic syndrome (MetS) is a cluster of common pathologies in which abdominal obesity is linked to an excess of visceral fat, insulin resistance, dyslipidemia and hypertension. The clustering of above factors increases the risk of type 2 diabetes mellitus and cardiovascular disease (Farooqui et al. 2012; Farooqui 2013; de la Monte and Tong 2013). At the molecular level, MetS is accompanied not only by dysregulation in the expression of adipokines and cytokines, but also by alterations in the levels of leptin, a peptide hormone released by white adipose tissue. These changes modulate immune response and inflammation that lead to alterations in the hypothalamic 'bodyweight/appetite/satiety set point' resulting in the initiation and development of the MetS (Farooqui et al. 2012; Farooqui 2013; de la Monte and Tong 2013). MetS is a risk factor for neurological disorders such as stroke, depression, and Alzheimer's disease. The molecular mechanism underlying the mirror relationship between metabolic syndrome and neurological disorders is not fully understood. However, cellular and neurochemical alterations observed in MetS like impairment of endothelial cell function, abnormality in essential fatty acid metabolism, and enhancement in phospholipid- and sphingolipid-derived lipid mediators along with abnormal insulin/leptin signaling may represent a pathological bridge between MetS and neurological disorders (stroke, AD, and depression) (Farooqui 2013). Long-term presence of hyperglycemia, insulin resistance, microvascular disease, glucose toxicity, oxidative stress, inflammation, and atherosclerosis may result in diabetes-mediated cerebrovascular disease, which may affect brain and accelerate cognitive decline and promote dementia and AD (Farooqui 2013). Long-term disturbance in insulin signaling may have a negative impact on memory formation and may promote the initiation of AD. Based on a marked decrease in CNS expression of genes encoding for insulin, insulin-like growth factor-1 and II (IGF-I, and IGF-II) receptors, and alterations in insulin signaling, it is proposed that AD may be called as a Type 3 Diabetes (de la Monte and Wands 2008; de la Monte 2009). There are similarities in neurochemical alterations between diabetes (a component of MetS) and AD. Both pathological conditions involve abnormal cholesterol levels, elevation in levels of phospholipid- and sphingolipid-derived lipid mediators, increase in expression of adipokines/cytokines, aggregation of Aβ, stimulation in GSK3β activity, and dysregulation of protein physiological and abnormal processes (Farooqui 2013). In addition, diabetes and AD also share other common abnormalities including impairment in glucose metabolism, elevation in oxidative stress, alterations in amyloidogenesis as well as peripheral oxidative and inflammatory stress, neural atrophy and/or degeneration, and cognitive decline. Another common feature of diabetes and AD are increased advanced glycation endproduct-modified proteins, which are found in diabetes and in the AD brain. The receptor for advanced glycation endproducts (RAGE) plays a prominent role in the pathogenesis of both diseases (Kojro and Postina 2009). In addition, a major role for insulin degrading enzyme in the degradation of Aβ peptide has been identified (Zhao et al. 2009).

Furthermore, soluble Aβ oligomer, which accumulates in AD brains and is recognized as potent synaptotoxins, has been proposed to play an important role in synapse failure in AD (Mucke and Selkoe 2012). Accumulation of Aβ oligomers in the brain not only produces oxidative stress, synapse deterioration and loss, but also inhibits synaptic plasticity. The accumulation of Aβ oligomer is also linked with impairment in hippocampal insulin signaling. These oligomers not only cause internalization and cellular redistribution of insulin receptors by blocking downstream hippocampal insulin signaling (Bomfim et al. 2012), but also induce hippocampal endoplasmic reticulum stress (Lourenco et al. 2013), establishing molecular parallels between AD and diabetes. Collectively, these studies suggest that Aβ oligomer may be a pathogenic link between AD and dysregulated peripheral glucose homeostasis that occurs in diabetes (Clarke et al. 2015).

1.6   Metal Ions and Pathogenesis of Alzheimer Disease

Human brain is extremely susceptible to oxidative damage not only due to the presence of polyunsaturated fatty acids in the neural membrane phospholipids, but also due to low levels of antioxidants and antioxidant enzymes. In addition, neurons of certain regions of the brain, such as the hippocampus, are particularly vulnerable to oxidative stress because of their low endogenous levels of vitamin E and glutathione relative to other brain regions. Low levels of antioxidants and antioxidant enzymes may be adequate under normal circumstances. However, high levels of ROS production have been reported to contribute to the pathogenesis of AD (Farooqui 2010a). It is well known that metal homeostasis in our body is strictly controlled by the interplay among transporters, channels, chaperones and metalloregulatory sensors. In neurodegenerative diseases and/or aging, this tightly control of this interplay is lost, resulting in imbalance of metal ion homeostasis (Bolognin et al. 2009; Breydo and Uversky 2011). In AD, oxidative stress is induced not only by the generation of high levels of lipid peroxidation, DNA, and protein oxidation products (4-HNE, 8-hydroxy-2′-deoxyguanosine, 8-oxo-7,8-dihydro-2′-deoxyguanosine, and protein carbonyls respectively), but also by the increased brain content of iron (Fe^{3+}) and copper (Cu^{2+}) (Fig. 1.11). Both ions interact with APP and are capable of stimulating free radical formation (e.g. hydroxyl radicals via Fenton reaction) and

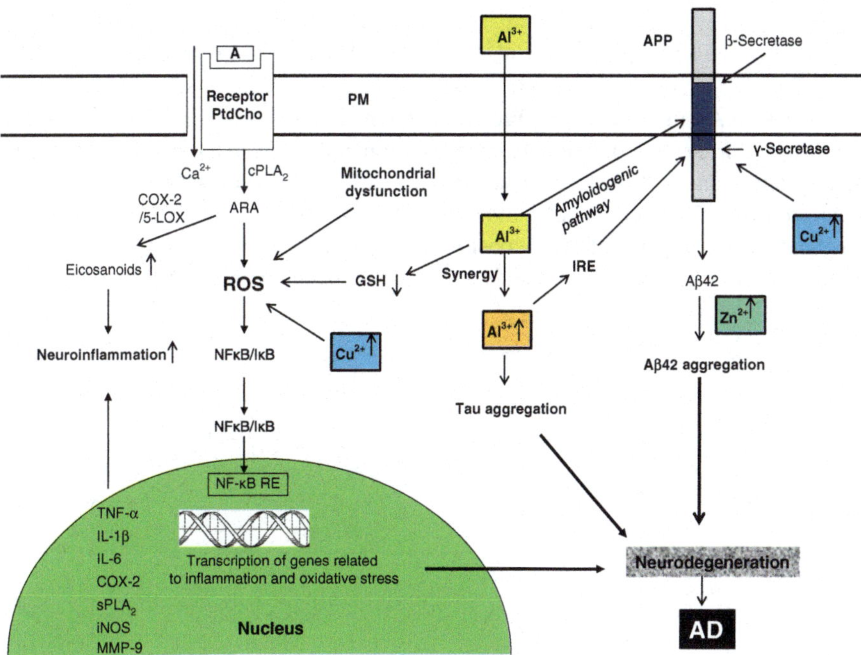

Fig. 1.11 Hypothetical diagram showing the involvement of metal ions and induction of oxidative stress and neuroinflammation in the pathogenesis of Alzheimer disease. Amyloid precursor protein (*APP*); β-amyloid (*Aβ*); agonist (*A*); Receptor (*R*); phosphatidylcholine (*PtdCho*); cytosolic phospholipase A₂ (*cPLA₂*); cyclooxygenase (*COX*); lipoxygenase (*LOX*); arachidonic acid (*ARA*); reactive oxygen species (*ROS*); nuclear factor-κB (*NF-κB*); nuclear factor-κB-response element (*NF-κB-RE*); inhibitory subunit of NF-κB (*I-κB*); tumor necrosis factor-α (*TNF-α*); interleukin-1β (*IL-1β*); interleukin-6 (*IL-6*); inducible nitric oxide synthase (*iNOS*); secretory phospholipase A₂ (*sPLA₂*); γ-secretase; β-secretase (*BACE-1*); iron response element (*IRE*)

increased protein and DNA oxidation. Elevations in levels of Fe^{3+}, Cu^{2+}, and Zn^{2+} have been reported in the normal brain during aging. Levels of Fe^{3+}, Cu^{2+}, and Zn^{2+} in normal brain parenchyma are 340 μM, 70 μM, and 350 μM, respectively. These levels are increased in brain parenchyma of AD patients by 700 μM, 300 μM, and 800 μM, respectively (Lovell et al. 1998). These metals are also enriched in both senile plaques and NFTs (Ayton et al. 2013). The presence of transition metals within the amyloid deposits in AD patients indicates that transition metals may directly interact with Aβ (Lovell et al. 1998).

Cu^{2+} produces oxidative stress by two mechanisms. First, it can directly catalyze the formation of ROS via a Fenton-like reaction. Second, exposure to elevated levels of copper significantly decreases glutathione levels (Speisky et al. 2009). Cu^{2+} in the presence of superoxide anion radical ($O_2^{\cdot-}$) or biological reductants such as ascorbic acid or GSH, is reduced to cuprous ion Cu^+ which is capable of catalyzing the formation of 0020 reactive hydroxyl radicals ($^{\cdot}OH$) through the decomposition of hydrogen peroxide via the Fenton reaction (Barbusinski 2009):

$$Cu^{2+} + O_2^{\cdot-} \rightarrow Cu^+ + O_2$$

$$Cu^+ + H_2O_2 \rightarrow Cu^{2+} + OH^- + {}^{\cdot}OH \quad \left(\text{Fenton reaction}\right)$$

Cu^{2+} binds to Aβ monomers via three histidine residues (His^6, His^{13}, and His^{14}) and a tyrosine residue (Tyr^{10}), inducing conformational changes in the peptide that facilitate its aggregation (Fig. 1.12) (Curtain et al. 2001). Based on several studies, it is suggested that Cu^{2+}-mediated disturbance in metal homeostasis may be directly involved in the process of Aβ deposition in AD brains. In addition, the aberrant interaction between transition metals and Aβ may generate ROS. Aβ interacts with

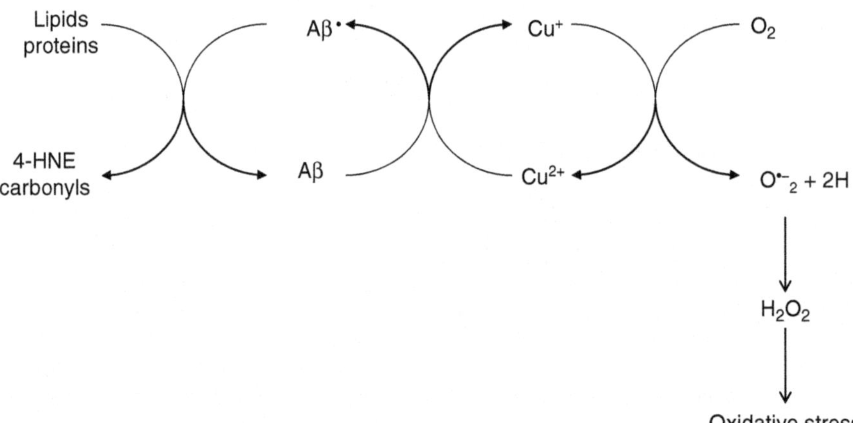

Fig. 1.12 Generation of H_2O_2 through the reaction between Aβ and Cu^{2+}. Aβ reduction of Cu^{2+} ions generates Aβ radicals (Aβ$^{\cdot}$) that extract protons from surrounding lipids and proteins generating 4-hydroxy-2-nonenal (4-HNE) and carbonyls respectively. Cu^+ reacts with molecular oxygen (O_2) forming H_2O_2

Cu^{2+} forming a cuproenzyme-like complex (Curtain et al. 2001). Electrons are then transferred from Aβ to Cu^{2+}, reducing Cu^{2+} to Cu^{+} and forming positively charged Aβ radical (Opazo et al. 2002). Cu^{+} then donates two electrons to oxygen, generating H_2O_2 (Opazo et al. 2002), producing conditions for further production of hydroxyl radicals (Fenton-type reaction) (Huang et al. 1999). After electron donation to O_2, the radicalized $\cdot$ Cu^{2+} complex is probably be restored to Aβ $\cdot$ Cu^{2+} by electron transfer from biological reducing agents such as cholesterol, catecholamines, and vitamin C (Opazo et al. 2002). Cu^{2+} also binds to Tau (Martic et al. 2013) and regulates its aggregation and phosphorylation (Zhou et al. 2007). Cu^{2+} plays an important role in amyloid plaques and NFT formation which are two key pathological features of AD. Pathological mechanisms contributing to Cu^{2+} and APP interactions are not fully understood. However, it is proposed that Cu^{2+}-mediated mediated amyloidogenic processing may involve a sub-population of APP molecules in cholesterol-rich lipid rafts in copper deficient AD brains (Hung et al. 2013). Co-localization of Aβ and a paradoxical high concentration of Cu^{2+} in lipid rafts promotes the formation of Aβ $\cdot$ Cu^{2+} complexes. These complexes can catalytically oxidize cholesterol to generate H_2O_2, oxysterols and other lipid peroxidation products that accumulate in brains of AD cases and transgenic mouse models (Kitazawa et al. 2009; Mao et al. 2012). Tau, the core protein component of NFTs, also interacts with Cu^{2+} and cholesterol facilitating the hyperphosphorylation and aggregation of Tau preceding the generation of NFTs. Collective evidence suggests that elevations in Cu^{2+} levels may contribute to the formation of senile plaques and NFTs (Hung et al. 2013).

$$Fe^{3+} + O_2^{\cdot-} \rightarrow Fe^{2+} + O_2$$

$$Fe^{2+} + H_2O_2 \rightarrow Fe^{3+} + OH^- + {}^{\cdot}OH \quad (\text{Fenton reaction})$$

$$Fe^{3+} + HO^- \rightarrow Fe^{2+} + {}^{\cdot}OH$$

Fe^{3+} is essential for many brain functions, including energy production, DNA synthesis and repair, phospholipid metabolism, myelination and neurotransmitter synthesis (Crichton et al. 2011). It also plays a crucial role in oxygen sensing and transport. The accumulation of iron has been reported to occur in the hippocampus and the cortex, brain regions that are affected in AD patients (Crichton et al. 2011; Smith et al. 2010). In these regions, Fe^{3+} and proteins that bind Fe^{3+} accumulate in the amyloid plaques (Silvestri and Camaschella 2008). This accumulation has been reported to play a role in the degree of tau phosphorylation (Egana et al. 2003). It is suggested that dysregulation of the level of iron in the brain may enhance the amyloidogenic pathway and potentiate the aggregation and toxicity of Aβ (Fig. 1.11). Like Cu^{2+}, Aβ also catalyzes the reduction of Fe^{3+} into Fe^{2+} with the generation H_2O_2, which is converted to OH$^{\cdot}$ (Opazo et al. 2002; Jiang et al. 2009). Three histidine residues at positions 6, 13, and 14 of Aβ may play a role in Fe^{3+} binding. Fe^{3+} has been shown to induce the aggregation of hyper-phosphorylated Tau, leading to the deposition of NFTs. Thus, iron contributes to Aβ aggregation as well as aggregation and deposition of Tau.

$$A\beta\text{-}Fe^{3+} + AscH \rightarrow A\beta\text{-}Fe^{2+} + Asc^- + H^+$$

$$A\beta\text{-}Fe^{3+} + Asc^- \rightarrow A\beta Fe^{2+} + Asc$$

$$A\beta\text{-}Fe^{2+} + H_2O_2 \rightarrow A\beta\text{-}Fe^{2+} + OH^- + \,{}^{\cdot}OH \quad (\text{Fenton reaction})$$

$$A\beta\text{-}Fe^{2+} + O_2 \rightarrow A\beta\text{-}Fe^{3+} + O_2^{\cdot-}$$

Converging evidence suggests that there is a relationship between iron, copper, and amyloid plaques in AD tissue (Meadowcroft et al. 2009; Ayton et al. 2013) and to a lesser extent, in the APP/PS1 model (Meadowcroft et al. 2009; Chamberlain et al. 2011; Bourassa et al. 2013). As stated above, iron and copper are essential elements required as a cofactor for numerous metabolic processes due to its ability to receive and donate electrons during redox cycling. Homeostasis of iron is tightly regulated under normal physiological conditions as excessive amounts of iron are known to cause cellular susceptibility to oxidative stress. Accumulation of iron throughout cortical tissue and focal deposition within $A\beta$ plaques are both known to occur within the AD brain (Connor et al. 1992a). In addition, altered regulation of iron management proteins are observed around $A\beta$ plaques and in AD cortical tissue. Specifically, robust staining of intracellular ferritin and extracellular transferrin is observed in the vicinity and periphery of AD $A\beta$ plaques (Connor et al. 1992b). The data strongly suggest that there is a disruption in brain iron homeostasis associated with AD and that the misregulation of iron plays a central role in disease pathology.

Zn^{2+}, an essential trace element and second in abundance in mammalian tissues (Paoletti et al. 2009), is not only critical for immunity, growth and development (Nolte et al. 2004), but is a cofactor for more than 300 enzymes and is essential for the correct functioning of over 2000 transcription factors (Takeda 2000; Jeong and Eide 2013). Zn^{2+} is redox inert and an essential trace element found in the brains of all animals including humans. In the brain, Zn is present at the highest concentrations in the hippocampus, amygdala, cerebral cortex, thalamus, and olfactory cortex (Frederickson et al. 2000). It plays important roles in various physiological functions such as in mitotic cell division, immune system activity, the synthesis of proteins, nucleic acids, gene expression, neurotransmission, hormonal storage and release, tissue repair, memory, the visual processes, and acts as a co-factor of more than 300 enzymes or metalloproteins (Hambidge 2000). Recent studies have revealed that Zn signaling plays crucial roles in various human biological systems (Hirano et al. 2008). Zn deficiency in human childhood is known to cause dwarfism, the retardation of mental and physical development, immune dysfunction, and learning disabilities (Prasad 2009).

Zn^{2+} is involved in at least three crucial events associated with the development of AD. First, Zn^{2+} interacts with the $A\beta$ monomer and then allows aggregation of monomers of $A\beta$ to soluble $A\beta$ oligomers and next to insoluble $A\beta$ plaques. Aggregation of NFTs proceeds in a similar way. Zn^{2+} binds to a tau protein, allowing the production of a tau complex. Additionally, in AD, Zn^{2+} not only contributes to autophagic dysfunction, but also promote the deregulation of intraneuronal calcium equilibrium (Cannon and Greenamyre 2011; McCord and Aizenman 2014; Szewczyk 2013). All of these events are correlated to the activation of many different signaling pathways involved in neuronal deterioration. Zn^{2+} inhibits cellular respiration and facilitates the leakage of electrons from the mitochondrial electron transport chain by inhibiting complex I and III (Frazzini et al. 2006). Zn^{2+} is also able to increase production of $O_2^{-\bullet}$ (via activation of protein kinase C (PKC) and increase activity of NADPH oxidase) and of nitric oxide (NO) (via induction of NO synthase) (Frazzini et al. 2006). NO can react with $O_2^{-\bullet}$ to produce peroxynitrite ($ONOO^-$), a free radical that can cause oxidation and nitration of proteins, lipids, and DNA (Mocchegiani et al. 2005; Frazzini et al. 2006). An important regulator of synaptic zinc is the zinc transporter-3 (ZnT3) protein which is essential for loading zinc into synaptic vesicles (Linkous et al. 2008). Levels of ZnT3 are decreased with aging in the brains of both mice and humans and are reduced even further in the brains of AD patients (Adlard et al. 2010). ZnT3 KO mice show defects in learning and memory at 6 months of age, and it is suggested that these mice provide a phenocopy for the synaptic and memory deficits of AD (Adlard et al. 2010). Zn^{2+} deficiency not only increases neuroinflammation, but also affects BDNF maturation and leading to cognitive dysfunction, aberrant intracellular Zn^{2+} mobilization or accumulation resulting in mitochondrial failure and ROS production. Extracellular Zn^{2+} overload within senile plaques also inhibits the iron-export ferroxidase activity further increasing ROS production and ultimately neuronal death (Duce and Bush 2010). Collective evidence suggests that Zn^{2+}, along with Cu^{2+}, and Fe^{3+} (released during neural transmission), directly bind to Aβ and accelerate its aggregation and accumulation into amyloid plaques (Morante 2008; Altamura and Muckenthaler 2009), These observations support the view that Fe^{3+}, Cu^{2+}, and Zn^{2+} may contribute to the pathogenesis of AD not only through the increased amyloid-β affinity, but also through changes in the redox status (Deibel et al. 1996). It also remains to be seen whether changes in metal ion-mediated oxidative stress are the cause or consequence of AD, there is a growing body of evidence showing a direct correlation between metal ions and key AD-related key proteins.

Magnesium ions (Mg^{2+}) are known to play a critical role in controlling synapse density and plasticity (Slutsky et al. 2004). In APPswe/PS1dE9 mice, elevation of brain Mg^{2+} by the treatment of Mg-L-threonate (MgT), not only upregulates NMDA-R signaling, but also reduces the expression of BACE1, an enzyme, which is closely associated with APP processing (Fig. 1.13) (Li et al. 2014). The upregulation of NMDAR may be due to the inhibition of calcineurin leading to downregulation of NMDAR signaling (Li et al. 2014). The reductions in BACE1 expression by MgT treatment may contribute to the prevention of synapse loss and improvement of cognitive function in the APPswe/PS1dE9 mice (Li et al. 2014). Levels of Mg^{2+} in the brain and serum are significantly decreased in AD patients compared with age-matched normal subjects (Andrasi et al. 2005; Barbagallo et al. 2011). If so, then simply restoring brain Mg^{2+} may produce beneficial effects in AD patients (Andrasi et al. 2005; Barbagallo et al. 2011). Thus, the intraperitoneal

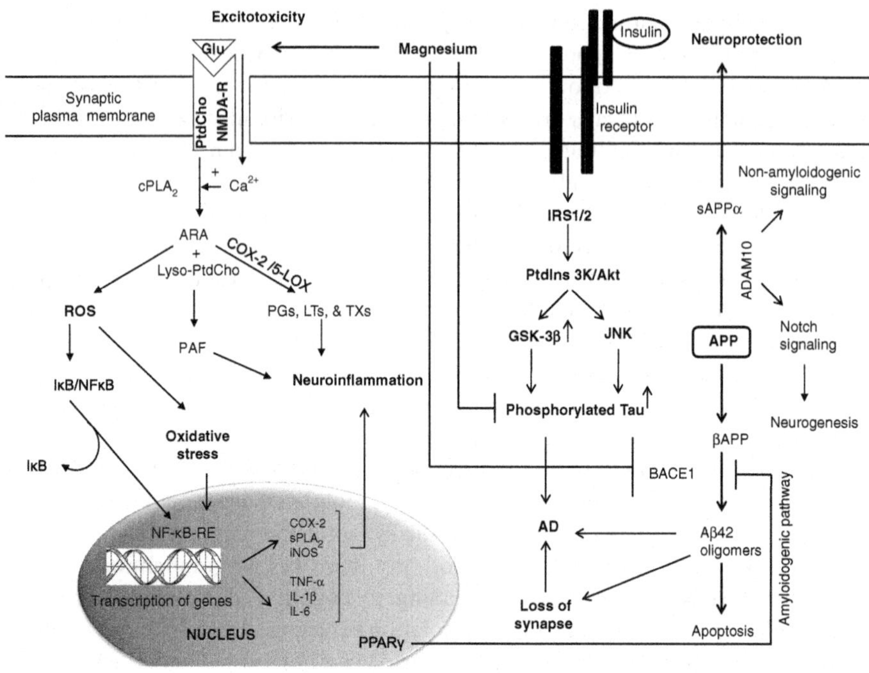

Fig. 1.13 Effect of magnesium on APP processing by magnesium. Notice the inhibition of BACE1 and hyperphosphorylated Tau by magnesium. Amyloid precursor protein (*APP*); β-amyloid (*Aβ*); glutamate (*Glu*); NMDA receptor (*NMDA-R*); phosphatidylcholine (*PtdCho*); cytosolic phospholipase A₂ (*cPLA₂*); lyso-phosphatidylcholine (*lyso-PtdCho*); cyclooxygenase (*COX*); lipoxygenase (*LOX*); arachidonic acid (*ARA*); platelet-activating factor (*PAF*); reactive oxygen species (*ROS*); nuclear factor-κB (*NF-κB*); nuclear factor-κB-response element (*NF-κB-RE*); inhibitory subunit of NF-κB (*I-κB*); tumor necrosis factor-α (*TNF-α*); interleukin-1β (*IL-1β*); interleukin-6 (*IL-6*); inducible nitric oxide synthase (*iNOS*); secretory phospholipase A₂ (*sPLA₂*); α-secretase (*ADAM10*); β-secretase (*BACE-1*); Insulin receptor substrate (*IRS*); phosphatidylinositol 3 kinase (*PtdIns 3K*); Glycogen synthase kinase 3 (*GSK-3*); C83 fragment and soluble (*sAPPα*); c-Jun NH2-terminal kinase (*JNK*); and poly(ADP)ribose polymerase (*PARP*)

injections of magnesium sulfate have been reported to elevate the brain magnesium levels and protect learning and memory capacities in streptozotocin-induced sporadic AD model rats (Xu et al. 2014). Magnesium sulfate has also been shown to reverse impairments in long-term potentiation (LTP), dendritic abnormalities, and the abnormal recruitment of synaptic proteins. Magnesium sulfate treatment decreases Tau hyperphosphorylation by increasing the inhibitory phosphorylation of GSK-3β at serine 9, thereby increasing the activity of Akt at Ser473 and PtdIns 3K at Tyr458/199, and improving insulin sensitivity (Xu et al. 2014) (Fig. 1.13). Collective evidence suggests that Mg²⁺ treatment protects cognitive function and synaptic plasticity by inhibiting GSK-3β in sporadic AD model rats (Xu et al. 2014; Li et al. 2014). More information is needed on neuroprotective effects of Mg²⁺ in animal models and AD patients.

1.7   Involvement of Neurotrophins in AD

BDNF is an important regulator of synaptic transmission and long-term potentiation (LTP) in the hippocampus and in other brain regions, playing a role in the formation of certain forms of memory. It is synthesized as a precursor, the proBDNF, which is proteolytically cleaved intracellularly either by enzymes like furin or pro-convertases, and secreted as the 14 kDa mature BDNF (Lessmann et al. 2003). Marked reduction in the levels of BDNF has been reported to occur in AD (Murer et al. 2001; Cotman 2005). Reduction in levels of BDNF and its receptor, tropomyosin receptor kinase B (TrkB) is accompanied by BDNF-mediated alterations in signaling related to synaptic dysfunction, neurodegeneration, and cognitive deficits (Murer et al. 2001; Cotman 2005). In AD transgenic mice (Tg2576), the accumulation of Aβ aggregates and increases in TNF-α and IL-1β signaling interference with BDNF signaling by impairing the axonal transport of BDNF in neurons (Poon et al. 2011, 2013). The high affinity binding of BDNF to TrkB leads to the activation of phosphoinositide-3-kinase (PtdIns 3K), Ras/MAPK, and PLCγ/PKC pathways, which are not only involved in local activation, but also associated with long-distance retrograde transport of the BDNF/TrkB signal (Fig. 1.14) (Du and Poo 2004).

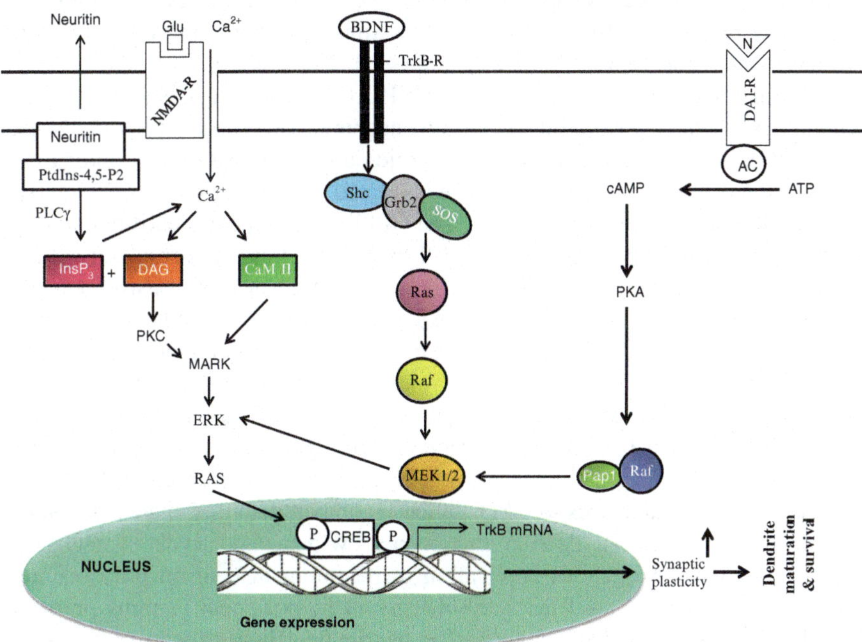

Fig. 1.14 BDNF and TrkB-R mediated signaling in the brain. Glutamate (*Glu*); NMDA receptor (*NMDA-R*); Brain-derived neurotrophic factor (*BDNF*); tyrosine kinase receptor B (*TrkB*); receptors phospholipase Cgama (*PLCγ*); diacylglycerol (*DAG*); protein kinase C (*PKC*); phosphatidylinositol 4,5-bisphosphate (*PtdIns4,5-P₂*); inositol 1,4,5-trisphosphate (*InsP₃*); calmodulin kinase II (*CaMKII*); and cyclic AMP response element-binding (*CREB*) protein

For axonal retrograde transport, TrkB is internalized from the cell surface into signaling endosomes that require the dynein motor protein to activate downstream signaling cascades (Delcroix et al. 2003; Heerssen et al. 2004). The PtdIns 3K pathway plays a critical role in retrograde trafficking (Nielsen et al. 1999) and because Aβ interferes with BDNF-mediated retrograde transport by impairing endosomal vesicle velocities leading to impairment in downstream signaling driven by BDNF/TrkB, including ERK5 activation, and CREB-dependent gene regulation, which are involved in neuroprotective and neuromodulatory effects (Poon et al. 2013). Aβ-mediated deficits in BDNF trafficking and signaling are mimicked by LDN (an inhibitor of Ubiquitin C-terminal hydrolase) and reversed by increasing cellular ubiquitin C-terminal hydrolase (UCH-L1) levels. These studies support the view that UCH-L1 is closely associated not only with BDNF receptor sorting, but also with endosomes signaling, and supporting retrograde transport. It is proposed that decrease in UCH-L1 mRNA levels in the hippocampi of AD brains may be involved in impaired BDNF/TrkB-mediated retrograde signaling, which may compromise synaptic plasticity and neuronal survival (Poon et al. 2013). In cultured neurons, sublethal doses of Aβ inhibit BDNF-MAPK/ERK and BDNF-PtdIns 3K signaling by interfering with insulin receptor substrate-1 and Shc, two docking proteins that link TrkB to the activation of downstream kinase cascades (Tong et al. 2004). Collective evidence suggests that the functional consequences of Aβ accumulation in AD may responsible for impaired BDNF axonal transport and vulnerability of neurons to apoptosis (Tong et al. 2004; Poon et al. 2011, 2013).

Neuritin or candidate plasticity gene 15 (CPG 15), which encodes a small, extracellular glycosylphosphatidylinositol (GPI)-anchored cell surface protein (Naeve et al. 1997), is a critical regulator for dendritic outgrowth, maturation, and axonal regeneration (Fujino et al. 2008). Its gene is located within the 6p2424-p25 interval on chromosome 6 (Fujino et al. 2003). The GPI linkage enables neuritin to anchor at cell surfaces, and upon cleavage of GPI by phospholipase the resultant soluble neuritin is released into the extracellular space. 3 cyclic AMP responsive elements present in the promoter region of the gene supporting the view that neuritin expression may be mediated by cyclic AMP responsive element binding protein (CREB). In vivo studies have shown an important role for CREB in activity-dependent neuritin expression in the barrel cortex of control and CREB α, Δ-knockout mice supporting the view that CREB is necessary for the regulation of neuritin-mediated neuroplasticity. The expression of neuritin is significantly decreased in the hippocampus and cerebral cortex of AD patients compared to age-matched control subjects (An et al. 2014). Tg2576 mice also show decreased levels of neutrin in hippocampus. The exogenous application of recombinant neuritin fully restored dendritic complexity as well as spine density in hippocampal neurons prepared from Tg2576 mice, whereas it did not affect neurite branching of neurons from their wild-type littermates. Furthermore, chronically infusion of neutrin into the brains of Tg2576 mice normalizes synaptic plasticity in acute hippocampal slices, leading to intact long-term potentiation. It is proposed that neutrin may act by increasing the levels of synaptophysin, a presynaptic marker, which represents an increase in synaptogenesis (An et al. 2014).

In brain, BDNF interacts with other neurotrophins such as TGF-β1, which is an anti-inflammatory and pleiotropic cytokine that regulates the proliferation, differentiation and survival of various cell types (Pál et al. 2012). TGF-β1 is highly expressed in the cerebral cortex, hippocampus, central amygdaloid nucleus, medial preoptic area, substantia nigra and brainstem (Vincze et al. 2010). It is a key regulator of the brain's responses to injury and inflammation (Shen et al. 2014). TGF-β1 protects neurons against β-amyloid toxicity (Caraci et al. 2008). TGF-β1 produces its effects by binding with a high-affinity transmembrane receptor complex consisting of the activin-like kinase 5 (ALK5)/TGF-β type I receptor as well as the TGF-β type II receptor (TβRII) subunits, which contains a serine/threonine kinase domain (Ten Dijke and Hill 2004). The binding of TGF-β1 with TβRII induces transphosphorylation of type I receptor by the type II receptor kinase. The consequent activation of type I receptor results in phosphorylation and translocation of Smads into the nucleus, where they regulate the expression of different target genes involved cell proliferation, differentiation, immune suppression and repair after injury (Ten Dijke and Hill 2004). Besides Smad-mediated gene transcription, TGF-β1 also activates Smad-independent pathways, including the extracellular-regulated kinase (ERK) pathway (Derynck and Zhang 2003), the NF-κB pathway (König et al. 2005), and the PtdIns 3K/Akt pathway (Bakin et al. 2000; Caraci et al. 2008). TGF-β/Smad-independent pathways have a key role in mediating different biological effects of TGF-β1 such as cell cycle inhibition, immune suppression and neuroprotective effects (Derynck and Zhang 2003; Caraci et al. 2008; Zhu et al. 2004). In addition, TGF-β1 also increases synaptic plasticity by enhancing the expression of BDNF and TrkB (Sometani et al. 2001). Significant decrease in TGF-β1 expression and signaling occurs in very early stages of AD. This impairment TGF-β1 signaling may facilitate and support to the pathogenesis of AD not only by decreasing BDNF and increasing the accumulation of Aβ but also by promoting Aβ-induced neurodegeneration in different animal models of AD (Wyss-Coray 2006; Caraci et al. 2008).

1.8 Wnt Signaling in the Progression of Alzheimer Disease

It is becoming increasingly evident that Wnt proteins are not only involved in regulating axon guidance and dendrite morphogenesis, but also contribute to synapse formation, synaptic plasticity, and long term potentiation (Inestrosa and Arenas 2010). These observations support the view that Wnt signaling may play an important role in regulating the formation and function of neuronal circuits (Ciani and Salinas 2005). Accumulation of Aβ in AD not only dysfunctions Wnt signaling, but also contributes to neuronal degeneration and synaptic impairment (Silva-Alvarez et al. 2013). Glycogen synthase kinase-3β (GSK-3β) is an important serine/threonine (Ser/Thr) kinases, which not only phosphorylates MAP tau, leading to NFT formation in AD, but is also regulates Wnt signaling (Rosso and Inestrosa 2013). Accumulating evidence suggests that Aβ pathology precedes hyperphosphorylated tau pathology (Folwell et al. 2010). However, the regulatory mechanism whereby Aβ induces hyperphosphorylated tau is still elusive. It is suggested that Wnt signaling through GSK3β may bridge the gap between Aβ and accumulation of Tau (Vargas et al. 2014). Thus, in vivo activation of Wnt signaling has been reported to improve episodic memory, increase in excitatory synaptic transmission, and enhancement in long-term potentiation in adult wild-type mice. Moreover, the activation of Wnt signaling may also rescue memory loss and improves synaptic dysfunction in APP/PS1-transgenic mice that model the amyloid pathology of Alzheimer disease. Based on detailed investigations, it is suggested that persistent activation of Wnt signaling through Wnt ligands, or inhibition of its negative regulators, such as Dickkopf-1 (DKK-1) and GSK-3β may protect against Aβ toxicity and ameliorate cognitive performance in AD (Shruster et al. 2011; Purro et al. 2012).

1.9   Leptin Signaling and Alzheimer Disease

Leptin is a 146 amino acid protein with a molecular weight of 16 kDa. It is encoded by the *ob* gene in the white adipose tissue (WAT) and is implicated in obesity, food intake, and energy homeostasis. Leptin circulates in the blood as a 16-kDa peptide and is transported into the brain via two distinct mechanisms. One involving a saturable transport system (receptor-mediated transcytosis) and the other promotes the transport of leptin in the brain through the cerebrospinal fluid (Farooqui 2013). Leptin acts through leptin receptors (obR), which are found in the brain (hypothalamus and hippocampus) and are encoded by the db gene (Farooqui 2013). In hypothalamus leptin signaling regulates food intake and energy homeostasis through the JAK/STAT, MAPK, PI3K and mTOR signaling, where as in hippocampus leptin increases synaptogenesis and aids in memory formation (Harvey et al. 2005). In addition, leptin not only increases neurogenesis in the dentate gyrus of adult mice (Garza et al. 2008), but also plays a critical role in hippocampal neuronal survival through the activation of PtdIns 3K/Akt and JAK2/STAT3 signal transduction pathways (Fig. 1.13) (Guo et al. 2008). Leptin also upregulates the expression of potent endogenous antioxidant enzyme Mn-SOD (manganese superoxide dismutase) and the anti-apoptotic protein Bcl-xL (B-cell lymphoma xL) in a STAT3-dependent manner in the hippocampus (Guo et al. 2008). The binding of leptin with leptin receptor results in Janus kinase 2 (Jak2)-mediated phosphorylation of two tyrosine residues (Tyr^{985} and Tyr^{1138}) in the cytoplasmic tail of the receptor (Münzberg and Myers 2005). Phosphorylated Tyr^{985} of leptin receptor recruits SH2-containing protein-tyrosine phosphatase-2, leading to activation of the mitogen-activated protein kinase/extracellular signal-regulated kinase (ERK) pathway (Bjørbaek et al. 2001). Phosphorylated Tyr^{1138} of leptin receptor activates signal transducer and activator of transcription 3 (STAT3), which dimerizes and is translocated to the nucleus where it acts as a transcription factor (Münzberg and Myers 2005). Jak2 also phosphorylates insulin receptor substrate-1 and -2, resulting in activation of the PtdIns 3K/Akt pathway (Niswender et al. 2001; Xu et al. 2005).

Leptin modulates Aβ production and metabolism (Fig. 1.15). Thus, chronic peripheral leptin administration in Tg2576 mice reduces levels of Aβ in the brain (Fewlass et al. 2004). The molecular mechanism associated with the effect of leptin is not fully understood. However, it is suggested that leptin decreases the activity of β-secretase (BACE-1) in the SH-SY5Y cells (Fewlass et al. 2004). In addition, leptin has been reported to decrease Tau phosphorylation explicitly at residues Ser[202], Ser[396], and Ser[404] in retinoic acid induced differentiated SH-SY5Y cells, differentiated human NT2 cells (NT2N), and rat primary cortical neurons (Greco et al. 2008, 2009). Aβ and leptin interact with the Akt/mammalian target of rapamycin complex1 (mTORC1) signaling pathway (Fig. 1.14) (Marwarha et al. 2010a, b; Marwarha and Ghribi 2012). Akt/mTORC1 activation retards tau phosphorylation by blocking the downstream enzyme GSK-3β. Collective evidence suggests that leptin downregulation precedes and promotes the accumulation of Aβ and phosphorylation of Tau. Leptin also protects cortical neurons from Aβ-mediated cell death by a signal transducer and activator of transcription-3 (STAT-3)-dependent mechanism (Doherty et al. 2012; Marwarha and Ghribi 2012). Based on the importance of leptin signaling in the brain, it is suggested that leptin signaling is associated with modulation of Aβ and its involvement in the pathogenesis of AD. It is

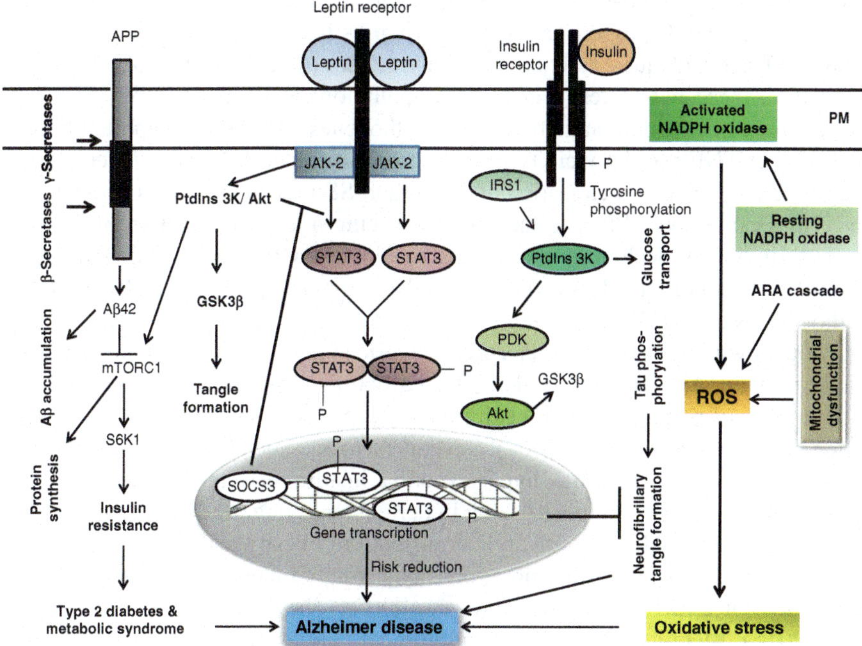

Fig. 1.15 Interactions between leptin and Aβ signaling and their relationship with development of Alzheimer disease. Arachidonic acid (*ARA*); reactive oxygen species (*ROS*); cyclooxygenase-2 (*COX-2*); phosphatidylinositol 3 kinase (*PtdIns 3K*); Glycogen synthase kinase 3 (*GSK-3*); amyloid precursor protein (*APP*); beta amyloid (*Aβ*); Janus kinase (*JAK*); Signal Transducer and Activator of Transcription (*STAT*); and suppressors of cytokine signaling (*SOCS*)

suggested that individuals with higher leptin levels have a much lower risk of developing AD (Holden et al. 2009; Lieb et al. 2009). Leptin deficiency, which may either be caused either by a defect in leptin transport across the BBB or by suppression of leptin signaling (Flier 2004), may result in increased risk of AD due to leptin resistance (Flier 2004; Doherty 2011).

1.10 Other Neurochemical Changes and Progression Toward AD

Above mentioned processes and pathways at some point may cause such cognitive impairments that they become classified as AD. Other AD symptoms include memory loss, poor sleep quality, emotional changes, anxiety and depression (Burns and Iliffe 2009; Landry and Liu-Ambrose 2014). This cognitive deterioration coincides with the neuropathological staging of AD (Braak and Braak 1991; Schroeter et al. 2009). It should be noted that the onset of neurochemical alterations in AD brain occurs many years or even a decade before the symptoms of AD occur (frequently above 65 years of age) (Brookmeyer et al. 1998). The progression of AD differs substantially between patients. Advancing dementia and severity of AD correlates not only with the disproportional loss of synapses between neurons, but also with number of senile plaques, and neurofibrillary tangles (Querfurth and LaFerla 2010). Therefore, these parameters can serve as pathological biomarkers for AD. The involvement of neuroinflammation in the pathogenesis of AD is supported by the presence of TNFα in the vicinity of SPs in post-mortem AD brains (McCoy and Tansey 2008). This co-localization of TNFα and SP is seen in brains of transgenic mice that expressed APP. The fact that transgenic mice with deletion of TNFR1 show a significant decline in microglia activation, BACE1 activity, Aβ pathology, less memory deficits and less neuron loss indeed suggesting a role for TNRF1 in the pathogenesis of AD (He et al. 2007).

Patients with diabetes mellitus and metabolic syndrome have an increased risk of developing AD (Brands et al. 2005; Farooqui et al. 2012; Farooqui 2013). About 80 % of patients with AD exhibit symptoms of glucose intolerance or diabetes mellitus (Janson et al. 2004). Several factors associated with diabetes and metabolic syndrome including hyperglycemia, insulin resistance, glucose intolerance, atherosclerosis, adiposity and hypertension, long term consumption of western diet, and sedentary life style contribute to obesity and risk of AD (Haan 2006; Farooqui 2015). Furthermore, obesity strongly increases the risk for developing AD later in life, than being overweight or obese at late life (Emmerzaal et al. 2015). Meta-analysis on epidemiological studies also strengthens the link between obesity and AD. It is concluded that "current trends of more overweight and obesity in childhood and adolescence may translate to longer exposure of the brain to potentially detrimental vascular and metabolic effects of adipose tissue (Emmerzaal et al. 2015)".

1.11 Conclusion

AD is the most prevalent and progressive neurodegenerative disease, afflicting 10 % of the population over the age of 65 and 50 % of the population over the age of 85. Clinically, AD is accompanied by a slow global decline in cognitive function, including deterioration of memory, reasoning, abstraction, language and emotional stability, culminating in a patient with end-stage disease, totally dependent on custodial care. Approximately 5 % of patients with AD have familial form of AD, which is related to a genetic predisposition, including mutations in the amyloid precursor protein, presenilin 1, and presenilin 2 genes. Rest of cases (95 %) of AD are sporadic. The pathogenesis of AD develops over many years before clinical symptoms appear, providing the opportunity to develop therapy that can slow or stop disease progression well before any clinical manifestation develops. The severity of AD pathology is clearly associates with the number of reactive astrocytes and activated microglia. Both neurons and glial cells contribute to the induction, maintenance, and progression of neuroinflammation and oxidative stress in AD by releasing proinflammatory cytokines, producing high levels of phospholipid-, sphingolipid-, and cholesterol-derived lipid mediators, and generating reactive oxygen and nitrogen species. AD also involves impairment in glucose metabolism and insulin resistance in the brain. Impairment in insulin signaling also plays an important role in the pathogenesis of AD. Accumulation of oligomeric Aβ may be the key step in the induction of cell death in AD. Amyloid plaque burden correlates poorly with memory deficits in AD patients; however, synaptic loss is a strong predictor of the clinical symptoms of AD. The mechanism by which oligomeric Aβ leads to synaptic degeneration and loss is not clear. However, Aβ oligomers are known to produce H_2O_2 not only through the activation of cupper-dependent superoxide dismutase and NADPH-oxidase in astrocytes, but also through the modulation of mitochondrial ROS generation. It is proposed that Aβ-mediated generation of ROS protein misfolding, induction of cytokines, and abnormal increase in intracellular calcium along with abnormal protein clearance defects through the ubiquitin-proteasome system may contribute to the pathogenesis of AD.

References

Abramov AY, Canevari L, Duchen MR (2004) Beta-amyloid peptides induce mitochondrial dysfunction and oxidative stress in astrocytes and death of neurons through activation of NADPH oxidase. J Neurosci 24:565–575

Adlard PA, Parncutt JM, Finkelstein DI, Bush AI (2010) Cognitive loss in zinc transporter-3 knock-out mice: a phenocopy for the synaptic and memory deficits of Alzheimer's disease? J Neurosci 30:1631–1636

Agostinho P, Cunha RA, Oliveira C (2010) Neuroinflammation, oxidative stress and the pathogenesis of Alzheimer's disease. Curr Pharm Des 16:2766–2778

Aizenstein HJ, Nebes RD, Saxton JA, Price JC, Mathis CA, Tsopelas ND, Ziolko SK, James JA, Snitz BE, Houck PR, Bi W, Cohen AD, Lopresti BJ, DeKosky ST, Halligan EM, Klunk WE

(2008) Frequent amyloid deposition without significant cognitive impairment among the elderly. Arch Neurol 65:1509–1517

Albensi BC, Mattson MP (2000) Evidence for the involvement of TNF and NF-κB in hippocampal synaptic plasticity. Synapse 35:151–159

Alonso AD, Grundke-Iqbal I, Barra HS, Iqbal K (1997) Abnormal phosphorylation of Tau and the mechanism of Alzheimer neurofibrillary degeneration: sequestration of microtubule-associated proteins 1 and 2 and the disassembly of microtubules by the abnormal Tau. Proc Natl Acad Sci U S A 94:298–303

Altamura S, Muckenthaler MU (2009) Iron toxicity in diseases of aging: Alzheimer's disease, Parkinson's disease and atherosclerosis. J Alzheimers Dis 16:879–895

Alzheimer's Association (2010) Alzheimer's disease facts and figures. Alzheimers Dement 6:158–194

Alzheimer's Association (2012) 2011 Alzheimer's diseases facts and figures: prevalence. Alzheimers Dement 7:12–13

Alzheimer's Association (2013) Alzheimer's disease facts and figures. Alzheimer Dementia 9:208–245

Amaducci L, Tesco G (1994) Aging as a major risk for degenerative diseases of the central nervous system. Curr Opin Neurol 7:283–286

An K, Jung JH, Jeong AY, Kim HG, Jung SY, Lee K, Kim HJ, Kim SJ, Jeong TY, Son Y, Kim HS, Kim JH (2014) Neuritin can normalize neural deficits of Alzheimer's disease. Cell Death Dis 5, e1523

Andrasi E, Pali N, Molnar Z, Kosel S (2005) Brain aluminum, magnesium and phosphorus contents of control and Alzheimer-diseased patients. J Alzheimers Dis 7:273–284

Andreyev AY, Fahy E, Guan Z, Kelly S, Li X, McDonald JG, Milne S, Myers D, Park H, Ryan A, Thompson BM, Wang E, Zhao Y, Brown HA, Merrill AH, Raetz CR, Russell DW, Subramaniam S, Dennis EA (2010) Subcellular organelle lipidomics in TLR-4-activated macrophages. J Lipid Res 51:2785–2797

Arboleda G, Morales LC, Benítez B, Arboleda H (2009) Regulation of ceramide-induced neuronal death: cell metabolism meets neurodegeneration. Brain Res Rev 59:333–346

Ariga T, McDonald MP, Yu RK (2008) Thematic review series: sphingolipids. Role of ganglioside metabolism in the pathogenesis of Alzheimer's disease—a review. J Lipid Res 49:1157–1175

Arioka M, Tsukamoto M, Ishiguro K, Kato R, Sato K, Imahori K, Uchida T (1993) Tau protein kinase II is involved in the regulation of the normal phosphorylation state of Tau protein. J Neurochem 60:461–468

Arnaud L, Robakis NK, Figueiredo-Pereira ME (2006) It may take inflammation, phosphorylation and ubiquitination to 'tangle' in Alzheimer's disease. Neurodegener Dis 3:313–319

Arnaud LT, Myeku N, Figueiredo-Pereira ME (2009) Proteasome-caspase-cathepsin sequence leading to Tau pathology induced by prostaglandin J2 in neuronal cells. J Neurochem 110:328–342

Artavanis-Tsakonas S, Rand MD, Lake RJ (1999) Notch signaling: cell fate control and signal integration in development. Science 284:770–776

Augustinack JC, Schneider A, Mandelkow EM, Hyman BT (2002) Specific Tau phosphorylation sites correlate with severity of neuronal cytopathology in Alzheimer's disease. Acta Neuropathol 103:26–35

Avila J, Gomez de Barreda E, Engel T, Lucas JJ, Hernandez F (2010) Tau phosphorylation in hippocampus results in toxic gain-of-function. Biochem Soc Trans 38:977–980

Ayton S, Lei P, Bush AI (2013) Metallostasis in Alzheimer's disease. Free Radic Biol Med 62:76–89

Bai Y, Li Q, Yang J, Zhou X, Yin X, Zhao D (2008) p75(NTR) activation of NF-kappaB is involved in PrP106–126-induced apoptosis in mouse neuroblastoma cells. Neurosci Res 62:9–14

Bakin AV, Tomlinson AK, Bhowmick NA, Moses HL, Arteaga CL (2000) Phosphatidylinositol 3-kinase function is required for transforming growth factor beta-mediated epithelial to mesenchymal transition and cell migration. J Biol Chem 275:36803–36810

Ballard C, Gauthier S, Corbett A, Brayne C, Aarsland D, Jones E (2011) Alzheimer's disease. Lancet 377:1019–1031

Bansal MP, Shalini S (2005) Effect of cholesterol and 7-beta hydroxycholesterol on glutathione status and nitric oxide production in murine peritoneal macrophages. Indian J Exp Biol 43:503–508

Barbagallo M, Belvedere M, Di Bella G, Dominguez LJ (2011) Altered ionized magnesium levels in mild-to-moderate Alzheimer's disease. Magnes Res 24:S115–S121

Barbusinski K (2009) Fenton reaction—controversy concerning the chemistry. Ecol Chem Eng 16:347–358

Barnes K, Matsas R, Hooper NM, Turner AJ, Kenny AJ (1988a) Endopeptidase-24.11 is strio-somally ordered in pig brain, and in contrast to aminopeptidase N and peptidyl dipeptidase A ("angiotensin converting enzyme"), is a marker for a set of striatal efferent fibres. Neuroscience 27(3):799–817

Barnes K, Turner AJ, Kenny AJ (1988b) Electronmicroscopic immunocytochemistry of pig brain shows that endopeptidase-24.11 is localized in neuronal membranes. Neurosci Lett 94:64–69

Barnes K, Doherty S, Turner AJ (1995) Endopeptidase-24.11 is the integral membrane peptidase initiating degradation of somatostatin in the hippocampus. J Neurochem 64:1826–1832

Barnham KJ, McKinstry WJ, Multhaup G, Galatis D, Morton CJ, Curtain CC, Williamson NA, White AR, Hinds MG, Norton RS, Beyreuther K, Masters CL, Parker MW, Cappai R (2003) Structure of the Alzheimer's disease amyloid precursor protein copper binding domain. A regulator of neuronal copper homeostasis. J Biol Chem 278:17401–17407

Barrier L, Ingrand S, Fauconneau B, Page G (2008) Gender-dependent accumulation of ceramides in the cerebral cortex of the APP(SL)/PS1Ki mouse model of Alzheimer's disease. Neurobiol Aging 31:1843–1853

Beraldo FH, Arantes CP, Santos TG, Queiroz NG, Young K, Rylett RJ, Markus RP, Prado MA, Martins VR (2010) Role of alpha7 nicotinic acetylcholine receptor in calcium signaling induced by prion protein interaction with stress-inducible protein 1. J Biol Chem 285:36542–36550

Berger SL, Kouzarides T, Shiekhattar R, Shilatifard A (2009) An operational definition of epi-genetics. Genes Dev 23:781–783

Bertram L (2009) Alzheimer's disease genetics current status and future perspectives. Int Rev Neurobiol 84:167–184

Bezprozvanny I, Mattson MP (2008) Neuronal calcium mishandling and the pathogenesis of Alzheimer's disease. Trends Neurosci 31:454–463

Bhat NR (2010) Linking cardiometabolic disorders to sporadic Alzheimer's disease: a perspective on potential mechanisms and mediators. J Neurochem 115:551–562

Biasini E, Turnbaugh JA, Unterberger U, Harris DA (2012) Prion protein at the crossroads of physiology and disease. Trends Neurosci 35:92–103

Bjørbaek C, Buchholz RM, Davis SM, Bates SH, Pierroz DD, Gu H, Neel BG, Myers MG Jr, Flier JS (2001) Divergent roles of SHP-2 in ERK activation by leptin receptors. J Biol Chem 276:4747–4755

Björkhem I, Lütjohann D, Diczfalusy U, Ståhle L, Ahlborg G, Wahren J (1998) Cholesterol homeostasis in human brain: turnover of 24S-hydroxycholesterol and evidence for a cerebral origin of most of this oxysterol in the circulation. J Lipid Res 39:1594–1600

Björkhem I, Andersson U, Ellis E, Alvelius G, Ellegard L, Diczfalusy U, Sjövall J, Einarsson C (2001) From brain to bile. Evidence that conjugation and omega-hydroxylation are important for elimination of 24S-hydroxycholesterol (cerebrosterol) in humans. J Biol Chem 276:37004–37010

Blennow K, Hampel H (2003) CSF markers for incipient Alzheimer's disease. Lancet Neurol 2:605–613

Bloom GS (2014) Amyloid-β and Tau: the trigger and bullet in Alzheimer disease pathogenesis. JAMA Neurol 71:505–508

Blurton-Jones M, Laferla FM (2006) Pathways by which Abeta facilitates Tau pathology. Curr Alzheimer Res 3:437–448

Bogdanovic N, Bretillon L, Lund EG, Diczfalusy U, Lannfelt L, Winblad B, Russell DW, Björkhem I (2001) On the turnover of brain cholesterol in patients with Alzheimer's disease. Abnormal induction of the cholesterol-catabolic enzyme CYP46 in glial cells. Neurosci Lett 314:45–48

Bohr VA (2002) Repair of oxidative DNA damage in nuclear and mitochondrial DNA, and some changes with aging in mammalian cells. Free Radic Biol Med 32:804–812

Boland K, Behrens M, Choi D, Manias K, Perlmutter DH (1996) The serpin-enzyme complex receptor recognizes soluble, non toxic amyloid-β peptide but not aggregated, cytotoxic amyloid-β peptide. J Biol Chem 271:18032–18044

Bolognin S, Messori L, Zatta P (2009) Metal ion physiopathology in neurodegenerative disorders. Neuromolecular Med 11:223–238

Bomfim TR, Forny-Germano L, Sathler LB, Brito-Moreira J, Houzel JC, Decker H, Silverman MA, Kazi H, Melo HM, McClean PL, Holscher C, Arnold SE, Talbot K, Klein WL, Munoz DP, Ferreira ST, De Felice FG (2012) An anti-diabetes agent protects the mouse brain from defective insulin signaling caused by Alzheimer's disease- associated Abeta oligomers. J Clin Invest 122:1339–1353

Borger E, Aitken L, Du H, Zhang W, Gunn-Moore FJ, Yan SS (2013) Is amyloid binding alcohol dehydrogenase a drug target for treating Alzheimer's disease? Curr Alzheimer Res 10:21–29

Bosco D, Fava A, Plastino M, Montaecini T, Puja A (2011) Possible implications of insulin resistance and glucose metabolism in Alzheimer's disease pathogenesis. J Cell Mol Med 15:1807–1821

Bourassa MW, Leskovjan AC, Tappero RV, Farquhar ER, Colton CA, Van Nostrand WE, Miller LM (2013) Elevated copper in the amyloid plaques and iron in the cortex are observed in mouse models of Alzheimer's disease that exhibit neurodegeneration. Biomed Spectrosc Imaging 2:129–139

Boyle PA, Wilson RS, Yu L, Barr AM, Honer WG, Schneider JA, Bennett DA (2013) Much of late life cognitive decline is not due to common neurodegenerative pathologies. Ann Neurol 74:478–489

Braak H, Braak E (1991) Neuropathological stageing of Alzheimer-related changes. Acta Neuropathol 82:239–259

Bramham CR, Messaoudi E (2005) BDNF function in adult synaptic plasticity: the synaptic consolidation hypothesis. Prog Neurobiol 76:99–12510

Brands AM, Biessels GJ, de Haan EH, Kappelle LJ, Kessels RP (2005) The effects of type 1 diabetes on cognitive performance: a meta-analysis. Diabetes Care 28:726–735

Brasnjevic I, Hof PR, Steinbusch HW, Schmitz C (2008) Accumulation of nuclear DNA damage or neuron loss: molecular basis for a new approach to understanding selective neuronal vulnerability in neurodegenerative diseases. DNA Repair (Amst) 7:1087–1097

Brazil DP, Hemmings BA (2001) Ten years of protein kinase B signalling: a hard Akt to follow. Trends Biochem Sci 26:657–664

Brewer GJ (1998) Age-related toxicity to lactate, glutamate, and beta-amyloid in cultured adult neurons. Neurobiol Aging 19:561–568

Brewer GJ (2010) Epigenetic oxidative redox shift (EORS) theory of aging unifies the free radical and insulin signaling theories. Exp Gerontol 45:173–179

Breydo L, Uversky VN (2011) Role of metal ions in aggregation of intrinsically disordered proteins in neurodegenerative diseases. Metallomics 3:1163–1180

Brookmeyer R, Gray S, Kawas C (1998) Projections of Alzheimer's disease in the United States and the public health impact of delaying disease onset. Am J Public Health 88:1337–1342

Brooksbank BW, McGovern J (1989) Gangliosides in the brain in adult Down's syndrome and Alzheimer's disease. Mol Chem Neuropathol 11:143–156

Bu G (2009) Apolipoprotein E and its receptors in Alzheimer's disease: pathways, pathogenesis and therapy. Nat Rev Neurosci 10:333–344

Burns A, Iliffe S (2009) Alzheimer's disease. Br Med J 338:b158

Cannon JR, Greenamyre JT (2011) The role of environmental exposures in neurodegeneration and neurodegenerative diseases. Toxicol Sci 124:225–250

Caraci F, Battaglia G, Busceti C, Biagioni F, Mastroiacovo F, Bosco P, Drago F, Nicoletti F, Sortino MA, Copani A (2008) TGF-beta 1 protects against Abeta-neurotoxicity via the phosphatidylinositol-3-kinase pathway. Neurobiol Dis 30:234–242

Carulla P, Bribián A, Rangel A, Gavín R, Ferrer I, Caelles C, Del Río JA, Llorens F (2011) Neuroprotective role of PrPC against kainate-induced epileptic seizures and cell death depends on the modulation of JNK3 activation by GluR6/7-PSD-95 binding. Mol Biol Cell 22:3041–3054

Cencioni C, Spallotta F, Martelli F, Valente S, Mai A, Zeiher AM, Gaetano C (2013) Oxidative stress and epigenetic regulation in ageing and age-related diseases. Int J Mol Sci 14:17643–17663

Chabrier MA, Cheng D, Castello NA, Green KN, LaFerla FM (2014) Synergistic effects of amyloid-beta and wild-type human Tau on dendritic spine loss in a floxed double transgenic model of Alzheimer's disease. Neurobiol Dis 64:107–117

Chakravarthy B, Menard M, Ito S, Gaudet C, Dal Pra I, Armato U, Whitfield J (2012) Hippocampal membrane-associated p75NTR levels are increased in Alzheimer's disease. J Alzheimers Dis 30:675–684

Chamberlain R, Wengenack TM, Poduslo JF, Garwood M, Jack CR Jr (2011) Magnetic resonance imaging of amyloid plaques in transgenic mouse models of Alzheimer's disease. Curr Med Imaging Rev 7:3–7

Chauhan NB, Siegel GJ, Feinstein DL (2005) Propentofylline attenuates Tau hyperphosphorylation in Alzheimer's Swedish mutant model Tg2576. Neuropharmacology 48:93–104

Chen F, Yu G, Arawaka S, Nishimura M, Kawarai T, Yu H, Tandon A, Supala A, Song YQ, Rogaeva E, Milman P, Sato C, Yu C, Janus C, Lee J, Song L, Zhang L, Fraser PE, St George-Hyslop PH (2001) Nicastrin binds to membrane-tethered Notch. Nat Cell Biol 3:751–754

Chen L, Lee HM, Greeley GH Jr, Englander EW (2007) Accumulation of oxidatively generated DNA damage in the brain: a mechanism of neurotoxicity. Free Radic Biol Med 42:385–393

Chen S, Yadav SP, Surewicz WK (2010) Interaction between human prion protein and amyloid-beta (Abeta) oligomers: role of N-terminal residues. J Biol Chem 285:26377–26383

Chiarini A, Dal Pra I, Whitfield JF, Armato U (2006) The killing of neurons by beta-amyloid peptides, prions, and pro-inflammatory cytokines. Ital J Anat Embryol 111:221–246

Chin J, Palop JJ, Puoliväli J, Massaro C, Bien-Ly N, Gerstein H, Scearce-Levie K, Masliah E, Mucke L (2005) Fyn kinase induces synaptic and cognitive impairments in a transgenic mouse model of Alzheimer's disease. J Neurosci 25:9694–9703

Choi J, Levey AI, Weintraub ST, Rees HD, Gearing M, Chin LS, Li L (2004) Oxidative modifications and down-regulation of ubiquitin carboxyl-terminal hydrolase L1 associated with idiopathic Parkinson's and Alzheimer's diseases. J Biol Chem 279:13256–13264

Chouliaras L, Rutten BP, Kenis G, Peerbooms O, Visser PJ, Verhey F, van Os J, Steinbusch HW, van den Hove DL (2010) Epigenetic regulation in the pathophysiology of Alzheimer's disease. Prog Neurobiol 90:498–510

Chouliaras L, Mastroeni D, Delvaux E, Grover A, Kenis G, Hof PR, Steinbusch HW, Coleman PD, Rutten BP, van den Hove DL (2013) Consistent decrease in global DNA methylation and hydroxymethylation in the hippocampus of Alzheimer's disease patients. Neurobiol Aging 34:2091–2099

Chuang JC, Jones PA (2007) Epigenetics and microRNAs. Pediatr Res 61:24R–29R

Ciani L, Salinas PC (2005) WNTs in the vertebrate nervous system: from patterning to neuronal connectivity. Nat Rev Neurosci 6:351–362

Clark CM, Schneider JA, Bedell BJ, Beach TG, Bilker WB, Mintun MA, Pontecorvo MJ, Hefti F, Carpenter AP, Flitter ML, Krautkramer MJ, Kung HF, Coleman RE, Doraiswamy PM, Fleisher AS, Sabbagh MN, Sadowsky CH, Reiman EP, Zehntner SP, Skovronsky DM; AV45-A07 Study Group (2011) Use of florbetapir-PET for imaging beta-amyloid pathology. JAMA 305:275–283

Clarke JR, Lyra E, Silva NM, Figueiredo CP, Frozza RL, Ledo JH, Beckman D, Katashima CK, Razolli D, Carvalho BM, Frazão R, Silveira MA, Ribeiro FC, Bomfim TR, Neves FS, Klein

WL, Medeiros R, LaFerla FM, Carvalheira JB, Saad MJ, Munoz DP, Velloso LA, Ferreira ST, De Felice FG (2015) Alzheimer-associated Aβ oligomers impact the central nervous system to induce peripheral metabolic deregulation. EMBO Mol Med 7:190–210

Cole SL, Vassar R (2007) The Alzheimer's disease β-secretase enzyme, BACE1. Mol Neurodegener 2:22

Conejo R, Lorenzo M (2001) Insulin signaling leading to proliferation, survival, and membrane ruffling in C2C12 myoblasts. J Cell Physiol 187:96–108

Connor JR, Snyder BS, Beard JL, Fine RE, Mufson EJ (1992a) Regional distribution of iron and iron-regulatory proteins in the brain in aging and Alzheimer's disease. J Neurosci Res 31:327–335

Connor JR, Menzies SL, St Martin SM, Mufson EJ (1992b) A histochemical study of iron, transferrin, and ferritin in Alzheimer's diseased brains. J Neurosci Res 31:75–83

Coppedè F, Migliore L (2009) DNA damage and repair in Alzheimer's disease. Curr Alzheimer Res 6:36–47

Cordy JM, Hooper NM, Turner AJ (2003) Exclusively targeting beta-secretase to lipid rafts by GPI-anchor addition up-regulates beta-site processing of the amyloid precursor protein. Proc Natl Acad Sci U S A 100:11735–11740

Costa RM, Drew J, Silva AJ (2005) Notch to remember. Trends Neurosci 28:429–435

Costantini C, Weindruch R, Della Valle G, Puglielli L (2005) A TrkA-to-p75NTR molecular switch activates amyloid β-peptide generation during aging. Biochem J 391:59–67

Costantini C, Scrable H, Puglielli L (2006) An aging pathway controls the TrkA to p75NTR receptor switch and amyloid β peptide generation. EMBO J 25:1997–2006

Cotman CW (2005) The role of neurotrophins in brain aging: a perspective in honor of Regino Perez-Polo. Neurochem Res 30:877–881

Crichton RR, Dexter DT, Ward RJ (2011) Brain iron metabolism and its perturbation in neurological diseases. J Neural Transm 118:301–314

Cui W, Sun Y, Wang Z, Xu C, Xu L, Wang F, Chen Z, Peng Y, Li R (2011) Activation of liver X receptor decreases BACE1 expression and activity by reducing membrane cholesterol levels. Neurochem Res 36:1910–1921

Curtain CC, Ali F, Volitakis I, Cherny RA, Norton RS, Beyreuther K, Barrow CJ, Masters CL, Bush AI, Barnham KJ (2001) Alzheimer's disease amyloid-β binds copper and zinc to generate an allosterically ordered membrane-penetrating structure containing superoxide dismutase-like subunits. J Biol Chem 276:20466–20473

Cutler RG, Kelly J, Storie K, Pedersen WA, Tammara A, Hatanpaa K, Troncoso JC, Mattson MP (2004) Involvement of oxidative stress-induced abnormalities in ceramide and cholesterol metabolism in brain aging and Alzheimer's disease. Proc Natl Acad Sci U S A 17:2070–2075

Dandona P (2002) Endothelium, inflammation, and diabetes. Curr Diab Rep 2:311–315

Dasari B, Prasanthi JR, Marwarha G, Singh BB, Ghribi O (2010) The oxysterol 27-hydroxycholesterol increases b-amyloid and oxidative stress in retinal pigment epithelial cells. BMC 10:22

Davison AN (1965) Brain sterol metabolism. Adv Lipid Res 3:171–196

Dawkins E, Small DH (2014) Insights into the physiological function of the β-amyloid precursor protein: beyond Alzheimer's disease. J Neurochem 129:756–769

Day JJ, Sweatt JD (2011) Epigenetic mechanisms in cognition. Neuron 70:813–829

de Calignon A, Fox LM, Pitstick R, Carlson GA, Bacskai BJ, Spires-Jones TL, Hyman BT (2010) Caspase activation precedes and leads to tangles. Nature 464:1201–1204

De Felice FG, Velasco PT, Lambert MP, Viola K, Fernandez SJ, Ferreira ST, Klein WL (2007) A beta oligomers induce neuronal oxidative stress through an *N*-methyl-D-aspartate receptor-dependent mechanism that is blocked by the Alzheimer drug memantine. J Biol Chem 282:11590–11601

De Felice FG, Wu D, Lambert MP, Fernandez SJ, Velasco PT, Lacor PN, Bigio EH, Jerecic J, Acton PJ, Shughrue PJ, Chen-Dodson E, Kinney GG, Klein WL (2009a) Alzheimer's disease-type neuronal Tau hyperphosphorylation induced by A beta oligomers. Neurobiol Aging 29:1334–1347

De Felice FG, Vieira MN, Bomfim TR, Decker H, Velasco PT, Lambert MP, Viola KL, Zhao WQ, Ferreira ST, Klein WL (2009b) Protection of synapses against Alzheimer's-linked toxins: insulin signaling prevents the pathogenic binding of Abeta oligomers. Proc Natl Acad Sci U S A 106:1971–1976

de la Monte SM (2013) Intranasal insulin therapy for cognitive impairment and neurodegeneration: current state of the art. Expert Opin Drug Deliv 10:1699–1709

de la Monte SM (2009) Insulin resistance and Alzheimer's disease. BMB Rep 42:475–481

de la Monte SM, Tong M (2013) Insulin resistance and metabolic failure underlie Alzheimer disease. In: Farooqui T, Farooqui AA (eds) Metabolic syndrome and neurological disorders. Wiley, Oxford, UK, pp 1–30

de la Monte SM, Tong M (2014) Brain metabolic dysfunction at the core of Alzheimer's disease. Biochem Pharmacol 88:548–559

de la Monte SM, Wands JR Jr (2008) Alzheimer's disease is type 3 diabetes-evidence reviewed. J Diabetes Sci Technol 2:1101–1113

Deaton AM, Bird A (2011) CpG islands and the regulation of transcription. Genes Dev 25:1010–1022

Deibel MA, Ehmann WD, Markesbery WR (1996) Copper, iron, and zinc imbalances in severely degenerated brain regions in Alzheimer's disease: possible relation to oxidative stress. J Neurol Sci 143:137–142

Delcroix JD, Valletta JS, Wu C, Hunt SJ, Kowal AS, Mobley WC (2003) NGF signaling in sensory neurons: evidence that early endosomes carry NGF retrograde signals. Neuron 39:69–84

Demuro A, Smith M, Parker I (2011) Single-channel Ca^{2+} imaging implicates Aβ1-42 amyloid pores in Alzheimer's disease pathology. J Cell Biol 195:515–524

Derynck R, Zhang YE (2003) Smad-dependent and Smad-independent pathways in TGF-beta family signalling. Nature 425:577–584

Desbène C, Malaplate-Armand C, Youssef I, Garcia P, Stenger C, Sauvée M, Fischer N, Rimet D, Koziel V, Escanyé MC, Oster T, Kriem B, Yen FT, Pillot T, Olivier JL (2012) Critical role of cPLA2 in Aβ oligomer-induced neurodegeneration and memory deficit. Neurobiol Aging 33:1123.e17–e29

Dickson JR, Kruse C, Montagna DR, Finsen B, Wolfe MS (2013) Alternative polyadenylation and miR-34 family members regulate Tau expression. J Neurochem 127:739–749

Doherty GH (2011) Obesity and the ageing brain: could leptin play a role in neurodegeneration? Curr Gerontal Geriatr Res 2011:708154

Doherty GH, Beccano-Kelly D, Yan SD, Gunn-Moore FJ, Harvey J (2012) Leptin prevents hippocampal synaptic disruption and neuronal cell death induced by amyloid β. Neurobiol Aging 34:226–237

Du JL, Poo MM (2004) Rapid BDNF-induced retrograde synaptic modification in a developing retinotectal system. Nature 429:878–883

Du H, Guo L, Fang F, Chen D, Sosunov AA, McKhann GM, Yan Y, Wang C, Zhang H, Molkentin JD, Gunn-Moore FJ, Vonsattel JP, Arancio O, Chen JX, Yan SD (2008) Cyclophilin D deficiency attenuates mitochondrial and neuronal perturbation and ameliorates learning and memory in Alzheimer's disease. Nat Med 14:1097–1105

Duce JA, Bush AI (2010) Biological metals and Alzheimer's disease: implications for therapeutics and diagnostics. Prog Neurobiol 92:1–18

Duce JA, Tsatsanis A, Cater MA, James SA, Robb E, Wikhe K, Leong SL, Perez K, Johanssen T, Greenough MA, Cho HH, Galatis D, Moir RD, Masters CL, McLean C, Tanzi RE, Cappai R, Barnham KJ, Ciccotosto GD, Rogers JT, Bush AI (2010) Iron-export ferroxidase activity of beta-amyloid precursor protein is inhibited by zinc in Alzheimer's disease. Cell 142:857–867

Egana JT, Zambrano C, Nunez MT, Gonzalez-Billault C, Maccioni RB (2003) Iron-induced oxidative stress modify Tau phosphorylation patterns in hippocampal cell cultures. Biometals 16:215–223

El Khoury J, Hickman SE, Thomas CA, Cao L, Silverstein SC, Loike JD (1996) Scavenger receptor-mediated adhesion of microglia to β-amyloid fibrils. Nature 382:716–719

Emmerzaal TL, Kiliaan AJ, Gustafson DR (2015) 2003-2013: a decade of body mass index, Alzheimer's disease, and dementia. J Alzheimers Dis 43:739–755

Englander EW (2008) Brain capacity for repair of oxidatively damaged DNA and preservation of neuronal function. Mech Ageing Dev 129:475–482

Escribano L, Simón AM, Gimeno E, Cuadrado-Tejedor M, López de Maturana R, García-Osta A, Ricobaraza A, Pérez-Mediavilla A, Del Río J, Frechilla D (2010) Rosiglitazone rescues memory impairment in Alzheimer's transgenic mice: mechanisms involving a reduced amyloid and Tau pathology. Neuropsychopharmacology 35:1593–1604

Evans MC, Barnes J, Nielsen C, Kim LG, Clegg SL, Blair M, Leung KK, Douiri A, Boyes RG, Ourselin S, Fox NC (2010) Volume changes in Alzheimer's disease and mild cognitive impairment: cognitive associations. Eur Radiol 20:674–682

Extance A (2010) Alzheimer's failure raises questions about disease-modifying strategies. Nat Rev Drug Discov 9:749–751

Fahrenholz F (2007) Alpha-secretase as a therapeutic target. Curr Alzheimer Res 4:412–417

Famer D, Meaney S, Mousavi M, Nordberg A, Bjorkhem I, Crisby M (2007) Regulation of alpha- and beta-secretase activity by oxysterols: cerebrosterol stimulates processing of APP via the a-secretase pathway. Biochem Biophys Res Commun 20:46–50

Fang CL, Wu WH, Liu Q, Sun X, Ma Y, Zhao YF, Li YM (2010) Dual functions of beta-amyloid oligomer and fibril in Cu(II)-induced H_2O_2 production. Regul Pept 163:1–6

Farooqui AA (2009a) Beneficial effects of fish oil on human brain. Springer, New York

Farooqui AA (2009b) Hot topics in neural membrane lipidology. Springer, New York, NY

Farooqui AA (2010a) Neurochemical aspects of neurotraumatic and neurodegeneratine diseases. Springer, New York

Farooqui AA (2010b) Studies on plasmalogen-selective phospholipase A_2 in brain. Mol Neurobiol 41:267–273

Farooqui AA (2011) Lipid mediators and their metabolism in the brain. Springer, New York

Farooqui AA (2013) Metabolic syndrome: an important risk factor for Stroke, Alzheimer, and depression. Springer, New York

Farooqui AA (2014) Inflammation and oxidative stress in neurological disorders. Springer, New York

Farooqui AA (2015) High calorie diet and human brain: metabolic consequences of long term consumption. Springer, New York

Farooqui AA, Horrocks LA (2007) Glycerophospholipids in the brain: phospholipases A_2 in neurological disorders. Springer, New York, NY

Farooqui AA, Rapoport SI, Horrocks LA (1997) Membrane phospholipid alterations in Alzheimer's disease: deficiency of ethanolamine plasmalogens. Neurochem Res 22:523–527

Farooqui AA, Horrocks LA, Farooqui T (2000) Deacylation-reacylation of neural membrane glycerophospholipids, a matter of life and death. J Mol Neurosci 14:123–133

Farooqui AA, Ong WY, Horrocks LA (2003) Plasmalogens, docosahexaenoic acid and neurological disorders. Adv Exp Med Biol 544:335–354

Farooqui AA, Horrocks LA, Farooqui T (2007) Modulation of inflammation in brain: a matter of fat. J Neurochem 101:577–599

Farooqui AA, Ong WY, Horrocks LA (2008) Neurochemical aspects of excitotoxicity. Springer, New York

Farooqui AA, Ong WY, Farooqui T (2010) Lipid mediators in the nucleus: their potential contribution to Alzheimer's disease. Biochim Biophys Acta 1801:906–916

Farooqui AA, Farooqui T, Panza F, Frisardi V (2012) Metabolic syndrome as a risk factor for neurological disorders. Cell Mol Life Sci 69:741–762

Feng J, Zhou Y, Campbell SL, Le T, Li E, Sweatt JD, Silva AJ, Fan G (2010) Dnmt1 and Dnmt3a maintain DNA methylation and regulate synaptic function in adult forebrain neurons. Nat Neurosci 13:423–430

Ferrer I (2009) Altered mitochondria, energy metabolism, voltage-dependent anion channel, and lipid rafts converge to exhaust neurons in Alzheimer's disease. J Bioenerg Biomembr 41:425–431

Fewlass DC, Noboa K, Pi-Sunyer FX, Johnston JM, Yan SD, Tezapsidis N (2004) Obesity-related leptin regulates Alzheimer's Abeta. FASEB J 18:1870–1878

Figueiredo-Pereira ME, Rockwell P, Schmidt-Glenewinkel T, Serrano P (2015) Neuroinflammation and J2 prostaglandins: linking impairment of the ubiquitin-proteasome pathway and mitochondria to neurodegeneration. Front Mol Neurosci 7:104

Fischer A, Sananbenesi F, Wang X, Dobbin M, Tsai LH (2007) Recovery of learning and memory is associated with chromatin remodelling. Nature 447:178–182

Flier JS (2004) Obesity wars: molecular progress confronts an expanding epidemic. Cell 116:337–350

Folwell J, Cowan CM, Ubhi KK, Shiabh H, Newman TA, Shepherd D, Mudher A (2010) Aβ exacerbates the neuronal dysfunction caused by human Tau expression in a Drosophila model of Alzheimer's disease. Exp Neurol 223:401–409

Fonteh AN, Chiang J, Cipolla M, Hale J, Diallo F, Chirino A, Arakaki X, Harrington MG (2013) Alterations in cerebrospinal fluid glycerophospholipids and phospholipase A_2 activity in Alzheimer's disease. J Lipid Res 54:2884–2897

Fox NC, Schott JM (2004) Imaging cerebral atrophy: normal ageing to Alzheimer's disease. Lancet 363:392–394

Frazzini V, Rockabrand E, Mocchegiani E, Sensi SL (2006) Oxidative stress and brain aging: is zinc the link? Biogerontology 7:307–314

Frederickson CJ, Suh SW, Silva D, Frederickson CJ, Thompson RB (2000) Importance of zinc in the central nervous system: the zinc-containing neuron. J Nutr 130:1471S–1483S

Frisoni GB, Rozzini L, Gozzetti A, Binetti G, Zanetti O, Bianchetti A, Trabucchi M, Cummings JL (1999) Behavioral syndromes in Alzheimer's disease: description and correlates. Dement Geriatr Cogn Disord 10:130–138

Fu X, Menke JG, Chen Y, Zhou G, MacNaul KL, Wright SD, Sparrow CP, Lund EG (2001) 27-Hydroxycholesterol is an endogenous ligand for liver X receptor in cholesterol-loaded cells. J Biol Chem 276:38378–38387

Fujino T, Lee WC, Nedivi E (2003) Regulation of cpg15 by signaling pathways that mediate synaptic plasticity. Mol Cell Neurosci 24:538–554

Fujino T, Wu Z, Lin WC, Phillips MA, Nedivi E (2008) cpg15 and cpg15-2 constitute a family of activity-regulated ligands expressed differentially in the nervous system to promote neurite growth and neuronal survival. J Comp Neurol 507:1831–1845

Gamblin TC, Chen F, Zambrano A, Abraha A, Lagalwar S, Guillozet al, Lu M, Fu Y, Garcia-Sierra F, LaPointe N, Miller R, Berry RW, Binder LI, Cryns VL (2003) Caspase cleavage of Tau: linking amyloid and neurofibrillary tangles in Alzheimer's disease. Proc Natl Acad Sci U S A 100:10032–10037

Garza JC, Guo M, Zhang W, Lu XY (2008) Leptin increases adult hippocampal neurogenesis in vivo and in vitro. J Biol Chem 283:18238–18247

Gasparini L, Gouras GK, Wang R, Gross RS, Beal MF, Greengard P, Xu H (2001) Stimulation of beta-amyloid precursor protein trafficking by insulin reduces intraneuronal beta-amyloid and requires mitogen-activated protein kinase signaling. J Neurosci 21:2561–2570

Giasson BI, Forman MS, Higuchi M, Golbe LI, Graves CL, Kotzbauer PT, Trojanowski JQ, Lee VM (2003) Initiation and synergistic fibrillization of Tau and α-synuclein. Science 300:636–640

Gleichmann M, Mattson MP (2011) Neuronal calcium homeostasis and dysregulation. Antioxid Redox Signal 14:1261–1273

Goate A, Chartier-Harlin MC, Mullan M, Brown J, Crawford F, Fidani L, Giuffra L, Haynes A, Irving N, James L (1991) Segregation of a missense mutation in the amyloid precursor protein gene with familial Alzheimer's disease. Nature 349:704–706

Golde TE, Dickson D, Hutton M (2006) Filling the Gaps in the abeta; cascade hypothesis of Alzheimer's disease. Curr Alzheimer Res 3:421–430

Gong CX, Iqbal K (2008) Hyperphosphorylation of microtubule-associated protein Tau: a promising therapeutic target for Alzheimer disease. Curr Med Chem 15:2321–2328

Gong Y, Chang L, Viola K, Lacor P, Lambert M, Finch C, Krafft GA, Klein WL (2003) Alzheimer's disease-affected brain: presence of oligomeric A beta ligands (ADDLs) suggests a molecular basis for reversible memory loss. Proc Natl Acad Sci U S A 100:10417–10422

Gräff J, Kim D, Dobbin MM, Tsai L-H (2011) Epigenetic regulation of gene expression in physiological and pathological brain processes. Physiol Rev 91:603–649

Grant WB, Campbell A, Itzhaki RF, Savory J (2002) The significance of environmental factors in the etiology of Alzheimer's disease. J Alzheimers Dis 4:179–189

Greco SJ, Sarkar S, Johnston JM, Zhu X, Su B, Casadesus G, Ashford JW, Smith MA, Tezapsidis N (2008) Leptin reduces Alzheimer's disease-related Tau phosphorylation in neuronal cells. Biochem Biophys Res Commun 376:536–541

Greco SJ, Sarkar S, Johnston JM, Tezapsidis N (2009) Leptin regulates Tau phosphorylation and amyloid through AMPK in neuronal cells. Biochem Biophys Res Commun 380:98–104

Grimm MO, Grimm HS, Patzold AJ, Zinser EG, Halonen R, Duering M, Tschäpe JA, De Strooper B, Müller U, Shen J (2005) Regulation of cholesterol and sphingomyelin metabolism by amyloid-β and presenilin. Nat Cell Biol 27:1118–1123

Grimm MO, Grimm HS, Hartmann T (2007) Amyloid beta as a regulator of lipid homeostasis. Trends Mol Med 13:337–344

Grimm MO, Rothhaar TL, Grösgen S, Burg VK, Hundsdörfer B, Haupenthal VJ, Friess P, Kins S, Grimm HS, Hartmann T (2011a) Trans fatty acids enhance amyloidogenic processing of the Alzheimer amyloid precursor protein (APP). J Nutr Biochem 23:1214–1223

Grimm MO, Kuchenbecker J, Grösgen S, Burg VK, Hundsdörfer B, Rothhaar TL, Friess P, de Wilde MC, Broersen LM, Penke B, Péter M, Vígh L, Grimm HS, Hartmann T (2011b) Docosahexaenoic acid reduces amyloid beta production via multiple pleiotropic mechanisms. J Biol Chem 286:14028–14039

Grimm MO, Zinser EG, Grösgen S, Hundsdörfer B, Rothhaar TL, Burg VK, Kaestner L, Bayer TA, Lipp P, Müller U, Grimm HS, Hartmann T (2012) Amyloid precursor protein (APP) mediated regulation of ganglioside homeostasis linking Alzheimer's disease pathology with ganglioside metabolism. PLoS One 7, e34095

Grimm MO, Zimmer VC, Lehmann J, Grimm HS, Hartmann T (2013) The impact of cholesterol, DHA, and sphingolipids on Alzheimer's disease. Biomed Res Int 2013:814390

Gu X, Sun J, Li S, Wu X, Li L (2013) Oxidative stress induces DNA demethylation and histone acetylation in SH-SY5Y cells: potential epigenetic mechanisms in gene transcription in Aβ production. Neurobiol Aging 34:1069–1079

Guan Z, Wang Y, Cairns NJ, Lantos PL, Dallner G, Sindelar PJ (1999) Decrease and structural modifications of phosphatidylethanolamine plasmalogen in the brain with Alzheimer disease. J Neuropathol Exp Neurol 58:740–747

Guo Z, Jiang H, Xu X, Duan W, Mattson MP (2008) Leptin-mediated cell survival signaling in hippocampal neurons mediated by JAK STAT3 and mitochondrial stabilization. J Biol Chem 283:1754–1763

Haan MN (2006) Therapy insight: type 2 diabetes mellitus and the risk of late-onset alzheimer's disease. Nat Clin Pract Neurol 2:159–166

Haass C, Mandelkow E (2010) Fynτ-amyloid: a toxic triad. Cell 142:356–358

Haass C, Selkoe DJ (2007) Soluble protein oligomers in neurodegeneration: lessons from the Alzheimer's amyloid beta-peptide. Nat Rev Mol Cell Biol 8:101–112

Hambidge M (2000) Human zinc deficiency. J Nutr 130:1344S–1349S

Han X (2007) Potential mechanisms contributing to sulfatide depletion at the earliest clinically recognizable stage of Alzheimer's disease: a tale of shotgun lipidomics. J Neurochem 103(Suppl 1):171–179

Han X (2010) The pathogenic implication of abnormal interaction between apolipoprotein E isoforms, amyloid-beta peptides, and sulfatides in Alzheimer's disease. Mol Neurobiol 41:97–106

Han X, Holtzman DM, McKeel DW Jr (2001) Plasmalogen deficiency in early Alzheimer's disease subjects and in animal models: molecular characterization using electrospray ionization mass spectrometry. J Neurochem 77:1168–1180

Han X, Holtzman DM, McKeel DW Jr, Kelley J, Morris JC (2002) Substantial sulfatide deficiency and ceramide elevation in very early Alzheimer's disease: potential role in disease pathogenesis. J Neurochem 82:809–818

Hanger DP, Anderton BH, Noble W (2009) Tau phosphorylation: the therapeutic challenge for neurodegenerative disease. Trends Mol Med 15:112–119

Hardy J, Selkoe DJ (2002) The amyloid hypothesis of Alzheimer's disease: progress and problems on the road to therapeutics. Science 297:353–356

Hartmann D, Tournoy J, Saftig P, Annaert W, De Strooper B (2001) Implication of APP secretases in notch signaling. J Mol Neurosci 17:171–181

Harvey J, Shanley LJ, O'Malley D, Irving AJ (2005) Leptin: a potential cognitive enhancer? Biochem Soc Trans 33:1029–1032

He P, Zhong Z, Lindholm K, Berning L, Lee W, Lemere C, Staufenbiel M, Li R, Shen Y (2007) Deletion of tumor necrosis factor death receptor inhibits amyloid beta generation and prevents learning and memory deficits in Alzheimer's mice. J Cell Biol 178:829–841

He X, Huang Y, Li B, Gong CX, Schuchman EH (2010) Deregulation of sphingolipid metabolism in Alzheimer's disease. Neurobiol Aging 31:398–408

Heerssen HM, Pazyra MF, Segal RA (2004) Dynein motors transport activated Trks to promote survival of target-dependent neurons. Nat Neurosci 7:596–604

Heneka MT, O'Banion MK (2007) Inflammatory processes in Alzheimer's disease. J Neuroimmunol 184:69–91

Henneman WJ, Sluimer JD, Barnes J, van der Flier WM, Sluimer IC, Fox NC, Scheltens P, Vrenken H, Barkhof F (2009) Hippocampal atrophy rates in Alzheimer disease: added value over whole brain volume measures. Neurology 72:999–1007

Herholz K (2012) Use of FDG PET as an imaging biomarker in clinical trials of Alzheimer's disease. Biomark Med 6:431–439

Hirai K, Aliev G, Nunomura A, Fujioka H, Russell RL, Atwood CS, Johnson AB, Kress Y, Vinters HV, Tabaton M, Shimohama S, Cash AD, Siedlak SL, Harris PL, Jones PK, Petersen RB, Perry G, Smith MA (2001) Mitochondrial abnormalities in Alzheimer's disease. J Neurosci 21:3017–3023

Hirano T, Murakami M, Fukada T, Nishida K, Yamasaki S, Suzuki T (2008) Roles of zinc and zinc signaling in immunity: zinc as an intracellular signaling molecule. Adv Immunol 97:149–176

Ho A, Sudhof TC (2004) Binding of F-spondin to amyloid-beta precursor protein: a candidate amyloid-beta precursor protein ligand that modulates amyloid-beta precursor protein cleavage. Proc Natl Acad Sci U S A 101:2548–2553

Holden KF, Lindquist K, Tylavsky FA, Rosano C, Harris TB, Yaffe K (2009) Serum leptin level and cognition in the elderly: findings from the Health ABC Study. Neurobiol Aging 30:1483–1489

Holland WL, Summers SA (2008) Sphingolipids, insulin resistance, and metabolic disease: new insights from in vivo manipulation of sphingolipid metabolism. Endocr Rev 29:381–402

Hong S, Ostaszewski BL, Yang T, O'Malley TT, Jin M, Yanagisawa K, Li S, Bartels T, Selkoe DJ (2014) Soluble Aβ oligomers are rapidly sequestered from brain ISF in vivo and bind GM1 ganglioside on cellular membranes. Neuron 82:308–319

Hooper C, Killick R, Lovestone S (2008) The GSK3 hypothesis of Alzheimer's disease. J Neurochem 104:1433–1439

Hsieh RH, Hou JH, Hsu HS, Wei YH (1994) Age-dependent respiratory function decline and DNA deletions in human muscle mitochondria. Biochem Mol Biol Int 32:1009–1022

Hu N, Yu JT, Tan L, Wang YL, Sun L, Tan L (2013) Nutrition and the risk of Alzheimer's disease. Biomed Res Int 2013:524820

Hu NW, Nicoll AJ, Zhang D, Mably AJ, O'Malley T, Purro SA, Terry C, Collinge J, Walsh DM, Rowan MJ (2014) mGlu5 receptors and cellular prion protein mediate amyloid-β-facilitated synaptic long-term depression in vivo. Nat Commun 5:3374

Huang X, Atwood CS, Hartshorn MA, Multhaup G, Goldstein LE, Scarpa RC, Cuajungco MP, Gray DN, Lim J, Moir RD, Tanzi RE, Bush AI (1999) The Aβ peptide of Alzheimer's disease directly produces hydrogen peroxide through metal ion reduction. Biochemistry 38:7609–7616

Huang Y, Tanimukai H, Liu F, Iqbal K, Grunake-Iqbal I, Gong CX (2004) Elevation of the level and activity of acid ceramidase in Alzheimer's disease brain. Eur J Neurosci 20:3489–3497

Hughes TM, Lopez OL, Evans RW, Kamboh MI, Williamson JD, Klunk WE, Mathis CA, Price JC, Cohen AD, Snitz BE, Dekosky ST, Kuller LH (2014) Markers of cholesterol transport are associated with amyloid deposition in the brain. Neurobiol Aging 35:802–807

Hung YH, Bush AI, La Fontaine S (2013) Links between copper and cholesterol in Alzheimer's disease. Front Physiol 4:111

Illenberger S, Zheng-Fischhofer Q, Preuss U, Stamer K, Baumann K, Trinczek B, Biernat J, Godemann R, Mandelkow EM, Mandelkow E (1998) The endogenous and cell cycle-dependent phosphorylation of Tau protein in living cells: implications for Alzheimer's disease. Mol Biol Cell 9:1495–1512

Illi B, Colussi C, Grasselli A, Farsetti A, Capogrossi MC, Gaetano C (2009) NO sparks off chromatin: tales of a multifaceted epigenetic regulator. Pharmacol Ther 123:344–352

Inestrosa NC, Arenas E (2010) Emerging roles of Wnts in the adult nervous system. Nat Rev Neurosci 11:77–86

Iqbal K, Liu F, Gong CX, Alonso AC, Grundke-Iqbal I (2009) Mechanisms of Tau-induced neurodegeneration. Acta Neuropathol 118:53–69

Iwata A, Nagata K, Hatsuta H, Takuma H, Bundo M, Iwamoto K, Tamaoka A, Murayama S, Saido T, Tsuji S (2014) Altered CpG methylation in sporadic Alzheimer's disease is associated with APP and MAPT dysregulation. Hum Mol Genet 23:648–656

Jahn H (2013) Memory loss in Alzheimer's disease. Dialogues Clin Neurosci 15:445–454

Jana A, Pahan K (2004) Fibrillar amyloid-beta peptides kill human primary neurons via NADPH oxidase-mediated activation of neutral sphingomyelinase. Implications for Alzheimer's disease. J Biol Chem 279:51451–51459

Janson J, Laedtke T, Parisi JE, O'Brien P, Petersen RC, Butler PC (2004) Increased risk of type 2 diabetes in Alzheimer disease. Diabetes 53:474–481

Jeong J, Eide DJ (2013) The SLC39 family of zinc transporters. Mol Aspects Med 34:612–619

Jetten AM (2009) Retinoid-related orphan receptors (RORs): critical roles in development, immunity, circadian rhythm, and cellular metabolism. Nucl Recept Signal 7:45–59

Jiang D, Li X, Williams R, Patel S, Men L, Wang Y, Zhou F (2009) Ternary complexes of iron, amyloid-β, and nitrilotriacetic acid: binding affinities, redox properties, and relevance to iron-induced oxidative stress in Alzheimer's disease. Biochemistry 48:7939–7947

Jick H, Zornberg GL, Jick SS, Seshadri S, Drachman DA (2000) Statins and the risk of dementia. Lancet 356:1627–1631

Johnston AM, Pirola L, Van Obberghen E (2003) Molecular mechanisms of insulin receptor substrate protein-mediated modulation of insulin signalling. FEBS Lett 546:32–36

Kaether C, Haass C (2004) A lipid boundary separates APP and secretases and limits amyloid beta-peptide generation. J Cell Biol 167:809–812

Kalanj S, Kracun I, Rosner H, Cosovic C (1991) Regional distribution of brain gangliosides in Alzheimer's disease. Neurol Croat 40:269–281

Kalaria RN, Maestre GE, Arizaga R, Friedland RP, Galasko D, Hall K, Luchsinger JA, Ogunniyi A, Perry EK, Potocnik F, Prince M, Stewart R, Wimo A, Zhang ZX, Antuono P; World Federation of Neurology Dementia Research Group (2008) Alzheimer's disease and vascular dementia in developing countries: prevalence, management, and risk factors. Lancet Neurol 7:812–826

Kandalepas PC, Vassar R (2012) Identification and biology of β-secretase. J Neurochem 120(Suppl 1):S55–S61

Kang J, Muller-Hill B (1990) Differential splicing of Alzheimer's disease amyloid A4 precursor RNA in rat tissues: PreA4(695) mRNA is predominantly produced in rat and human brain. Biochem Biophys Res Commun 166:1192–1200

Kang J, Lemaire HG, Unterbeck A, Salbaum JM, Masters CL, Grzeschik KH, Multhaup G, Beyreuther K, Muller-Hill B (1987) The precursor of Alzheimer's disease amyloid A4 protein resembles a cell-surface receptor. Nature 325:733–736

Kang HS, Angers M, Beak JY, Wu X, Gimble JM, Wada T, Xie W, Collins JB, Grissom SF, Jetten AM (2007) Gene expression profiling reveals a regulatory role for ROR alpha and ROR gamma in phase I and phase II metabolism. Physiol Genomics 31:281–294

Kanoski SE, Davidson TL (2011) Western diet consumption and cognitive impairment: links to hippocampal dysfunction and obesity. Physiol Behav 103:59–68

Katsel P, Li C, Haroutunian V (2007) Gene expression alterations in the sphingolipid metabolism pathways during progression of dementia and Alzheimer's disease: a shift toward ceramide accumulation at the earliest recognizable stages of Alzheimer's disease? Neurochem Res 32:845–856

Katura T, Moriya T, Nakahata N (2010) 15-Deoxy-delta 12,14-prostaglandin J2 biphasically regulates the proliferation of mouse hippocampal neural progenitor cells by modulating the redox state. Mol Pharmacol 77:601–611

Katzman R, Saitoh T (1991) Advances in Alzheimer's disease. FASEB J 5:278–286

Kawahara M (2010) Neurotoxicity of beta-amyloid protein: oligomerization, channel formation and calcium dyshomeostasis. Curr Pharm Des 16:2779–2789

Kessels HW, Nguyen LN, Nabavi S, Malinow R (2010) The prion protein as a receptor for amyloid-beta. Nature 466:E3–E4

Khosravani H, Zhang Y, Tsutsui S, Hameed S, Altier C, Hamid J, Altier C, Hamid J, Chen L, Villemaire M, Ali Z, Jirik FR, Zamponi GW (2008) Prion protein attenuates excitotoxicity by inhibiting NMDA receptors. J Cell Biol 181:551–565

Kim J, Basak JM, Holtzman DM (2009) The role of apolipoprotein E in Alzheimer's disease. Neuron 63:287–303

Kim J, Lee HJ, Lee KW (2010) Naturally occurring phytochemicals for the prevention of Alzheimer's disease. J Neurochem 112:1415–1430

Kim J, Yoon H, Basak J, Kim J (2014) Apolipoprotein E in synaptic plasticity and Alzheimer's disease: potential cellular and molecular mechanisms. Mol Cells 37:767–776

Kimberly WT, LaVoie MJ, Ostaszewski BL, Ye W, Wolfe MS, Selkoe DJ (2003) γ-Secretase is a membrane protein complex comprised of presenilin, nicastrin, aph-1, and pen-2. Proc Natl Acad Sci U S A 100:6382–6387

Kitazawa M, Cheng D, Laferla FM (2009) Chronic copper exposure exacerbates both amyloid and Tau pathology and selectively dysregulates cdk5 in a mouse model of AD. J Neurochem 108:1550–1560

Kleene R, Loers G, Langer J, Frobert Y, Buck F, Schachner M (2007) Prion protein regulates glutamate-dependent lactate transport of astrocytes. J Neurosci 27:12331–12340

Klewpatinond M, Davies P, Bowen S, Brown DR, Viles JH (2008) Deconvoluting the Cu^{2+} binding modes of full-length prion protein. J Biol Chem 283:1870–1881

Klunk WE, Engler H, Nordberg A, Wang Y, Blomqvist G, Holt DP, Bergström M, Savitcheva I, Huang GF, Estrada S, Ausén B, Debnath ML, Barletta J, Price JC, Sandell J, Lopresti BJ, Wall A, Koivisto P, Antoni G, Mathis CA, Långström B (2004) Imaging brain amyloid in Alzheimer's disease with Pittsburgh compound-B. Ann Neurol 55:306–319

Knowles JK, Rajadas J, Nguyen TV, Yang T, LeMieux MC, Vander Griend L, Ishikawa C, Massa SM, Wyss-Coray T, Longo FM (2009) The p75 neurotrophin receptor promotes amyloid-β(1–42)-induced neuritic dystrophy in vitro and in vivo. J Neurosci 29:10627–10637

Kojro E, Postina R (2009) Regulated proteolysis of RAGE and AbetaPP as possible link between type 2 diabetes mellitus and Alzheimer's disease. J Alzheimers Dis 16:865–878

Kong Y, Wu J, Yuan L (2014) MicroRNA expression analysis of adult-onset Drosophila Alzheimer's disease model. Curr Alzheimer Res 11:882–891

Konietzko U (2012) AICD nuclear signaling and its possible contribution to Alzheimer's disease. Curr Alzheimer Res 9:200–216

König HG, Kögel D, Rami A, Prehn JH (2005) TGF-{beta}1 activates two distinct type I receptors in neurons implications for neuronal NF-{kappa}B signaling. J Cell Biol 168:1077–1086

Kopan R, Goate A (2000) A common enzyme connects notch signaling and Alzheimer's disease. Genes Dev 14:2799–2806

Koponen E, Lakso M, Castrén E (2004) Overexpression of the full-length neurotrophin receptor trkB regulates the expression of plasticity-related genes in mouse brain. Brain Res Mol Brain Res 130:81–94

Kotti T, Head DD, McKenna CE, Russell DW (2008) Biphasic requirement for geranylgeraniol in hippocampal long-term potentiation. Proc Natl Acad Sci U S A 105:11394–11399

Krogh-Madsen R, Plomgaard P, Keller P, Keller C, Pedersen BK (2004) Insulin stimulates interleukin-6 and tumor necrosis factor-alpha gene expression in human subcutaneous adipose tissue. Am J Physiol Endocrinol Metab 286:E234–E238

Kumar S, Walter J (2011) Phosphorylation of amyloid beta (Aβ) peptides – a trigger for formation of toxic aggregates in Alzheimer's disease. Aging (Albany, NY) 3:PMC3184981

Kuner PR, Schubenel RC, Hertel C (1998) β-amyloid binds to p75^{NTR} and activates NF-κB in human neuroblastoma cells. J Neurosci Res 54:798–804

Labouèbe G, Liu S, Dias C, Zou H, Wong JC, Karunakaran S, Clee SM, Phillips AG, Boutrel B, Borgland SL (2013) Insulin induces long-term depression of ventral tegmental area dopamine neurons via endocannabinoids. Nat Neurosci 16:300–308

Lacor PN, Buniel MC, Chang L, Fernandez SJ, Gong Y, Viola KL, Lambert MP, Velasco PT, Bigio EH, Finch CE, Krafft GA, Klein WL (2004) Synaptic targeting by Alzheimer's-related amyloid beta oligomers. J Neurosci 24:10191–10200

Lacor PN, Buniel MC, Furlow PW, Clemente AS, Velasco PT, Wood M, Viola KL, Klein WL (2007) Abeta oligomer-induced aberrations in synapse composition, shape, and density provide a molecular basis for loss of connectivity in Alzheimer's disease. J Neurosci 27:796–807

Laird FM, Cai H, Savonenko AV, Farah MH, He K, Melnikova T, Wen H, Chiang HC, Xu G, Koliatsos VE, Borchelt DR, Price DL, Lee HK, Wong PC (2005) BACE1, a major determinant of selective vulnerability of the brain to amyloid-beta amyloidogenesis, is essential for cognitive, emotional, and synaptic functions. J Neurosci 25:11693–11709

Lambert MP, Barlow AK, Chromy BA, Edwards C, Freed R, Liosatos M, Morgan TE, Rozovsky I, Trommer B, Viola KL, Wals P, Zhang C, Finch CE, Krafft GA, Klein WL (1998) Diffusible, nonfibrillar ligands derived from Abeta1-42 are potent central nervous system neurotoxins. Proc Natl Acad Sci U S A 95:6448–6453

Landry GJ, Liu-Ambrose T (2014) Buying time: a rationale for examining the use of circadian rhythm and sleep interventions to delay progression of mild cognitive impairment to Alzheimer's disease. Front Aging Neurosci 6:325

Lasagna-Reeves CA, Sengupta U, Castillo-Carranza D, Gerson JE, Guerrero-Munoz M, Troncoso JC, Jackson GR, Kayed R (2014) The formation of Tau pore-like structures is prevalent and cell specific: possible implications for the disease phenotypes. Acta Neuropathol Commun 2:56

Lauren J, Gimbel D, Nygaard H, Gilbert J, Strittmatter S (2009) Cellular prion protein mediates impairment of synaptic plasticity by amyloid-beta oligomers. Nature 457:1128–1132

Leandro P, Gomes CM (2008) Protein misfolding in conformational disorders: rescue of folding defects and chemical chaperoning. Mini Rev Med Chem 8:901–911

Lee HG, Zhu X, Castellani RJ, Nunomura A, Perry G, Smith MA (2007) Amyloid-β in Alzheimer disease: the null versus the alternate hypotheses. J Pharmacol Exp Ther 321(3):823–829

Lee H-K, Kumar P, Fu Q, Rosen KM, Querfurth HW (2009) The insulin/Akt signaling pathway is targeted by intracellular β-amyloid. Mol Biol Cell 20:1533–1544

Lee JC, Simonyi A, Sun AY, Sun GY (2010) Phospholipases A_2 and neural membrane dynamics: implications for Alzheimer's disease. J Neurochem 116:813–819

Leoni V (2009) Oxysterols as markers of neurological disease – a review. Scan J Clin Lab Invest 69:22–25

Leoni V, Solomon A, Kivipelto M (2010) Links between ApoE, brain cholesterol metabolism, Tau and amyloid beta-peptide in patients with cognitive impairment. Biochem Soc Trans 38:1021–1025

Lessmann V, Gottmann K, Malcangio M (2003) Neurotrophin secretion: current facts and future prospects. Prog Neurobiol 69:341–374

Li YM, Lai MT, Xu M, Huang Q, DiMuzio-Mower J, Sardana MK, Shi XP, Yin KC, Shafer JA, Gardell SJ (2000) Presenilin 1 is linked with gamma -secretase activity in the detergent solubilized state. Proc Natl Acad Sci U S A 97:6138–6143

Li T, Ma G, Cai H, Price DL, Wong PC (2003) Nicastrin is required for assembly of presenilin/γ-secretase complexes to mediate notch signaling and for processing and trafficking of β-amyloid precursor protein in mammals. J Neurosci 23:3272–3277

Li Z, Melandri F, Berdo I, Jansen M, Hunter L, Wright S, Valbrun D, Figueiredo-Pereira ME (2004a) Delta12-Prostaglandin J2 inhibits the ubiquitin hydrolase UCH-L1 and elicits ubiquitin-protein aggregation without proteasome inhibition. Biochem Biophys Res Commun 319:1171–1180

Li R, Lindholm K, Yang LB, Yue X, Citron M, Yan R, Beach T, Sue L, Sabbagh M, Cai H, Wong P, Price D, Shen Y (2004b) Amyloid β peptide load is correlated with increased β-secretase activity in sporadic Alzheimer's disease patients. Proc Natl Acad Sci U S A 101:3632–3637

Li LM, Liu QH, Qiao JT, Zhang C (2009) Aβ_{31-35}-induced neuronal apoptosis is mediated by JNK-dependent extrinsic apoptosis pathway. Neurosci Bull 25:361–366

Li W, Yu J, Huang X, Abumaria N, Zhu Y, Huang X, Xiong W, Pen C, Liu X-C, Chui D, Liu G (2014) Elevation of brain magnesium prevents synaptic loss and reverses cognitive deficits in Alzheimer's disease mouse model. Mol Brain 7:65

Lieb W, Beiser AS, Vasan RS, Tan ZS, Au R, Harris TB, Roubenoff R, Auerbach S, DeCarli C, Wolf PA, Seshadri S (2009) Association of plasma leptin levels with incident Alzheimer disease and MRI measures of brain aging. JAMA 302:2565–2572

Linkous DH, Flinn JM, Koh JY, Lanzirotti A, Bertsch PM, Jones BF, Giblin LJ, Frederickson CJ (2008) Evidence that the ZNT3 protein controls the total amount of elemental zinc in synaptic vesicles. J Histochem Cytochem 56:3–6

Liu W, Dou F, Feng J, Yan Z (2001) RACK1 is involved in β-amyloid impairment of muscarinic regulation of GABAergic transmission. Neurobiol Aging 32:1818–1826

Liu Y, Qin L, Wilson BC, An L, Hong J-S, Liu B (2002) Inhibition by naloxone stereoisomers of β-amyloid peptide (1-42)-induced superoxide production in microglia and degeneration of cortical and mesencephalic neurons. J Pharmacol Exp Ther 302:1212–1219

Liu D, Xu Y, Feng Y, Liu H, Shen X, Chen K, Ma J, Jiang H (2006) Inhibitor discovery targeting the intermediate structure of beta-amyloid peptide on the conformational transition pathway: implications in the aggregation mechanism of beta-amyloid peptide. Biochemistry 45:10963–10972

Liu H, Li W, Ahmad M, Miller TM, Rose ME, Poloyac SM, Uechi G, Balasubramani M, Hickey RW, Graham SH (2011) Modification of ubiquitin-C-terminal hydrolase-L1 by cyclopentenone prostaglandins exacerbates hypoxic injury. Neurobiol Dis 41:318–328

Longo FM, Massa SM (2008) Small molecule modulation of p75 neurotrophin receptor functions. CNS Neurol Disord Drug Targets 7:63–70

Lorenzo A, Yuan M, Zhang Z, Paganetti PA, Sturchler-Pierrat C, Staufenbiel M, Mautino J, Vigo FS, Sommer B, Yankner BA (2000) Amyloid β interacts with the amyloid precursor protein: a potential toxic mechanism in Alzheimer's disease. Nat Neurosci 3:460–464

Lourenco MV, Clarke JR, Frozza RL, Bomfim TR, Forny-Germano L, Batista AF, Sathler LB, Brito-Moreira J, Amaral OB, Silva CA, Freitas-Correa L, Espírito-Santo S, Campello-Costa P, Houzel JC, Klein WL, Holscher C, Carvalheira JB, Silva AM, Velloso LA, Munoz DP, Ferreira ST, De Felice FG (2013) TNF-alpha mediates PKR-dependent memory impairment and brain IRS-1 inhibition induced by Alzheimer's beta-amyloid oligomers in mice and monkeys. Cell Metab 18:831–843

Lovell MA, Markesbery WR (2008) Oxidatively modified RNA in mild cognitive impairment. Neurobiol Dis 29:169–175

Lovell MA, Robertson JD, Teesdale WJ, Campbell JL, Markesbery WR (1998) Copper, iron and zinc in Alzheimer's disease senile plaques. J Neurol Sci 158:47–52

Lukiw WJ, Andreeva TV, Grigorenko AP, Rogaev EI (2013) Studying micro RNA function and dysfunction in Alzheimer's disease. Front Genet 3:327

Lund EG, Xie C, Kotti T, Turley SD, Dietschy JM, Russell DW (2003) Knockout of the cholesterol 24-hydroxylase gene in mice reveals a brain-specific mechanism of cholesterol turnover. J Biol Chem 278:22980–22988

Luo Y, Bolon B, Kahn S, Bennett BD, Babu-Khan S, Denis P, Fan W, Kha H, Zhang J, Gong Y, Martin L, Louis JC, Yan Q, Richards WG, Citron M, Vassar R (2001) Mice deficient in BACE1, the Alzheimer's β-secretase, have normal phenotype and abolished β-amyloid generation. Nat Neurosci 4:231–232

Lustbader JW, Cirilli M, Lin C, Xu HW, Takuma K, Wang N, Caspersen C, Chen X, Pollak S, Chaney M, Trinchese F, Liu S, Gunn-Moore F, Lue LF, Walker DG, Kuppusamy P, Zewier ZL, Arancio O, Stern D, Yan SS, Wu H (2004) ABAD directly links Abeta to mitochondrial toxicity in Alzheimer's disease. Science 304:448–452

Lutjohann D, Papassotiropoulos A, Bjorkhem I, Locatelli S, Bagli M, Oehring RD, Schlegel U, Jessen F, Rao ML, von Bergmann K, Heun R (2000) Plasma 24S-hydroxycholesterol (cerebrosterol) is increased in Alzheimer and vascular demented patients. J Lipid Res 41:195–198

Ma H, Lesne S, Kotilinek L, Steidl-Nichols JV, Sherman M, Younkin L, Younkin S, Forster C, Sergeant N, Delacourte A, Vassar R, Citron M, Kofuji P, Boland LM, Ashe KH (2007) Involvement of beta-site APP cleaving enzyme 1 (BACE1) in amyloid precursor protein-mediated enhancement of memory and activity-dependent synaptic plasticity. Proc Natl Acad Sci U S A 104:8167–8172

Mandelkow EM, Biernat J, Drewes G, Gustke N, Trinczek B, Mandelkow E (1995) Tau domains, phosphorylation, and interactions with microtubules. Neurobiol Aging 16:355–362, discussion 362–353

Mandelkow EM, Schweers O, Drewes G, Biernat J, Gustke N, Trinczek B, Mandelkow E (1996) Structure, microtubule interactions, and phosphorylation of Tau protein. Ann N Y Acad Sci 777:96–106

Mandelkow EM, Stamer K, Vogel R, Thies E, Mandelkow E (2003) Clogging of axons by Tau, inhibition of axonal traffic and starvation of synapses. Neurobiol Aging 24:1079–1085

Mantha AK, Sarkar B, Tell G (2013) A short review on the implications of base excision repair pathway for neurons: relevance to neurodegenerative diseases. Mitochondrion 16:38–49

Mao X, Ye J, Zhou S, Pi R, Dou J, Zang L, Chen X, Chao X, Li W, Liu M, Liu P (2012) The effects of chronic copper exposure on the amyloid protein metabolisim associated genes' expression in chronic cerebral hypoperfused rats. Neurosci Lett 518:14–18

Martic S, Rains MK, Kraatz HB (2013) Probing copper/Tau protein interactions electrochemically. Anal Biochem 442:130–137

Martin L, Latypova X, Wilson CM et al (2013) Tau protein kinases: involvement in Alzheimer's disease. Ageing Res Rev 12:289–309

Martins IJ, Berger T, Sharman MJ, Verdile G, Fuller SJ, Martins RN (2009) Cholesterol metabolism and transport in the pathogenesis of Alzheimer's disease. J Neurochem 111:1275–1308

Marwarha G, Ghribi O (2012) Leptin signaling and Alzheimer's disease. Am J Neurodegener Dis 1:245–265

Marwarha G, Dasari B, Prasanthi JR, Schommer J, Ghribi O (2010a) Leptin reduces the accumulation of Abeta and phosphorylated Tau induced by 27-hydroxycholesterol in rabbit organotypic slices. J Alzheimers Dis 19:1007–1019

Marwarha G, Dasari B, Prabhakara JP, Schommer J, Ghribi O (2010b) β-Amyloid regulates leptin expression and Tau phosphorylation through the mTORC1 signaling pathway. J Neurochem 115:373–384

Marwarha G, Raza S, Prasanthi JR, Ghribi O (2013) Gadd153 and NF-κB crosstalk regulates 27-hydroxycholesterol-induced increase in BACE1 and β-amyloid production in human neuroblastoma SH-SY5Y cells. PLoS One 8, e70773

Masliah E, Mallory M, Alford M, DeTeresa R, Hansen LA, McKeel DW Jr, Morris JC (2001) Altered expression of synaptic proteins occurs early during progression of Alzheimer's disease. Neurology 56:127–129

Mast N, Norcross R, Andersson U, Shou M, Nakayama L, Bjorkhem I, Pikuleva IA (2003) Broad substrate specificity of human cytochrome P450 46A1 which initiates cholesterol degradation in the brain. Biochemistry 42:14284–14292

Mateos L, Akterin S, Gil-Bea FJ, Spulber S, Rahman A, Björkhem I, Schultzberg M, Flores-Morales A, Cedazo-Minguez A (2009) Activity-regulated cytoskeleton-associated protein in rodent brain is down-regulated by high fat diet in vivo and by 27-hydroxycholesterol in vitro. Brain Pathol 19:69–80

Matsuzaki K, Kato K, Yanagisawa K (2010) Abeta polymerization through interaction with membrane gangliosides. Biochim Biophys Acta 1801:868–877

Mattson MP (2004) Metal-catalyzed disruption of membrane protein and lipid signaling in the pathogenesis of neurodegenerative disorders. Ann N Y Acad Sci 1012:37–50

Matus A (1994) Stiff microtubules and neuronal morphology. Trends Neurosci 17:19–22

McCord MC, Aizenman E (2014) The role of intracellular zinc release in aging, oxidative stress, and Alzheimer's disease. Front Aging Neurosci 6:77

McCoy MK, Tansey MG (2008) TNF signaling inhibition in the CNS: implications for normal brain function and neurodegenerative disease. J Neuroinflammation 5:45

McKhann G, Drachman D, Folstein M, Katzman R, Price D, Stadlan EM (1984) Clinical diagnosis of Alzheimer's disease: report of the NINCDS-ADRDA Work Group under the auspices of department of health and human services task force on Alzheimer's disease. Neurology 34:939–944

Mclauren Dorrance A, Graham D, Dominiczak A, Fraser R (2000) Inhibition of nitric oxide synthesis increases erythrocyte membrane fluidity and unsaturated fatty acid content. Am J Hypertens 13:1194–1202

McNay EC, Ong CT, McCrimmon RJ, Cresswell J, Bogan JS, Sherwin RS (2010) Hippocampal memory processes are modulated by insulin and high-fat-induced insulin resistance. Neurobiol Learn Mem 93:546–553

Meadowcroft MD, Connor JR, Smith MB, Yang QX (2009) MRI and histological analysis of beta-amyloid plaques in both human Alzheimer's disease and APP/PS1 transgenic mice. J Magn Reson Imaging 29:997–1007

Metcalfe MJ, Huang Q, Figueiredo-Pereira ME (2012) Coordination between proteasome impairment and caspase activation leading to TAU pathology: neuroprotection by cAMP. Cell Death Dis 3, e326

Meziane H, Dodart JC, Mathis C, Little S, Clemens J, Paul SM, Ungerer A (1998) Memory-enhancing effects of secreted forms of the beta-amyloid precursor protein in normal and amnestic mice. Proc Natl Acad Sci U S A 95:12683–12688

Mi S, Lee X, Shao Z, Thill G, Ji B, Relton J, Levesque M, Allaire N, Perrin S, Sands B, Crowell T, Cate RL, McCoy JM, Pepinsky RB (2004) LINGO-1 is a component of the Nogo-66 receptor/p75 signaling complex. Nat Neurosci 7:221–228

Millan MJ (2013) An epigenetic framework for neurodevelopmental disorders: from pathogenesis to potential therapy. Neuropharmacology 68:2–82

Mitani Y, Akashiba H, Saita K, Yarimizu J, Uchino H, Okabe M, Asai M, Yamasaki S, Nozawa T, Ishikawa N, Shitaka Y, Ni K, Matsuoka N (2014) Pharmacological characterization of the novel γ-secretase modulator AS2715348, a potential therapy for Alzheimer's disease, in rodents and nonhuman primates. Neuropharmacology 79:412–419

Mocchegiani E, Bertoni-Freddari C, Marcellini F, Malavolta M (2005) Brain, aging and neurodegeneration: role of zinc ion availability. Prog Neurobiol 75:367–390

Morante S (2008) The role of metals in beta-amyloid peptide aggregation: X-Ray spectroscopy and numerical simulations. Curr Alzheimer Res 5:508–524

Morishima-Kawashima M, Hasegawa M, Takio K, Suzuki M, Yoshida H, Titani K, Ihara K (1995) Proline-directed and non-proline-directed phosphorylation of PHF-Tau. J Biol Chem 270:823–829

Morris M, Maeda S, Vossel K, Mucke L (2011) The many faces of Tau. Neuron 70:410–426

Mucke L, Selkoe DJ (2012) Neurotoxicity of amyloid beta-protein: synaptic and network dysfunction. Cold Spring Harb Perspect Med 2:a006338

Münzberg H, Myers MG Jr (2005) Molecular and anatomical determinants of central leptin resistance. Nat Neurosci 8:566–570

Murer MG, Yan Q, Raisman-Vozari R (2001) Brain-derived neurotrophic factor in the control human brain, and in Alzheimer's disease and Parkinson's disease. Prog Neurobiol 63:71–124

Naeve GS, Ramakrishnan M, Kramer R, Hevroni D, Citri Y et al (1997) Neuritin: a gene induced by neural activity and neurotrophins that promotes neuritogenesis. Proc Natl Acad Sci U S A 94:2648–2653

Nalivaeva NN, Belyaev ND, Kerridge C, Turner AJ (2014) Amyloid-clearing proteins and their epigenetic regulation as a therapeutic target in Alzheimer's disease. Front Aging Neurosci 6:235

Nathlie P, Jean-Noel O (2008) Processing of amyloid precursor protein and amyloid peptide neurotoxicity. Curr Alzheimer Res 5:92–99

Nelson PT, Alafuzoff I, Bigio EH, Bouras C, Braak H, Cairns NJ, Castellani RJ, Crain BJ, Davies P, Del Tredici K, Duyckaerts C, Frosch MP, Haroutunian V, Hof PR, Hulette CM, Hyman BT, Iwatsubo T, Jellinger KA, Jicha GA, Kövari E, Kukull WA, Leverenz JB, Love S, Mackenzie IR, Mann DM, Masliah E, McKee AC, Montine TJ, Morris JC, Schneider JA, Sonnen JA, Thal DR, Trojanowski JQ, Troncoso JC, Wisniewski T, Woltjer RL, Beach TG (2012) Correlation of Alzheimer disease neuropathologic changes with cognitive status: a review of the literature. J Neuropathol Exp Neurol 71:362–381

Niedowicz DM, Studzinski CM, Weidner AM, Platt TL, Kingry KN, Beckett TL, Bruce-Keller AJ, Keller JN, Murphy MP (2013) Leptin regulates amyloid β production via the γ-secretase complex. Biochim Biophys Acta 1832:439–444

Nielsen E, Severin F, Backer JM, Hyman AA, Zerial M (1999) Rab5 regulates motility of early endosomes on microtubules. Nat Cell Biol 1:376–382

Nikolaev A, McLaughlin T, O'Leary DD, Tessier-Lavigne M (2009) APP binds DR6 to trigger axon pruning and neuron death via distinct caspases. Nature 457:981–989

Nistico R, Cavallucci V, Piccinin S, Macri S, Pignatelli M, Mehdawy B, Blandini F, Laviola G, Lauro D, Mercuri NB, D'Amelio M (2012) Insulin receptor beta-subunit haploinsufficiency impairs hippocampal late-phase LTP and recognition memory. Neuromolecular Med 14:262–269

Niswender KD, Morton GJ, Stearns WH, Rhodes CJ, Myers MG Jr, Schwartz MW (2001) Intracellular signalling. Key enzyme in leptin-induced anorexia. Nature 413:794–795

Nolte C, Gore A, Sekler I, Kresse W, Hershfinkel M, Hoffman A, Kettenmann H, Moran A (2004) ZnT1 expression in astroglial cells protects against zinc toxicity and slows the accumulation of intracellular zinc. Glia 43:145–155

Nunomura A, Hofer T, Moreira PT, Castellani RJ, Smith MA, Perry G (2009) RNA oxidation in Alzheimer disease and related neurodegenerative disorders. Acta Neuropathol 118:151–166

Nykjaer A, Lee R, Teng KK, Jansen P, Madsen P, Nielsen MS, Jacobsen C, Kliemannel M, Schwarz E, Willnow TE, Hempstead BL, Petersen CM (2004) Sortilin is essential for proNGF-induced neuronal cell death. Nature 427:843–848

Octave JN (2005) Alzheimer disease: cellular and molecular aspects Bull. Mem Acad R Med Belg 160:445–449

Oddo S (2008) The ubiquitin-proteasome system in Alzheimer's disease. J Cell Mol Med 12:363–373

Oddo S, Caccamo A, Kitazawa M, Tseng BP, LaFerla FM (2003) Amyloid deposition precedes tangle formation in a triple transgenic model of Alzheimer's disease. Neurobiol Aging 24:1063–10670

Oikawa N, Goto M, Ikeda K, Taguchi R, Yanagisawa K (2012) The γ-secretase inhibitor DAPT increases the levels of gangliosides at neuritic terminals of differentiating PC12 cells. Neurosci Lett 525:49–53

Olkkonen VM, Béaslas O, Nissilä E (2012) Oxysterols and their cellular effectors. Biomolecules 2:76–103

Ong WY, Kim J-H, He X, Chen P, Farooqui AA, Jenner AM (2010) Changes in brain cholesterol metabolome after kainate excitotoxicity. Mol Neurobiol 41:299–313

Opazo C, Huang X, Cherny RA, Moir RD, Roher AE, White AR, Cappai R, Masters CL, Tanzi RE, Inestrosa NC, Bush AI (2002) Metalloenzyme-like activity of Alzheimer's disease β-amyloid: Cu-dependent catalytic conversion of dopamine, cholesterol, and biological reducing agents to neurotoxic H_2O_2. J Biol Chem 277:40302–40308

Pál G, Vincze C, Renner É, Wappler EA, Nagy Z, Lovas G, Dobolyi A (2012) Time course, distribution and cell types of induction of transforming growth factor betas following middle cerebral artery occlusion in the rat brain. PLoS One 7, e46731

Paoletti P, Vergnano AM, Barbour B, Casado M (2009) Zinc at glutamatergic synapses. Neuroscience 158:126–136

Paudel HK, Lew J, Ali Z, Wang JH (1993) Brain proline-directed protein kinase phosphorylates Tau on sites that are abnormally phosphorylated in Tau associated with Alzheimer's paired helical filaments. J Biol Chem 268:23512–23518

Peleg S, Sananbenesi F, Zovoilis A, Burkhardt S, Bahari-Javan S, Agis-Balboa RC, Cota P, Wittnam JL, Gogol-Doering A, Opitz L, Salinas-Riester G, Dettenhofer M, Kang H, Farinelli L, Chen W, Fischer A (2010) Altered histone acetylation is associated with age-dependent memory impairment in mice. Science 328:753–756

Perini G, Della-Bianca V, Politi V, Della Valle G, Dal-Pra I, Rossi F, Armato U (2002) Role of p75 neurotrophin receptor in the neurotoxicity by beta-amyloid peptides and synergistic effect of inflammatory cytokines. J Exp Med 195:907–918

Pettegrew JW, Panchalingam K, Hamilton RL, McClure RJ (2001) Brain membrane phospholipid alterations in Alzheimer's disease. Neurochem Res 26:771–782

Pfrieger FW (2003) Outsourcing in the brain do neurons depend on cholesterol delivery by astrocytes? Bioessays 25:72–78

Pimenova AA, Thathiah A, De Strooper B, Tesseur I (2014) Regulation of amyloid precursor protein processing by serotonin signaling. PLoS One 9, e87014

Poon WW, Blurton-Jones M, Tu CH, Feinberg LM, Chabrier MA, Harris JW, Jeon NL, Cotman CW (2011) β-Amyloid impairs axonal BDNF retrograde trafficking. Neurobiol Aging 32:821–833

Poon WW, Carlos AJ, Aguilar BL, Berchtold NC, Kawano CK, Zograbyan V, Yaopruke T, Shelanski M, Cotman CW (2013) Levels of soluble apolipoprotein E/amyloid-β (Aβ) complex are reduced and oligomeric Aβ increased with APOE4 and Alzheimer disease in a transgenic mouse model and human samples. J Biol Chem 288:16937–16948

Postina R, Schroeder A, Dewachter I, Bohl J, Schmitt U, Kojro E, Prinzen C, Endres K, Hiemke C, Blessing M, Flamez P, Dequenne A, Godaux E, van Leuven F, Fahrenholz F (2004) A disintegrin-metalloproteinase prevents amyloid plaque formation and hippocampal defects in an Alzheimer disease mouse model. J Clin Invest 113:1456–1464

Powers ET, Morimoto RI, Dillin A, Kelly JW, Balch WE (2009) Biological and chemical approaches to diseases of proteostasis deficiency. Annu Rev Biochem 78:959–991

Prasad AS (2009) Impact of the discovery of human zinc deficiency on health. J Am Coll Nutr 28:257–265

Pratico D, Zhukareva V, Yao Y, Uryu K, Funk CD, Lawson JA, Trojanowski JQ, Lee VM (2004) 12/15-lipoxygenase is increased in Alzheimer's disease: possible involvement in brain oxidative stress. Am J Pathol 164:1655–1662

Price JL, McKeel DW Jr, Buckles VD, Roe CM, Xiong C, Grundman M, Hansen LA, Petersen RC, Parisi JE, Dickson DW, Smith CD, Davis DG, Schmitt FA, Markesbery WR, Kaye J, Kurlan R, Hulette C, Kurland BF, Higdon R, Kukull W, Morris JC (2009) Neuropathology of nondemented aging: presumptive evidence for preclinical Alzheimer disease. Neurobiol Aging 30:1026–1036

Puglielli L, Ellis BC, Saunders AJ, Kovacs DM (2003) Ceramide stabilizes β-site amyloid precursor protein-cleaving enzyme 1 and promotes amyloid β-peptide biogenesis. J Biol Chem 278:19777–19783

Purro SA, Dickins EM, Salinas PC (2012) The secreted Wnt antagonist dickkopf-1 is required for amyloid β-mediated synaptic loss. J Neurosci 32:3492–3498

Puzzo D, Privitera L, Leznik E, Fà M, Staniszewski A, Palmeri A, Arancio O (2008) Picomolar amyloid-beta positively modulates synaptic plasticity and memory in hippocampus. J Neurosci 28:14537–14545

Qiu WQ, Folstein MF (2006) Insulin, insulin-degrading enzyme and amyloid-beta peptide in Alzheimer's disease: review and hypothesis. Neurobiol Aging 27:190–198

Querfurth HW, LaFerla FM (2010) Alzheimer's disease. N Engl J Med 362:329–344

Radhakrishnan A, Goldstein JL, McDonald JG, Brown MS (2008) Switch-like control of SREBP-2 transport triggered by small changes in ER cholesterol: a delicate balance. Cell Metab 8:512–521

Ramasamy R, Yan SF, Schmidt AM (2009) RAGE therapeutic target and biomarker of the inflammatory response-the evidence mounts. J Leukoc Biol 86:505–512

Rao JS, Rapoport SI, Kim HW (2011) Altered neuroinflammatory, arachidonic acid cascade and synaptic markers in postmortem Alzheimer's disease brain. Transl Psychiatry 1, e31

Rao JS, Keleshian VL, Klein S, Rapoport SI (2012) Epigenetic modifications in frontal cortex from Alzheimer's disease and bipolar disorder patients. Transl Psychiatry 2, e132

Reddy PH (2009) Amyloid beta, mitochondrial structural and functional dynamics in Alzheimer's disease. Exp Neurol 218:286–292

Reddy PH, Beal MF (2008) Amyloid beta, mitochondrial dysfunction and synaptic damage: implications for cognitive decline in aging and Alzheimer's disease. Trends Mol Med 14:45–53

Reddy PH, Geethalakshmi M, Byung SP, Joline J, Geoffrey M, William W Jr, Jeffrey K, Maria M (2005) Differential loss of synaptic proteins in Alzheimer's disease: implications for synaptic dysfunction. J Alzheimers Dis 7:103–117

Reitz C, Mayeux R (2014) Alzheimer disease: epidemiology, diagnostic criteria, risk factors and biomarkers. Biochem Pharmacol 88:640–651

Renner M, Lacor PN, Velasco PT, Xu J, Contractor A, Klein WL, Triller A (2010) Deleterious effects of amyloid beta oligomers acting as an extracellular scaffold for mGluR5. Neuron 66:739–754

Reser JE (2009) Alzheimer's disease and natural cognitive aging may represent adaptive metabolism reduction programs. Behav Brain Funct 5:13

Rissman RA, Poon WW, Blurton-Jones M, Oddo S, Torp R, Vitek MP, LaFerla FM, Rohn TT, Cotman CW (2004) Caspase-cleavage of Tau is an early event in Alzheimer disease tangle pathology. J Clin Invest 114:121–130

Ritter A, Cummings J (2015) Fluid biomarkers in clinical trials of Alzheimer's disease therapeutics. Front Neurol 6:186

Roberson ED, Scearce-Levie K, Palop JJ, Yan F, Cheng IH, Wu T, Gerstein H, Yu GQ, Mucke L (2007) Reducing endogenous Tau ameliorates amyloid beta-induced deficits in an Alzheimer's disease mouse model. Science 316:750–754

Roberson ED, Halabisky B, Yoo JW, Yao J, Chin J, Yan F, Wu T, Hamto P, Devidze N, Yu G-Q, Palop JJ, Noebels JL, Mucke L (2011) Amyloid-β/Fyn-induced synaptic, network, and cognitive impairments depend on Tau levels in multiple mouse models of Alzheimer's disease. J Neurosci 31:700–711

Robertson KD, Wolffe AP (2000) DNA methylation in health and disease. Nat Rev Genet 1:11–19

Rohan de Silva HA, Jen A, Wickenden C, Jen LS, Wilkinson SL, Patel AJ (1997) Cell-specific expression of beta-amyloid precursor protein isoform mRNAs and proteins in neurons and astrocytes. Brain Res Mol Brain Res 47:147–156

Roselli F, Tirard M, Lu J, Hutzler P, Lamberti P, Livrea P, Morabito M, Almeida OF (2005) Soluble beta-amyloid1-40 induces NMDA-dependent degradation of postsynaptic density-95 at glutamatergic synapses. J Neurosci 25:11061–1070

Rosso SB, Inestrosa NC (2013) WNT signaling in neuronal maturation and synaptogenesis. Front Cell Neurosci 7:103

Russell DW, Halford RW, Ramirez DM, Shah R, Kotti T (2009) Cholesterol 24-hydroxylase: an enzyme of cholesterol turnover in the brain. Annu Rev Biochem 78:1017–1040

Sacchetti P, Sousa KM, Hall AC, Liste I, Steffensen KR, Theofilopoulos S, Parish CL, Hazenberg C, Richter LA, Hovatta O, Gustafsson JA, Arenas E (2009) Liver X receptors and oxysterols promote ventral midbrain neurogenesis in vivo and in human embryonic stem cells. Cell Stem Cell 5:409–419

Sala Frigerio C, Lau P, Salta E, Tournoy J, Bossers K, Vandenberghe R, Wallin A, Bjerke M, Zetterberg H, Blennow K, De Strooper B (2011) Reduced expression of hsa-miR27a-3p in CSF of patients with Alzheimer disease. Neurology 81:2103–2106

Santucci R, Sinibaldi F, Fiorucci L (2008) Protein folding, unfolding and misfolding: role played by intermediate states. Mini Rev Med Chem 8:57–62

Saravia FE, Revsin Y, Gonzalez Deniselle MC, Gonzalez SL, Roig P, Lima A, Homo-Delarche F, De Nicola AF (2002) Increased astrocyte reactivity in the hippocampus of murine models of type 1 diabetes: the nonobese diabetic (NOD) and streptozotocin-treated mice. Brain Res 957:345–353

Scheff SW, Price DA, Schmitt FA, DeKosky ST, Mufson EJ (2007) Synaptic alterations in CA1 in mild Alzheimer disease and mild cognitive impairment. Neurology 68:1501–1508

Schmidt HD, Duman RS (2007) The role of neurotrophic factors in adult hippocampal neurogenesis, antidepressant treatments and animal models of depressive-like behavior. Behav Pharmacol 18:391–418

Schmitt B, Bernhardt T, Moeller H-J, Heuser I, Frölich L (2004) Combination therapy in Alzheimer's disease. CNS Drugs 18:827–844

Schmitz A, Tikkanen R, Kirfel G, Herzog V (2002) The biological role of the Alzheimer amyloid precursor protein in epithelial cells. Histochem Cell Biol 117:171–180

Schroeter ML, Stein T, Maslowski N, Neumann J (2009) Neural correlates of Alzheimer's disease and mild cognitive impairment: a systematic and quantitative meta-analysis involving 1351 patients. Neuroimage 47:1196–1206

Selkoe DJ (2004) Cell biology of protein misfolding: the examples of Alzheimer's and Parkinson's diseases. Nat Cell Biol 6:1054–1061

Selkoe DJ (2008) Soluble oligomers of the amyloid beta-protein impair synaptic plasticity and behavior. Behav Brain Res 192:106–113

Selkoe DJ (2011) Alzheimer's disease. Cold Spring Harb Perspect Biol 3:a004457

Seneff S, Wainwright G, Mascitelli L (2011) Nutrition and Alzheimer's disease: the detrimental role of a high carbohydrate diet. Eur J Intern Med 22:134–140

Sethi P, Lukiw WJ (2009) Micro-RNA abundance and stability in human brain: specific alterations in Alzheimer's disease temporal lobe neocortex. Neurosci Lett 459:100–104

Shan X, Lin CLG (2006) Quantification of oxidized RNAs in Alzheimer's disease. Neurobiol Aging 27:657–662

Shan X, Tashiro H, Lin CL (2003) The identification and characterization of oxidized RNAs in Alzheimer's disease. J Neurosci 23:4913–4921

Shaw LM, Korecka M, Clark CM, Lee VM, Trojanowski JQ (2007) Biomarkers of neurodegeneration for diagnosis and monitoring therapeutics. Nat Rev Drug Discov 6:295–303

Shelat PB, Chalimoniuk M, Wang JH, Strosznajder JB, Lee JC, Sun AY, Simonyi A, Sun GY (2008) Amyloid beta peptide and NMDA induce ROS from NADPH oxidase and AA release from cytosolic phospholipase A_2 in cortical neurons. J Neurochem 106:45–55

Shen WX, Chen JH, Lu JH, Peng YP, Qiu YH (2014) TGF-β1 protection against Aβ1-42-induced neuroinflammation and neurodegeneration in rats. Int J Mol Sci 15: 22092–22108

Shruster A, Eldar-Finkelman H, Melamed E, Offen D (2011) Wnt signaling pathway overcomes the disruption of neuronal differentiation of neural progenitor cells induced by oligomeric amyloid β-peptide. J Neurochem 116:522–529

Siegel GJ, Chauhan NB (2000) Neurotrophic factors in Alzheimer's and Parkinson's disease brain. Brain Res Brain Res Rev 33:199–227

Silva-Alvarez C, Arrazola MS, Godoy JA, Ordenes D, Inestrosa NC (2013) Canonical Wnt signaling protects hippocampal neurons from Abeta oligomers: role of non-canonical Wnt-5a/Ca_{2+} in mitochondrial dynamics. Front Cell Neurosci 7:97

Silvestri L, Camaschella C (2008) A potential pathogenetic role of iron in Alzheimer's disease. J Cell Mol Med 12:1548–1550

Simons K, Ikonen E (2000) How cells handle cholesterol? Science 290:1721–1726

Skovronsky DM, Moore DB, Milla ME, Doms RW, Lee VM (2000) Protein kinase C-dependent alpha-secretase competes with beta-secretase for cleavage of amyloid-beta precursor protein in the trans-golgi network. J Biol Chem 275:2568–2575

Slutsky I, Sadeghpour S, Li B, Liu G (2004) Enhancement of synaptic plasticity through chronically reduced Ca^{2+} flux during uncorrelated activity. Neuron 44:835–849

Smith MA, Zhu X, Tabaton M, Liu G, McKeel DW Jr, Cohen ML, Wang X, Siedlak SL, Dwyer BE, Hayashi T, Nakamura M, Nunomura A, Perry G (2010) Increased iron and free radical generation in preclinical Alzheimer disease and mild cognitive impairment. J Alzheimers Dis 19:363–372

Smith PY, Delay C, Girard J, Papon MA, Planel E, Sergeant N, Buée L, Hébert SS (2011) MicroRNA-132 loss is associated with Tau exon 10 inclusion in progressive supranuclear palsy. Hum Mol Genet 20:4016–4024

Söderberg M, Edlund C, Kristensson K, Dallner G (1991) Fatty acid composition of brain phospholipids in aging and in Alzheimer's disease. Lipids 26:421–425

Sometani A, Kataoka H, Nitta A, Fukumitsu H, Nomoto H, Furukawa S (2001) Transforming growth factor-beta1 enhances expression of brain-derived neurotrophic factor and its receptor, TrkB, in neurons cultured from rat cerebral cortex. J Neurosci Res 66:369–376

Soscia SJ, Kirby JE, Washicosky KJ, Tucker SM, Ingelsson M, Hyman B, Burton MA, Goldstein LE, Duong S, Tanzi RE, Moir RD (2010) The Alzheimer's disease-associated amyloid beta-protein is an antimicrobial peptide. PLoS One 5, e9505

Sotthibundhu A, Sykes AM, Fox B, Underwood CK, Thangnipon W, Coulson EJ (2008) β-amyloid$_{1-42}$ induces neuronal death through the p75 neurotrophin receptor. J Neurosci 28:3941–3946

Speisky H, Gomez M, Burgos-Bravo F, Lopez-Alarcon C, Jullian C, Olea-Azar C, Aliaga ME (2009) Generation of superoxide radicals by copper-glutathione complexes: redox-consequences associated with their interaction with reduced glutathione. Bioorg Med Chem 17:1803–1810

Stanyon HF, Patel K, Begum N, Viles JH (2014) Copper(II) sequentially loads onto the N-terminal amino group of the cellular prion protein before the individual octarepeats. Biochemistry 53:3934–3939

Starkov AA, Beal FM (2008) Portal to Alzheimer's disease. Nat Med 14:1020–1021

Stephenson D, Rash K, Smalstig B, Roberts E, Johnstone E, Sharp J, Panetta J, Little S, Kramer R, Clemens J (1999) Cytosolic phospholipase A_2 is induced in reactive glia following different forms of neurodegeneration. Glia 27:110–128

Stoothoff WH, Johnson GVW (2005) Tau phosphorylation: physiological and pathological consequences. Biochim Biophys Acta 1739:280–297

Stranahan AM, Norman ED, Lee K, Cutler RG, Telljohann RS, Egan JM, Mattson MP (2008) Diet-induced insulin resistance impairs hippocampal synaptic plasticity and cognition in middle-aged rats. Hippocampus 18:1085–1088

Sultana R, Perluigi M, Butterfield DA (2009) Oxidatively modified proteins in Alzheimer's disease (AD), mild cognitive impairment and animal models of AD: role of Abeta in pathogenesis. Acta Neuropathol 118:131–150

Sun GY, Horrocks LA, Farooqui AA (2007) The roles of NADPH oxidase and phospholipases A_2 in oxidative and inflammatory responses in neurodegenerative diseases. J Neurochem 103:1–16

Sun GY, He Y, Chuang DY, Lee JC, Gu Z, Simonyi A, Sun AY (2012) Integrating cytosolic phospholipase A_2 with oxidative/nitrosative signaling pathways in neurons: a novel therapeutic strategy for AD. Mol Neurobiol 46:85–95

Svennerholm L, Gottfries CG (1994) Membrane lipids, selectively diminished in Alzheimer brains, suggest synapse loss as a primary event in early-onset form (type I) and demyelination in late-onset form (type II). J Neurochem 62:1039–1047

Swaminathan A, Jicha GA (2014) Nutrition and prevention of Alzheimer's dementia. Front Aging Neurosci 6:282

Szewczyk B (2013) Zinc homeostasis and neurodegenerative disorders. Front Aging Neurosci 5:33

Tabaton M, Zhu X, Perry G, Smith MA, Giliberto L (2010) Signaling effect of amyloid-beta(42) on the processing of AbetaPP. Exp Neurol 221:18–25

Takami M, Funamoto S (2012) gamma-Secretase-dependent proteolysis of transmembrane domain of amyloid precursor protein: successive tri- and tetrapeptide release in amyloid beta-protein production. Int J Alzheimers Dis 2012, 591392

Takeda A (2000) Zinc homeostasis and functions of zinc in the brain. Biometals 14:343–351

Takeuchi M, Yamagishi S (2008) Possible involvement of advanced glycation end-products (AGEs) in the pathogenesis of Alzheimer's disease. Curr Pharm Des 14:973–978

Tan Z, Shi L, Schreiber SS (2009) Differential expression of redox factor-1 associated with β-amyloid-mediated neurotoxicity. Open Neurosci J 3:26–34

Ten Dijke P, Hill CS (2004) New insights into TGF-beta-Smad signalling. Trends Biochem Sci 29:265–273

Teng J, Takei Y, Harada A, Nakata T, Chen J, Hirokawa N (2001) Synergistic effects of MAP2 and MAP1B knockout in neuronal migration, dendritic outgrowth, and microtubule organization. J Cell Biol 155:65–76

Terry RD, Masliah E, Salmon DP, Butters N, DeTeresa R, Hill R, Hansen LA, Katzman R (1991) Physical basis of cognitive alterations in Alzheimer's disease: synapse loss is the major correlate of cognitive impairment. Ann Neurol 30:572–580

Thinakaran G, Koo EH (2008) Amyloid precursor protein trafficking, processing and function. J Biol Chem 283:29615–29619

Thounaojam MC, Kaushik DK, Basu A (2013) MicroRNAs in the brain: it's regulatory role in neuroinflammation. Mol Neurobiol 47:1034–1044

Tian Y, Crump CJ, Li YM (2010) Dual role of alpha-secretase cleavage in the regulation of gamma-secretase activity for amyloid production. J Biol Chem 285:32549–32556

Tiffany HL, Lavigne MC, Cui YH, Wang JM, Leto TL, Gao JL, Murphy PM (2001) Amyloid-β induces chemotaxis and oxidant stress by acting at formylpeptide receptor 2 (FPR2), a G protein-coupled receptor expressed in phagocytes and brain. J Biol Chem 276:23645–23652

Tong L, Balazs R, Thornton PL, Cotman CW (2004) β-amyloid peptide at sublethal concentrations downregulates brain-derived neurotrophic factor functions in cultured cortical neurons. J Neurosci 24:6799–6809

Townsend M, Mehta T, Selkoe DJ (2007) Soluble Abeta inhibits specific signal transduction cascades common to the insulin receptor pathway. J Biol Chem 282:33305–33312

Tremblay ML, Giguere V (2008) Phosphatases at the heart of FoxO metabolic control. Cell Metab 7:101–103

Trousson A, Bernard S, Petit PX, Liere P, Pianos A, El Hadri K, Lobaccaro JM, Ghandour MS, Raymondjean M, Schumacher M, Massaad C (2009) 25-hydroxycholesterol provokes oligodendrocyte cell line apoptosis and stimulates the secreted phospholipase A2 type IIA via LXR beta and PXR. J Neurochem 109:945–958

Tyan SH, Shih AY, Walsh JJ, Maruyama H, Sarsoza F, Ku L, Eggert S, Hof PR, Koo EH, Dickstein DL (2012) Amyloid precursor protein (APP) regulates synaptic structure and function. Mol Cell Neurosci 51:43–52

Um JW, Kaufman AC, Kostylev M, Heiss JK, Stagi M, Takahashi H, Kerrisk ME, Vortmeyer A, Wisniewski T, Koleske AJ, Gunther EC, Nygaard HB, Strittmatter SM (2013) Metabotropic glutamate receptor 5 is a coreceptor for Alzheimer aβ oligomer bound to cellular prion protein. Neuron 79:887–902

Unger J, McNeill TH, Moxley RT 3rd, White M, Moss A, Livingston JN (1989) Distribution of insulin receptor-like immunoreactivity in the rat forebrain. Neuroscience 31:143–157

Upadhya SC, Hegde AN (2007) Role of the ubiquitin proteasome system in Alzheimer's disease. BMC Biochem 8(Suppl 1):S12 van der Heide LP, Ramakers GMJ, Smidt MP (2006) Insulin signaling in the central nervous system: learning to survive. Prog Neurobiol 79:205–221

van Helmond Z, Miners JS, Kehoe PG, Love S (2010) Higher soluble amyloid beta concentration in frontal cortex of young adults than in normal elderly or Alzheimer's disease. Brain Pathol 20:787–793

van Meer G, Voelker DR, Feigenson GW (2008) Membrane lipids: where they are and how they behave. Nat Rev Mol Cell Biol 9:112–124

Vargas JY, Fuenzalida M, Inestrosa NC (2014) In vivo activation of Wnt signaling pathway enhances cognitive function of adult mice and reverses cognitive deficits in an Alzheimer's disease model. J Neurosci 34:2191–2202

Vassar R (2004) BACE1: the beta-secretase enzyme in Alzheimer's disease. J Mol Neurosci 23:105–114

Vazquez A (2013) Metabolic states following accumulation of intracellular aggregates: implications for neurodegenerative diseases. PLoS One 8, e63822

Vega GL, Weiner MF, Lipton AM, von Bergmann K, Lutjohann D, Moore C, Svetlik D (2003) Reduction in levels of 24S-hydroxycholesterol by statin treatment in patients with Alzheimer disease. Arch Neurol 60:510–515

Velez-Pardo C, Garcia Ospina G, Jimenez del Rio M (2002) Aβ[25-35] peptide and iron promote apoptosis in lymphocytes by an oxidative stress mechanism: involvement of H_2O_2, caspase-3, NF-κB, p53 and c-Jun. Neurotoxicology 23:351–365

Verdile G, Fuller S, Atwood CS, Laws SM, Gandy SE, Martins RN (2004) The role of beta amyloid in Alzheimer's disease: still a cause of everything or the only one who got caught? Pharmacol Res 50:397–409

Vincze C, Pál G, Wappler EA, Szabó ER, Nagy ZG, Lovas G, Dobolyi A (2010) Distribution of mRNAs encoding transforming growth factors-beta1, -2, and -3 in the intact rat brain and after experimentally induced focal ischemia. J Comp Neurol 518:3752–3770

Wagner SL, Zhang C, Cheng S, Nguyen P, Zhang X, Rynearson KD, Wang R, Li Y, Sisodia SS, Mobley WC, Tanzi RE (2014) Soluble γ-secretase modulators selectively inhibit the production of the 42-amino acid amyloid β peptide variant and augment the production of multiple carboxy-truncated amyloid β species. Biochemistry 53:702–713

Walsh DM, Klyubin I, Fadeeva JV, Cullen WK, Anwyl R, Wolfe MS, Rowan MJ, Selkoe DJ (2002) Naturally secreted oligomers of amyloid beta protein potently inhibit hippocampal long-term potentiation in vivo. Nature 416:535–539

Wang L-J, Colella R, Roisen FJ (1998a) Ganglioside GM1 alters neuronal morphology by modulating the association of MAP2 with microtubules and actin filaments. Brain Res Dev Brain Res 105:227–239

Wang D, Kreutzer DA, Essigmann JM (1998b) Mutagenicity and repair of oxidative DNA damage: insights from studies using defined lesions. Mutat Res 400:99–115

Wang HY, Lee DH, D'Andrea MR, Peterson PA, Shank RP, Reitz AB (2000) β-amyloid(1-42) binds to α-7 nicotinic acetylcholine receptor with high affinity. Implications for Alzheimer's disease pathology. J Biol Chem 275:5626–5632

Wang Y, Kumar N, Crumbley C, Griffin PR, Burris TP (2010) A second class of nuclear receptors for oxysterols: regulation of RORalpha and RORgamma activity by 24S-hydroxycholesterol (cerebrosterol). Biochim Biophys Acta 1801:917–923

Watt NT, Taylor DR, Kerrigan TL, Griffiths HH, Rushworth JV, Whitehouse IJ, Hooper NM (2012) Prion protein facilitates uptake of zinc into neuronal cells. Nat Commun 3:1134

Wells K, Farooqui AA, Liss L, Horrocks LA (1995) Neural membrane phospholipids in Alzheimer disease. Neurochem Res 20:1329–1333

West MJ, Coleman PD, Flood DG, Troncoso JC (1994) Differences in the pattern of hippocampal neuronal loss in normal aging and Alzheimer's disease. Lancet 344:769–772

Winoto A, Littman DR (2002) Nuclear hormone receptors in T lymphocytes. Cell 109:S57–S66

Wolozin B (2004) Cholesterol and the biology of Alzheimer's disease. Neuron 41:7–10

Wolozin B, Kellman W, Ruosseau P, Celesia GG, Siegel G (2000) Decreased prevalence of Alzheimer disease associated with 3-hydroxy-3-methylglutaryl coenzyme A reductase inhibitors. Arch Neurol 57:1439–1443

Wyss-Coray T (2006) Tgf-Beta pathway as a potential target in neurodegeneration and Alzheimer's. Curr Alzheimer Res 3:191–195

Xie H, Johnson GV (1997) Ceramide selectively decreases Tau levels in differentiated PC12 cells through modulation of calpain I. J Neurochem 8:1020–1030

Xie S, Xiao JX, Gong GL, Zang YF, Wang YH, Wu HK, Jiang XX (2006) Voxel-based detection of white matter abnormalities in mild Alzheimer disease. Neurology 66:1845–1849

Xu P, Guo M, Hay BA (2004) MicroRNAs and the regulation of cell death. Trends Genet 20:617–624

Xu AW, Kaelin CB, Takeda K, Akira S, Schwartz MW, Barsh GS (2005) PI3K integrates the action of insulin and leptin on hypothalamic neurons. J Clin Invest 115:951–958

Xu ZP, Li L, Bao J, Wang ZH, Zeng J, Liu EJ, Li XG, Huang RX, Gao D, Li MZ, Zhang Y, Liu GP, Wang JZ (2014) Magnesium protects cognitive functions and synaptic plasticity in streptozotocin-induced sporadic Alzheimer's model. PLoS One 9, e108645

Yaar M, Zhai S, Pilch PF, Doyle SM, Eisenhauer PB, Fine RE, Gilchrest BA (1997) Binding of β-amyloid to the p75 neurotrophin receptor induces apoptosis. A possible mechanism for Alzheimer's disease. J Clin Invest 100:2333–2340

Yaar M, Zhai S, Fine RE, Eisenhauer PB, Arble BL, Stewart KB, Gilchrest BA (2002) Amyloid β binds trimers as well as monomers of the 75-kDa neurotrophin receptor and activates receptor signaling. J Biol Chem 277:7720–7725

Yan SD, Chen X, Fu J, Chen M, Zhu H, Roher A, Slattery T, Zhao L, Nagashima M, Morser J, Migheli A, Nawroth P, Stern D, Schmidt AM (1996) RAGE and amyloid-beta peptide neurotoxicity in Alzheimer's disease. Nature 382:685–691

Yao ZX, Papadopoulos V (2002) Function of beta-amyloid in cholesterol transport: a lead to neurotoxicity. FASEB J 16:1677–1679

Yu G, Nishimura M, Arawaka S, Levitan D, Zhang L, Tandon A, Song YQ, Rogaeva E, Chen F, Kawarai T, Supala A, Levesque L, Yu H, Yang DS, Holmes E, Milman P, Liang Y, Zhang DM, Xu DH, Sato C, Rogaev E, Smith M, Janus C, Zhang Y, Aebersold R, Farrer LS, Sorbi S, Bruni A, Fraser P, St George-Hyslop P (2000) Nicastrin modulates presenilin-mediated notch/glp-1 signal transduction and βAPP processing. Nature 407:48–54

Zempel H, Thies E, Mandelkow E, Mandelkow EM (2010) Abeta oligomers cause localized Ca(2+) elevation, missorting of endogenous Tau into dendrites, Tau phosphorylation, and destruction of microtubules and spines. J Neurosci 30:11938–11950

Zha Q, Ruan Y, Hartmann T, Beyreuther K, Zhang D (2004) GM1 ganglioside regulates the proteolysis of amyloid precursor protein. Mol Psychiatry 9:946–952

Zhang Y, Schuff N, Du AT, Rosen HJ, Kramer JH, Gorno-Tempini ML, Miller BL, Weiner MW (2009) White matter damage in frontotemporal dementia and Alzheimer's disease measured by diffusion MRI. Brain 132:2579–2592

Zhang R, Zhang Q, Niu J, Lu K, Xie B, Cui D, Xu S (2014) Screening of microRNAs associated with Alzheimer's disease using oxidative stress cell model and different strains of senescence accelerated mice. J Neurol Sci 338:57–64

Zhao WQ, Alkon DL (2001) Role of insulin and insulin receptor in learning and memory. Mol Cell Endocrinol 177:125–134

Zhao W, Chen H, Xu H, Moore E, Meiri N, Quon MJ, Alkon DL (1999) Brain insulin receptors and spatial memory. Correlated changes in gene expression, tyrosine phosphorylation, and signaling molecules in the hippocampus of water maze trained rats. J Biol Chem 274:34893–34902

Zhao WQ, De Felice FG, Fernandez S, Chen H, Lambert MP, Quon MJ, Krafft GA, Klein WL (2008) Amyloid beta oligomers induce impairment of neuronal insulin receptors. FASEB J 22:246–260

Zhao WQ, Lacor PN, Chen H, Lambert MP, Quon MJ, Krafft GA, Klein WL (2009) Insulin receptor dysfunction impairs cellular clearance of neurotoxic oligomeric a{beta}. J Biol Chem 284:18742–18753

Zhao W, Wu X, Xie H, Ke Y, Yung WH (2010) Permissive role of insulin in the expression of long-term potentiation in the hippocampus of immature rats. Neurosignals 18:236–245

Zhao Y, Bhattacharjee S, Jones BM, Hill J, Dua P, Lukiw WJ (2014) Regulation of neurotropic signaling by the inducible, NF-κB-sensitive miRNA-125b in Alzheimer's disease (AD) and in primary human neuronal-glial (HNG) cells. Mol Neurobiol 50:97–106

Zhou LX, Du JT, Zeng ZY, Wu WH, Zhao YF, Kanazawa K, Ishizuka Y, Nemoto T, Nakanishi H, Li YM (2007) Copper (II) modulates in vitro aggregation of a Tau peptide. Peptides 28:2229–2234

Zhu Y, Culmsee C, Klumpp S, Krieglstein J (2004) Neuroprotection by transforming growth factor-beta1 involves activation of nuclear factor-kappaB through phosphatidylinositol-3-OH kinase/Akt and mitogen-activated protein kinase-extracellular-signal regulated kinase1,2 signaling pathways. Neuroscience 123:897–906

Zhu D, Bungart BL, Yang X, Zhumadilov Z, Lee JC, Askarova S (2015) Role of membrane biophysics in Alzheimer's-related cell pathways. Front Neurosci 9:186

Ziegenhorn AA, Schulte-Herbrüggen O, Danker-Hopfe H, Malbranc M, Hartung HD, Anders D, Lang UE, Steinhagen-Thiessen E, Schaub RT, Hellweg R (2007) Serum neurotrophins—a study on the time course and influencing factors in a large old age sample. Neurobiol Aging 28:1436–1445

Zou K, Gong JS, Yanagisawa K, Michikawa M (2002) A novel function of monomeric amyloid beta-protein serving as an antioxidant molecule against metal-induced oxidative damage. J Neurosci 22:4833–4841

Zuccato C, Cattaneo E (2009) Brain-derived neurotrophic factor in neurodegenerative diseases. Nat Rev Neurol 5:311–322

Chapter 2
Potential Animal Models of Alzheimer Disease and Their Importance in Investigating the Pathogenesis of Alzheimer Disease

2.1 Introduction

As stated in Chap. 1, AD is a progressive and irreversible neurodegenerative disease characterized by progressive loss of memory and cognitive function. Risk factors for AD include old age, positive family history, unhealthy life style, consumption of high fat diet, and exposure to toxic environment (Farooqui 2015). Clinically, AD is characterized by deterioration of memory and cognitive function, progressive impairment of activities of daily living, and several neuropsychiatric symptoms. Neuropathologically, AD is characterized by the accumulation of beta-amyloid (Aβ) protein that forms plaques and tau protein phosphorylation that promote the formation and deposition neurofibrillary tangles (NFT) (Farooqui 2010). Many biochemical mechanisms have been proposed to explain the pathogenesis of AD including production of reactive oxygen species, disruption of calcium homeostasis, activation of Wnt pathway, excitotoxicity, activation of apoptotic pathways, neuronal degeneration, and neurotransmitter deficits, the precise role of abnormal protein aggregates in the pathogenesis of AD remains to be clarified (Huang and Jiang 2009; Welsh-Bohmer and White 2009; Querfurth and LaFerla 2010). Human autopsies and animal models studies have indicated that both senile plaques and NFT are co-localized with activated glial cells, supporting the view that reactive gliosis may be closely associated with the pathogenetic role of AD (Craft et al. 2006; Farooqui 2013). Increased generation of Aβ peptides not promotes neuroinflammation through the upregulation of different cytokines, and pro-inflammatory mediators (Tuppo and Arias 2005). It is well known that astrocytes play an important role in the controlling the cerebral homeostasis. Accumulation of Aβ and activation of astrocytes in AD initially (for a short time) is a neuroprotective response aimed at removing injurious stimuli. However, uncontrolled and prolonged activation of astrocytes produces detrimental effects that override the beneficial effects due to upregulation of different cytokines and proinflammatory mediators leading to neurodegeneration directly as well as in an autocrine/paracrine manner expanding the

© Springer International Publishing Switzerland 2016
A.A. Farooqui, *Therapeutic Potentials of Curcumin for Alzheimer Disease*,
DOI 10.1007/978-3-319-15889-1_2

neuropathological damage in AD (Mrak and Griffin 2001; Pekny et al. 2014). Among above mentioned hypothesis, Aβ hypothesis has a big support among researchers. According to Aβ hypothesis the accumulation of senile plaques and neurofibrillary tangles is accompanied by neuronal atrophy and progressive synaptic failure, which initially appears in the entorhinal region and the temporal lobe, before progressing to the limbic system and subsequently to major areas of the neocortex, severely damaging the brain (Braak and Braak 1995). Aβ is a peptide (4 kDa) generated by proteolytic processing of the amyloid precursor protein (APP), a transmembrane glycoprotein βAPP (~770 amino acids), which has been implicated in the regulation of neuronal cytoarchitecture, synaptic plasticity, axon guidance, and cell–cell interactions in the brain (Hardy and Selkoe 2002; Zhang et al. 2007, 2011; Haass and Selkoe 2007). To explain neurodegeneration in AD, amyloid cascade hypothesis has been proposed (Tanzi and Bertram 2005). According to this hypothesis, amyloid precursor protein (APP) is processed either by the non-amyloidogenic pathway, or the amyloidogenic pathway (Fig. 2.1) (Chow et al. 2010; Zhang et al. 2012a). In the non-amyloidogenic pathway, α-secretase cleaves APP in the ectodomain within the Aβ region of the APP protein, which precludes the generation of the Aβ peptide (Chow et al. 2010; Zhang et al. 2012a). In the amyloidogenic pathway, APP is processed by the β-site APP-cleaving enzyme (BACE), releasing a soluble APP fragment (sAPPβ), which is secreted outside the cell, leaving behind a membrane-associated C-terminal fragment of 99 or 89 amino acids [C99 or C89 (CTFβ)]. The CTFβ is then broken down by γ-secretase, generating the Aβ peptide and a cytoplasmic APP intracellular domain (AICD) (Chow et al. 2010; Zhang et al. 2012b). Aβ42 peptide oligomerizes, and readily forms aggregates that accumulate in the brain to form plaques whose recognition by brain cell microglial cells instigate a pro-inflammatory microglial response and the release of ROS and pro-inflammatory cytokines (Small et al. 2001; Fu et al. 2014). In addition, neurodegenerative process in AD is associated with alterations in neurogenesis leading to memory dysfunction (Donovan et al. 2006). Aβ accumulation is the consequence of an altered balance between protein synthesis, aggregation rate, and clearance. Accumulation of Aβ plaques contributes not only to the alterations in cellular activities, but also to disrupted communication in the brain, leading to neurotoxic inflammation and neuronal death. NMDA receptors play an important role in the production of Aβ42. Activation of synaptic NMDA receptors promotes the non-amyloidogenic pathway, which not only reduces the generation of Aβ42, but also upregulates extracellular signal-regulated kinase (ERK) and Ca^{2+}/calmodulin-dependent protein kinase (CAMK) pathways. These processes promote cyclic AMP (cAMP) signaling pathway, which is closely associated with the formation of long-term memory (Lonze and Ginty 2002). The cAMP-dependent protein kinase A (PKA), when allosterically activated by cAMP, can phosphorylate cAMP response element binding protein (CREB), a basic leucine zipper transcription factor at serine 133 (Gonzalez and Montminy 1989). Phosphorylated CREB then interacts with the transcription coactivator CREB-binding protein to initiate the transcription and translation of CREB target genes, which are required for the synaptic plasticity mediating long-term memory formation. Recent studies have demonstrated that CREB enhances

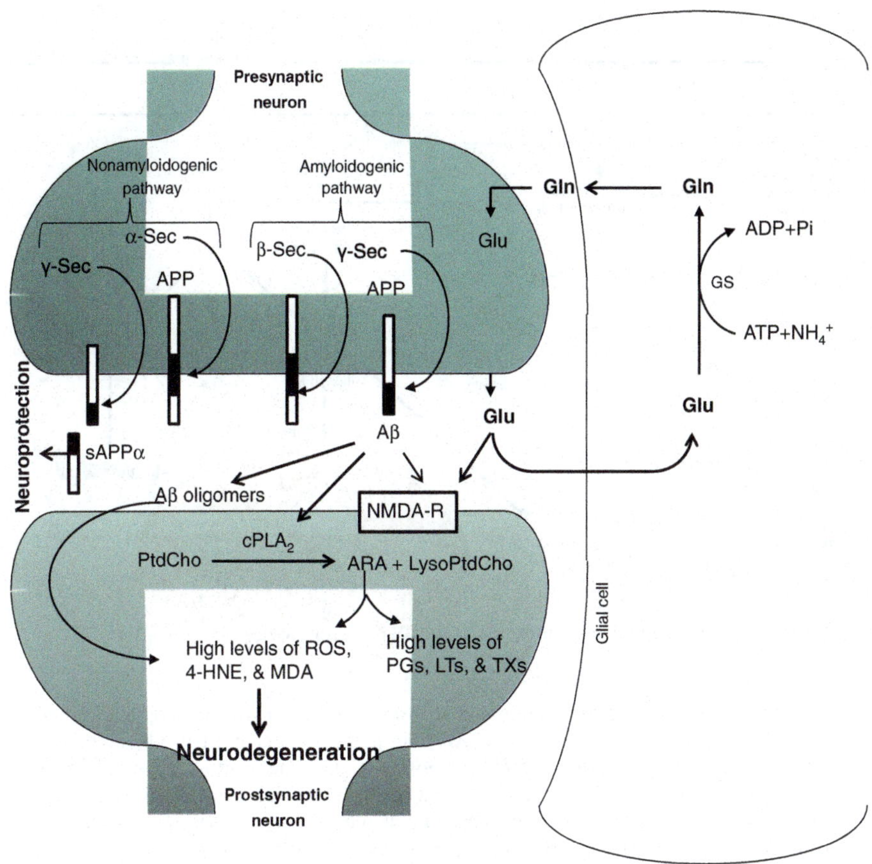

Fig. 2.1 Diagram showing β-Amyloid hypothesis and related molecular events associated with the pathogenesis of Alzheimer disease. Glutamate (*Glu*); glutamine (*Gln*); NMDA receptor (*NMDA-R*); phosphatidylcholine (*PtdCho*); lyso-phosphatidylcholine (*lyso-PtdCho*); cytosolic phospholipase A₂ (*cPLA₂*); arachidonic acid (*ARA*); α-secretase (*α-Sec*); β-secretase (*β-Sec*); γ-secretase (*γ-Sec*); amyloid precursor protein (*APP*); β-Amyloid (*Aβ*); Prostaglandins (*PGs*); leukotrienes (*LTs*); thromboxanes (*TXs*); 4-hydroxynonenal (*4-HNE*); and malondialdehyde (*MDA*)

short-term memory by up-regulating brain-derived neurotrophic factor (BDNF), suggesting that CREB signaling is involved in the formation of both short- and long-term memory (Suzuki et al. 2011). CREB-mediated gene expression is impaired in the brains of both AD mouse models and patients (Gong et al. 2004; Phillips et al. 1991), as well as in cultured neurons insulted with Aβ (Tong et al. 2001). Conversely, activation of extrasynaptic NMDA receptors promotes the amyloidogenic pathway leading to increased production of Aβ42 and loss of Ca²⁺ homeostasis. Increased production of Aβ42 not only downregulates the phosphorylation of CREB and enhances LTD, but also induces mitochondrial dysfunction leading to apoptotic cell death (Fig. 2.2) (Hardingham et al. 2002; Bordji et al. 2010). Subsequent activation of downstream signal transduction pathways (such as

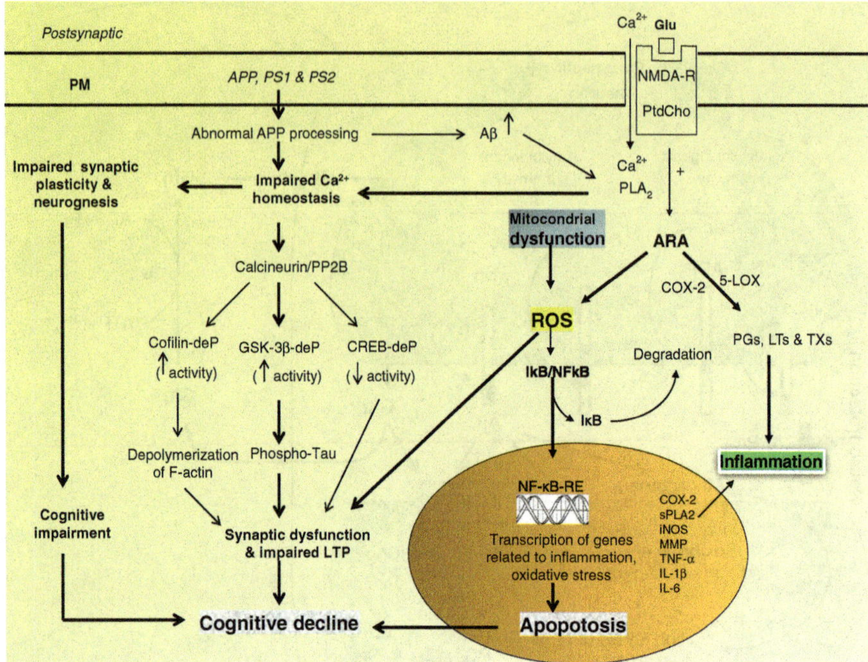

Fig. 2.2 Involvement of NMDA receptor and abnormal APP processing in apoptotic cell death and cognitive decline in Alzheimer disease. Amyloid precursor protein (*APP*); β-amyloid (*Aβ*); glutamate (*Glu*); NMDA receptor (*NMDA-R*); phosphatidylcholine (*PtdCho*); phospholipase A₂ (*PLA₂*); cyclooxygenase-2 (*COX-2*); 5-lipoxygenase (*5-LOX*); arachidonic acid (*ARA*); prostaglandins (*PGs*); leukotrienes (*LTs*); thromboxanes (*TXs*); reactive oxygen species (*ROS*); nuclear factor-κB (*NF-κB*); nuclear factor-κB-response element (*NF-κB-RE*); inhibitory subunit of NF-κB (*I-κB*); tumor necrosis factor-α (*TNF-α*); interleukin-1β (*IL-1β*); interleukin-6 (*IL-6*); inducible nitric oxide synthase (*iNOS*); secretory phospholipase A₂ (*sPLA₂*); death domain (*DD*); nitric oxide (*NO*); long term potentiation (*LTP*); genes for APP, PS1, and PS2 (*APP; PS1, and PS2*, respectively); Ca²⁺/calmodulin-dependent protein phosphatase (*calcineurin*); protein phosphatase II (*PP2B*); cofilin (actin binding and modulating proteins); glycogen synthase kinase 3 beta (*GSK-3β*); and cAMP response element-binding protein (*CREB*)

dephosphorylation and activation of the actin filament severing protein cofilin by calcineurin) induce a cascade of pathological events causing synaptic disruption and neuronal loss through mitochondrial dysfunction, induction of oxidative stress, neuroinflammation and alterations in bioenergetic, leading to dysregulation of synaptic neurotransmission and abnormal neuronal network activity (Fig. 2.2) (De Felice et al. 2007; Selkoe 2008; Palop and Mucke 2010; Sakono and Zako 2010; Tomiyama et al. 2010; Farooqui 2010).

Despite of many criticisms against the amyloid cascade hypothesis, it is becoming increasingly evident that this hypothesis can explain not all, but many molecular and cellular aspects of AD including Aβ and Tau pathology. As stated in Chap. 1, most AD cases (more than 95 %) are sporadic with over 65 years old and only less than 5 % cases are of genetic (familial, FAD) origin-that is, related to a genetic predisposition with mutations in the amyloid precursor protein, presenilin 1, and presenilin 2 genes. Apolipoprotein E (APOE) polymorphisms, sometimes referred to as familial late-onset AD, are not mutations per se, but are a significant predisposing factor (Adalbert et al. 2007). In particular, the *APOE* gene has three isoforms (ϵ2, ϵ3 and ϵ4), with the ϵ4 isoform being the strongest predisposing allele (Bu 2009). APOE ϵ3/ϵ4 heterozygotes have two- to threefold higher risk of developing AD compared with ϵ3/ϵ3 homozygotes, and ϵ4/ϵ4 homozygotes have more than twofold the risk of the ϵ3/ϵ4 genotype, while the presence of ϵ2 is somewhat protective (Aggarwal et al. 2005). Though sporadic and FAD forms of AD reflect similar pathologies, the underlying causes of pathogenic may vary considerably. As stated above, FAD is linked to specific mutations in APP or PS1 or PS2, located at chromosomes 21, 14, and 1, respectively leading to accumulation of toxic Aβ species in the brain by mid-life. Sporadic AD manifests later in life (over the age of 65 years), and is triggered by more complex neurochemical mechanisms along with genetic components and lifestyle factors (e.g. diet, exercise, and sleep). The histopathological similarity between sporadic and early familial cases has been taken as evidence for a common etiology of the disease. Because in vitro and in vivo data indicated that early onset FAD mutations give rise to the generation of more Aβ peptides and their accumulation has been proposed to be involved in the pathogenesis of FAD. In contrast, the pathogenesis of sporadic AD is very complex and multifactorial involving complex interactions among multiple genetic, epigenetic, and environmental factors. Clinical and epidemiological studies indicate that aging, stress, long term consumption of high calorie diet, aluminum, and viral infections may contribute to the risk of AD (Grant et al. 2002). At the neurochemical level pathogenesis of sporadic AD not only involves the accumulation of Aβ and hyperphosphorylated Tau, but also excitotoxicity, disruption of intracellular calcium homeostasis, oxidative stress, neuroinflammation, loss of memory formation along with reduction in the expression of trophic factors, impairments of axonal transport, and mitochondrial dysfunction (Leuner et al. 2007; Farooqui 2010).

2.2 Potential Animal Models and Alzheimer Disease

Animal models of AD are needed to study the signal transduction mechanisms underlying AD pathogenesis and learning about the effect of genetic and environmental risk factors involved in the pathogenesis of AD. In addition, animal models are also used for developing diagnostic tests and investigating the therapeutic effects of drugs on neuropathology and cognitive function in AD. Animal models are needed for the establishment of pharmacodynamics and pharmacokinetic parameters, the toxicity analysis of new drugs for the treatment of AD. Collective evidence suggests that animal models of AD are not only a cornerstone for studying the pathogenesis of AD, but also for developing and studying pharmacokinetics of drugs.

2.2.1   Invertebrate Models of Alzheimer Disease

Invertebrate models have several advantages over vertebrate models (Link 2005; Wu and Luo 2005). The genes and pathways of invertebrate organisms are well-suited to the study of human disease because both pharmacological and genetic manipulations can be performed easily to understand the function of their orthologs in vivo (Table 2.1). Invertebrate models also have other advantages. They are inexpensive, easy to work with, have short lifespans, and often have very well-characterized in terms of stereotypical development and behavior. Two invertebrate model organisms: roundworm *(Caenorhabditis elegan)* and fruit fly *(Drosophila melanogaster)* (Saraceno et al. 2013) qualify for the above criteria. These models are useful tools for studying human AD not only because genes contributing to human AD are homologues in invertebrates, but also because many signaling pathways are conserved and display similar activities (Li and Le 2013). *C. elegans* has been used a fundamental tool for dissecting the pathways that link lifespan to AD. Specifically, one of the major pathways that regulate lifespan is the insulin/IGF-1 signaling (IIS) pathway—a pathway that has been validated in nematodes, flies and mice and strongly implicated in humans (Kenyon et al. 1993; Holzenberger et al. 2003). In *C. elegans* model of AD, knockdown of the insulin/IGF-1 receptor DAF-2 results not only in longevity, but also retardation of Aβ toxicity by delaying the onset of paralysis, supporting the view that there may be a link between the mechanisms of aging and proteotoxicity (Cohen et al. 2006). Modulation of lifespan by

Table 2.1 Listing of invertebrate orthogenes and vertebrate genes contributing to the pathogenesis of AD

Protein	Caenorhabditis elegans	Drosophila	Zebrafish	Mouse
APP	Apl-1	appl	Appa, appb	APP
ADAM10	Sup-17	kuzbanian	No α-secretase	ADAM 10 gene present
ADAM17	Adm-4	dBACE	Absent	ADAM 10
β-Secretase	Absent	Absent	Absent	β-secretase
γ-Secretase complex	γ-Secretase complex	γ-Secretase complex	Incomplete γ-Secretase complex	Complete γ-secretase
Tau	Ptl-1	dtau	Mapta/maptb	Tau
APOE	Absent	Absent	Present	APOE
Presenilins	Absent	Absent	Psen1 and psen2	PS1 and PS2
APLP2	Absent	Absent	Absent	Present
MAPT	Absent	Absent	Absent	Present
PSEN1	Absent	Absent	Absent	Present

Most of the proteins associated with AD are evolutionarily conserved in *Drosophila* and *Caenorhabditis elegans* making these organisms attractive model systems for understand the conserved molecular functions of these genes linked to AD. zebrafish (*Danio rerio*) is a promising model organism for studying molecular events in AD (Saraceno et al. 2013)

DAF-2 is highly dependent on HSF-1 and DAF-16, two transcription factors, which have been reported to drive the expression of longevity genes (Hsu et al. 2003). Both transcription factors block proteotoxicity, but they did so through opposing effects. HFS-1 promotes disaggregation, while DAF-16 enhances aggregation forward, possibly as a means of sequestering the amyloidogenic protein from the cellular milieu (Cohen et al. 2006). *C. elegans* model of AD also expresses the tau homologue Ptl-1. Similarly, in *Drosophila*, the expression of human wild-type and mutant forms of Tau and Aβ has provided useful information on the role of Tau and Aβ proteins under physiological and pathological conditions (Wittmann et al. 2001). Collective evidence suggests that invertebrate animal models provide an in vivo system useful for dissecting the molecular mechanisms underlying neurodegeneration in AD. Significantly important information has been obtained on molecular and neurochemical aspects of AD using *Caenorhabditis elegans* and *Drosophila melanogaster* models (Saraceno et al. 2013; Li and Le 2013). Despite of above mentioned advantages invertebrate models, transgenic approaches in *Caenorhabditis elegans* and *Drosophila melanogaster* models suffers from several unphysiological features, such as (a) high protein levels due to the integration of multiple transgene copies into the genome, (b) alterations in brain area specificity and subcellular expression pattern of the transgene compared with the endogenous gene because of the use of an exogenous promoter, and (c) disruption of endogenous gene expression due to the insertion of transgene into the host genome (Baker and Götz 2015). Consequently, alternative strategies such as knock-in approach (P301L mutation of tau into the murine *MAPT* locus) and development od senescence-accelerated SAMP (senescence-accelerated mouse-prone) strain.

2.2.2 *Vertebrate Models for Alzheimer Disease*

Use of mice (*Mus musculus*) for the development of animal models offer several advantages over invertebrate models. Mice are vertebrates, which more closely related to humans than invertebrate models such as yeast, worms, or flies (Saraceno et al. 2013). Whole genome of mouse has been mapped (Waterston et al. 2002). The proportion of mouse genes with a single identifiable ortholog in the human genome is ~80 %. This makes the mouse an ideal model for investigating environmental and genetic manipulations, which are not feasible in higher primates and humans. The small size and short gestation and life span makes mice amenable animals for rapid breeding in large and, consequently, the feasibility of many studies in a relatively short period. In addition, preclinical experiments with mice model of human diseases can thus be performed in relative short time periods, enabling the chronic study on the effects of drugs in these models. A valid mice model for AD should not only exhibit progressive AD-like neuropathology and cognitive deficits, but like humans it should manifest some memory loss and cognitive deficits with advancing age.

Studies on transgenic (Tg) mice have provided useful information into the chronology of events leading to the pathogenesis of AD. For example, double-Tg mice, which over-express human mutant APP and tau (Tg line APP^{sw}-tau^{vlw}) mimic several characteristics of the AD phenotype such as deposition of Aβ, hyperphosphorylation of Tau, formation of NFT, glial cell proliferation, and significant neuronal loss in the entorhinal cortex (EC) and CA1 subfield of the hippocampus (Perez et al. 2005; Ribe et al. 2005). All the above phenotypic traits of AD develop in these mice in an age-dependent manner and are accompanied by progressive hippocampus-dependent memory impairment. However, neurodegeneration in these mice predates overt deposition of Aβ, supporting the view that extracellular fibrillar amyloid may not be causing neuronal death. Furthermore, the extent of neurodegeneration in these mice does not correlate well with total immunostained amyloid plaque burden (Ribe et al. 2005). Thus, studies on mice models of AD have provided us an excellent opportunity to track the natural history of oligomeric Aβ (also known as ADDLs) accumulation in their brains and to study the relationships of these Aβ species to AD-related neuropathological changes and cognition (Perez et al. 2005; Ribe et al. 2005). Oligomeric forms of Aβ have been reported to instigate memory loss through their ability to target synapses and disrupt synaptic plasticity (Wang et al. 2002), including inhibition of long-term potentiation (Walsh et al. 2002; Townsend et al. 2006) and prolonged maintenance of long-term depression (Wang et al. 2002). This suggests that soluble oligomeric forms, not fibrillar deposits of Aβ are pathologically important for the synaptic dysfunction of AD (Li et al. 2009a; Koffie et al. 2009). Using microdialysis technique on interstitial fluid (ISF) samples from Alzheimer model APP/PS1 Tg mice at 3 different age stages of AD-like amyloid plaque development, it is shown that high molecular weight (HMW) and low-molecular-weight (LMW) Aβ oligomers are present in brain ISF samples and that levels of ISF Aβ oligomers become elevated with age in the brain of APP/PS1 Tg mice (Takeda et al. 2013). The clearance of HMW Aβ oligomers is slower than LMW Aβ after acute inhibition of γ-secretase activity to stop Aβ synthesis supporting the view that the rate of clearance of various Aβ oligomers from the brain is different from each other (Takeda et al. 2013). As stated in Chap. 1, Aβ oligomers interact with a number of postsynaptic receptors including ionotropic and metabotropic glutamate receptors, the cellular prion protein (PrP^C), neuroligin, the Wnt receptor, and insulin receptors (Krafft and Klein 2010; Ferreira and Klein 2011; Viola and Klein 2015). Many neurotoxic effects have been described as resulting from the interaction of Aβ oligomers with several receptors or co-receptors (Velasco et al. 2012).

Extensive investigations on mice models of AD have indicated that unlike the human AD neuropathology, which displays massive neurodegeneration, only very few transgenic animal models show neuronal death and on a scale that does not compare to what is seen on postmortem human brains (Elder et al. 2010). In addition, the way the genetic manipulation translates into the histological and clinical recapitulation of the AD highly depends on the promoter used to insert the transgene and on the genetic background of the recipient animal (Elder et al. 2010). This actually makes any comparison between transgenic mouse models difficult. Furthermore, many mice models do not show cognitive dysfunction despite overexpression of APP (Masliah et al. 2001). The formation of neurofibrillary tangles (NFT) is not observed in most of the APP overexpressing models (Ribeiro et al.

2013). Studies on generation of AD transgenic mice models using Tau protein have revealed that only minor motor impairments with little Tau protein accumulation (mostly in brain and spinal cord), however classic NFT are not observed (Eriksen and Janus 2007; Wiedlocha et al. 2012). Another important issue is that many different mice strains or hybrid strains have been used for developing transgenic mice models (Joseph et al. 2001). The strain heterogeneity makes it difficult to compare transgenic models, as there are strain specific differences in the performance of behavioral tasks have been observed (Joseph et al. 2001). Hybrid mouse strains can also have vision problems that confound any results obtained from behavioral testing (Joseph et al. 2001; Brown 2007). Another aspect of AD pathology, such as the location of Aβ plaques and neurofibrillary tangles, vary depending on the promoter region used for the incorporation of transgene into the animal's genome (Braidy et al. 2012; Lecanu and Papadopoulos 2013). Therefore, different models using similar genetic mutations can produce very different brain pathologies and cognitive deficits. Collective evidence suggests that presently available mice models do not fulfill above mentioned criteria. Thus, at the present time an ideal animal model for AD is not available (Cuadrado-Tejedor and García-Osta 2014). It is worth noting that almost all transgenic models only related to the familial early onset form of AD, which represents a mere 5 % of AD cases. The remaining 95 % are sporadic late-onset forms, the causes and pathogenesis of this form remain elusive. Converging evidence thus suggest that at present mouse models display some neurochemical, neuropathological, and behavioral alterations of AD. However, they do not recapitulate all aspects of human AD. Furthermore, failure of AD immunotherapy in mouse models indicates that there is a need for developing superior models of the AD pathology with cognitive dysfunction.

The ideal transgenic model should mimic multiple aspects of the disease including its etiology and a time dependent progression of the pathology, involving similar structures and cells similar to the human pathology. Identifying and targeting the cognitive deficits that occur early in the course of the human AD are critical for producing the maximum impact of treatment on cognitive function and quality of life in AD patients. Earliest neuropathological changes in human AD occur in hippocampus and entorhinal cortex, followed by changes in the medial temporal lobe. In human AD the earliest detectable deficits in cognition are seen in medial temporal lobe-dependent episodic memory (Schmitt et al. 2000; Smith et al. 2007). These early deficits in episodic memory are followed closely by deficits in semantic memory, and both are developed before other deficits in cognitive domains such as attention, visuospatial memory, or executive function (Bondi et al. 2008). These observations support the view that cognitive functions such as episodic and semantic memory that depend heavily on the neural circuitry of the medial and lateral temporal lobes may be impaired earlier than cognitive abilities depending on the circuitry of other brain regions. The development of cognitive deficits in mouse models of AD shows similar, but not identical patterns of progression suggesting that mouse transgenic models do not fully recapitulate the inevitable neuronal loss. Some transgenic mice fail to even demonstrate the phenotypic alterations associated with the modeled diseases, providing further evidence that humans and primates can be more vulnerable than rodents to the same triggers inducing neurodegeneration, a phenomenon also observed in pharmacological models (Przedborski et al. 2001).

Rats offer numerous advantages over mice for the development of animal models. The rats are physiologically, genetically and morphologically closer to humans than mice (Jacob and Kwitek 2002). Their larger body and brain size facilitates intrathecal administration of drugs, microdialysis, multiple sampling of cerebrospinal fluid, in vivo electrophysiology, as well as neurosurgical and neuroimaging procedures (Tesson et al. 2005). Like humans, the rat contains 6 isoforms of Tau (Hanes et al. 2009; Tran et al. 2013), although the ratio of 4R/3R Tau isoforms is different (9:1 in rats; 1:1 in humans). In addition, rats not only share a good homology with humans in apoE amino acid sequences (73.5 % with human apoE3, 73.9 % with apoE4), but also show finer and more accurate motor coordination than mice and exhibit a richer behavioral display (McLean et al. 1983). Based on these advantages, it is suggested that rats can be used for developing better animal models of AD than mice (Carmo and Claudio Cuello 2013).

The earliest transgenic rat models of AD show accumulation of intracellular Aβ but no senile plaques. Lack of senile plaques may be due to inadequate Aβ levels, since higher concentrations are required to initiate the Aβ deposition. Some of these models also show synaptic dysfunction supporting the view that cognitive deficits are independent of plaque formation but correlate better with Aβ oligomers and other Aβ species (Millington et al. 2014). In contrast, UKUR25 and UKUR28 transgenic rat strains show an accumulation of intracellular Aβ-immunoreactive material in pyramidal neurons of the neocortex and in CA2 and CA3 regions of the hippocampus. These rat models not only support the role of Aβ in the amyloid cascade at the early and pre-plaque phase of the amyloid pathology, but also show dysregulation of ERK2 activation in the brain (Echeverria et al. 2004a) (Table 2.2). Furthermore, it is also reported that accumulation of Aβ is sufficient to trigger the initial steps of the tau-phosphorylation cascade, which may be responsible for impairments in learning and alterations in the MWM task (Echeverria et al. 2004a). Collective evidence suggests that rat models of AD in rats show significant changes in synaptic proteins and memory formation (Vercauteren et al. 2004).

Table 2.2 Animal models of AD in rat that have been used for obtaining information on AD pathogenesis

Name of animal models	Reference
McGill-R-Thy1-APP	Leon et al. (2010)
UKUR25	Echeverria et al. (2004b)
UKUR28	Echeverria et al. (2004b)
Tg6590	Kloskowska et al. (2010)
Tg478	Flood et al. (2009)
Tg1116	Flood et al. (2009)
Tg11587	Liu et al. (2008)
APP21	Agca et al. (2008)
APP31	Agca et al. (2008)

Several transgenic mouse models expressing mutated forms of human Tau containing neurofibrillary degeneration have also been developed (Mocanu et al. 2008; Ramsden et al. 2005). Transgenic mouse model only show minor motor impairments and tau protein accumulation (mostly in brain and spinal cord), however classic NFT are not observed. Behavior analysis of mice and rat models has indicated that rats not only show more progressive cognitive decline in spatial navigation, but also display disturbances in sensorimotor and reflex responses (Hrnkova et al. 2007) than mice. These impairments correlate with the progressive accumulation of argyrophilic NFTs, mature sarcosyl-insoluble Tau complexes, and extensive axonal damage in the brain stem and spinal cord. Although, hyperphosphorylated Tau is present in cortex and hippocampus, but no neuronal loss or occurrence of neurofibrillary tangles has been observed in the brain (Hrnkova et al. 2007). These rats also show a decrease in lifespan (Zilka et al. 2006; Koson et al. 2008).

Infusion of low doses of LPS into rat brain ventricular system results in an animal model with neuroinflammation. This animal model has several parallels characteristics of human AD, including increase in microglial cell activation, onset of astrogliosis, and elevation in tissue levels of IL-1β and TNF-α, elevation in levels of APP (Hauss-Wegrzyniak et al. 1998; Wenk, et al. 2000), along with deficit in the working memory (Hauss-Wegrzyniak et al. 1998, 1999a, b). Above mentioned neurochemical and immunochemical changes have been quantified by Magnetic Resonance Imaging (MRI) in the animal model and AD patients (Bobinski et al. 1999; Forloni et al. 1992). It is also reported that like human AD, the chronic LPS infusion into the ventral forebrain in animal model also results in chronic IL-1β or TNF-α increase and selectively degeneration of cholinergic cells in a time- but not dose-dependent manner (Wenk and Willard 1998; Willard et al. 1999). In the LPS infusion animal model, behavioral, biochemical, and pathological deficits induced by chronic LPS infusion are reversible with chronic administration of either an NSAID (Hauss-Wegrzyniak et al. 1999a, b) or an IL-1RA (Bluthe et al. 1992).

It should be noted that NSAID-mediated beneficial effect is observed only in young rats, with no significant attenuation of the deficits in old rats (Hauss-Wegrzyniak et al. 1999b). NSAID therapy does not have any effect in human AD patients.

There are fundamental differences gene expression, neural circuitry, brain size, proportions of gray and white matters, and neurochemical responses between rodent and human brains. Nonhuman primates (great apes, baboons, macaques, and marmosets) due to genetic lineages share many structural and functional features with humans. So they may provide better animal model for AD than rodents (Finch and Austad 2012). It is realized that the management and care of nonhuman primates are more complicated and the related costs are much higher. Despite of these complications, use of nonhuman primate animal models may provide information on higher intellectual functions such as planning of complex cognitive behaviors, personality expression, decision-making and moderating social behavior (Sutcliffe and Hutcheson 2012).

2.3   Neurotoxin-Based Animal Models for Alzheimer Disease

Neurotoxin-based models involve the disruption of multiple neurotransmitter systems, which partially contribute to the pathophysiology of neurochemical, cognitive, and behavioral disturbances associated with AD. The majority of animal models within this category are based Aon the cholinergic hypothesis of AD (Craig et al. 2011), which states that loss of cholinergic function in the brain contributes significantly to the cognitive decline associated with advanced age and AD (Bartus 2000). Degeneration of cholinergic neurons in the nucleus basalis of Meynert, situated in the basal forebrain and primarily projecting to the neocortex, occurs early in the course of AD (Whitehouse et al. 1982; Dournaud et al. 1995). Intraparenchymal or intracerebroventricular microinjections of glutamate analogs (quinolic, kainic, N-methyl-D-aspartic, ibotenic and quisqualic acids) and the cholinotoxin (AF64A) have been used to generate animal models for AD (Fig. 2.3) (Toledana and Álvarez 2010). Glutamate analogs induce degeneration of glutamatergic neurons, where as AF64A preferentially triggers degeneration of cholinergic neurotransmission (Stephens et al. 1987; Nakahara et al. 1988).

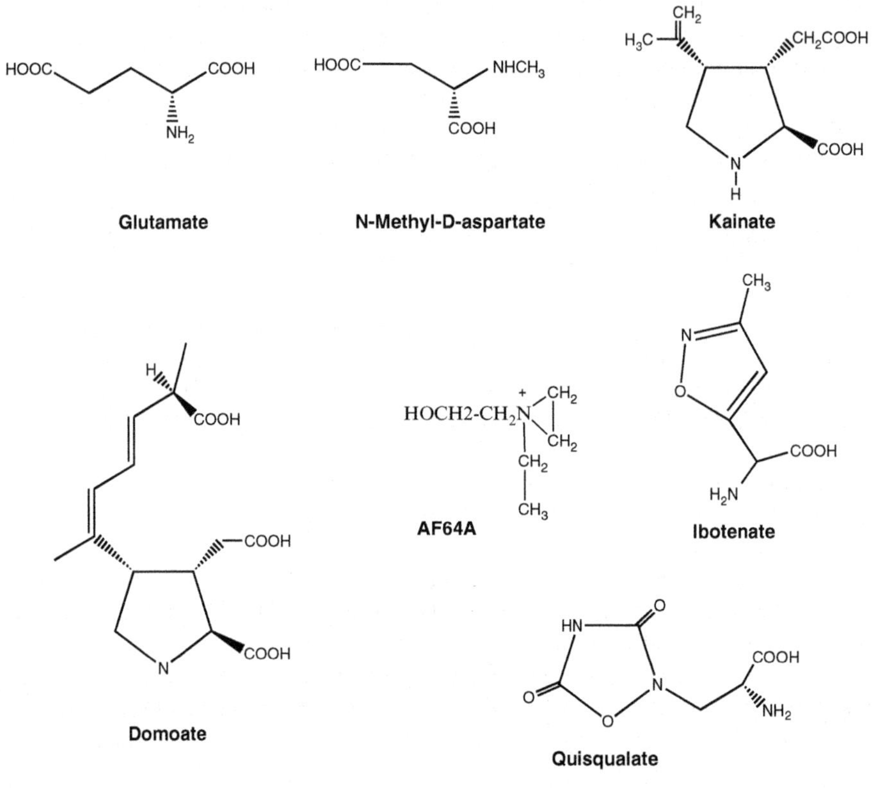

Fig. 2.3 Chemical structures of neurotoxins used for developing animal models of Alzheimer disease

2.3.1 Cholinergic and Glutamatergic Signaling Animal Models of Alzheimer Disease

It is well known that both cholinergic and glutamatergic neurons are located in the hippocampus and in the frontal, temporal and parietal cortex are severely affected in AD, whereas similar neurons in the motor and sensory cortex are relatively spared (Francis 2003). Since the hippocampus and cortex are essential for learning and memory, it is possible that degeneration of cholinergic and glutamatergic neurons may be an early event in the pathogenesis of AD (Kar et al. 2004; Morris 2002). Studies on animal models of AD have indicated that upregulation of cholinergic presynaptic boutons occurs before the involvement of glutamatergic terminals, thus raising the possibility that a compromised cholinergic system may affect the functioning/survival of glutamatergic neurons in the brain (Bell and Cuello 2006). Indeed, pyramidal neurons of the cortex that use glutamate as their primary transmitter are known to possess both cholinergic and glutamatergic receptors and receive inputs from the basal forebrain cholinergic neurons (Francis 2003).

Neurochemical investigations on tissues from biopsy and autopsy of the brains of individuals with AD have indicated that a profound reduction in the activity of the ACh-synthesizing enzyme, choline acetyltransferase (ChAT), in the neocortex, which correlates positively with the severity of dementia (Geula and Mesulam 1994; Lander and Lee 1998; Davies and Maloney 1976). Reduced choline uptake, ACh release and loss of cholinergic neurons from the basal forebrain region further indicate a selective presynaptic cholinergic deficit in the hippocampus and neocortex of brains of individuals with AD. ACh exerts effects on the central nervous system by interacting with G-protein-coupled muscarinic and ligand-gated cation channel nicotinic receptors. It is generally believed that M2 receptors, most of which are located on presynaptic cholinergic terminals, are reduced in the brains of individuals with AD (Lander and Lee 1998; Nordberg, et al. 1992). The density of postsynaptic M1 receptors remains unaltered, but there is some evidence for disruption of the coupling between the receptors, their G-proteins and second messengers (Nordberg et al. 1992; Warpman et al. 1993). Administration of acetylcholine agonists (pilocarpine and nicotine) increases learning and memory levels, but acetylcholine antagonists (scopolamine and succinylcholine) decreases learning and memory. Some studies have shown that during learning, the level of acetylcholine is increased in the amygdala, which plays an important role in memory consolidation (McGaugh 2004). It appears that the cholinergic system is involved in mediating this process (McGaugh 2004). The perfect performance of central cholinergic systems (nicotinic and muscarinic systems) is important for consolidation with shuttle box. Administration of acetylcholine agonist and antagonist via ICV affects the consolidation, in a dose-dependent manner (Eidi et al. 2006).

Stimulation of glutamate receptor results in breakdown of neural membrane phospholipids (phosphatidylcholine and plasmalogen) by the stimulation of cytosolic phospholipase A_2 ($cPLA_2$) and plasmalogen-selective phospholipase A_2 ($PlsEtn-PLA_2$). Stimulation of $cPLA_2$ increases the levels of arachidonate-derived

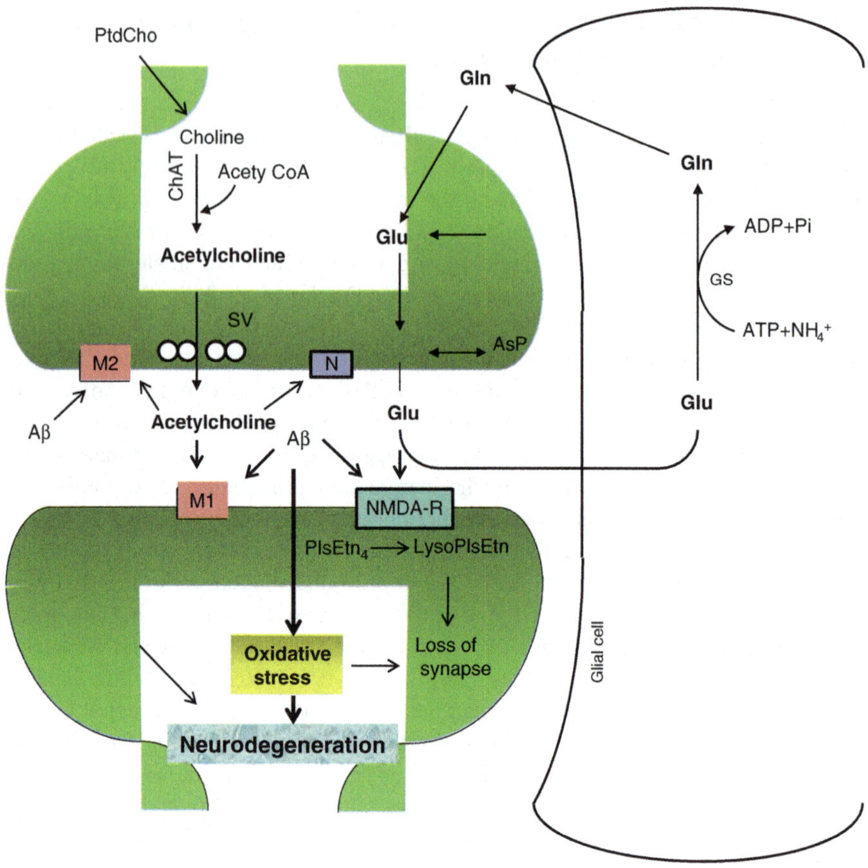

Fig. 2.4 Activities of cholinergic and glutamatergic neurons in the pathogenesis of Alzheimer disease. Amyloid precursor protein (*APP*); β-amyloid (*Aβ*); glutamate (*Glu*); glutamine (*Gln*); NMDA receptor (*NMDA-R*); plasmalogen (*PlsEtn*); 4 (plasmalogen-selective phospholipase A₂); lyso- plasmalogen (*lyso-PlsEtn*); neural membrane phosphatidylcholine (*PtdCho*); muscarinic M1 receptor (*M1*); muscarinic M2 receptor (*M2*); synaptic vesicles (*SV*); and presynaptic nicotinic receptor (*N*)

enzymic and non-enzymic lipid mediators (eicosanoids and 4-HNE, malonaldehyde, respectively), whereas activation of PlsEtn-PLA₂ catabolizes plasmalogen, which are major component of synaptic plasma membrane leading to the loss of synapse (Figs. 2.1 and 2.4). These observations support the view that there is a neurochemical basis of interactions between cholinergic and glutamatergic systems and their potential implications in triggering pathological abnormalities in Alzheimer disease (Revett et al. 2013).

Overstimulation of NMDA receptors for longer time period (i.e., more than 24 h) increases amyloidogenic APP processing and formation of high levels of Aβ (Bordji et al. 2010; Lesné et al. 2005). In AD, the accumulation of Aβ not only enhances neuronal sensitivity to glutamate, but also increases the activity of synaptic networks, resulting in excitatory potentials and Ca^{2+} influx (Brorson et al. 1995). Aβ-mediates its toxic effect either by facilitating Ca^{2+} influx into neurons leading to the activation of Ca^{2+}-dependent enzymes or by forming an oligomeric pore in the membrane. These processes may stimulate more glutamate release from glutamatergic axon terminals and/or increase intracellular calcium concentration in dendrites, thus rendering neurons vulnerable to excitotoxicity (Bobich et al. 2004; Bezprozvanny and Mattson 2008). Aβ oligomers can also promote the generation of ROS, which may trigger membrane-associated oxidative stress leading to impairment in the functions of ion-motive ATPases and glutamate and glucose transporters rendering neurons vulnerable to excitotoxicity (Camandola and Mattson 2011). Overstimulation of glutamate receptors may not only result in the collapse of mitochondrial potential and deregulation of calcium homeostasis, but also production of high levels of ROS, 4-hydroxynonenal (4-HNE), and other arachidonic acid-derived lipid mediators (Farooqui and Horrocks 2006). 4-HNE forms adducts with membrane proteins including those crucial for maintaining ATP levels, resting membrane potential and extracellular glutamate levels (Esterbauer et al. 1991; Farooqui 2011).

Changes in Tau metabolism are also related with NMDA receptor function. Tau has a dendritic function in postsynaptic targeting of the Src kinase Fyn, which phosphorylates the NMDA receptor (Suzuki and Okumura-Noji 1995). Missorting of Tau in transgenic mice expressing truncated Tau or absence of Tau in Tau knockout mice disrupt postsynaptic targeting of Fyn. Reduced expression of Tau uncouples NMDA-mediated excitotoxicity and mitigates Aβ toxicity (Ittner et al. 2010). Reducing endogenous Tau levels prevent behavioral deficits in transgenic mice expressing human APP, and protect both transgenic and nontransgenic mice against excitotoxicity (Roberson et al. 2007). Collective evidence suggests that chronic neuronal excitotoxicity may contribute to AD via promoting abnormal hyperphosphorylation of tau (Liang et al. 2009).

2.3.2   Aluminum in the Development of Animal Models of Alzheimer Disease

Aluminum is the most common metal and the third most abundant element in the earth's crust (Exley 2012). Humans get exposed to toxic levels of aluminum via common products such as antiperspirants, antacids, food, water, aluminum-based household products, cosmatics, and vaccines. In vitro and in vivo studies have indicated that aluminum produces oxidative stress though it is devoid of redox capacity in biological systems (Sharma et al. 2013; Satoh et al. 2005). Aluminum produces apoptotic cell death through the involvement of mitochondrial and endoplastic reticulum-mediated oxidative stress processes associated with caspase 9, caspase 12, and caspase 3 activation (Rizvi et al. 2014).

Levels of aluminum are significantly increased in brains of patients with AD. The molecular mechanisms associated with neurotoxic action of aluminum in AD are not fully understood. However, in vitro studies indicate that at low levels aluminum

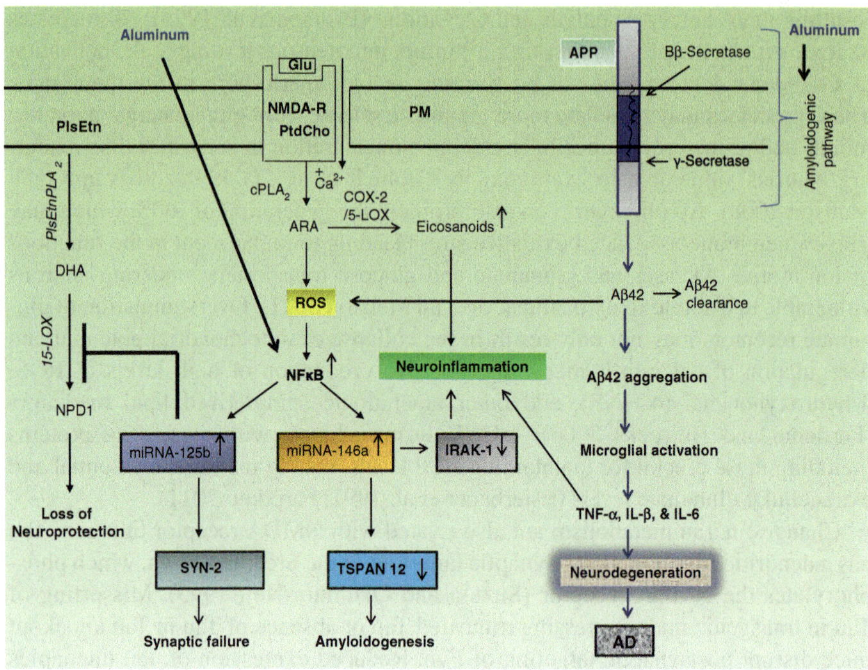

Fig. 2.5 Contribution of aluminum in the pathogenesis of Alzheimer disease. Amyloid precursor protein (*APP*); β-amyloid (*Aβ*); glutamate (*Glu*); NMDA receptor (*NMDA-R*); phosphatidylcholine (*PtdCho*); cytosolic phospholipase A₂ (*cPLA₂*); cyclooxygenase-2 (*COX-2*); lipoxygenase (*LOX*); arachidonic acid (*ARA*); docosahexaenoic acid (*DHA*); eicosanoids (prostaglandins, leukotrienes, and thromboxanes); neuroprotectin D1 (*NPD1*); reactive oxygen species (*ROS*); nuclear factor-κB (*NF-κB*); tumor necrosis factor-α (*TNF-α*); interleukin-1β (*IL-1β*); interleukin-6 (*IL-6*); interleukin-1 receptor-associated kinase (*IRAK*); synapsin-2 (*SYN-2*), and tetraspanin-12 (*TSPAN12*)

induces Tau aggregation (Mizoroki et al. 2007). Aluminum may also modulate Aβ aggregation, oligomerization, and ROS-mediated neurotoxicity (Fig. 2.5) (Bharathi et al. 2008; Rondeau et al. 2009a, b; Rodella et al. 2008; Walton and Wang 2009; Yumoto et al. 2009). Aluminum not only alters normal processing of Aβ precursor protein (Drago et al. 2008), but also stimulates amyloidogenesis. In addition, aluminum inhibits the proteolytic degradation of Aβ peptide via cathepsin D, triggering the intracellular accumulation of Aβ peptide (Sakamoto et al. 2006). Therefore, many primary therapeutic goals are targeted at reducing the metal-induced Aβ aggregation into toxic components. One of the therapeutic strategies is development of the agents that can chelate metal ions (Zatta et al. 2009) and to prevent the metal ions from the interaction with Aβ peptide as well as to attenuate the metal-induced redox activity and neurotoxicity of the peptides (Rodríguez-Rodríguez et al. 2009).

Chronic intragastric (i.g.) administration of aluminium gluconate (Al³⁺ 200 mg/ kg per day) not only results in significant increase of hippocampal metal ion levels (Al, Fe, Mn, Cu and Zn), but also causes learning and memory function disorders in rats (Yu et al. 2014). Aluminium gluconate administration-mediated chronic brain damage in rats can be prevented by meloxicam, a COX-2 inhibitor (Su et al. 2009) suggesting that the over-expression of COX-2 may play an important role in the neurodegeneration, and the inhibitors of COX-2 may prevent the acute and chronic brain damages mediated by aluminium gluconate. In addition, aluminum stimulates NF-κB, which is involved in IL-1 receptor-associated kinase (IRAK)-mediated neuroinflammation (Zhao et al. 2014). Aluminum has also been reported to inhibit brain carbohydrate metabolizing enzymes and utilization of carbohydrates. This may be one potential mechanism by which aluminum may act as a neurotoxicant (Lai and Blass 1984). Contribution of aluminum in the pathogenesis of AD is supported by several recently described observations: (a) aluminum promotes inflammatory signaling through the activation of NF-κB (Bondy 2013; Walton 2013) and (b) aluminum induces strikingly similar messenger RNA (mRNAs) and micro RNAs (miRNAs) to those found to be increased in AD. These miRNAs (miRNA-9, miRNA-34a, miRNA-125b, miRNA-146a, and miRNA-155) are under transcriptional control by the pro-inflammatory transcription factor NF-κB. Among these miRNAs subfamily, miRNA-125b occurs abundantly in human brain. Bioinformatics analysis has demonstrated that an up-regulated miRNA-125b may potentially target the 3′ untranslated region (3′-UTR) of the messenger RNA (mRNA) encoding (a) a 15-lipoxygenase (15-LOX) (Zhao et al. 2014), the enzyme that oxidizes and facilitates the conversion of docosahexaneoic acid into neuroprotectin D1 (NPD1), a docosanoid, which is closely associated with neuroprotective effects of docosahexaenoic acid (Farooqui 2009, 2011). In addition, dietary aluminum enhances lipid peroxidation, oxidative stress, apoptosis, and gene expression deficits in transgenic animal models of AD (Praticò et al. 2002; Bharathi et al. 2008; Zhang et al. 2012b). Finally, like human AD, the administration of aluminum in animal models contributes to alterations in chromatin, impairment in ATP production and utilization (Lukiw and Pogue 2007; Pogue et al. 2012; Bhattacharjee et al. 2013). Furthermore, in aged rats, aluminum treatment alters levels of copper, zinc, and manganese in certain brain regions and results in an enlargement of hippocampal mossy fibers (Fattoretti et al. 2004). In rat brain, aluminum -induced damages to the brain include neuropathological, neurochemical, neurophysiological, and neurobehavioral alterations. Among the alterations, the most notable are poor learning and behavioral functions, which involve changes in acetylcholinesterase, an enzyme, which is closely associated with deterioration of the learning ability of rats (Kawahara and Kato-Negishi 2011). The animal models show that subcutaneous injections of aluminum hydroxide induce apoptotic neuronal death, decrease in motor function, and increase in anxiety in mice (Shaw et al. 2013). Rabbits have been reported to very sensitive to aluminum exposure, with intracerebral and intravenous infusions reproducing some of the pathological features consistent with AD (Savory et al. 2006). However, oral administration of aluminum has proven less successful in inducing pathological features of AD. AD models mentioned above have been used to gain knowledge not only on molecular mechanism of action of neurotoxins, but also on the neural mechanisms underlying memory dysfunction caused by neurotoxins. This has resulted in better understanding of cholinergic innervations in the aetiology and treatment of AD. The suitability of neurotoxin models has been questioned because conflicting and controversial results due to the chemical nature of lesion-inducing neurotoxins, concentration of neurotoxin used, and even the morphological, histochemical, biochemical and cognitive methods used to produce phenotypes in the model (Toledana and Álvarez 2010). Neurotoxin-based models produce neurodegeneration in hippocampus and cortical areas in animal, but neurotoxin models have failed to replicate the classic pathological hallmarks and the insidious and progressive nature of the human AD (Toledana and Álvarez 2010).

2.3.3   *Transgenic Models of Alzheimer Disease*

Most transgenic mouse models are generated by microinjecting complementary DNA (cDNA), containing a transgene of interest into the pronucleβ of a large number of zygotes (Cho et al. 2009). Resulting embryos are then implanted into pseudo-pregnant dams for normal gestation. Generating gene targeted mice is a complex process (Cho et al. 2009; Platt et al. 2013). Creating viable mice takes many attempts, and consumes a significant amount of resources. After the initial genetic modification has been introduced, a new mouse line can be crossed into a pre-existing mouse line that already displays one or more other aspects of the disease neuropathology. Hence, given sufficient time, funding, and resources one can build increasingly complex mice models of the AD. Using transgenic mice many AD models have been developed. These mice not only overexpress mutant forms of human APP, presenilins, and/or tau protein in the brain, but also show many neurochemical characteristics. Thus, knockout mice have been designed and developed for alterations in APP, secretases, i.e., BACE, PSEN1 and PSEN2, ADAM10 (Shen et al. 1997; Luo et al. 2001; Lee et al. 2003) as well as for APP and Tau proteins. Examples these models are Tg2576, PDAPP, TgAPP23, Tg-APPswe/PS1dE9, 3xTg-AD, and 5XFAD mice. The list of transgenic AD models is available at the web site of the Alzheimer Research Forum (http://www.alzforum.org/res/com/tra/default.asp). Many of the transgenic AD models show accumulation of Aβ, plaque pathogenesis, gliosis, neuronal loss, Tau pathology, and/or cognitive impairments, but no single transgenic AD model recapitulates all aspects of AD neurochemistry and pathology. Using above mouse transgenic models, most investigators have focused their attention on understanding the molecular mechanism related to suppression of genes that encode proteins that contribute to the pathogenesis AD along with neurobehavioral and pathological changes. The comparative analysis of these AD models suggests that AD models can be classified into two distinct plaque deposition groups. Early plaque depositing models such as APPswe/PS1dE9, 3xTg-AD and 5XFAD, which may be useful to study the biochemical aspects of APP metabolism, whereas late plaque depositing models such as Tg2576, PDAPP, and TgAPP23, which can provide useful information on physiological and environmental aspects of AD pathogenesis, which occur on a longer time scale (Shen et al. 1997; Luo et al. 2001; Lee et al. 2003; Lee and Han 2013). More than 20 autosomal dominant APP mutations linked to AD have been discovered (http://www.molgen.ua.ac.be/ADMutations). These mutations show enhancement in the aggregation of Aβ by several mechanisms such as Swedish mutation, Arctic mutation, and a mutation near the γ-secretase site. Swedish mutation promotes APP cleavage near the β-secretase site (Mullan et al. 1992) leading to enhancement in overall production of all forms of Aβ. The Arctic mutation (a mutation within Aβ) enhances protofibril formation (Nilsberth et al. 2001). Several mutations near the γ-secretase site increase the relative production of the Aβ42 (Goate et al. 1991; Murrell et al. 2000). The impact of β-secretase deletion in wild-type mice produces subtle changes in anxiety and sensorimotor abilities (Kobayashi et al. 2008) leading to enhancement in long-term depression (Wang et al. 2008). In contrast, β-secretase manipulations in APP overexpression models not only prevent amyloid pathology, neurodegeneration, and astrogliosis, but also restore cognitive deficits (Ohno et al. 2007). Restoration of long-term potentiation and improved cognitive performance are also reported after partial reduction of β-secretase in 5xFAD animals (Kimura et al. 2010). Conversely, human *bace1* (*hbace1*) coexpression in mice carrying human *app*$_{swe}$ (Mohajeri et al. 2004) or *app*$_{695}$ (Chiocco et al. 2004) elevates APP processing and the release of toxic Aβ42, sAPPβ, C99, and C89 terminal fragments. These findings support the view that β-secretase is the key enzyme in amyloidosis, and its inhibition can be used as a target for the treatment of AD.

To avoid the complications of transgenic protein overexpression, attempts have also been to generate more physiologically relevant animal models of AD. Thus, AD knock-in models are generated by introducing human *APP* and/or *PSEN1* FAD mutations and humanized Aβ to the endogenous mouse gene (Guo et al. 2012; Flood et al. 2002; Köhler et al. 2005). Knock-in mice with human *APP* have several advantages over the traditional transgenic models. Due to the presence of native promoter control mice containing human *APP* show physiological levels of protein expression without any changes in the temporal and spatial expression patterns. In contrast to transgenic models in which the existence of mouse proteins may complicate the phenotypes, the mouse gene products are replaced with the humanized mutant proteins in knock-in models. In contrast to human AD, Knock-in mice show the expression of human Aβ, but no tau abnormality has been reported. Duplications of the APP gene also lead to the induction of all forms of Aβ (Sleegers et al. 2006). Recently, an autosomal recessive mutation involving the deletion of glutamate at Aβ residue 22 has been discovered in a woman with dementia who apparently lacks amyloid plaques imaged with PiB (Tomiyama et al. 2008). This discovery raises possibility that amyloid plaques may not be required for the onset of neurodegeneration in AD. Studies on the effect of genetic ablation of Nrf2 on APP/Aβ processing and/or aggregation as well as changes in autophagic dysfunction in APP/PS1 mice indicate that there is a significant increase in inflammatory response in APP/PS1 mice lacking Nrf2. These changes are accompanied by increase in intracellular levels of APP, Aβ (1-42), and Aβ (1-40) without a change in the total full-length APP. APP/PS1 mouse with Nrf2 deficiency not only displays a shift in APP and Aβ levels in the insoluble fraction, but also show an increase in poly-ubiquitin conjugated proteins. APP/PS1-mediated autophagic dysfunction is also enhanced in Nrf2-deficient mice. Finally, neurons in the APP/PS1/Nrf2-/- mice display an increase in the accumulation of multivesicular bodies, endosomes, and lysosomes (Joshi et al. 2015).

In vitro and in vivo studies strongly indicate that high level of Aβ peptide is the primary causative agents in the pathogenesis of AD (Tanzi and Bertram 2005). Reduction in Aβ clearance and its deposition is one potential mechanism leading to increased cerebral Aβ levels in AD. However, it is also possible that small increases in Aβ production over time may tip the balance toward Aβ accumulation. *APP* mutant mice show an age-dependent extracellular plaque deposition primarily in neocortex and hippocampus, accompanied by severe gliosis. Most *APP* mutants contain no neurofibrillary tangles. However, they do contain amyloid deposits and hyperphosphorylated Tau but without tangles (Tiraboschi et al. 2004). One exception is the transgenic model expressing *APP, PS1,* and *Mapt* (3xTg-AD) characterized by Aβ plaques and neurofibrillary tangles (Oddo et al. 2003). The number of CA1 neurons is inversely correlated with CA1 plaque load and neuron loss was observed primarily in the vicinity of extracellular plaques. The molecular mechanisms linking Aβ and tau pathologies remain elusive. According to the Aβ cascade hypothesis, excessive amount of Aβ peptides generated by abnormal APP metabolism initiates the pathogenesis of AD, which leads to Aβ plaque formation, tau hyperphosphorylation, and neurodegeneration (Karran et al. 2011). This hypothesis is supported by AD genetics (Golde et al. 2011), but not by mouse AD model studies. In APP transgenic line J20 model the aggressive deposition of Aβ is not accompanied by enhanced Tau phosphorylation (Roberson et al. 2007). Aβ and Tau hyperphosphorylation coexist but in an independent manner in a double transgenic mouse model of human mutant APP (APP23) and wild type tau (ALZ17) (Clavaguera et al. 2013). Thus, more studies are needed on mechanisms linking Aβ and Tau pathologies. Collective evidence suggests that transgenic models have provided some valuable information on the molecular mechanism and understanding of AD progression, but they still do not recapitulate all aspects of human AD (Zheng et al. 1996; Takei et al. 2000).

Presenilin knockout mice have also been reported to display marked neurodegeneration in cerebral cortex along with loss of memory and induction of synaptic dysfunction (Shen et al. 1997; Saura et al. 2004). Thus far, over 200 autosomal dominant mis-sense mutations have been reported in the genes for APP and presenilin (the γ-secretase catalytic subunit). These mutations may contribute to FAD, which are found very near to the β- and γ-secretase cleavage sites. They may not only contribute to increase APP processing, but also mediate the elevation in levels of total Aβ as well as Aβ42. BACE is the exclusive β-secretase, which controls the production of Aβ and has an essential role in the etiology of AD. Knockout mice for β-secretase have also been generated. They do not produce Aβ and are perfectly viable tool for understanding the neurochemical mechanisms of pathogenesis of AD (Luo et al. 2001; Roberds et al. 2001). However, BACE knockout mice show significant decrease in the intensity of myelination and reduction in myelin thickness (Hu et al. 2006; Willem et al. 2006), supporting the view that BACE may play an important role in myelinogenesis and brain development.

Several transgenic or gene-targeted mouse lines expressing human apoE3 or apoE4 have also been developed, without co-expression of mutant hAPP. Transgenic mice expressing apoE4 in neurons on a murine *Apoe* knockout background show age- and female gender-dependent spatial learning and memory deficits, which are not seen in neuron-specific apoE3 mice (Raber et al. 2000). Morphological studies have shown that neuronal apoE3, but not apoE4, retards the age-dependent neuronal death in apoE-null mice (Buttini et al. 1999, 2010). ApoE4 not only impairs synaptogenesis, but also decreases dendritic spine density in vivo in apoE transgenic and gene-targeted mice as well as in primary neuronal cultures (Brodbeck et al. 2011; Dumanis et al. 2009). In addition, neural stem cells in adult mice express apoE and apoE4 impairs adult hippocampal neurogenesis (Li et al. 2009b), which may contribute to apoE4-mediated impairment in learning and memory and cognitive function. Since there is no Aβ accumulation in any of these apoE4 mouse models, which support the view that an Aβ-independent role of apoE4 in inducing neuronal and behavioral deficits in vivo. While many of the above mentioned transgenic mice accumulate Aβ and develop Aβ plaque pathology along with cognitive impairment, they are unable to induce NFT formation. To determine the contribution of tau protein hyperphosphorylation in the pathogenesis of AD, several mouse models have been established that overexpress either wild-type or mutated human tau protein. It is reported that Tau protein mutations are associated with frontotemporal dementia, but not with AD (Duyckaerts et al. 2008). Introduction of human Tau proteins containing FTD mutations result in NFT formation (Gotz et al. 2001; Lewis et al. 2000; Tanemura et al. 2002; Allen et al. 2002). Tau protein containing G272V and P301S mutations produce both NFT formation and severe cognitive deficits (Schindowski et al. 2006). In an effort to model NFT pathology that is relevant to AD rather than FTD, tau knockout mice were crossed with mice expressing human genomic tau protein, resulting in mice expressing human but not murine tau protein (hTau). However, these mice express minimal NFT pathology (Andorfer et al. 2003).

It is becoming increasingly evident that type 2 diabetes mellitus and metabolic syndrome are risk factors for stroke, AD, and depression (Farooqui et al. 2012; Farooqui 2013). Due to improved treatments, type 2 diabetes mellitus patients are living longer, putting them at increased risk for age-related complications along with risk of stroke, AD, and depression. Recent studies have described the generation of a novel mouse model combining the key features of obesity, diabetes, and AD. In these studies, the obese and diabetic *db/db* mouse (Srinivasan and Ramarao 2007) is crossed with the APP$^{\Delta NL/\Delta NL}$ × PS1$^{P264L/P264L}$ knock-in model of AD (Reaume et al. 1996; Siman et al. 2000; Niedowicz et al. 2014). The resulting mice are called *db/AD*. These mice are morbidly obese, have glucose intolerant, show insulin resistance, and display parenchymal amyloid plaques similar to the parental lines. In addition, these mice show profound cognitive impairment and marked cerebrovascular abnormalities, which are Aβ/tau-independent mechanism. Long term consumption of high-fat diet is known to induce the accumulation of Aβ not only in the brain of wild type rabbits, rodents, and APP Tg mice (Sparks et al. 1994; Refolo et al. 2000; Ho et al. 2004), but also in humans (Farooqui 2015). The molecular mechanisms associated with high-fat mediated Aβ accumulation in the brain are not fully understood. However, autophagosome-mediated enhancement in amyloidogenic APP processing (Son et al. 2012) or up-regulation of BACE1 (Guglielmotto et al. 2012) may increase high-fat induced Aβ generation by above mentioned mechanisms. Furthermore, soluble Aβ itself is believed to reduce endothelial function and vascular reactivity in mice (Niwa et al. 2000) and humans (Dumas et al. 2012; den Abeelen et al. 2014). Collective evidence suggests that *db/AD* model is a unique. It can be used to study overlap among molecular mechanisms of obesity, type 2 diabetes mellitus, and AD in old animals.

2.4   Animal Models of Alzheimer Disease in Cell Culture

Attempts have been to establish Aβ-pathologies such as production, secretion, oligomerization and aggregation of Aβ peptides utilizing a novel platform to model the pathological processing of mutant human APPswe protein for Aβ genesis, oligomerization and aggregation, the initial events of AD pathogenesis (Ghate et al. 2014). Neurosphere cultures have been prepared from AD transgenic (*APPswe, PSEN1dE9*) mice embryos. These cultures not only show positive expression for both transgenes at the mRNA level and express humanized APP and its proteolytic products including Aβ peptides. Analysis of Tg+ve neurosphere lysates the presence of both monomeric and various oligomeric Aβ peptides similar to an 18-month old Tg+ve mouse brain homogenate. Tg+ve neurosphere cultures secrete a large amount of human Aβ peptides that consist of Aβ40 and Aβ42 with a very high Aβ42/Aβ40 ratio comparable to that of human AD brain homogenates and more than any cellular model of AD. Tg+ve culture supernatants also contain monomeric and various pathogenic Aβ peptide oligomers (ranging from 2-mer to 12-mer; the Aβ star oligomer) (Ghate et al. 2014). In addition, conformation-dependent immunocytochemistry demonstrated the presence of intracellular and extracellular Aβ peptides within neurospheres. The neurosphere culture system has many advantages over existing cellular models. Thus, (a) neurosphere cultures contain both brain stem and progenitor like cells, which can differentiate towards mature brain cells like neurons and astrocytes that are not possible in transformed cell lines, (b) these cultures can synthesize and secrete both Aβ peptides, (c) these cultures show high Aβ42/Aβ40 ratio, (d) produce pathogenic Aβ peptide oligomerization, which is comparable with the animal models of AD and much higher than existing cellular models of AD, including iPSC based models of AD (Israel et al. 2012) and (e) demonstrate intracellular and extracellular aggregation of Aβ peptides. It is proposed that studies with neurosphere cell culture may advance not only our understanding of pathogenesis of AD, but may provide better understanding of therapeutic agents on decreasing the beta amyloid synthesis and aggregation within neural cells (Ghate et al. 2014).

2.5 The Gap Between Mouse Models and Human Patients of Alzheimer Disease

The lack of an ideal animal model and specific biomarkers for AD progression makes it not only difficult to learn molecular mechanism associated with the pathogenesis of AD, but also complicate to discover the drugs that can prevent neurodegeneration in AD. At present mice models of AD mimic only few aspects of the disease, which are neither enough to learn about the molecular mechanism, specific biomarkers, and develop new treatment (Elder et al. 2010). Another possibility is that senile plaques and neurofibrillary tangles are endpoints for different disease-driving mechanisms. Thus, achieving a successful inhibition of Aβ and tau pathologies may not result in the successful for treating AD. AD is a multifactorial disease so its treatment may require a multitarget approach. To generate better animal models for AD, one has to develop better understanding of the molecular neuropathological mechanisms not only associated with neurodegeneration, but also behavioral and memory losses. Another important point is either the lack or low of neurodegeneration in animal models compared to human subjects, who show slow and continuous neurodegeneration with the progression of the disease. This is tempting to speculate that more research breakthroughs in development of animal models are needed for the development of models reflecting the heterogeneity of the disease (Cuadrado-Tejedor and García-Osta 2014). Also, discovery of specific biomakers is necessary not only to identify AD progression in a large population, but also for monitoring clinical trials and responses to medication.

2.6 Conclusion

AD is a multifactorial disease characterized by the accumulation of senile plaques, which are composed of oligomers of Aβ and neurofibrillary tangles and hyperphosphorylated Tau protein. In addition, neurochemical changes in AD include slow excitotoxicity, mitochondrial dysfunction, oxidative stress, and neuroinflammation. Neurotoxin-induce animal models of AD show very little neuropathological changes, but they induce mitochondrial dysfunction, oxidative stress, and neuroinflammation. Animal models for AD have been developed in both invertebrates (fruit flies and roundworms) and vertebrates (mice, rats, and rabbits). Most mice models are based on familial AD mutations of genes involved in the amyloidogenic process, such as the APP, MAPT, PS1, PS2 tau protein and apoE. Some models also incorporate Tau mutations, which are known to cause frontotemporal dementia, a condition, which shares some elements of neuropathology with AD. Transgenic mice develop several lesions similar to those of AD, including diffuse and neuritic amyloid deposits, cerebral amyloid angiopathy, dystrophic neurites and synapses, and amyloid-associated neuroinflammation. However, other features of AD, such as neurofibrillary tangles, nerve cell loss, and significant memory deficits are not satisfactorily reproduced in these models. This suggests that despite various modifications specific to AD in the genome of animals, investigators have failed to create an ideal animal model, which can be fully characterized by all the pathological and neurochemical changes that can occur in AD. Nevertheless, the role of transgenic animals is undeniable, both in research on AD neuropathology and for testing new therapies, such as immunotherapy. It is well understood that it is difficult to reproduce all anatomical characteristics and cognitive ability of humans in mice because of lower-order of cognition found in mice. In addition, there are substantial anatomical differences between mouse and human brains, particularly that the mouse brain has a higher gray-to-white matter ratio. Still, transgenic mice have provided valuable genetic, neurochemical, and neuropathological information on AD. Better transgenic models of AD are needed for future research in higher animals, which are closer to humans not only in anatomy, but also in cognitive function, behavior and social responses.

References

Adalbert R, Gilley J, Coleman MP (2007) Aβ, tau and ApoE4 in Alzheimer's disease, the axonal connection. Trends Mol Med 13:135–142

Agca C, Fritz JJ, Walker LC, Levey AI, Chan AW, Lah JJ, Agca Y (2008) Development of transgenic rats producing human beta-amyloid precursor protein as a model for Alzheimer's disease: transgene and endogenous APP genes are regulated tissue-specifically. BMC Neurosci 9:28

Aggarwal NT, Wilson RS, Beck TL, Bienias JL, Berry-Kravis E, Bennett DA (2005) The apolipoprotein E ε4 allele and incident Alzheimer's disease in persons with mild cognitive impairment. Neurocase 11:3–7

Allen B, Ingram E, Takao M, Smith MJ, Jakes R, Virdee K, Yoshida H, Holzer M, Craxton M, Emson PC, Atzori C, Migheli A, Crowther RA, Ghetti B, Spillantini MG, Goedert M (2002) Abundant tau filaments and nonapoptotic neurodegeneration in transgenic mice expressing human P301S tau protein. J Neurosci 22:9340–9351

Andorfer C, Kress Y, Espinoza M, de Silva R, Tucker KL, Barde YA, Duff K, Davies P (2003) Hyperphosphorylation and aggregation of tau in mice expressing normal human tau isoforms. J Neurochem 86:582–590

Baker S, Götz J (2015) What we can learn from animal models about cerebral multi-morbidity. Alzheimers Res Ther 7:11

Bartus RT (2000) On neurodegenerative diseases, models and treatment strategies: lessons learned and lessons forgotten a generation following the cholinergic hypothesis. Exp Neurol 163:495–529

Bell KFS, Cuello AC (2006) Altered synaptic function in Alzheimer's disease. Eur J Pharmacol 545:11–21

Bezprozvanny I, Mattson MP (2008) Neuronal calcium mishandling and the pathogenesis of Alzheimer's disease. Trends Neurosci 31:454–463

Bharathi V, Vasudevaraju P, Govindaraju M, Palanisamy AP, Sambamurti K, Rao KS (2008) Molecular toxicity of aluminium in relation to neurodegeneration. Indian J Med Res 128:545–556

Bhattacharjee S, Zhao Y, Hill JM, Culicchia F, Kruck TP, Percy ME, Pogue AI, Walton JR, Lukiw WJ (2013) Selective accumulation of aluminum in cerebral arteries in Alzheimer's disease. J Inorg Biochem 126:35–37

Bluthe RM, Dantzer R, Kelley KW (1992) Effects of IL-1 receptor antagonist on the behavioral effects of lipopolysaccharide in rat. Brain Res 573:318–320

Bobich JA, Zheng Q, Campbell A (2004) Incubation of nerve endings with a physiological concentration of Abeta1-42 activates CaV2.2(N-Type)-voltage operated calcium channels and acutely increases glutamate and noradrenaline release. J Alzheimers Dis 6:243–255

Bobinski M, de Leon MJ, Convit A, De Santi S, Wegiel J, Tarshish CY, Saint Louis LA, Wisniewski HM (1999) MRI of entorhinal cortex in mild Alzheimer's disease. Lancet 353:38–40

Bondi MW, Jak AJ, Delano-Wood L, Jacobson MW, Delis DC, Salmon DP (2008) Neuropsychological contributions to the early identification of Alzheimer's disease. Neuropsychol Rev 18:73–90

Bondy SC (2013) Prolonged exposure to low levels of aluminum leads to changes associated with brain aging and neurodegeneration. Toxicology 315:1–7

Bordji K, Becerril-Ortega J, Nicole O, Buisson A (2010) Activation of extrasynaptic, but not synaptic, NMDA receptors modifies amyloid precursor protein expression pattern and increases amyloid-ss production. J Neurosci 30:15927–15942

Braak H, Braak E (1995) Staging of Alzheimer's disease-related neurofibrillary changes. Neurobiol Aging 16:271–278, discussion 278-284

Braidy N, Muñoz P, Palacios AG, Castellano-Gonzalez G, Inestrosa NC, Chung RS, Sachdev P, Guillemin GJ (2012) Recent models for Alzheimer's disease: clinical implications and basic research. J Neural Transm 119:173–195

Brodbeck J, McGuire J, Liu Z, Meyer-Franke A, Balestra ME, Jeong DE, Pleiss M, McComas C, Hess F, Witter D, Peterson S, Childers M, Goulet M, Liverton N, Hargreaves R, Freedman S, Weisgraber KH, Mahley RW, Huang Y (2011) Structure-dependent impairment of intracellular apolipoprotein E4 trafficking and its detrimental effects are rescued by small-molecule structure correctors. J Biol Chem 286:17217–17226

Brorson JR, Bindokas VP, Iwama T, Marcuccilli CJ, Chisholm JC, Miller RJ (1995) The Ca^{2+} influx induced by beta-amyloid peptide 25–35 in cultured hippocampal neurons results from network excitation. J Neurobiol 26:325–338

Brown RE (2007) Behavioural phenotyping of transgenic mice. Can J Exp Psychol 61:328–344

Bu G (2009) Apolipoprotein E and its receptors in Alzheimer's disease: pathways, pathogenesis and therapy. Nat Rev Neurosci 10:333–344

Buttini M, Orth M, Bellosta S, Akeefe H, Pitas RE, Wyss-Coray T, Mucke L, Mahley RW (1999) Expression of human apolipoprotein E3 or E4 in the brains of $Apoe^{-/-}$ mice: Isoform-specific effects on neurodegeneration. J Neurosci 19:4867–4880

Buttini M, Masliah E, Yu GQ, Palop JJ, Chang S, Bernardo A, Lin C, Wyss-Coray T, Huang Y, Mucke L (2010) Cellular source of apolipoprotein E4 determines neuronal susceptibility to excitotoxic injury in transgenic mice. Am J Pathol 177:563–569

Camandola S, Mattson MP (2011) Aberrant subcellular neuronal calcium regulation in aging and Alzheimer's disease. Biochim Biophys Acta 1813:965–973

Carmo SD, Claudio Cuello A (2013) Modeling Alzheimer's disease in transgenic rats. Mol Neurodegener 8:37

Chiocco MJ, Kulnane LS, Younkin L, Younkin S, Evin G, Lamb BT (2004) Altered amyloid-β metabolism and deposition in genomic-based β-secretase transgenic mice. J Biol Chem 279: 52535–52542

Cho A, Haruyama N, Kulkarni AB (2009) Generation of transgenic mice. Curr Protoc Cell Biol 19:11

Chow VW, Mattson MP, Wong PC, Gleichmann M (2010) An overview of APP processing enzymes and products. Neuromolecular Med 12:1–12

Clavaguera F, Akatsu H, Fraser G, Crowther RA, Frank S, Hench J, Probst A, Winkler DT, Reichwald J, Staufenbiel M, Ghetti B, Goedert M, Tolnay M (2013) Brain homogenates from human tauopathies induce tau inclusions in mouse brain. Proc Natl Acad Sci U S A 110:9535–9540

Cohen E, Bieschke J, Perciavalle RM, Kelly JW, Dillin A (2006) Opposing activities protect against age-onset proteotoxicity. Science 313:1604–1610

Craft JM, Watterson DM, Van Eldik LJ (2006) Human amyloid beta-induced neuroinflammation is an early event in neurodegeneration. Glia 53:484–490

Craig LA, Hong NS, McDonald RJ (2011) Revisiting the cholinergic hypothesis in the development of Alzheimer's disease. Neurosci Biobehav Rev 35:1397–1409

Cuadrado-Tejedor M, García-Osta A (2014) Current animal models of Alzheimer's disease: challenges in translational research. Front Neurol 5:182

Davies P, Maloney AJF (1976) Selective loss of central cholinergic neurons in Alzheimer's disease. Lancet 2:1403

De Felice FG, Velasco PT, Lambert MP, Viola K, Fernandez SJ, Ferreira ST, Klein WL (2007) Abeta oligomers induce neuronal oxidative stress through an N-methyl-D-aspartate receptor-dependent mechanism that is blocked by the Alzheimer drug memantine. J Biol Chem 282:11590–11601

den Abeelen AS, Lagro J, van Beek AH, Claassen JA (2014) Impaired cerebral autoregulation and vasomotor reactivity in sporadic Alzheimer's disease. Curr Alzheimer Res 11:11–17

Donovan MH, Yazdani U, Norris RD, Games D, German DC, Eisch AJ (2006) Decreased adult hippocampal neurogenesis in the PDAPP mouse model of Alzheimer's disease. J Comp Neurol 495:70–83

Dournaud P, Delaere P, Hauw JJ, Epelbaum J (1995) Differential correlation between neurochemical deficits, neuropathology, and cognitive status in Alzheimer's disease. Neurobiol Aging 16:817–823

Drago D, Bolognin S, Zatta P (2008) Role of metal ions in the abeta oligomerization in Alzheimer's disease and in other neurological disorders. Curr Alzheimer Res 5:500–507

Dumanis SB, Tesoriero JA, Babus LW, Nguyen MT, Trotter JH, Ladu MJ, Weeber EJ, Turner RS, Xu B, Rebeck GW, Hoe HS (2009) ApoE4 decreases spine density and dendritic complexity in cortical neurons in vivo. J Neurosci 29:15317–15322

Dumas A, Dierksen GA, Gurol ME, Halpin A, Martinez-Ramirez S, Schwab K, Rosand J, Viswanathan A, Salat DH, Polimeni JR, Greenberg SM (2012) Functional magnetic resonance imaging detection of vascular reactivity in cerebral amyloid angiopathy. Ann Neurol 72:76–81

Duyckaerts C, Potier MC, Delatour B (2008) Alzheimer disease models and human neuropathology: similarities and differences. Acta Neuropathol 115:5–38

Echeverria V, Ducatenzeiler A, Dowd E, Jänne J, Grant SM, Szyf M, Wandosell F, Avila J, Grimm H, Dunnett SB, Hartmann T, Alhonen L, Cuello AC (2004a) Altered mitogen-activated protein kinase signaling, tau hyperphosphorylation and mild spatial learning dysfunction in transgenic rats expressing the beta-amyloid peptide intracellularly in hippocampal and cortical neurons. Neuroscience 29:583–592

Echeverria V, Ducatenzeiler A, Alhonen L, Janne J, Grant SM, Wandosell F, Muro A, Baralle F, Li H, Duff K, Szyf M, Cuello A (2004b) Rat transgenic models with a phenotype of intracellular Abeta accumulation in hippocampus and cortex. J Alzheimers Dis 6:209–219

Eidi A, Eidi M, Mahmoodi G, Oryan SH (2006) Effect of vitamin E on memory retention in rats: possible involvement of cholinergic system. Eur Neuropsychopharmacol 16:101–106

Elder GA, Gama Sosa MA, De Gasperi R (2010) Transgenic mouse models of Alzheimer's disease. Mt Sinai J Med 77:69–81

Eriksen JL, Janus CG (2007) Plaques, tangles and memory loss in mouse models of neurodegeneration. Behav Genet 37:79–100

Esterbauer H, Schaur RJ, Zollner H (1991) Chemistry and biochemistry of 4-hydroxynonenal, malonaldehyde and related aldehydes. Free Radic Biol Med 11:81–128

Exley C (2012) The coordination chemistry of aluminium in neurodegenerative disease. Coordination Chem Rev 256:2142–2146

Farooqui AA (2009) Beneficial effects of fish oil on human brain. Springer, New York

Farooqui AA (2010) Neurochemical aspects of neurotraumatic and neurodegenerative diseases. Springer, New York

Farooqui AA (2011) Lipid mediators and their metabolism in the brain. Springer, New York

Farooqui AA (2013) Metabolic syndrome: an important risk factor for stroke, Alzheimer disease, and depression. Springer, New York

Farooqui AA (2015) High calorie diet and human brain: metabolic consequences of long term consumption. Springer, New York

Farooqui AA, Horrocks LA (2006) Phospholipase A_2-generated lipid mediators in the brain: the good, the bad, and the ugly. Neuroscientist 12:245–260

Farooqui AA, Farooqui T, Panza F, Frisardi V (2012) Metabolic syndrome as a risk factor for neurological disorders. Cell Mol Life Sci 69:741–762

Fattoretti P, Bertoni-Freddari C, Balietti M, Giorgetti B, Solazzi M, Zatta P (2004) Chronic aluminum administration to old rats results in increased levels of brain metal ions and enlarged hippocampal mossy fibers. Ann N Y Acad Sci 1019:44–47

Ferreira ST, Klein WL (2011) The Aβ oligomer hypothesis for synapse failure and memory loss in Alzheimer's disease. Neurobiol Learn Mem 96:529–543

Finch CE, Austad SN (2012) Primate aging in the mammalian scheme: the puzzle of extreme variation in brain aging. Age (Dordr) 34:1075–1091

Flood DG, Reaume AG, Dorfman KS, Lin YG, Lang DM et al (2002) FAD mutant PS-1 gene-targeted mice: increased A beta 42 and A beta deposition without APP overproduction. Neurobiol Aging 23:335–348

Flood DG, Lin YG, Lang DM, Trusko SP, Hirsch JD, Savage MJ, Scott RW, Howland DS (2009) A transgenic rat model of Alzheimer's disease with extracellular Abeta deposition. Neurobiol Aging 30:1078–1090

Forloni G, Demicheli F, Giorgi S, Bendotti C, Angeretti N (1992) Expression of amyloid precursor protein mRNAs in endothelial, neuronal and glial cells: modulation by IL-1. Mol Brain Res 16:128–134

Francis PT (2003) Glutamatergic systems in Alzheimer's disease. Int J Geriatr Psychiatry 18:S15–S21

Fu R, Shen Q, Xu P, Luo JJ, Tang Y (2014) Phagocytosis of microglia in the central nervous system diseases. Mol Neurobiol 49:1422–3410

Geula C, Mesulam MM (1994) Cholinergic system and related neuropathological predilection patterns in Alzheimer's disease. In: Terry RD, Katzman R, Bick KL (eds) Alzheimer disease. Raven, New York, pp 263–291

Ghate PS, Sidhar H, Carlson GA, Giri RK (2014) Development of a novel cellular model of Alzheimer's disease utilizing neurosphere cultures derived from B6C3 Tg(APPswe, PSEN1dE9)85Dbo/J embryonic mouse brain. Springerplus 3:161

Goate AM, Chartier-Harlin CM, Mullan M, Brown J, Crawford F, Fidani L, Giuffra L, Haynes A, Irving N, James L (1991) Segregation of a missense mutation in the amyloid precursor protein gene with familial Alzheimer's disease. Nature 349:704–706

Golde TE, Schneider LS, Koo EH (2011) Anti-abeta therapeutics in Alzheimer's disease: the need for a paradigm shift. Neuron 69:203–213

Gong B, Vitolo OV, Trinchese F, Liu S, Shelanski M, Arancio O (2004) Persistent improvement in synaptic and cognitive functions in an Alzheimer mouse model after rolipram treatment. J Clin Invest 114:1624–1634

Gonzalez GA, Montminy MR (1989) Cyclic AMP stimulates somatostatin gene transcription by phosphorylation of CREB at serine 133. Cell 59:675–680

Gotz J, Chen F, van Dorpe J, Nitsch RM (2001) Formation of neurofibrillary tangles in P301l tau transgenic mice induced by Abeta 42 fibrils. Science 293:1491–1495

Grant WB, Campbell A, Itzhaki RF, Savory J (2002) The significance of environmental factors in the etiology of Alzheimer's disease. J Alzheimers Dis 4:179–189

Guglielmotto M, Aragno M, Tamagno E, Vercellinatto I, Visentin S, Medana C, Catalano MG, Smith MA, Perry G, Danni O, Boccuzzi G, Tabaton M (2012) AGEs/RAGE complex upregulates BACE1 via NF-kappaB pathway activation. Neurobiol Aging 33(1), e113

Guo Q, Wang Z, Li H, Wiese M, Zheng H (2012) APP physiological and pathophysiological functions: insights from animal models. Cell Res 22:78–89

Haass C, Selkoe DJ (2007) Soluble protein oligomers in neurodegeneration: lessons from the Alzheimer's amyloid beta-peptide. Nat Rev Mol Cell Biol 8:101–112

Hanes J, Zilka N, Bartkova M, Caletkova M, Dobrota D, Novak M (2009) Rat tau proteome consists of six tau isoforms: implication for animal models of human tauopathies. J Neurochem 108:1167–1176

Hardingham GE, Fukunaga Y, Bading H (2002) Extrasynaptic NMDARs oppose synaptic NMDARs by triggering CREB shut-off and cell death pathways. Nat Neurosci 5:405–414

Hardy J, Selkoe DJ (2002) The amyloid hypothesis of Alzheimer's disease: progress and problems on the road to therapeutics. Science 297:353–356

Hauss-Wegrzyniak B, Dobrzanski P, Stoehr JD, Wenk GL (1998) Chronic neuroinflammation in rats reproduces components of the neurobiology of Alzheimer's disease. Brain Res 780: 294–303

Hauss-Wegrzyniak B, Vraniak P, Wenk GL (1999a) The effects of a novel NSAID upon chronic neuroinflammation are age dependent. Neurobiol Aging 20:305–313

Hauss-Wegrzyniak B, Willard LB, Del Soldato P, Pepeu G, Wenk GL (1999b) Peripheral administration of novel anti-inflammatories can attenuate the effects of chronic inflammation within the CNS. Brain Res 815:36–43

Ho L, Qin W, Pompl PN, Xiang Z, Wang J, Zhao Z, Peng Y, Cambareri G, Rocher A, Mobbs CV, Hof PR, Pasinetti GM (2004) Diet-induced insulin resistance promotes amyloidosis in a transgenic mouse model of Alzheimer's disease. FASEB J 18:902–904

Holzenberger M, Dupont J, Ducos B, Leneuve P, Géloën A, Even PC, Cervera P, Le Bouc Y (2003) IGF-1 receptor regulates lifespan and resistance to oxidative stress in mice. Nature 421:182–187

Hrnkova M, Zilka N, Minichova Z, Koson P, Novak M (2007) Neurodegeneration caused by expression of human truncated tau leads to progressive neurobehavioural impairment in transgenic rats. Brain Res 1130:206–213

Hsu A-L, Murphy CT, Kenyon C (2003) Regulation of aging and age-related disease by DAF-16 and heat-shock factor. Science 300:1142–1145

Hu X, Hicks CW, He W, Wong P, Macklin WB, Trapp BD, Yan R (2006) Bace1 modulates myelination in the central and peripheral nervous system. Nat Neurosci 9:1520–1525

Huang HC, Jiang ZF (2009) Accumulated amyloid-beta peptide and hyperphosphorylated tau protein: relationship and links in Alzheimer's disease. J Alzheimers Dis 16:15–27

Israel MA, Yuan SH, Bardy C, Reyna SM, Mu Y, Herrera C, Hefferan MP, Van Gorp S, Nazor KL, Boscolo FS, Carson CT, Laurent LC, Marsala M, Gage FH, Remes AM, Koo EH, Goldstein LS (2012) Probing sporadic and familial Alzheimer's disease using induced pluripotent stem cells. Nature 482:216–220

Ittner LM, Ke YD, Delerue F, Bi M, Gladbach A, van Eersel J, Wölfing H, Chieng BC, Christie MJ, Napier IA, Eckert A, Staufenbiel M, Hardeman E, Götz J (2010) Dendritic function of tau mediates amyloid-beta toxicity in Alzheimer's disease mouse models. Cell 142:387–397

Jacob HJ, Kwitek AE (2002) Rat genetics: attaching physiology and pharmacology to the genome. Nat Rev Genet 3:33–42

Joseph J, Shukitt-Hale B, Denisova NA, Martin A, Perry G, Smith MA (2001) Copernicus revisited: amyloid beta in Alzheimer's disease. Neurobiol Aging 22:131–146

Joshi G, Gan KA, Johnson DA, Johnson JA (2015) Increased Alzheimer's disease-like pathology in the APP/ PS1ΔE9 mouse model lacking Nrf2 through modulation of autophagy. Neurobiol Aging 36:664–679

Kar S, Slowikowski SP, Westaway D, Mount H (2004) Interactions between β-amyloid and central cholinergic neurons: implications for Alzheimer's disease. J Psychiatry Neurosci 29:427–441

Karran E, Mercken M, De Strooper B (2011) The amyloid cascade hypothesis for Alzheimer's disease: an appraisal for the development of therapeutics. Nat Rev Drug Discov 10:698–712

Kawahara M, Kato-Negishi M (2011) Link between aluminum and the pathogenesis of Alzheimer's disease: the integration of the aluminum and amyloid cascade hypotheses. Int J Alzheimer Dis 2011:276393

Kenyon C, Chang J, Gensch E, Rudner A, Tabtiang R (1993) A C. elegans mutant that lives twice as long as wild type. Nature 366:461–464

Kimura R, Devi L, Ohno M (2010) Partial reduction of BACE1 improves synaptic plasticity, recent and remote memories in Alzheimer's disease transgenic mice. J Neurochem 113:248–261

Kloskowska E, Pham TM, Nilsson T, Zhu S, Oberg J, Codita A, Pedersen LA, Pedersen JT, Malkiewicz K, Winblad B, Folkesson R, Benedikz E (2010) Cognitive impairment in the Tg6590 transgenic rat model of Alzheimer's disease. J Cell Mol Med 14:1816–1823

Kobayashi D, Zeller M, Cole T, Buttini M, McConlogue L, Sinha S, Freedman S, Morris RG, Chen KS (2008) BACE1 gene deletion: impact on behavioural function in a model of Alzheimer's disease. Neurobiol Aging 29:861–873

Koffie RM, Meyer-Luehmann M, Hashimoto T, Adams KW, Mielke ML, Garcia-Alloza M, Micheva KD, Smith SJ, Kim ML, Lee VM, Hyman BT, Spires-Jones TL (2009) Oligomeric amyloid beta associates with postsynaptic densities and correlates with excitatory synapse loss near senile plaques. Proc Natl Acad Sci U S A 106:4012–4017

Köhler C, Ebert U, Baumann K, Schröder H (2005) Alzheimer's disease-like neuropathology of gene-targeted APP-SLxPS1mut mice expressing the amyloid precursor protein at endogenous levels. Neurobiol Dis 20:528–540

Koson P, Zilka N, Kovac A, Kovacech B, Korenova M, Filipcik P, Novak M (2008) Truncated tau expression levels determine life span of a rat model of tauopathy without causing neuronal loss or correlating with terminal neurofibrillary tangle load. Eur J Neurosci 28:239–246

Krafft GA, Klein WL (2010) ADDLs and the signaling web that leads to Alzheimer's disease. Neuropharmacology 59:230–242

Lai JC, Blass JP (1984) Inhibition of brain glycolysis by aluminum. J Neurochem 42:438–446

Lander CJ, Lee JM (1998) Pharmacological drug treatment of Alzheimer disease: the cholinergic hypothesis revisited. J Neuropathol Exp Neurol 57:719–731

Lecanu L, Papadopoulos V (2013) Modeling Alzheimer's disease with non-transgenic rat models. Alzheimers Res Ther 5:1–9

Lee JE, Han PL (2013) An update of animal models of Alzheimer disease with a reevaluation of plaque depositions. Exp Neurobiol 22:84–95

Lee DC, Sunnarborg SW, Hinkle CL, Myers TJ, Stevenson MY, Russell WE, Castner BJ, Gerhart MJ, Paxton RJ, Black RA, Chang A, Jackson LF (2003) TACE/ADAM17 processing of EGFR ligands indicates a role as a physiological convertase. Ann N Y Acad Sci 995:22–38

Leon WC, Canneva F, Partridge V, Allard S, Ferretti MT, DeWilde A, Vercauteren F, Atifeh R, Ducatenzeiler A, Klein W, Szyf M, Alhonen L, Cuello AC (2010) A novel transgenic rat model with a full Alzheimer's-like amyloid pathology displays pre-plaque intracellular amyloid-beta-associated cognitive impairment. J Alzheimers Dis 20:113–126

Lesné S, Ali C, Gabriel C, Croci N, MacKenzie ET, Glabe CG, Plotkine M, Marchand-Verrecchia C, Vivien D, Buisson A (2005) NMDA receptor activation inhibits α-secretase and promotes neuronal amyloid-β production. J Neurosci 25:9367–9377

Leuner K, Hauptmann S, Hauptmann S (2007) Mitochondrial dysfunction: the first domino in brain aging and Alzheimer's disease? Antioxid Redox Signal 9:1659–1675

Lewis J, McGowan E, Rockwood J, Melrose H, Nacharaju P, Van Slegtenhorst M, Gwinn-Hardy K, Murphy PM, Baker M, Yu X, Duff K, Hardy J, Corral A, Lin WL, Yen SH, Dickson DW, Davies P, Hutton M (2000) Neurofibrillary tangles, amyotrophy and progressive motor disturbance in mice expressing mutant (P301L) tau protein. Nat Genet 25:402–405

Li J, Le W (2013) Modeling neurodegenerative diseases in Caenorhabditis elegans. Exp Neurol 250:94–103

Li S, Hong S, Shepardson NE, Walsh DM, Shankar GM, Selkoe D (2009a) Soluble oligomers of amyloid beta protein facilitate hippocampal long-term depression by disrupting neuronal glutamate uptake. Neuron 62:788–801

Li G, Bien-Ly N, Andrews-Zwilling Y, Xu Q, Bernardo A, Ring K, Halabisky B, Deng C, Mahley RW, Huang Y (2009b) GABAergic interneuron dysfunction impairs hippocampal neurogenesis in adult apolipoprotein E4 knockin mice. Cell Stem Cell 5:634–645

Liang Z, Liu F, Iqbal K, Grundke-Iqbal I, Gong CX (2009) Dysregulation of tau phosphorylation in mouse brain during excitotoxic damage. J Alzheimers Dis 17:531–539

Link CD (2005) Invertebrate models of Alzheimer's disease. Genes Brain Behav 4:147–156

Liu L, Orozco IJ, Planel E, Wen Y, Bretteville A, Krishnamurthy P, Wang L, Herman M, Figueroa H, Yu WH, Arancio O, Duff K (2008) A transgenic rat that develops Alzheimer's disease-like amyloid pathology, deficits in synaptic plasticity and cognitive impairment. Neurobiol Dis 31:46–57

Lonze BE, Ginty DD (2002) Function and regulation of CREB family transcription factors in the nervous system. Neuron 35:605–623

Lukiw WJ, Pogue AI (2007) Induction of specific micro RNA (miRNA) species by ROS-generating metal sulfates in primary human brain cells. J Inorg Biochem 101:1265–1269

Luo Y, Bolon B, Kahn S, Bennett BD, Babu-Khan S, Denis P, Fan W, Kha H, Zhang J, Gong Y, Martin L, Louis JC, Yan Q, Richards WG, Citron M, Vassar R (2001) Mice deficient in BACE1, the Alzheimer's beta-secretase, have normal phenotype and abolished beta-amyloid generation. Nat Neurosci 4:231–232

Masliah E, Sisk A, Mallory M, Games D (2001) Neurofibrillary pathology in transgenic mice overexpressing V717F beta-amyloid precursor protein. J Neuropathol Exp Neurol 60:357–368

McGaugh JL (2004) The amygdala modulates the consolidation of memories of emotionally arousing experiences. Annu Rev Neurosci 27:1–28

McLean JW, Fukazawa C, Taylor JM (1983) Rat apolipoprotein E mRNA. Cloning and sequencing of double-stranded cDNA. J Biol Chem 258:8993–9000

Millington C, Sonego S, Karunaweera N, Rangel A, Aldrich-Wright JR, Campbell IL, Gyengesi E, Münch G (2014) Chronic neuroinflammation in Alzheimer's disease: new perspectives on animal models and promising candidate drugs. Biomed Res Int 2014:309129

Mizoroki T, Meshitsuka S, Maeda S, Murayama M, Sahara N, Takashima A (2007) Aluminum induces tau aggregation in vitro but not in vivo. J Alzheimers Dis 11:419–427

Mocanu MM, Nissen A, Eckermann K, Khlistunova I, Biernat J, Drexler D, Petrova O, Schönig K, Bujard H, Mandelkow E, Zhou L, Rune G, Mandelkow EM (2008) The potential for beta-structure in the repeat domain of tau protein determines aggregation, synaptic decay, neuronal loss, and coassembly with endogenous Tau in inducible mouse models of tauopathy. J Neurosci 28:737–748

Mohajeri MH, Saini KD, Nitsch RM (2004) Transgenic BACE expression in mouse neurons accelerates amyloid plaque pathology. J Neural Transm 111:413–425

Morris JC (2002) Challenging assumptions about Alzheimer's disease: mild cognitive impairment and the cholinergic hypothesis. Ann Neurol 51:143–144

Mrak RE, Griffin WS (2001) The role of activated astrocytes and of the neurotrophic cytokine S100B in the pathogenesis of Alzheimer's disease. Neurobiol Aging 22:915–922

Mullan M, Crawford F, Axelman K, Houlden H, Lilius L, Winblad B, Lannfelt L (1992) A pathogenic mutation for probable Alzheimer's disease in the APP gene at the N-terminus of beta-amyloid. Nat Genet 1:345–347

Murrell JR, Hake AM, Quaid KA, Farlow MR, Ghetti B (2000) Early-onset Alzheimer disease caused by a new mutation (V717L) in the amyloid precursor protein gene. Arch Neurol 57:885–887

Nakahara N, Iga Y, Mizobe F, Kawanishi G (1988) Effects of intracerebroventricular injection of AF64A on learning behaviors in rats. Jpn J Pharmacol 48:121–130

Niedowicz DM, Reeves VL, Platt TL, Kohler K, Beckett TL, Powell DK, Lee TL, Sexton TR, Song ES, Brewer LD, Latimer CS, Kraner SD, Larson KL, Ozcan S, Norris CM, Hersh LB, Porter NM, Wilcock DM, Murphy MP (2014) Obesity and diabetes cause cognitive dysfunction in the absence of accelerated β-amyloid deposition in a novel murine model of mixed or vascular dementia. Acta Neuropathol Commun 2:64

Nilsberth C, Westlind-Danielsson A, Eckman CB, Condron MM, Axelman K, Forsell C, Stenh C, Luthman J, Teplow DB, Younkin SG, Näslund J, Lannfelt L (2001) The 'Arctic' APP mutation (E693G) causes Alzheimer's disease by enhanced Abeta protofibril formation. Nat Neurosci 4:887–893

Niwa K, Younkin L, Ebeling C, Turner SK, Westaway D, Younkin S, Ashe KH, Carlson GA, Iadecola C (2000) Abeta 1-40-related reduction in functional hyperemia in mouse neocortex during somatosensory activation. Proc Natl Acad Sci U S A 97:9735–9740

Nordberg A, Alafuzoff I, Winbald B (1992) Nicotinic and muscarinic receptor subtypes in the human brain: changes with aging and dementia. J Neurosci Res 31:103–111

Oddo S, Caccamo A, Shepherd JD, Murphy MP, Golde TE, Kayed R, Metherate R, Mattson MP, Akbari Y, LaFerla FM (2003) Triple-transgenic model of Alzheimer's disease with plaques and tangles: intracellular Aβ and synaptic dysfunction. Neuron 39:409–421

Ohno M, Cole SL, Yasvoina M, Zhao J, Citron M, Berry R, Disterhoft JF, Vassar R (2007) BACE1 gene deletion prevents neuron loss and memory deficits in 5XFAD APP/PS1 transgenic mice. Neurobiol Dis 26:134–145

Palop JJ, Mucke L (2010) Amyloid-β-induced neuronal dysfunction in Alzheimer's disease: from synapses toward neural networks. Nat Neurosci 13:812–818

Pekny M, Wilhelmsson U, Pekna M (2014) The dual role of astrocyte activation and reactive gliosis. Neurosci Lett 565:30–38

Perez M, Ribe E, Rubio A, Lim F, Morán MA, Ramos PG, Ferrer I, Isla MT, Avila J (2005) Characterization of a double (amyloid precursor protein-tau) transgenic: tau phosphorylation and aggregation. Neuroscience 130:339–347

Phillips HS, Hains JM, Armanini M, Laramee GR, Johnson SA, Winslow JW (1991) BDNF mRNA is decreased in the hippocampus of individuals with Alzheimer's disease. Neuron 7:695–702

Platt TL, Reeves VL, Murphy MP (2013) Transgenic models of Alzheimer's disease: better utilization of existing models through viral transgenesis. Biochim Biophys Acta 1832:1437–1448

Pogue AI, Jones BM, Bhattacharjee S, Percy ME, Zhao Y, Lukiw WJ (2012) Metal-sulfate induced generation of ROS in human brain cells: detection using an isomeric mixture of 5- and 6-carboxy-2′,7′-dichlorofluorescein diacetate (carboxy-DCFDA) as a cell permeant tracer. Int J Mol Sci 13:9615–9626

Praticò D, Uryu K, Sung S, Tang S, Trojanowski JQ, Lee VM (2002) Aluminum modulates brain amyloidosis through oxidative stress in APP transgenic mice. FASEB J 16:1138–1140

Przedborski S, Jackson-Lewis V, Naini AB, Jakowec M, Petzinger G, Miller R, Akram M (2001) The Parkinsonian toxin 1-methyl-4-phenyl-1,2,3,6-tetrahydropyridine (MPTP): a technical review of its utility and safety. J Neurochem 76:1265–1274

Querfurth HW, LaFerla FM (2010) Alzheimer's disease. N Engl J Med 362:329–344

Raber J, Wong D, Yu GQ, Buttini M, Mahley RW, Pitas RE, Mucke L (2000) Alzheimer's disease: Apolipoprotein E and cognitive performance. Nature 404:352–354

Ramsden M, Kotilinek L, Forster C, Paulson J, McGowan E, SantaCruz K, Guimaraes A, Yue M, Lewis J, Carlson G, Hutton M, Ashe KH (2005) Age-dependent neurofibrillary tangle formation, neuron loss, and memory impairment in a mouse model of human tauopathy (P301L). J Neurosci 25:10637–10647

Reaume AG, Howland DS, Trusko SP, Savage MJ, Lang DM, Greenberg BD, Siman R, Scott RW (1996) Enhanced amyloidogenic processing of the beta-amyloid precursor protein in gene-targeted mice bearing the Swedish familial Alzheimer's disease mutations and a "humanized" Abeta sequence. J Biol Chem 271:23380–23388

Refolo LM, Malester B, LaFrancois J, Bryant-Thomas T, Wang R, Tint GS, Sambamurti K, Duff K, Pappolla MA (2000) Hypercholesterolemia accelerates the Alzheimer's amyloid pathology in a transgenic mouse model. Neurobiol Dis 7:321–331

Revett TJ, Baker GB, Jhamandas J, Kar S (2013) Glutamate system, amyloid β peptides and tau protein: functional interrelationships and relevance to Alzheimer disease pathology. J Psychiatry Neurosci 38:6–23

Ribe EM, Perez M, Puig B, Gich I, Lim F, Cuadrado M, Sesma T, Catena S, Sánchez B, Nieto M, Gómez-Ramos P, Morán MA, Cabodevilla F, Samaranch L, Ortiz L, Pérez A, Ferrer I, Avila J, Gómez-Isla T (2005) Accelerated amyloid deposition, neurofibrillary degeneration and neuronal loss in double mutant APP/tau transgenic mice. Neurobiol Dis 20:814–822

Ribeiro FM, Camargos ER, de Souza LC, Teixeira AL (2013) Animal models of neurodegenerative diseases. Rev Bras Psiquiatr 35(Suppl 2):S82–S91

Rizvi SHM, Parveen A, Verma AK, Ahmad M, Arshad M, Mahdi AA (2014) Aluminium induced endoplasmic reticulum stress mediated cell death in SH-SY5Y neuroblastoma cell line is independent of p53. PLoS One 9, e98409

Roberds SL, Anderson J, Basi G, Bienkowski MJ, Branstetter DG, Chen KS, Freedman SB, Frigon NL, Games D, Hu K, Johnson-Wood K, Kappenman KE, Kawabe TT, Kola I, Kuehn R, Lee M, Liu W, Motter R, Nichols NF, Power M, Robertson DW, Schenk D, Schoor M, Shopp GM, Shuck ME, Sinha S, Svensson KA, Tatsuno G, Tintrup H, Wijsman J, Wright S, McConlogue L (2001) BACE knockout mice are healthy despite lacking the primary beta-secretase activity in brain: implications for Alzheimer's disease therapeutics. Hum Mol Genet 10:1317–1324

Roberson ED, Scearce-Levie K, Palop JJ, Yan F, Cheng IH, Wu T, Gerstein H, Yu GQ, Mucke L (2007) Reducing endogenous tau ameliorates amyloid beta-induced deficits in an Alzheimer's disease mouse model. Science 316:750–754

Rodella LF, Ricci F, Borsani E, Stacchiotti A, Foglio E, Favero G, Favero G, Rezzani R, Mariani C, Bianchi R (2008) Aluminium exposure induces Alzheimer's disease-like histopathological alterations in mouse brain. Histol Histopathol 23:433–439

Rodríguez-Rodríguez C, Sánchez de Groot N, Rimola A, Alvarez-Larena A, Lloveras V, Vidal-Gancedo J, Ventura S, Vendrell J, Sodupe M, González-Duarte P (2009) Design, selection, and characterization of thioflavin-based intercalation compounds with metal chelating properties for application in Alzheimer's disease. J Am Chem Soc 131:1436–1451

Rondeau V, Jacqmin-Gadda H, Commenges D, Helmer C, Dartigues JF (2009) Aluminum and silica in drinking water and the risk of Alzheimer's disease or cognitive decline: findings from 15-year follow-up of the PAQUID cohort. Am J Epidemiol 169:489–496

Sakamoto T, Saito H, Ishii K, Takahashi H, Tanabe S, Ogasawara Y (2006) Aluminum inhibits proteolytic degradation of amyloid beta peptide by cathepsin D: a potential link between aluminum accumulation and neuritic plaque deposition. FEBS Lett 580:6543–6549

Sakono M, Zako T (2010) Amyloid oligomers: formation and toxicity of Abeta oligomers. FEBS J 277:1348–1358

Saraceno C, Musardo S, Marcello E, Pelucchi S, Luca MD (2013) Modeling Alzheimer's disease: from past to future. Front Pharmacol 4:77

Satoh E, Okada M, Takadera T, Ohyashiki T (2005) Glutathione depletion promotes aluminum-mediated cell death of pc12 cells. Biol Pharm Bull 28:941–946

Saura CA, Choi SY, Beglopoulos V, Malkani S, Zhang D, Shankaranarayana Rao BS, Chattarji S, Kelleher RJ 3rd, Kandel ER, Duff K, Kirkwood A, Shen J (2004) Loss of presenilin function causes impairments of memory and synaptic plasticity followed by age-dependent neurodegeneration. Neuron 42:23–36

Savory J, Herman MM, Ghribi O (2006) Mechanisms of aluminum-induced neurodegeneration in animals: implications for Alzheimer's disease. J Alzheimers Dis 10:135–144

Schindowski K, Bretteville A, Leroy K, Bégard S, Brion JP, Hamdane M, Buée L (2006) Alzheimer's disease-like tau neuropathology leads to memory deficits and loss of functional synapses in a novel mutated tau transgenic mouse without any motor deficits. Am J Pathol 169:599–616

Schmitt FA, Davis DG, Wekstein DR, Smith CD, Ashford JW, Markesbery WR (2000) "Preclinical" AD revisited: neuropathology of cognitively normal older adults. Neurology 55:370–376

Selkoe DJ (2008) Biochemistry and molecular biology of amyloid beta-protein and the mechanism of Alzheimer's disease. Handb Clin Neurol 89:245–260

Sharma DR, Sunkaria A, Wani WY, Sharma RK, Kandimalla RJ, Bal A, Gill KD (2013) Aluminium induced oxidative stress results in decreased mitochondrial biogenesis via modulation of PGC-1α expression. Toxicol Appl Pharmacol 273:365–380

Shaw CA, Li Y, Tomljenovic L (2013) Administration of aluminium to neonatal mice in vaccine-relevant amounts is associated with adverse long term neurological outcomes. J Inorganic Biochem 128:237–244

Shen J, Bronson RT, Chen DF, Xia W, Selkoe DJ, Tonegawa S (1997) Skeletal and CNS defects in Presenilin-1-deficient mice. Cell 89:629–639

Siman R, Reaume AG, Savage MJ, Trusko S, Lin YG, Scott RW, Flood DG (2000) Presenilin-1 P264L knock-in mutation: differential effects on abeta production, amyloid deposition, and neuronal vulnerability. J Neurosci 20:8717–8726

Sleegers K, Brouwers N, Gijselinck I, Theuns J, Goossens D, Wauters J, Del-Favero J, Cruts M, van Duijn CM, Van Broeckhoven C (2006) APP duplication is sufficient to cause early onset Alzheimer's dementia with cerebral amyloid angiopathy. Brain 129:2977–2983

Small DH, Mok SS, Bornstein JC (2001) Alzheimer's disease and Abeta toxicity: from top to bottom. Nat Rev Neurosci 2:595–598

Smith GE, Pankratz VS, Negash S, Machulda MM, Petersen RC, Boeve BF, Knopman DS, Lucas JA, Ferman TJ, Graff-Radford N, Ivnik RJ (2007) A plateau in pre-Alzheimer memory decline: evidence for compensatory mechanisms? Neurology 69:133–139

Son SM, Song H, Byun J, Park KS, Jang HC, Park YJ, Jang HC, Park YJ, Mook-Jung I (2012) Accumulation of autophagosomes contributes to enhanced amyloidogenic APP processing under insulin-resistant conditions. Autophagy 8:1842–1844

Sparks DL, Scheff SW, Hunsaker JC III, Liu H, Landers T, Gross DR (1994) Induction of Alzheimer-like beta-amyloid immunoreactivity in the brains of rabbits with dietary cholesterol. Exp Neurol 126:88–94

Srinivasan K, Ramarao P (2007) Animal models in type 2 diabetes research: an overview. Indian J Med Res 125:451–472

Stephens PH, Tagari P, Cuello AC (1987) Ethylcholine mustard aziridinium ion lesions of the rat cortex result in retrograde degeneration of basal forebrain cholinergic neurons: implications for animal models of neurodegenerative disease. Neurochem Res 12:613–618

Su Q, Yang J, Zhang P (2009) Effect of meloxicam on cyclooxygenase 2 expression of chronic aluminium overload-induced nerve degeneration in rat hippocampus. Chin J Pharmacol Toxicol 23:1–6

Sutcliffe J, Hutcheson D (2012) Perspectives on the non-human primate touch-screen self ordered spatial search paradigm. In: Spink AJ, Grieco F, Krips OE, Loijens LWS, Noldus LPJJ, Zimmerman PH (eds) 8th International conference on methods and techniques in behavioral research. Measuring Behavior Conferences, Wageningen, pp 182–185

Suzuki T, Okumura-Noji K (1995) NMDA receptor subunits epsilon 1 (NR2A) and epsilon 2 (NR2B) are substrates for Fyn in the postsynaptic density fraction isolated from the rat brain. Biochem Biophys Res Commun 216:582–588

Suzuki A, Fukushima H, Mukawa T, Toyoda H, Wu LJ, Zhao MG, Xu H, Shang Y, Endoh K, Iwamoto T, Mamiya N, Okano E, Hasegawa S, Mercaldo V, Zhang Y, Maeda R, Ohta M, Josselyn SA, Zhuo M, Kida S (2011) Upregulation of CREB-mediated transcription enhances both short- and long-term memory. J Neurosci 31:8786–8802

Takeda S, Hashimoto T, Roe AD, Hori Y, Spires-Jones TL, Hyman BT (2013) Brain interstitial oligomeric amyloid β increases with age and is resistant to clearance from brain in a mouse model of Alzheimer's disease. FASEB J 20:3239–3248

Takei Y, Teng J, Harada A, Hirokawa N (2000) Defects in axonal elongation and neuronal migration in mice with disrupted tau and map1b genes. J Cell Biol 150:989–1000

Tanemura K, Murayama M, Akagi T, Hashikawa T, Tominaga T, Ichikawa M, Yamaguchi H, Takashima A (2002) Neurodegeneration with tau accumulation in a transgenic mouse expressing V337M human tau. J Neurosci 22:133–141

Tanzi RE, Bertram L (2005) Twenty years of the Alzheimer's disease amyloid hypothesis: a genetic perspective. Cell 120:545–555

Tesson L, Cozzi J, Ménoret S, Rémy S, Usal C, Fraichard A, Anegon I (2005) Transgenic modifications of the rat genome. Transgenic Res 14:531–546

Tiraboschi P, Sabbagh MN, Hansen LA, Salmon DP, Merdes A, Gamst A, Masliah E, Alford M, Thai LJ, Corey-Bloom J (2004) Alzheimer disease without neocortical neurofibrillary tangles: "a second look". Neurology 62:1141–1147

Toledana A, Álvarez MI (2010) Lesion-induced vertebrate models of Alzheimer dementia. In: De Deyn PP, Van Dam D (eds) Animal models of dementia. Springer Science + Business Media, New York, pp 295–345

Tomiyama T, Nagata T, Shimada H, Teraoka R, Fukushima A, Kanemitsu H, Takuma H, Kuwano R, Imagawa M, Ataka S, Wada Y, Yoshioka E, Nishizaki T, Watanabe Y, Mori H (2008) A new amyloid beta variant favoring oligomerization in Alzheimer's-type dementia. Ann Neurol 63:377–387

Tomiyama T, Matsuyama S, Iso H, Umeda T, Takuma H, Ohnishi K, Ishibashi K, Teraoka R, Sakama N, Yamashita T, Nishitsuji K, Ito K, Shimada H, Lambert MP, Klein WL, Mori H (2010) A mouse model of amyloid beta oligomers: their contribution to synaptic alteration, abnormal tau phosphorylation, glial activation, and neuronal loss in vivo. J Neurosci 30:4845–4856

Tong L, Thornton PL, Balazs R, Cotman CW (2001) β-Amyloid-(1–42) impairs activity-dependent cAMP response element-binding protein signaling in neurons at concentrations in which cell is not compromised. J Biol Chem 276:17301–17306

Townsend M, Shankar GM, Mehta T, Walsh DM, Selkoe DJ (2006) Effects of secreted oligomers of amyloid beta-protein on hippocampal synaptic plasticity: a potent role for trimers. J Physiol 572:477–492

Tran TN, Kim SH, Gallo C, Amaya M, Kyees J, Narayanaswami V (2013) Biochemical and biophysical characterization of recombinant rat apolipoprotein E: similarities to human apolipoprotein E3. Arch Biochem Biophys 529:18–25

Tuppo EE, Arias HR (2005) The role of inflammation in Alzheimer's disease. Int J Biochem Cell Biol 37:289–305

Velasco PT, Heffern MC, Sebollela A, Popova IA, Lacor PN, Lee KB, Sun X, Tiano BN, Viola KL, Eckermann AL, Meade TJ, Klein WL (2012) Synapse-binding subpopulations of Aβ oligomers sensitive to peptide assembly blockers and scFv antibodies. ACS Chem Neurosci 3:972–981

Vercauteren FG, Clerens S, Roy L, Hamel N, Arckens L, Vandesande F, Alhonen L, Janne J, Szyf M, Cuello AC (2004) Early dysregulation of hippocampal proteins in transgenic rats with Alzheimer's disease-linked mutations in amyloid precursor protein and presenilin 1. Brain Res Mol Brain Res 132:241–259

Viola KL, Klein WL (2015) Amyloid β oligomers in Alzheimer's disease pathogenesis, treatment and diagnosis. Acta Neuropathol 129:183–206

Walsh DM, Klyubin I, Fadeeva JV, Cullen WK, Anwyl R, Wolfe MS, Rowan MJ, Selkoe DJ (2002) Naturally secreted oligomers of amyloid beta protein potently inhibit hippocampal long-term potentiation in vivo. Nature 416:535–539

Walton JR (2013) Aluminum involvement in the progression of Alzheimer's disease. J Alzheimers Dis 35:7–43

Walton JR, Wang MX (2009) APP expression, distribution and accumulation are altered by aluminum in a rodent model for Alzheimer's disease. J Inorg Biochem 103:1548–1554

Wang HW, Pasternak JF, Kuo H, Ristic H, Lambert MP, Chromy B, Viola KL, Klein WL, Stine WB, Krafft GA, Trommer BL (2002) Soluble oligomers of beta amyloid (1–42) inhibit long-term potentiation but not long-term depression in rat dentate gyrus. Brain Res 924:133–140

Wang H, Song L, Laird F, Wong PC, Lee HK (2008) BACE1 knock-outs display deficits in activity-dependent potentiation of synaptic transmission at mossy fiber to CA3 synapses in the hippocampus. J Neurosci 28:8677–8681

Warpman U, Alafuzoff I, Nordberg A (1993) Coupling of muscarinic receptors to GTP proteins in postmortem human brain – alterations in Alzheimer's disease. Neurosci Lett 150:39–43

Waterston RH, Lindblad-Toh K, Birney E (2002) Mouse Genome Sequencing Consortium, Initial sequencing and comparative analysis of the mouse genome. Nature 420:520–562

Welsh-Bohmer KA, White CL III (2009) Alzheimer disease: what changes in the brain cause dementia? Neurology 72, e21

Wenk GL, Willard LB (1998) The neural mechanisms underlying cholinergic cell death within the basal forebrain. Int J Dev Neurosci 16:729–735

Wenk GL, Hauss-Wegrzyniak B, Willard LB (2000) Pathological and biochemical studies of chronic neuroinflammation may lead to therapies for Alzheimer's disease in research and perspectives in neurosciences: neuroimmune neurodegenerative and psychiatric disorders. Springer, Heidelberg, pp 73–77

Whitehouse PJ, Price DL, Clark AW, Coyle JT, DeLong MR (1982) Alzheimer disease: evidence for selective loss of cholinergic neurons in the nucleus basalis. Ann Neurol 10:122–126

Wiedlocha M, Stańczykiewicz B, Jakubik M, Rymaszewska J (2012) [Selected mice models based on APP, MAPT and presenilin gene mutations in research on the pathogenesis of Alzheimer's disease]. Postepy Hig Med Dosw (Online) 66:415–430

Willard LB, Hauss-Wegrzyniak B, Wenk GL (1999) Pathological and biochemical consequences of acute and chronic neuroinflammation within the basal forebrain cholinergic system of rats. Neuroscience 88:193–200

Willem M, Garratt AN, Novak B, Citron M, Kaufmann S, Rittger A, DeStrooper B, Saftig P, Birchmeier C, Haass C (2006) Control of peripheral nerve myelination by the beta-secretase BACE1. Science 314:664–666

Wittmann CW, Wszolek MF, Shulman JM, Salvaterra PM, Lewis J, Hutton M, Feany MB (2001) Tauopathy in Drosophila: neurodegeneration without neurofibrillary tangles. Science 293:711–714

Wu Y, Luo Y (2005) Transgenic C. elegans as a model in Alzheimer's research. Curr Alzheimer Res 2:37–45

Yu L, Jiang R, Su Q, Yu H, Yang J (2014) Hippocampal neuronal metal ion imbalance related oxidative stress in a rat model of chronic aluminium exposure and neuroprotection of meloxicam. Behav Brain Funct 10:6

Yumoto S, Kakimi S, Ohsaki A, Ishikawa A (2009) Demonstration of aluminum in amyloid fibers in the cores of senile plaques in the brains of patients with Alzheimer's disease. J Inorg Biochem 103:1579–1584

Zatta P, Hegde ML, Bharathi P, Suram A, Venugopal C, Jagannathan R, Poddar P, Srinivas P, Sambamurti K, Rao KJ, Scancar J, Messori L, Zecca L (2009) Challenges associated with metal chelation therapy in Alzheimer's disease. J Alzheimers Dis 17:457–468

Zhang C, Khandelwal PJ, Chakraborty R, Cuellar TL, Sarangi S, Patel SA, Cosentino CP, O'Connor M, Lee JC, Tanzi RE, Saunders AJ (2007) An AICD-based functional screen to identify APP metabolism regulators. Mol Neurodegener 2:15

Zhang YW, Thompson R, Zhang H, Xu H (2011) APP processing in Alzheimer's disease. Mol Brain 4:3

Zhang H, Ma Q, Zhang YW, Xu H (2012) Proteolytic processing of Alzheimer's β-amyloid precursor protein. J Neurochem 120(Suppl 1):9–21

Zhang QL, Jia L, Jiao X, Guo WL, Ji JW, Yang HL, Niu Q (2012b) APP/PS1 transgenic mice treated with aluminum: an update of Alzheimer's disease model. Int J Immunopathol Pharmacol 25:49–58

Zhao Y, Bhattacharjee S, Jones BM, Hill J, Dua P, Lukiw WJ (2014) Regulation of neurotropic signaling by the inducible, NF-kB-sensitive miRNA-125b in Alzheimer's disease (AD) and in primary human neuronal-glial (HNG) cells. Mol Neurobiol 50:97–106

Zheng H, Jiang M, Trumbauer ME, Hopkins R, Sirinathsinghji DJ, Stevens KA, Conner MW, Slunt HH, Sisodia SS, Chen HY, Van der Ploeg LH (1996) Mice deficient for the amyloid precursor protein gene. Ann N Y Acad Sci 777:421–426

Zilka N, Filipcik P, Koson P, Fialova L, Skrabana R, Zilkova M, Rolkova G, Kontsekova E, Novak M (2006) Truncated tau from sporadic Alzheimer's disease suffices to drive neurofibrillary degeneration in vivo. FEBS Lett 580:3582–3588

Chapter 3
Metabolism, Bioavailability, Biochemical Effects of Curcumin in Visceral Organs and the Brain

3.1 Introduction

Curcumin ($C_{21}H_{20}O_6$) or diferuloylmethane (bis-α,β-unsaturated β-diketone) is a hydrophobic polyphenolic compound (mol mass of 368.38) present in the Indian spice turmeric (curry powder). It is derived from the rhizomes of *Curcuma longa*, which belongs to family Zingiberaceae (Anand et al. 2008a). It has been used for centuries in Chinese traditional medicine and Indian medicine (Ayurvedic medicine) systems as a nociceptive, anti-inflammatory, and antishock drug to relieve pain and inflammation in muscles (Anand et al. 2008a) and for the treatment of many pathological conditions, such as rheumatism, digestive and inflammatory disorders, intermittent fevers, urinary discharges, leukoderma and amenorrhoea as part of traditional medicine (Anand et al. 2008a). Turmeric products have been declared as safe not only by the FDA (http://www.accessdata.fda.gov/scripts/fcn/gras_notices/GRN000460.pdf) in the USA, and the Natural Health Products Directorate of Canada, but also by the Joint Expert Committee of the Food and Agriculture Organization/World Health Organization (FAO/WHO) (National Cancer Institute 1996). Thus, curcumin is a safe and non-toxic compound, which exhibits a wide range of pharmacological activities, such as anti-inflammatory, antioxidant, anticarcinogenic, antimutagenic, anticoagulant, antifertility, antidiabetic, antibacterial, antifungal, antiprotozoal, antiviral, antifibrotic, antivenom, antiulcer, hypotensive and hypocholesteremic activities (Anand et al. 2008a; Aggarwal et al. 2014).

The chemical structure of curcumin consists of two methoxyl groups, two phenolic hydroxyl groups, and three double conjugated bonds. The two aryl rings containing ortho-methoxy phenolic OH^- groups are symmetrically linked to a β-diketone moiety.

The presence of intramolecular hydrogen atoms transfer at the β-diketone chain of curcumin results in the existence of keto and enol tautomeric conformations in equilibrium (Fig. 3.1). Keto-enol tautomers of curcumin also exist in several *cis* and *trans* forms. The relative concentrations of *cis* and *trans* forms vary according to

© Springer International Publishing Switzerland 2016

A.A. Farooqui, *Therapeutic Potentials of Curcumin for Alzheimer Disease*,

DOI 10.1007/978-3-319-15889-1_3

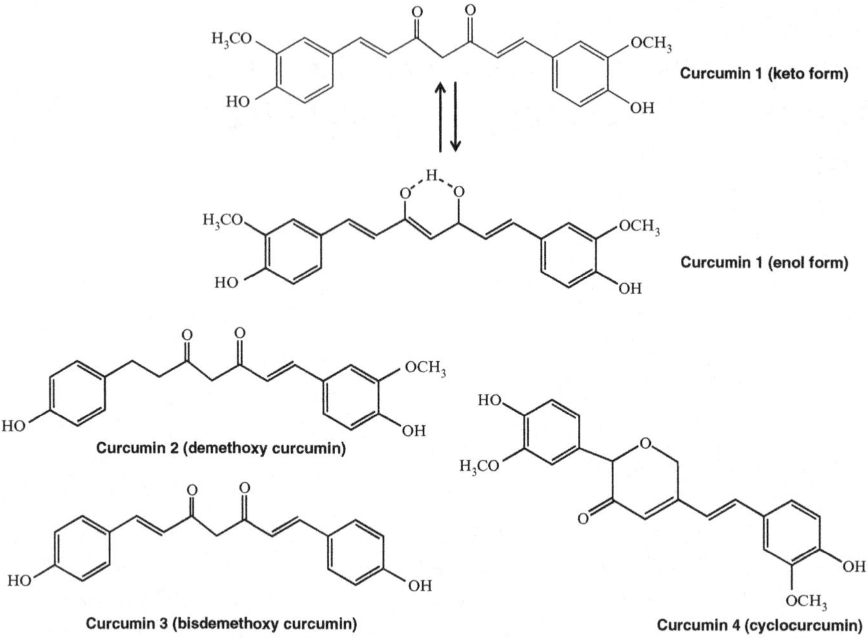

Fig. 3.1 Chemical structures of tautomers of curcumin along with curcumin 2 (methoxycurcumin), and curcumin 3 (demethoxycurcumin)

temperature, polarity of solvent, pH and substitution of the aromatic rings (Cornago et al. 2008; Bertolasi et al. 2008). Thus, a predominant keto form occurs in acidic and neutral solutions and as a stable enol form, which occurs in alkaline media. The amount of keto-enol-enolate of the heptadienone moiety in equilibrium plays a crucial role in the physicochemical properties of curcumin. Chemical structural features of curcumin dictate the biochemical and biophysical activities of curcumin. The o-methoxyphenol group and methylenic hydrogen are responsible for the anti-oxidant activity of curcumin, and curcumin donates an electron/hydrogen atom to reactive oxygen species. Curcumin interacts with a number of biomolecules through non-covalent and covalent binding. As stated above, the hydrogen bonding and hydrophobicity of curcumin, arising from the aromatic and tautomeric structures along with the flexibility of the linker group are responsible for the non-covalent interactions. The α,β-unsaturated β-diketone moiety covalently binds with protein thiols, through Michael reaction (Priyadarsini 2013). The β-diketo group is capable of chelating transition metals, thereby reducing the metal induced toxicity and some of the metal complexes exhibit improved antioxidant activity as enzyme mimics (Priyadarsini 2013).

The alcoholic extract of turmeric contains three curcuminoids, viz., curcumin (also referred as curcumin I, 77 %), desmethoxycurcumin (curcumin II, 17 %), and bisdesmethoxycurcumin (curcumin III, 3 %) (Fig. 3.1). In addition, turmeric also contains volatile oils (natlantone, tumerone and zingiberone), proteins, sugars and

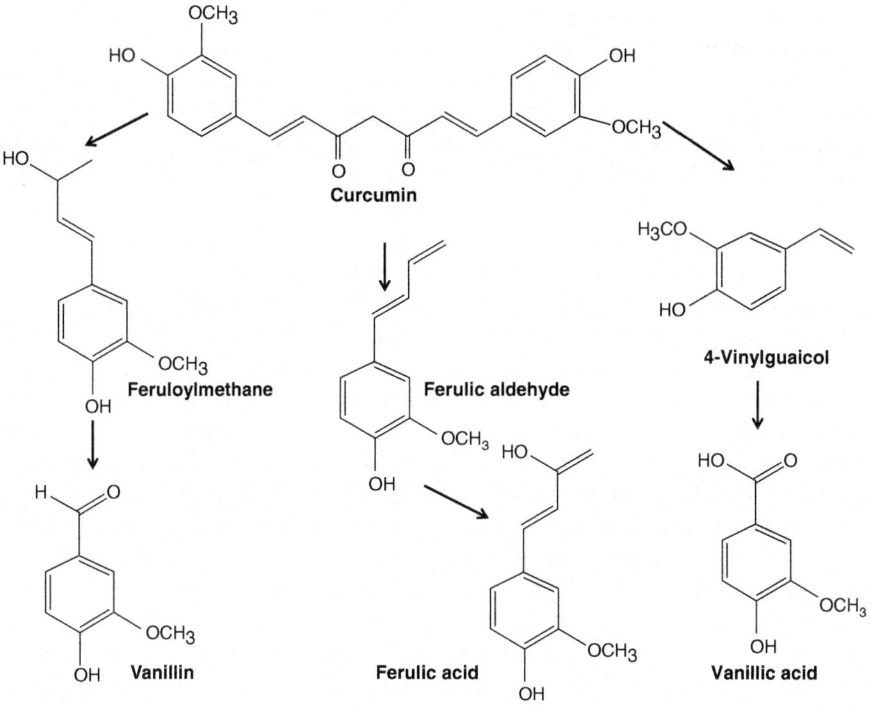

Fig. 3.2 Degradation products of curcumin under physiological conditions

resins. Curcumin is insoluble in water. It is readily soluble in organic solvents such as dimethylsulfoxide, acetone, and ethanol. More than 90 % decomposition of curcumin occurs in a serum free 0.1 M phosphate buffer (pH 7.2) at 37 °C. The degraded products have been identified as trans-6-(4′-hydroxy-3′-methoxyphenyl)-2,4,-dioxo-5-hexenal (major products), vanillin, vanillic acid, ferulic acid (4-hydroxy-3-methoxy cinnamic acid) and feruloylmethane using high performance liquid chromatography (HPLC) and gas chromatography mass spectrophotometry (GC-MS) analyses (Wang et al. 1997) (Fig. 3.2). Among degradation products of curcumin, ferulic acid possesses three distinctive structural motifs, which contribute to its free radical scavenging activity. The presence of electron donating groups on the benzene ring (3 methoxy and more importantly 4-hydroxyl) of ferulic acid gives the additional property of terminating free radical chain reactions. Furthermore the functionality-the carboxylic acid group in ferulic acid with an adjacent unsaturated C–C double bond provides additional attack sites for free radicals and thus preventing them from attacking the membrane. In addition, the carboxylic acid group in ferulic acid also acts as an anchor for binding with the lipid bilayer and providing some protection against lipid peroxidation (Srinivasan et al. 2007; Kanski et al. 2002). The addition of ascorbic acid, *N*-acetylcysteine and glutathione slows the decomposition of curcumin in both cell culture medium and 0.1 M phosphate buffer under alkaline conditions (Wang et al. 1997). The complexation of curcumin with cyclodextrin improves its solubility to more than 10,000 folds in aqueous solution at pH 5 (Tomren et al. 2007).

At neuropharmacological level, the analgesic effects of curcumin are mediated by the suppression of nitrite, TNF-α, and capsaicin-induced TRPV1 activity, and through the descending noradrenergic and serotonergic systems (Yeon et al. 2010; Zhao et al. 2012). Antiinflammatory and antioxidant activities of curcumin are mediated by the inhibition of cytosolic phospholipase A_2 (cPLA$_2$), cyclooxygenase 2 (COX-2), 5-lipoxygenase (5-LOX) glutathione S-transferases, down-regulation of inflammatory transcription factors, and upregulation of heme oxygenase-1 respectively (Jeong et al. 2009). Curcumin also interacts and modulates activities of DNA (cytosine-5)-methyltransferase-1, inducible nitric oxide synthase (iNOS), thioredoxin reductase, and protein kinases (such as protein kinase C, mammalian target of rapamycin, mitogen-activated protein kinases, and Akt). Many of above mentioned enzymes contribute not only to its ability to interfere with multiple signaling cascades, such as cell cycle regulators, apoptotic proteins, pro-inflammatory cytokines, but also to its modulatory effects on proliferative regulators and transcription factors such as nuclear factor-kappa B (NF-κB), Nrf2, Stat3, tumor necrosis factor-α, forkhead box O3a, transcription factor activator protein-1 (AP-1), and CRAC (Shishodia et al. 2007; Shishodia 2013). It also inhibits secretion of interleukins and cytokines as well as an expression of human epidermal growth factor receptor (EGFR) (Chen et al. 2006). Curcumin also inhibits the development of cancer cells and tumor growth, suppresses proliferation, and blocks angiogenesis and inflammation (Shehzad and Lee 2013).

3.2   Bioavailability of Curcumin and Its Analogs in Visceral Organs and Brain

Curcumin can be delivered to the body through oral, nasal, intraperitoneal, and intravenous injections. Several studies have indicated that curcumin is remarkably well tolerated, but bioavailability of curcumin is very low (Anand et al. 2008a; Goel et al. 2008; McClure et al. 2015). There are three major reasons for the low bioavailability: (a) its poor absorption, (b) its rapid metabolism, and (c) its rapid systemic elimination. In addition, curcumin has a short biological half life in circulation and visceral organs (Anand et al. 2008a; Bisht and Maitra 2009). In rodents, curcumin undergoes rapid metabolism by conjugation and reduction, and its disposition after oral dosing. However, information about comprehensive pharmacokinetic data is not available. It is reported that 10 mg/kg of curcumin given intravenously to rats yielded a maximum serum level of 0.36 ± 0.05 µg/mL, whereas 500 mg/kg of curcumin administered orally only yielded a 0.06 ± 0.01 µg/mL maximum serum level (Yang et al. 2007). Furthermore, oral administration of curcumin at a dose of 2 g/kg results in a maximum serum concentration of 1.35 ± 0.23 µg/mL at time 0.83 h, while in humans the dose of 2 g of curcumin results in either undetectable or very low (0.006 ± 0.005 µg/mL at 1 h) serum levels (Shoba et al. 1998). Very little information is available on pharmacokinetic of curcumin in humans. First phase I and II clinical trial of curcumin have been performed in patients with advanced colorectal cancer for up to 4 months at several doses (500, 1000, 2000, 4000, 8000, and 12,000 mg/day) without any toxicity (Sharma et al. 2001; Cheng et al. 2001). The serum concentration of curcumin usually peaks at 1–2 h after oral intake of curcumin and gradually declines within 12 h. The average peak serum concentrations after taking 4000 mg, 6000 mg and 8000 mg of curcumin were 0.51 ± 0.11 µM, 0.63 ± 0.06 µM and 1.77 ± 1.87 µM, respectively. In this study serum levels of curcumin peak at one and two hours post-administration declined rapidly. So far, an upper level of toxicity has not been established for curcumin. Studies have shown that a dosage as high as 12 g/day is safe and tolerable to humans with a few reporting mild side-effects (Goel et al. 2008; Jiao et al. 2009). Studies on curcumin metabolism in animals indicate that it is rapidly metabolized rapidly through glucuronidation (Fig. 3.3) and sulfation or it is reduced to hexahydrocurcumin in

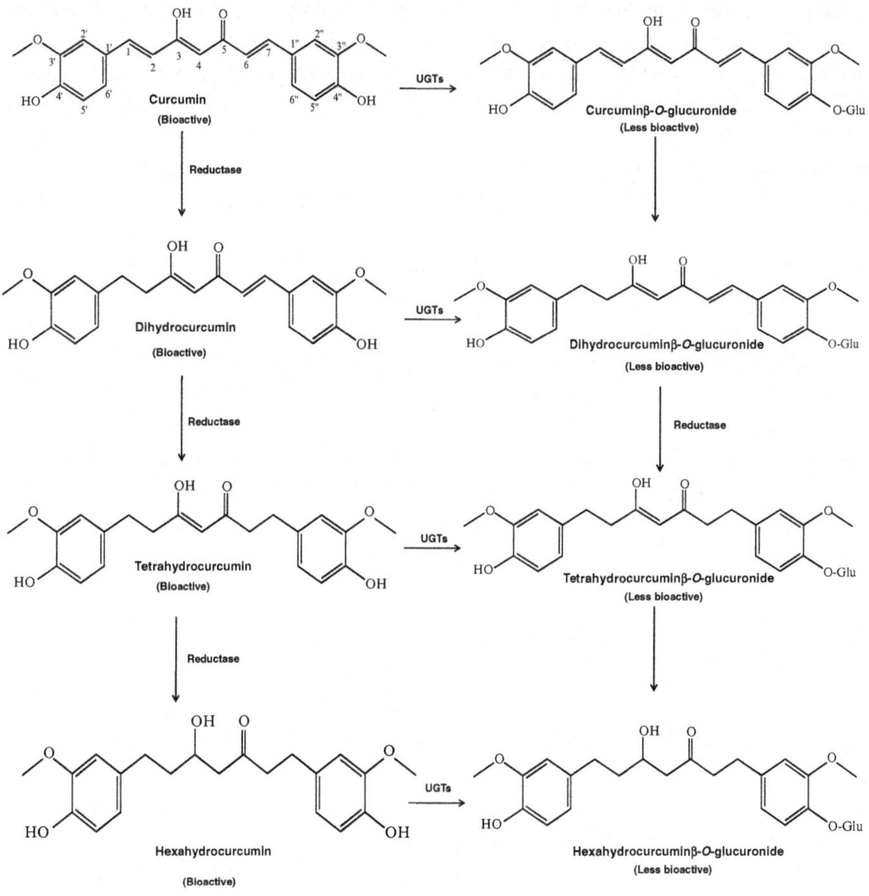

Fig. 3.3 Detoxification products of curcumin

liver, and intestine (Wahlstrom and Blennow 1978). Glucuronidation and sulfation of curcumin is catalyzed by UDP-Glucuronosyltransferases (UGTs) and sulfotransferases in liver and intestine (Hoehle et al. 2007; Asai and Miyazawa 2000).

3.3   Different Approaches for the Delivery of Curcumin to the Brain

It is well known that the brain and spinal cord are protected by the blood–brain barrier (BBB), a protective mechanism that controls cerebral homeostasis and provides the central nervous system with unique protection against all foreign matter (Roney et al. 2005). The BBB retards the entry of 98 % of small molecules and 100 % of large molecules in the brain. It is located at the level of the capillaries between the blood and cerebral tissue, and is characterized by the presence of tight intracellular junctions and polarized expression of many transport systems (Weiss et al. 2009). BBB reduces the penetration of curcumin. In this case, the use of certain delivery systems (formulations) may prove particularly effective. The bioavailability of curcumin can be enhanced by numerous approaches including the use of adjuvant like piperine (Shoba et al. 1998; Suresh and Srinivasan 2010), use of structural analogs of curcumin (e.g., FLL32, EF-24, EF4, FMeC1 and FMeC2) (Fig. 3.4) (Thomas et al. 2008; Olivera et al. 2012; Yanagisawa et al. 2015), use of chemical complexation agents, such as cyclodextrin (Tønnesen et al. 2002), phospholipid (Liu et al. 2006), polysaccharides, and protein (Todd 1991) and preparation of liposomes and nanoparticles (Ghalandarlaki et al. 2014; Cheng et al. 2013a). These approaches have not only resulted in longer circulation and increase in the cellular permeability, but also resistance against metabolic processing leading to efficient delivery to tissues and higher half life in the circulation. Some of these approaches have also allowed curcumin to penetrate the blood-brain barrier (BBB) effectively (Shoba et al. 1998; Suresh and Srinivasan 2010; Yanagisawa et al. 2011; Ghalandarlaki et al. 2014). The delivery of curcumin to the brain can enhance the adult hippocampus neurogenesis in the dentate gyrus region of hippocampus.

3.3.1   *Piperine and Curcumin Delivery*

As stated above, curcumin has a poor absorption rate and undergoes rapid metabolism which severely curtails its bioavailability. Piperine, a major alkaloidal constituent of black pepper (*Piper nigrum*), is a powerful inhibitor of hepatic and intestinal glucuronidation. Several studies have indicated that piperine increases the bioavailability and bioefficacy of curcumin by inhibiting glucuronidation (Atal et al. 1985; Shoba et al. 1998). In addition, administration of piperine and curcumin

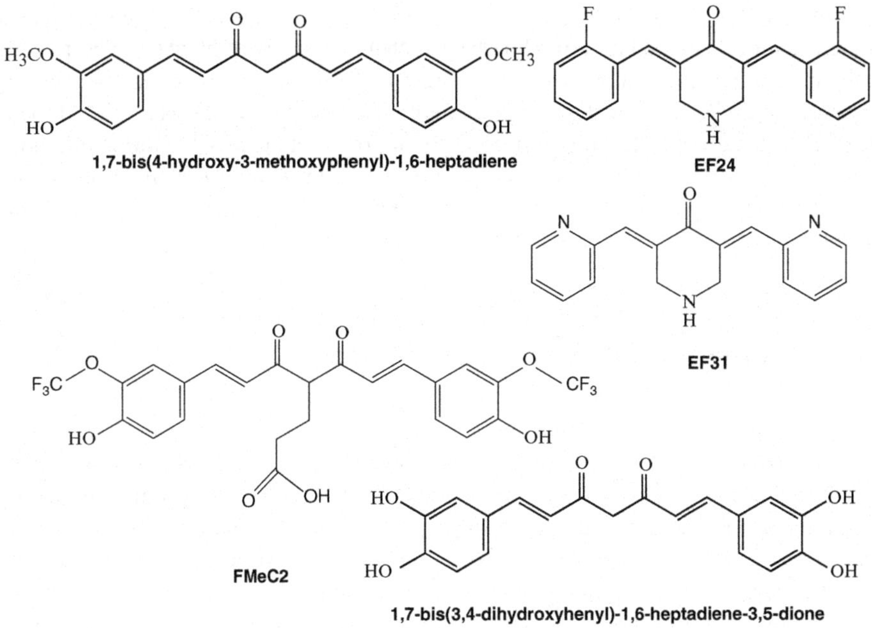

Fig. 3.4 Chemical structures of structural analogs of curcumin

also protect against the chronic unpredictable stress-induced cognitive impairment and associated oxidative damage in mice (Rinwa and Kumar 2012; Sehgal et al. 2012). Thus, co-administration of 2 g/kg curcumin and 20 mg/kg piperine to rats results in higher serum curcumin concentrations at 1 and 2 h after the oral administration. Pharmacokinetic analysis have indicated that piperine not only increases Tmax (time to reach the maximum serum concentration), but also increases half-life and total clearance of curcumin leading to increase in bioavailability (154 %) (Shoba et al. 1998; Suresh and Srinivasan 2010). In humans, the bioavailability of curcumin is increased by 2000 % in 45 min after the co-administration with piperine (Arcaro et al. 2014). Based on above data it is suggested that piperine increases the bioavailability of curcumin not only through the inhibition of glucuronidation process (Anand et al. 2007), but also through the efflux transport (Singh et al. 2013).

Administration of curcumin phospholipid conjugates show higher absorption and better activity without any adverse complications (Marczylo et al. 2007; Begum et al. 2008; Gota et al. 2010). Curcumin phospholipid conjugates enhance the adult hippocampus neurogenesis by increasing the number of newly generated cells in the dentate gyrus region of hippocampus (Kim et al. 2008). Another formulation called NanoCurc™ is also developed recently (Ray et al. 2011). This formulation (a polymeric nanoparticle encapsulated curcumin) is completely water soluble. Treatment of neuronally differentiated human SK-N-SH cells with NanoCurc™ protects against ROS (H₂O₂) mediated insults. In vivo, intraperitoneal (IP) injections of

NanoCurc™ at a dose of 25 mg/kg twice daily in athymic mice produce significant curcumin levels in the brain (0.32 µg/g). Biochemical study of NanoCurc™-treated athymic mice not only show reduction in levels of H_2O_2, but also decrease in caspase 3 and caspase 7 activities in the brain, along with elevation in levels of glutathione (GSH). Increase in reduced to oxidized glutathione (GSH:GSSH) ratio in athymic mice brain versus control brain indicates a favorable redox intracellular environment. These results support the view that NanoCurc™ may represent an optimized formulation to deliver the curcumin to the brain (Ray et al. 2011).

3.3.2 Nanocarriers for Curcumin Delivery

Nanocarriers are small size (typically 10e100 nm), which are commonly used for the targeted drug delivery (Torchilin 2009; Lammers et al. 2011). Nanocarriers may not only improve the circulation time of the loaded drug, but also promote its accumulation at the pathological site exploiting the so-called 'enhance permeation and retention (EPR) effect' (Torchilin 2012). During the last decades, various types of nanocarriers, such as polymeric micelles and nanoparticles, liposomes, conjugates, and peptide carriers have been developed for drug delivery/targeting and some systems have reached clinical evaluations and applications (Svenson 2012; Duncan and Gaspar 2011).

To improve the bioavailability of curcumin, Poly (lactic-*co*-glycolic acid) (PLGA) encapsulated curcumin has been developed (Fig. 3.5). PLGA can be hydrolyzed in the body to yield biodegradable lactic acid and glycolic acid. Thus, PLGA microparticles and nanoparticles provide a biodegradable carrier platform for sustained release of curcumin with improved bio-availability (Anand et al. 2010; Chan et al. 2009; Yang et al. 2012). Originally approved by FDA for the use in USA, PLGA has been explored as a drug delivery carrier in many oncologic, ophthalmologic, and other medical applications (Anand et al. 2010; Chan et al. 2009; Cohen-Sela et al. 2009). In vitro studies have shown that PLGA-curcumin has very rapid and more efficient cellular uptake than aqueous curcumin. Intravenous administration of either curcumin or PLGA-curcumin (2.5 mg/kg), exhibits almost two to four times as high serum concentration of PLGA-curcumin than curcumin in visceral tissues (Anand et al. 2010; Khalil et al. 2013). For delivery of curcumin into the brain PLGA nanoparticles with Tet-1 peptide have been used (Mathew et al. 2012). It is shown that curcumin encapsulated-PLGA nanoparticles are able to enter brain, destroy amyloid aggregates, and exhibit anti-oxidative property. PLGA nanoparticles are non-cytotoxic. Curcumin-encapsulated PLGA nanoparticles also induce neural stem cells NSC proliferation and neuronal differentiation in vitro, and in the hippocampus and subventricular zone of adult rats, curcumin-encapsulated PLGA nanoparticles stimulate neurogenesis. Detailed investigations have shown that curcumin-encapsulated PLGA nanoparticles significantly increase expression of genes involved in cell proliferation (reelin, nestin, and Pax6) and neuronal differentiation (neurogenin, neuroD1, neuregulin, neuroligin,

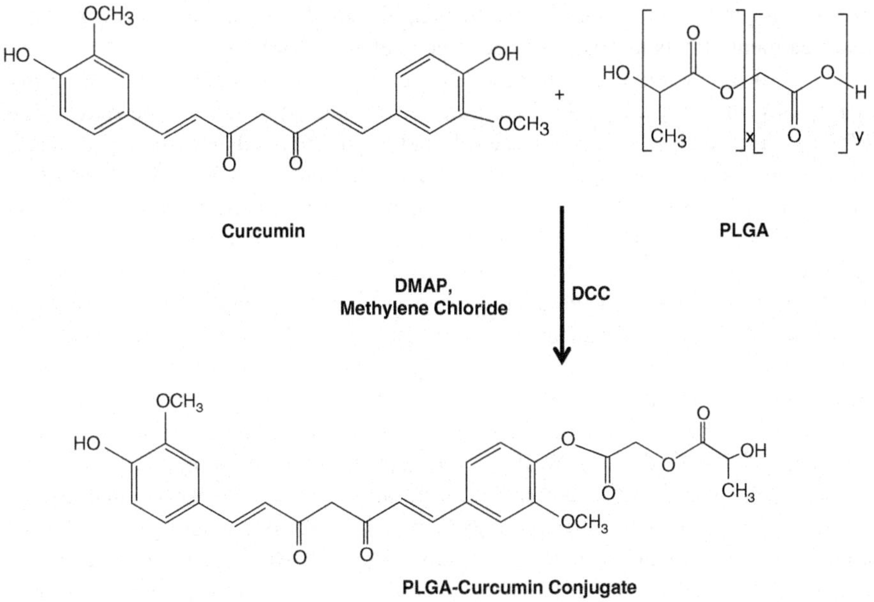

Fig. 3.5 Reactions showing the synthesis of curcumin PLGA conjugate

and Stat3). Curcumin nanoparticles increase neuronal differentiation by activating the Wnt/β-catenin pathway, involved in the regulation of neurogenesis (Tiwari et al. 2014). Collectively, these studies support the view that the encapsulation of curcumin in PLGA does not destroy its inherent properties and so, the PLGA-curcumin nanoparticles can be used for the treatment of neurological disorders (Mathew et al. 2012; Doggui et al. 2012; Tiwari et al. 2014). However, this way of encapsulation suffers from several disadvantages, such as the low encapsulation efficiency, the broad particle size distribution, and the process-induced denaturation of the drug cargos.

The latest studies on nanoparticle technology in AD model Tg2576 mice have indicated that the bioavailability of curcumin, particle size, and stability can be increased by preparing nanoparticles using polyethylene glycol-polylactic acid co-block polymer, and polyvinylpyrrolidone in a multi-inlet vortex mixer, followed by freeze drying with β-cyclodextrin (Cheng et al. 2013a). This nanocurcumin powder, unformulated curcumin, or placebo is orally administered to Tg2576 mice for 3 months. Before and after treatment, studies on memory are performed by radial arm maze and contextual fear conditioning tests. Treatment with nanocurcumin results in significantly better memory in the contextual fear conditioning test than placebo and tendencies toward better working memory in the radial arm maze test than ordinary curcumin (Cheng et al. 2013a). Nanocurcumin treatment also results in significantly higher curcumin concentration in plasma and six times higher area under the curve and mean residence time in brain than ordinary curcumin indicating that this novel nanocurcumin formulation produce highly stabilized nanoparticles with positive treatment effects in Tg2576 mice (Cheng et al. 2013a).

Recently a new microparticle-based delivery system is designed curcumin using poly(propylene sulfide) (PPS). Curcumin-loaded microspheres are prepared using PPS to target and scavenge intracellular ROS in activated macrophages. These microspheres reduce not only in vitro cell death caused by cytotoxic levels of ROS, but also decrease tissue-level ROS in vivo in the diabetic mouse hind limb ischemia model of peripheral arterial disease (Poole et al. 2015). Interestingly, due to the ROS scavenging properties of PPS, the microparticles without curcumin (Control or blank) also show inherent therapeutic properties leading to a synergistic effect in these assays. Functionally, local delivery of curcumin-PPS microspheres accelerate recovery from hind limb ischemia in diabetic mice, as demonstrated using non-invasive imaging techniques (Poole et al. 2015). These studies suggest curcumin-loaded microspheres can be used for improving local curcumin bioavailability for treatment of chronic diseases with oxidative stress and neuroinflammation.

Liposomes consisting of bovine brain sphingomyelin, cholesterol, and 1,2-stearoyl-sn-glycero-3-phosphoethanolamine-N-[maleimide(poly(ethylene glycol)-2000)] and surface functionalized with the apolipoprotein E (ApoE) peptide as targeting ligand can be loaded with curcumin (Re et al. 2011). In another study, a cationic liposome-polyethylene glycol (PEG)epolyethylenimine (PEI) complex (lipoePEGe-PEI complex, LPPC) has been used for the encapsulation of curcumin (Lin et al. 2012). These curcumin loaded liposomes have been used for the treatment of various types of cancers (Prasad et al. 2014). Liposomes have been bifunctionalized with a peptide derived from the apolipoprotein-E receptor-binding domain for BBB targeting and with phosphatidic acid for Aβ binding (Balducci et al. 2014). Electron microscopic studies have indicated that bifunctionalized liposomes have ability to hinder the aggregation of Aβ in vitro (EM experiments). Administration of bifunctionalized liposomes to APP/presenilin 1 transgenic mice (aged 10 months) for 3 weeks (three injections per week) reduces total brain-insoluble Aβ1-42 (−33 %) as assessed by ELISA, and the number and total area of plaques (−34 %) detected by histological procedures. Furthermore, brain Aβ oligomers are also reduced (−70.5 %) as assessed by SDS-PAGE. Plaque reduction is confirmed in APP23 transgenic mice (aged 15 months) either histologically or by PET imaging with [^{11}C]Pittsburgh compound B (PIB). Most importantly liposomes can enter the brain in an intact form, as determined by confocal microscopy experiments with fluorescently labeled liposomes. These data support the view that bifunctionalized liposomes destabilize brain Aβ aggregates and promote peptide removal across the BBB (Balducci et al. 2014). This preparation can be loaded with curcumin and used for the treatment of AD.

Cyclodextrin (CD) has been also used for improving the bioavailability of curcumin. It is reported that encapsulated curcumin (CDC) not only has a greater cellular uptake and longer half-life in the cancer cells compared with free curcumin, but also is more active than free curcumin in inhibiting TNF-α-mediated activation of the inflammatory transcription factor NF-kappaB and in suppressing gene products regulated by NF-kappaB, including those involved in cell proliferation (cyclin

D1), invasion (MMP-9), and angiogenesis (VEGF) (Yadav et al. 2010). Furthermore, CDC is also more active than free curcumin in inducing the death receptors DR4 and DR5. Annexin V staining, cleavage of caspase-3 and PARP, and DNA fragmentation also indicates that CDC is more potent than free curcumin in inducing apoptosis of leukemic cells (Yadav et al. 2010). Oral administration of curcumin in CD containing nanoparticles and unformulated curcumin to Tg2576 mice for 3 months indicates that CD containing nanoparticles treated mice performed better on radial arm maze and contextual fear conditioning tests (Cheng et al. 2013b). Administration of nanocurcumin results in significantly higher curcumin concentration in plasma and better delivery to the brain of Tg2576 mice than unformulated curcumin (Cheng et al. 2013b).

Recently inhalable forms of curcumin analogs (perfluoro curcumin analog, FMeC1) have been developed. These analogs have been used for ^{19}F NMR imaging to study the binding of curcumin to Aβ (Yanagisawa et al. 2011). FMeC1 compound can be aerosolized using a center-flow atomizer, diluted with air and subsequently delivered by nose through inhalation (McClure et al. 2015). Preliminary data indicate that drug delivery using atomization approach alleviates toxicity concerns and efficiently deposits FMeC1 compound in the brain slightly better than intravenous injection. Notably, delivery of the FMeC1 compound reaches concentrations in the brain detectable by ^{19}F NMR (McClure et al. 2015). Using this procedure it is possible to examine the colocalize FMeCl with Aβ plaques in the cortex and hippocampal regions of the 5XFAD mouse brain under fluorescence microscopy (McClure et al. 2015).

3.4 Biochemical Activities and Targets of Curcumin Action

Curcumin mediates its effects through the modulation of transcription factors (NF-κB, AP-1, STAT, and Nrf2), inflammatory cytokines (TNF-α, IL-1β, and IL-6), enzymes (COX-2, LOX, MMP9, MAPK, mTOR, Akt), growth factors (VEGF, EGF, and FGF), growth factor receptors (EGFR, HER-2, and AR), adhesion molecules (ELAM-1, ICAM-1, VCAM-1), and apoptosis related proteins (Bcl-2, caspases, DR, Fas) (Goel et al. 2008; Zhou et al. 2011) (Fig. 3.6). These targets are associated with regulation of signal transduction processes involved in antioxidant, anti-inflammatory, anti-depressant like, antidiabetic, anti-hyperalgesic, anticarcinogenic, antimicrobial, hepatoprotective, thrombosuppressive, and antinociceptive effects of curcumin (Ghorbani et al. 2014; Zhu et al. 2014). It should be noted that most above mentioned activities of curcumin have been assigned to methoxy, hydroxyl, α,β-unsaturated carbonyl moiety or to diketone groups present in the curcumin structure. A major metabolite of curcumin is called as tetrahydrocurcumin (THC), which lacks α,β-unsaturated carbonyl moiety and is white in color. It differs with curcumin in its chemical and biochemical activities (Aggarwal et al. 2014). Some investigators have indicated that curcumin exhibits both pro-oxidant and anti-oxidant properties while THC mediates superior antioxidant activities than

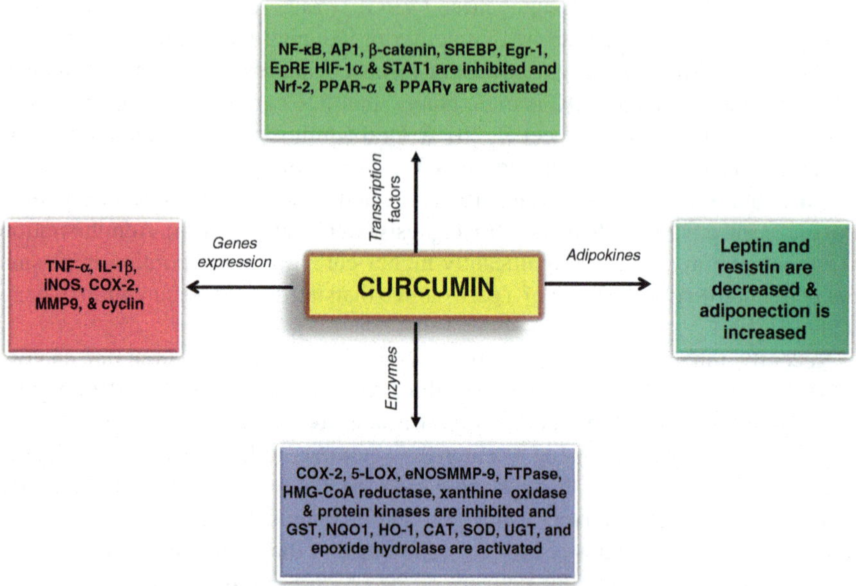

Fig. 3.6 Effect of curcumin on gene expression, adipokines, transcription factors and enzymes

curcumin (Aggarwal et al. 2014). Curcumin increases the expression of heme oxygenase-1 (HO-1) through the activation of the nuclear factor-erythroid-2-related factor 2 (Nrf2). In contrast, THC has no effect on HO-1 expression (Jeong et al. 2006, 2009; Farombi et al. 2008). Curcumin is reported to modulate the mobility of membranes, inhibits activities of many enzymes (DNA (cytosine-5)-methyltransferase-1, heme oxygenase-1, COX-2, inducible nitric oxide synthase, 5-lipoxygenase), transcription factors (Nrf2, β-catenin, NF-κB, and forkhead box O3a), and growth factors (vascular endothelial growth factor) (Anand et al. 2008a). In addition, curcumin also retards the formation of nitric oxide, and Aβ plaques, and ROS. Other investigators have reported that THC mediates higher antioxidant activity due to its ability to better activate GSH peroxidase, glutathione-S-transferase, NADPH: quinone reductase as well as superior free radical quenching activities than curcumin (Aggarwal et al. 2014).

3.4.1   Antioxidant Properties of Curcumin

Oxidative stress, a process that overwhelms the antioxidant defenses of the cells, is caused by over production of reactive oxygen species (ROS), such as superoxide anions ($O_2{}^{\cdot-}$), hydroxyl ($^{\cdot}OH$), alkoxyl ($RO^{\cdot}$), and peroxyl radicals ($ROO^{\cdot}$), and hydrogen peroxide (H_2O_2). These mediators contain unpaired electron. In general, $^{\cdot}OH$ is considered to be most active and harmful by-product of oxidative metabolism, which can cause molecular damage to neural and nonneural cells (Farooqui 2014). It shows an average lifetime of 10^{-9} s and can react with nearly every biomolecule such as nuclear DNA, mitochondrial DNA, proteins and membrane lipids. ˙OH can also be artificially generated by the Fenton-reaction Aoyagi et al. 2001).

Copper (Cu^{2+}) is an integral part of many important enzymes involved in a number of vital biological processes. Although normally bound to proteins, Cu may be released and become free to catalyze the formation of highly reactive ˙OH. In the brain, Cupric (Cu^{2+}) is an important metal ion, which is not only required for the formation and maintenance of myelin, but is also associated with the formation of melanin pigment in skin, hair, and eyes. Several enzymes, such as cytochrome c oxidase and copper, zinc-superoxide dismutase contain Cu^{2+}. Cytochrome c oxidase catalyzes the reduction of oxygen to water, an essential step in cellular respiration and copper, zinc-superoxide dismutase (Cu, Zn-SOD) scavenges the free radical superoxide. Non-specific Cu^{2+}-binding to thiol enzymes may modify the catalytic activities of cytochrome P450 monooxygenase. Cu^{2+} not only oxidizes, but can also bind to some amino acid residues of the P450 monooxygenase (Letelier et al. 2009). Additionally, Cu^{2+} is a constituent of dopamine-beta-hydroxylase, a critical enzyme in the catecholamine biosynthetic pathway (Uauy et al. 1998). Converging evidence therefore suggests that Cu-enzyme/protein-related malfunctions may contribute to the development of many visceral and neurological disorders (Gaetke et al. 2014). Cu^{2+} produces oxidative stress by catalyzing the production of ROS via a Fenton-like reaction (Liochev and Fridovich 2002). The cupric ions (Cu^{2+}), in the presence of superoxide anion radical or ascorbic acid or GSH are reduced to cuprous ion (Cu^{+}), which is capable of catalyzing the formation of reactive ˙OH through the decomposition of hydrogen peroxide (H_2O_2) via the Fenton reaction (Aruoma et al. 1991; Prousek 2007; Barbusinski 2009).

$$Cu^{2+} + O_2 \rightarrow Cu^{+} + O_2$$

$$Cu^{+} + H_2O_2 \rightarrow Cu^{2+} + \text{˙OH} + OH \quad \left(\text{Fenton reaction}\right)$$

Generation of ˙OH can ultimately leads to the oxidative damage to not only to proteins, but also to lipids and DNA.

Iron occurs in cells in ferrous (Fe^{2+}) and ferric (Fe^{3+}) forms. In biochemical reactions Fe^{2+} acts as an electron donor and Fe^{3+} acts as an electron acceptor. Iron contributes to the production of ˙OH and $O_2^{\cdot -}$ through the following reaction sequence.

$$Fe^{2+} + H_2O_2 \rightarrow Fe^{3+} + OH + \text{˙OH} \quad \left(\text{Fenton reaction}\right)$$

$$Fe^{3+} + O_2 \rightarrow Fe^{2+} + O_2^{\cdot -}$$

Net reaction

$$H_2O_2 + O_2^{\cdot -} \rightarrow OH + \text{˙OH} + O_2 \quad \left(\text{Haber-Weiss reaction}\right)$$

Under normal physiological conditions, the generation of ROS can be controlled and neutralized by a variety of antioxidant enzymes, such as superoxide dismutase (Monari et al. 2009), as well as several non-enzymatic molecules, such as glutathione and vitamins A, C and E (Martindale and Holbrook 2002). As stated above, the generation of low levels of ROS is necessary for normal cell signaling associated with protein phosphorylation (Rhee et al. 2000), response to growth factors (Bae et al. 1997), and activation of antigen-specific T-cell activation (Sena et al. 2013) and maintenance of hippocampal stem and progenitor cells (Dickinson et al. 2011). However, inactivity of antioxidant enzymes and deficiency of cellular antioxidants (glutathione and vitamins A, C and E) or an excess of ROS, the balance between oxidant and antioxidant is lost and the cell goes into a state of oxidative stress, which can disrupt normal cell signaling, as well as induce damage to cell organelles and the genome (Garcia-Garcia et al. 2012).

The mammalian brain is particularly sensitive to ROS-mediated oxidative damage not only due to the high oxygen consumption, but also due to the presence of high levels of unsaturated fatty acids, which are esterified at sn-2 position of glycerol moiety of neural membrane phospholipids. Major sources of ROS in brain are mitochondrial respiratory chain, uncontrolled arachidonic acid (ARA) cascade, and activation of NADPH oxidase (Sun et al. 2007). At their high concentrations, ROS react with different macromolecules, thereby causing damage to, DNA, proteins, and lipids (Stoner et al. 2008). An important feature of neurological diseases, which include neurotraumatic, neurodegenerative, and neuropsychiatric diseases is the induction of oxidative stress (Barnham et al. 2004; Maes et al. 2011; Farooqui 2010). At the molecular level, ROS not only damage neural cells through their oxidative effects, but also through the regulation of cellular redox, which plays an important role in the modulation of critical cellular functions (mainly in astrocytes and microglia), such as mitogen-activated protein (MAP) kinase cascade activation, ion transport, calcium mobilization, and apoptosis program activation (Emerit et al. 2004; Farooqui and Horrocks 2007; Farooqui 2010).

Curcumin is a bifunctional antioxidant, which not only scavenges free radicals ($O_2^{\cdot-}$, $\cdot OH$, and singlet oxygen), but also up-regulates various cytoprotective and antioxidant enzymes and proteins (Ak and Gülçin 2008; Barzegar and Moosavi-Movahedi 2011). Antioxidant properties of curcumin are strongly associated with the phenolic hydroxyls, which can capture or scavenge free radicals. Thus, treatment with curcumin increases the expression of cytoprotective proteins such as superoxide dismutase (SOD), catalase (CAT) (Panchal et al. 2008), glutathione reductase (GR), glutathione peroxidase (GPx) (Yarru et al. 2009), heme oxygenase 1 (HO-1) (Jeong et al. 2009), glutathione-S-transferase (GST), NAD(P)H: quinone oxidoreductase 1 (NQO1) (Fig. 3.7) (Ye et al. 2007) and γ-glutamylcysteine ligase (γGCL) (Rushworth et al. 2006). The molecular mechanism involved in the induction of cytoprotective proteins by curcumin is not fully understood. However, it is suggested that stimulation of cytoprotective protein by curcumin may involve the activation of nuclear factor erythroid-derived 2 (Nrf2) (Cuadrado et al. 2009; Rojo et al. 2012). Nrf2 is a member of Cap 'n' collar/basic-leucine zipper family transcription factor, which regulates the ARE containing genes. Nrf2 has six erythroid-

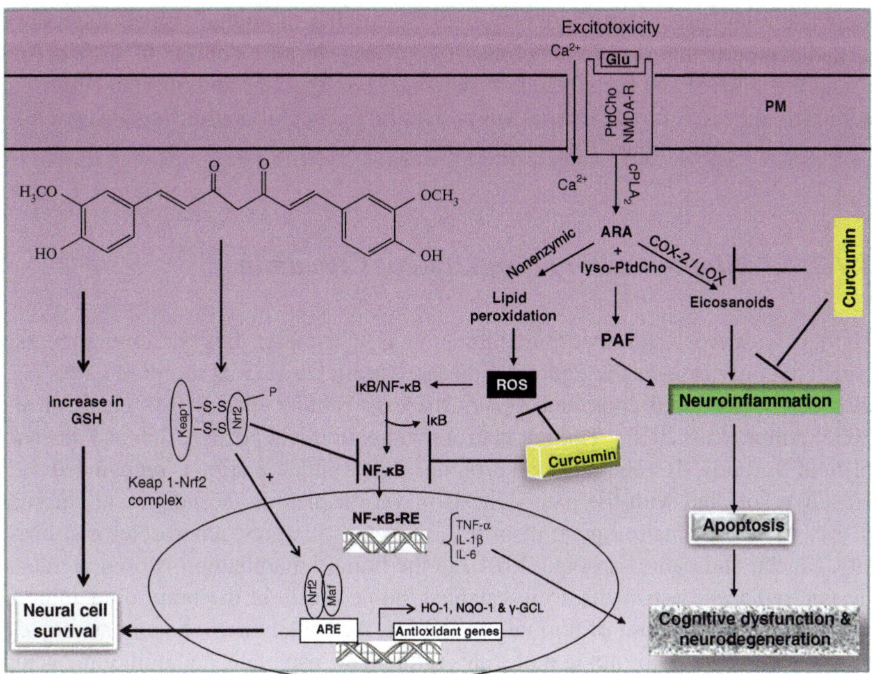

Fig. 3.7 Hypothetical diagram showing signal transduction processes associated with the effect of curcumin on Nrf2 and NF-κB. *N*-Methyl-D-aspartate (*NMDA-R*); glutamate (*Glu*); Phosphatidylcholine (*PtdCho*); cytosolic phospholipase A₂ (*cPLA₂*); cyclooxygenase-2 (*COX-2*); lipoxygenase (*LOX*); arachidonic acid (*ARA*); reactive oxygen species (*ROS*); nuclear factor-κB (*NF-κB*); nuclear factor-kappaB-response element (*NF-κB-RE*); inhibitory subunit of NF-κB (*I-κB*); tumor necrosis factor-α (*TNF-α*); interleukin-1β (*IL-1β*); interleukin-6 (*IL-6*); inducible nitric oxide synthase (*iNOS*); secretory phospholipase A₂ (*sPLA₂*); nuclear factor E2-related factor 2 (*Nrf2*); kelch-like ECH-associated protein 1 (*Keap1*); antioxidant response element (*ARE*); small leucine zipper proteins (*Maf*); heme oxygenase (*HO-1*); NADPH quinine oxidoreductase (*NQO-1*); γ-glutamate cystein ligase (*γ-GCL*)

derived CNC homology protein (ECH) domains. Under normal conditions, Nrf2 resides in the cytoplasm where it binds with the actin-binding protein, Kelch like ECH-associating protein 1 (Keap1), and is rapidly degraded (half life ~ 20 min) by ubiquitin-proteasome pathway. Keap 1 acts as negative regulator of Nrf2 (Joshi and Johnson 2012). Treatment with curcumin disrupts Nrf2-Keap 1 complex. Free Nrf2 is translocated into the nucleus, where it binds to small Maf proteins and increase the transcription rate of antioxidant response elements (ARE) driven cytoprotective genes, numerous protective enzymes and scavengers. Furthermore, curcumin is reported to increase the synthesis and concentration of reduced glutathione (GSH) in astrocytes and neurons by induction of γGCL (Lavoie et al. 2009). Recent studies have indicated that curcumin can reduce oxidative damage and cognitive deficits associated with aging (Shishodia et al. 2005). Studies of animal models have suggested that curcumin may be beneficial in neurodegenerative conditions such as AD (Yang et al. 2005) and focal cerebral ischemia (Thiyagarajan and Sharma 2004). In addition, the curcumin treatment can protect hippocampal neurons against excitotoxic and traumatic injury (Wu et al. 2006).

3.4.2 Antiinflammatory Properties of Curcumin

Unlike oxidative stress, neuroinflammation is a complex host defense process, which not only isolates the injured or diseased brain tissue from uninjured area, but also destroys injured cells and repairs the extracellular matrix (Minghetti et al. 2005; Amor et al. 2010). Though neuroinflammation is a positive defense mechanism of the body, dysregulated and prolonged neuroinflammation is recognized and closely associated with the pathogenesis of various neural and nonneural chronic diseases, such as neurological disorders, diabetes, allergies, atherosclerosis, obesity, cancer, and pain (Farooqui 2014). In the brain, inflammation is orchesterated by microglia and astrocytes to re-establish homeostasis in the brain after injury-mediated disequilibrium of normal physiology. Activated microglia and astrocytes release an array of neurotoxic molecules (glutamate, aspartate, and quinolinic acid, nitric oxide, and ROS), inflammatory cytokines and chemokines (TNF-α, interleukin (IL)-1β, and IL-6, intercellular adhesion molecule (ICAM)-5, and neural cell adhesion molecule (NCAM) (Dong and Benveniste 2001). These cytokines/chemokines stimulate phospholipases A_2 and COXs, which degrade neural membrane phospholipids and release of arachidonic acid and lyso-phosphatidylcholine. Arachidonic acid is oxidized to pro-inflammatory eicosanoid (prostaglandins, leukotrienes, and thromboxanes) by COXs and LOXs and lyso-phosphatidylcholine is utilized for the synthesis of pro-inflammatory platelet-activating factor. Three isoforms of COXs have been reported to occur in mammalian tissues. While COX-1 is involved in housekeeping functions and is constitutively and stably expressed in cells and in tissues, and COX-3 which appears expressed only in some specific tissues including brain and spinal cord (Phillis et al. 2006), COX-2 is normally low in most cells but is constitutively elevated in 80–90 % of colorectal and other cancers (Phillis et al. 2006). Eicosanoids, the oxygenated C18 to C22 compounds derived from n-6 fatty acids along with inflammatory cytokines and chemokines initiate, intensify, and maintain neuroinflammation. Resolution of inflammation involves the generation of IL-10, arachidonic acid-derived lipoxins, and docosahexaenoic acid-derived resolvins and neuroprotectins) (Farooqui 2010). Resolution of neuroinflammation is essential for the termination of the beneficial effects of neuroinflammation (Serhan and Savill 2005). Oxidative stress and neuroinflammation are interlinked with neurodegenerative processes in neurological disorders although it is difficult to establish the temporal sequence of their relationship. Proinflammatory transcription factor, NF-κB is redox sensitive. ROS stimulates NF-κB, which exists in the cytoplasm of neural cells as a heteromeric p50/p65 complexed with an inhibitory subunit, IκB. High oxidative stress promotes the dissociation NF-κB from NF-κB- IκB complex. This frees NF-κB and allows it to translocate to the nucleus, where it binds to genome through NF-κB response element, facilitating the expression of more than 200 target genes including genes for inflammatory cytokines resulting in the onset of neuroinflammation. Many of genes contribute to cell proliferation, invasion, metastasis, and chemoresistance. Therefore, ROS trigger the release of inflammatory cytokines, which through the activation of phospholipases A_2, COX-I2, and LOXs in turn enhance ROS production, thus establishing a vicious circle (Farooqui 2013).

Curcumin retards neuroinflammation not only by inhibiting p65 translocation to the nucleus and suppressing IκBα degradation in numerous cell types (Singh and Aggarwal 1995), but also by blocking COXs and LOXs and the uptake of arachidonic acid by macrophages. These processes limit the availability of arachidonic acid for eicosanoid production (Menon and Sudheer 2007; Rajasekaran 2011). By inhibiting NF-κB activation, curcumin suppresses the expression of various cell survival and proliferative genes, including Bcl-2, Bcl-xL, cyclin D1, interleukin (IL)-6, cyclooxygenase 2 (COX-2) and matrix metallopeptidase (MMP)-9, and subsequently arrests cell cycle, inhibits proliferation, and induces apoptosis (Aggarwal et al. 2004). In addition, Curcumin may also act by increasing the level and activity of proteins involved in antioxidative defense. Thus, curcumin enhances activities of protein kinases, and GST, CAT, SOD, UDP-glucuronosyl transferase (UGT), HO-1, and sirtuins. Activation of these enzymes not only contributes to neuroprotection, but is also essential for homeostasis in the vascular system (Maines 2000; Yang et al. 2013).

3.4.3 Anti-Excitotoxic Activities of Curcumin

Excitotoxicity is defined as a process by which high levels of glutamate excites neurons leading to their demise (Farooqui et al. 2008). Glutamate and its analog interact with excitatory amino acid receptors on the cell membrane and produce their effects on neurons. Excitotoxicity not only induces calcium influx, but initiates a cascade of events involving increase in free radical production, mitochondrial dysfunction, and activation of many calcium-dependent enzymes, including those involved in the generation and metabolism of arachidonic acid (Farooqui et al. 2008). These enzymes include isoforms of phospholipase A_2 (PLA_2), COX-2, 5-LOX, and epoxygenases (EPOX). Generation of high levels oxygenated arachidonic acid metabolites is called as uncontrolled "arachidonic acid cascade", which contributes to neural cell injury and death in neurotraumatic, neurodegenerative, and neuropsychiatric diseases (Farooqui 2010).

Treatment of SH-SY5Y cells with glutamate not only induces ROS formation and endoplasmic reticulum stress (ER stress), but also activates thioredoxin-interacting protein/NOD-like receptor pyrin domain containing-3 (TXNIP/NLRP3) inflammasome, leading to damage in the hippocampus (Li et al. 2015). Treatment of hippocampus or SH-SY5Y cells with curcumin inhibits the ER transmembrane kinase/endoRNase (RNase) (IRE1α) and PKR-like ER-resident kinase (PERK) phosphorylation with suppression of intracellular ROS generation (Li et al. 2015). Curcumin also increases AMPK activity and knockdown of AMPKα with specific siRNA abrogates its inhibitory effects on IRE1α and PERK phosphorylation, supporting the view that AMPK activity is essential for the suppression of ER stress (Li et al. 2015). As a result, curcumin not only reduces TXNIP expression, but also inhibits NLRP3 inflammasome activation through the downregulation of NLRP3 and induction of caspase-1 induction. These processes contribute to curcumin-mediated reduction of IL-1β secretion. Specific fluorescent probe and flow cytometry analysis demonstrate that curcumin retards mitochondrial malfunction and protects cell survival from glutamate neurotoxicity (Li et al. 2015).

Exposure of rat cortical neurons to glutamate and its analogs produce a significant decrease in brain-derived neurotrophic factor (BDNF) level, reduction in cell viability, and increase in cellular apoptosis (Wang et al. 2008). This neurotrophin mediates its action through TrkB (tropomyosin-related kinase B) receptors, which are coupled with the activation of the Ras/ERK, phosphatidylinositol 3-kinase/Akt and phospholipase C-γ (PLC-γ) pathways. In hippocampus, BDNF modulates LTP by acting at pre- and post-synaptic levels. BDNF also modulates the transport of mRNAs along dendrites and their translation at the synapse in cortical region by modulating the initiation and elongation phases of protein synthesis, and by acting on specific miRNAs (Genheden et al. 2015). Pretreatment of neurons with curcumin reverses the BDNF expression and cell viability in a dose- and time-dependent manner. However, K252a, a Trk receptor inhibitor which is known to inhibit the activity of BDNF, can inhibit the survival-inducing effect of curcumin. In addition, the up-regulation of BDNF levels by curcumin is also suppressed by K252a. Pretreatment of neurons with tyrosine kinase B (TrkB) antibody not only inhibits the activity of BDNF, but also blocks the protective effect of curcumin. Collectively, these studies suggest that the neuroprotective effect of curcumin may be mediated via BDNF/TrkB-MAPK/PtdIns 3 K-CREB signaling pathway (Wang et al. 2010).

3.4.4 Anti-Diabetic Activities of Curcumin

Diabetes, a complex endocrine and metabolic condition caused by long term hyperglycemia, impaired insulin secretion, insulin resistance (reduction in insulin ability to stimulate glucose uptake from body peripheral tissues), and β-cell dysfunction (Stumvoll et al. 2005) along with the onset of oxidative stress, chronic low-grade inflammation, and ER stress (Garg et al. 2012). It is proposed that oxidative stress, inflammatory stress, and ER stress may interact with each other during pathogenesis of diabetes. Improper management of diabetes may result in a number of health problems, including visceral diseases (heart and kidney diseases, blindness, and peripheral vascular disease) and neurological disorders (stroke, Alzheimer disease,

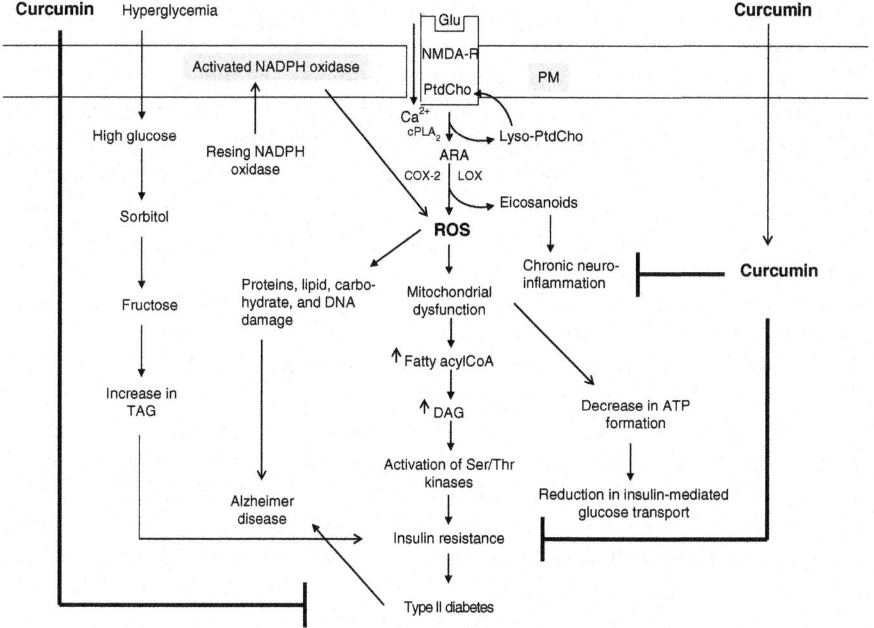

Fig. 3.8 Hypothetical diagram showing signal transduction processes associated with the antidiabetic effects of curcumin. Plasma membrane (*PM*); *N*-methyl-D-aspartate (*NMDA-R*); glutamate (*Glu*); Phosphatidylcholine (*PtdCho*); lyso-phosphatidylcholine (*lyso-PtdCho*); cytosolic phospholipase A₂ (*cPLA₂*); cyclooxygenase-2 (*COX-2*); lipoxygenase (*LOX*); arachidonic acid (*ARA*); reactive oxygen species (*ROS*); diacylglycerol (*DAG*); and triacylglycerol (*TAG*)

and depression, nerve damage). There are two types of diabetes, namely type 1 and type II diabetes. The molecular mechanisms involved in the pathogenesis insulin resistance in type 1 and type II diabetes are not fully understood. However, accumulation of lipids in the liver is considered to be one of the primary mechanisms that contribute to insulin resistance and type II diabetes (Fig. 3.8). Type 1 diabetes is caused by the inability of insulin-secreting pancreatic β cells to produce insulin (Stumvoll et al. 2005), where as type II diabetes is caused by long term consumption of western diet (high calorie diet), which is enriched in saturated fats and simple carbohydrates (Farooqui 2015). In type II diabetes elevations in nonesterified saturated fatty acids (palmitic and stearic acids), triacylglycerol, diacylglycerol, acylcarnitines, and ceramide are known to contribute to the molecular mechanism of insulin resistance and type II diabetes (Itani et al. 2002; Adams et al. 2004, 2009; Holland et al. 2007). In addition, elevation in inflammatory cytokines, adipokines, and mitochondrial dysfunction also contribute to insulin resistance. Type II diabetes also has a strong genetic component and many genes including genes for calpain 10, potassium inward-rectifier 6.2, peroxisome proliferator-activated receptor gamma, and insulin receptor substrate-1 may also contribute to type II diabetes (Stumvoll et al. 2005).

It is becoming increasingly evident that in addition to peripheral complications, type II diabetes also produces complications in the brain. One of the potential neurological complications of type II diabetes is cognitive deficits. Interesting to note that the structural, electrophysiological, neurochemical and anatomical complications of cognitive deficits in diabetes are strikingly similar to chronic stress observed in animals as well as in patients with stress-related psychiatric illnesses such as major depressive disorder (Reagan 2012). The molecular mechanisms associated with cognitive deficits are not fully understood. However, it is well known that insulin enters the brain via a saturable transport system and once inside the brain insulin stimulates and enhances neuroplasticity (Reger and Craft 2006; Reagan 2007).

Decrease in insulin activity, development of insulin resistance and hyperglycemia in the brain of diabetes/MetS patients may decrease neuroplasticity leading to the neurological deficits (Reagan et al. 2008). In addition, hyperglycemia and insulin resistance may also alter the balance between pro-oxidant and anti-oxidant cascades in the brain contributing to neuroplasticity deficits in diabetes/MetS. Thus, high levels of ROS and lipid peroxidation products, as well as decreases in antioxidant systems are characteristic features of brain tissue from experimental models of diabetes (Gispen and Biessels 2000). Metabolic syndrome (MetS), which is a complex pathological condition characterized by the abdominal obesity, dyslipidemia, hypertension, insulin resistance and type II diabetes has been reported to be an important risk factor for stroke, Alzheimer disease and depression (Farooqui et al. 2012; Farooqui 2013). Family history, age, environmental and lifestyle factors (diet and physical inactivity, and exposure to toxins) contribute to predisposition for the development of MetS as well as neurological disorders. Thus, incidences of stroke are two- to fourfold higher in patients with diabetes/MetS and cardiovascular diseases compared to normal subjects of the same age. Similarly, patients with diabetes/MetS have a two- to threefold increased risk for developing dementia and AD and MetS doubles the risk of depression (Farooqui 2013).

Curcumin not only reduces blood glucose level by reducing the hepatic glucose production, suppressing hyperglycemia-mediated inflammatory state, and stimulating glucose uptake through the up-regulation of GLUT4, GLUT2 and GLUT3 genes expressions, but also by activating AMP kinase, promoting the PPAR ligand-binding activity, and inducing peroxisome proliferator-activated receptor-gamma (PPAR-γ) activation (Nishiyama et al. 2005; Ghorbani et al. 2014). In addition, the administration of curcumin to diabetic rats not only increases GSH, but also increases activities of antioxidant enzymes activities (HO-1, GST, GPx, γGCL, CAT, SOD) and even gene expression of these enzymes to counteract the oxidative stress (Hussein and Abu-Zinadah 2010). Among above antioxidant enzymes, HO-1 has attracted considerable attention by virtue of its antioxidant, anti-inflammatory, and antiapoptotic properties (Pae et al. 2008). As stated above, Curcumin activates HO-1 expression through nuclear translocation of Nrf2 and its upstream kinases (ERK and p38 MAPK pathways). In addition, curcumin inhibits TNF-α (El-Azab et al. 2011) and plasma free fatty acids levels (FFA) (El-Moselhy et al. 2011). Curcumin also inhibits NF-κB activation (Soetikno et al. 2011), protein carbonyl formation (Suryanarayana et al. 2007), lipid peroxidation (El-Azab et al. 2011), and activities of lysosomal enzymes (N-acetyl-β-D-glucosaminidase, β-D-glucuronidase, β-D-galactosidase) (Chougala et al. 2012). Curcumin also decreases the levels of thiobarbituric acid reactive substances, and the activity of sorbitol dehydrogenase (Murugan and Pari 2007). Curcumin also can increase plasma insulin level and increase lipoprotein lipase (LPL) activity. Finally, curcumin is involved in activating of enzymes in liver, which are associated with glycolysis, gluconeogenic, and lipid metabolic process (Seo et al. 2008), and activating transcription factor Nrf2 function as well (He et al. 2012). Most of this information has been confirmed with studies on genetically obese *ob/ob* mice and obese mice consuming a high-fat diet have indicated that oral consumption of curcumin not only inhibits NF-κB activation in the liver, but increases the production of the adipokine adiponectin in adipose tissue; and ameliorates inflammation, hyperglycemia, and insulin resistance (Weisberg et al. 2008). Curcumin has also been reported to Improve lipidemia through the induction of PPAR-γ activity (Nishiyama et al. 2005), which is linked to adipogenesis (Deng et al. 2011). This improvement may also be implicated in the normalization of enzymatic activities (Seo et al. 2008) involved in lipid peroxidation (Mahesh et al. 2004) and glucose metabolism, including antioxidant enzymes (SOD and CAT) and GPx, hepatic glucose regulating enzymes glucose-6-phosphatase, phosphoenolpyruvate carboxykinase, hepatic lipid regulating enzymes (fatty acid synthase, 3-hydroxy-3-methylglutaryl coenzyme reductase, and acyl-CoA: cholesterol acyltransferase), and malondialdehyde (MDA) (Jain et al. 2009).

AMP-activated protein kinase (AMPK) is a strong energy regulator that controls whole-body glucose homeostasis in the liver and other key tissues in type 2 diabetes (Schultze et al. 2012; Kim et al. 2009). AMPK stimulates glucose uptake and mediates suppression of hepatic gluconeogenesis. Curcumin not only inhibits glucose-6-phosphatase, phosphoenolpyruvate carboxykinase in H4IIE rat hepatoma and Hep3B human hepatoma cells, but increases the phosphorylation of AMPK (Fujiwara et al. 2008) and its downstream target acetyl-CoA carboxylase (ACC) (Shehzad et al. 2011) in H4IIE and Hep3B cells. Accumulating evidences thus suggest that curcumin-mediated changes in carbohydrate metabolism along with the stimulation of insulin secretion from pancreatic tissues, improvement in pancreatic cell function, reduction of insulin resistance, antioxidant and anti-inflammatory properties of curcumin may be responsible for the beneficial effects of curcumin in diabetes (Ghorbani et al. 2014).

3.4.5 *Antinociceptive Effects of Curcumin*

Nociception (pain) is a complex multidimensional experience comprising sensory-discriminative, affective-motivational and cognitive-evaluative components (Melzack and Casey 1968) involving peripheral and central mechanisms (Svensson and Yaksh 2002; Broom et al. 2004). Based on etiology, nociception may be classified into tissue damage-induced inflammatory/nociception and nerve damage-mediated neuropathy. The former is mediated by a painful stimulus on nociceptors, the later by a primary lesion or dysfunction in the brain and spinal cord. The pathophysiology of the nociception is extremely complex. Currently, analgesics (opiates, nonsteroidal anti-inflammatory drugs, and benzodiazepine), antispasmodics, antidepressants are the most common medications for acute as well as peripheral and central chronic pain conditions; yet most of these medications are not always optimal and have drawbacks of addiction and constipation. Furthermore, the paradoxical development of analgesic tolerance, inadequate pain relief, and nociceptive sensitization with prolonged opioid use has also proved an unfortunate obstacle for their clinic applications (Chu et al. 2008).

Curcumin has been reported to relieve from central and peripheral pain. Thus, acute curcumin treatment not only attenuates formalin-induced defensive behaviors, visceral pain as measured by acetic acid induced writhing response, but also relieves capsaicin-induced thermal hyperalgesia and reserpine-induced fibromyalgia-like behaviors (Tajik et al. 2007, 2008; Mittal et al. 2009). Repeated curcumin treatments are needed for relieving peripheral neuropathic pain. Similarly, repeated curcumin treatment is also necessary to get relief from the preexisting hyperalgesia. In a chronic constriction injury (CCI) model of neuropathic pain, acute curcumin treatment has no effect on mechanical and thermal hyperalgesia, but repeated curcumin treatment progressively and completely reverse the CCI-induced hypersensitivity (Tajik et al. 2007, 2008; Mittal et al. 2009). The primary mechanisms associated with curcumin induced antinociceptive actions in postoperative pain are not fully understood. However, numerous potential biochemical and pharmacological mechanisms have been linked to the antinociceptive actions of curcumin. These include monoamine system (Zhao et al. 2012), nitrogen oxidase signaling (Brouet and Ohshima 1995), GABAa receptors (Zhao et al. 2013), opioid receptors, and interleukin receptors (Abe et al. 1999). The molecular mechanisms involved in antinociceptive activities of curcumin on above receptors are not clearly understood. However, some information is available on attenuation of opioid tolerance by curcumin has been recently published. It is reported that curcumin acts by inhibiting Ca^{2+}/calmodulin-dependent protein kinase II α (CaMKIIα) (Hu et al. 2015), a protein kinase that plays acritical for opioid tolerance and dependence. In addition, involvement of spinal β_2 adrenoceptors and 5-HT$_{1A}$ receptors has also been reported in curcumin-mediated antinociception in a mouse model of neuropathic pain. These two receptor systems are involved in spinal pain processing (Zhao et al. 2012). These mechanisms may also be involved in curcumin mediated relief of post-operative pain. Curcumin administration also reduces reserpine induced pain and depression by reversing the changes in serotonin and substance P levels induced by reserpine (Arora et al. 2011). In facial pain model, curcumin exhibits pain ameliorating, synergistic effects when administered alongside NSAIDs (Mittal et al. 2009). Furthermore, the analgesic effects of curcumin are mediated by the suppression of nitrite, TNF-α, and capsaicin-induced TRPV1 activity, through the descending noradrenergic and serotonergic systems (Sharma et al. 2006; Yeon et al. 2010; Zhao et al. 2012).

3.4.6 Anticancer Effects of Curcumin

Cancer is a group of over 200 neoplastic diseases caused by the dysregulation of multiple cell signaling pathways (Vogelstein and Kinzler 2004). Development of cancer is supported by the dysregulation of as many as 500 different genes. The dysregulation of various genes may occur over a period as long as 20–30 years before a given cancer begins to manifest its symptoms (Hasima and Aggarwal 2012). Multiple factors contribute to the development of cancers. Only 5–10 % of all cancers are of genetic origin, whereas the remaining 90–95 % may be linked to lifestyle factors (smoking, unhealthy diet, and obesity) and the environment (sun exposure, radiation and environmental pollutants) (Anand et al. 2008b). In addition, epigenetic alterations are associated with changes in DNA methylation, covalent modifications of histones, and alterations in miRNA expression. These three mechanisms are interconnected and appear to be key players in tumor progression and failure of conventional chemotherapy (Jones 2002). The underlying molecular mechanisms by which these risk factors induce cancer are not fully understood. However, most cancers are accompanied by the onset of oxidative stress and inflammation (Anand et al. 2008b). Although, significant progress has been made on the treatment of cancers, but road to studies for the treatment of cancer has not been smooth. Because of its ability to scavenge free radicals and inhibit inflammation, curcumin has been investigated for cancer chemoprevention and tumor growth suppression.

Curcumin shows beneficial effects in several types of cancer in patients. Recent studies have indicated that oral administration of curcumin (6.0 g daily) during radiotherapy reduces the severity of radiation dermatitis in breast cancer patients (Ryan et al. 2013). In animal model oral administration of curcumin inhibits lung cancer (Cheng et al. 2013b), skin (Phillips et al. 2013), head and neck (Clark et al. 2010), oral (Lin et al. 2010), hepatocellular carcinoma (Yoysungnoen et al. 2006), mammary tumors, lymphomas, leukemias (Huang et al. 1998), and familial adenomatous polyposis (Perkins et al. 2002). The anticancer effects of curcumin are not only due to its ability to suppress proliferation of a wide variety of tumor cells, inhibition of transcription factors (NF-κB, AP-1 and Egr-1), but also due to the down-regulation of enzymes (COX2, LOX, NOS, MMP-9), and cytokines (TNF-α and chemokines) (Aggarwal et al. 2003; Jurenka 2009). It is also known that through down-regulation of vascular endothelial growth factor (VEGF), basic fibroblast growth factor (bFGF) and epidermal growth factor (EGF), as well as angiopoietin and hypoxia-inducible factors (HIF)-1α, curcumin suppresses angiogenesis and restricts the growth of tumors (Gururaj et al. 2002). In addition, curcumin also decreases the expression of cell surface adhesion molecules and cyclin D1 and inhibits the activity of c-Jun N-terminal kinase, protein tyrosine kinases and protein serine/threonine kinases (Aggarwal et al. 2003; Jurenka 2009). All these enzymes support the growth of cancer cells. Furthermore, as stated above, curcumin also induces potent antioxidant and anti-inflammatory activities.

Curcumin not only reduces the expression of positive regulators of DNA methyltransferase 1 (DNMT1), p65 and Sp1, but also acts as a modulator of histone deacetylase (HDAC) and histone acetyltransferase (HAT) (Balasubramanyam et al. 2004). Both of these properties can significantly alter the epigenome of cancer cells by reactivating prometastatic genes and proto-oncogenes (Ravindran et al. 2009). Curcumin promotes the balance between histone acetyl transferase and histone deacetylase activity by selectively activating or inactivating the expression of genes implicated in cancer death and progression, respectively. Finally, curcumin modulates miRNAs (miR-15a, miR-16, miR-21, miR-22, miR-26, miR-101, miR-146, miR-200, miR-203, and let-7) and their multiple target genes, supporting the view that curcumin not only restore the epigenetic regulation balance, but also suppresses tumor initiation, promotion, invasion, metastasis, and angiogenesis (Teiten et al. 2013; Tuorkey 2014).

3.5   Curcumin and Iron Chelation

Iron is vital for almost all living organisms. This essential element is necessary for human life. Iron containing proteins are not only involved in oxygen transport, oxygen sensing, electron transfer, energy metabolism and DNA synthesis, but also contribute to the regulation of cell growth and differentiation (Lieu et al. 2001; Richardson 2002; Hentze et al. 2004). Excess of iron can cause heart disease and cancer where as deficiency of iron contributes to fatigue, decrease in immunity or anemia, which can have serious health consequences (Rajapurkar et al. 2012). Iron is a powerful pro-oxidant and can cause cellular damage by producing reactive oxygen species in different tissues of the body (Basuli et al. 2014). Iron plays an important role in various neurodegenerative diseases such as Parkinson and Alzheimer diseases are associated with iron metabolism deregulation (Bush 2013; Jenner 1989). In biological fluids iron is found in ferric or ferrous forms. The presence of ferric or ferrous form depends on the reductive power of the medium (commonly dictated by the GSH-NADPH/NADH levels). In extracellular fluids, essentially all the iron is transported by the protein transferrin, which shields the Fe^{3+} from the environment and renders it virtually redox-inactive as well as non-exchangeable with, or displaceable by, physiological substances or metals (Gozzelino and Arosio 2015).

Several studies have indicated that treatment of mice with curcumin for 6 months results in symptoms of iron deficiency. Curcumin has also been reported to modulate proteins of iron metabolism in cells and in tissues supporting the view that curcumin may act as an iron chelator (Bernabe-Pineda et al. 2004; Jiao et al. 2009). Thus, it is important to monitor the iron status in human subjects undergoing clinical trials when treatment involves high doses of native curcumin (Chin et al. 2014). Since chelation of iron may be the underlying mechanism, encapsulation of curcumin in liposomes or nanocarrier may help to reduce its direct interaction with transition metals and thereby their chelation. Micelles and liposomes have been used as vehicles to improve the bioavailability of curcumin (Kakkar et al. 2010; Schiborr et al. 2014) and provide the additional advantage that lower oral doses can be administered, which would lead to a smaller number of potential curcumin:iron complexes and thus a reduced potential for adverse effects on iron status.

3.6 Neuroprotective Activities of Curcumin

It is well known that the brain hippocampus plays an important role in memory formation and spatial navigation. Adult hippocampal neurogenesis is closely related to memory formation (Sahay et al. 2011; Stone et al. 2011). Progenitor cells located in the subgranular zone of the hippocampal dentate gyrus divide, proliferate, differentiate, and give rise to new neurons (Duman 2004). Overall decline in cellular proliferation has been reported to occur during brain aging (Cameron and McKay 1999; Choi et al. 2010). Neurogenesis is negatively regulated by age and stress (Cameron and McKay 1999; Choi et al. 2010). In the dentate gyrus of aged mice neurogenesis is promoted by low levels of ROS, expression of BDNF, caloric restriction, exercise, and physiological activation. Wnt-catenin signaling pathway contributes to neurogenesis through the maintenance of synaptic plasticity, survival, proliferation, and differentiation in embryonic and adult brains (van Praag et al. 2002; Clevers et al. 2014).

Curcumin mediates its neuroprotective effects and memory restoring effects through the prevention of oxidative stress. Thus, chronic curcumin oral treatment significantly improves the colchicine-mediated cognitive impairment in mice by reducing lipid peroxidation (Kumar et al. 2007a, b). Furthermore, curcumin not only reverses 3-nitropropionic acid and ethanol-mediated memory deficit by suppressing the nitric oxide synthase/nitric oxide pathway (Yu et al. 2013a), but also protects against ethanol-, aluminum-, and iron-mediated neurotoxicity in rodents (Kumar et al. 2009; Yu et al. 2013b). In the okadaic acid-mediated neural impairment model, oral administration of curcumin significantly improves the memory function as assessed by both Morris water maze and passive avoidance tests (Rajasekar et al. 2013). Curcumin not only reverses Aβ-mediated cognitive deficits and neuropathological alterations (Frautschy et al. 2001), but also promotes hippocampal neurogenesis (Xu et al. 2007; Zhao et al. 2008) through activation of the canonical Wnt pathway (Tiwari et al. 2014), which can be knocked down by stereotaxically injecting Wnt3a siRNA into the hippocampus. Wnt pathway can be restored by the treatment with curcumin nanoparticles (Tiwari et al. 2014). Based on these findings, it is proposed that the dose-dependent neurogenic activity of curcumin resembles the hormesis, a process characterized by low dose activation and high dose inhibition of biochemical activity (Lichtenwalner and Parent 2006; Park and Lee 2011; Farooqui 2015; Pluta et al. 2015).

3.7 Conclusion

Curcumin (diferuloylmethane), an orange-yellow component of turmeric or curry powder has been used in some medicinal preparation as well as used as a food-coloring agent. In recent years, both in vitro and in vivo studies have indicated that curcumin produces antioxidant, anti-inflammatory, anticancer, antiviral, and antiarthritic properties. The poor bioavailability of curcumin is the major hurdle for its more widespread use in animals and humans. However, encapsulation of curcumin into liposomes, cyclodextrin, curcumin conjugate with PLGA, complexation with phospholipids, and synthesis of curcumin analogs have made it easy to bypass this problem. It has been demonstrated that this new ways of delivering curcumin have resulted in increased absorption and allowing increase in levels of curcumin to various body tissues. The underlying mechanisms of these effects are diverse and appear to involve the regulation of various molecular targets, including transcription factors (NF-κB), growth factors (vascular endothelial cell growth factor), inflammatory cytokines (TNF-α, IL1, and IL-6), and enzymes (protein kinases such as MAPK, Akt, COX-2, and 5-LOX). Accumulating evidence suggests that due to its efficacy and regulation of multiple targets, as well as its safety for human use, curcumin can be used for the treatment of neurological disorders, visceral diseases and various types of cancers.

References

Abe Y, Hashimoto S, Horie T (1999) Curcumin inhibition of inflammatory cytokine production by human peripheral blood monocytes and alveolar macrophages. Pharmacol Res 39:41–47

Adams JM II, Pratipanawatr T, Berria R, Wang E, DeFronzo RA, Sullards MC, Mandarino LJ (2004) Ceramide content is increased in skeletal muscle from obese insulin-resistant humans. Diabetes 53:25–31

Adams SH, Hoppel CL, Lok KH, Zhao L, Wong SW, Minkler PE, Hwang DH, Newman JW, Garvey WT (2009) Plasma acylcarnitine profiles suggest incomplete long-chain fatty acid beta-oxidation and altered tricarboxylic acid cycle activity in type 2 diabetic African-American women. J Nutr 139:1073–1081

Aggarwal BB, Kumar A, Bharti AC (2003) Anticancer potential of curcumin: preclinical and clinical studies. Anticancer Res 23:363–398

Aggarwal S, Takada Y, Singh S, Myers JN, Aggarwal BB (2004) Inhibition of growth and survival of human head and neck squamous cell carcinoma cells by curcumin via modulation of nuclear factor-kappaB signaling. Int J Cancer 111:679–692

Aggarwal BB, Deb L, Prasad S (2014) Curcumin differs from tetrahydrocurcumin for molecular targets, signaling pathways and cellular responses. Molecules 20:185–205

Ak T, Gülçin I (2008) Antioxidant and radical scavenging properties of curcumin. Chem Biol Interact 174:27–37

Amor S, Puentes F, Baker D, van der Valk P (2010) Inflammation in neurodegenerative diseases. Immunology 129:154–169

Anand P, Kunnumakkara AB, Newman RA, Aggarwal BB (2007) Bioavailability of curcumin: problems and promises. Mol Pharm 4:807–818

Anand P, Thomas SG, Kunnumakkara AB, Sundaram C, Harikumar KB, Sung B, Tharakan ST, Misra K, Priyadarsini IK, Rajasekharan KN, Aggarwal BB (2008a) Biological activities of curcumin and its analogues (Congeners) made by man and Mother Nature. Biochem Pharmacol 76:1590–1611

Anand P, Kunnumakkara AB, Sundaram C, Harikumar KB, Tharakan ST, Lai OS, Sung B, Aggarwal BB (2008b) Cancer is a preventable disease that requires major lifestyle changes. Pharm Res 25:2097–2116

Anand P, Nair HB, Sung B, Kunnumakkara AB, Yadav VR, Tekmal RR, Aggarwal BB (2010) Design of curcumin-loaded PLGA nanoparticles formulation with enhanced cellular uptake, and increased bioactivity in vitro and superior bioavailability in vivo. Biochem Pharmacol 79:330–338

Aoyagi S, Yamazaki M, Miyasaka T (2001) Clarification of enhanced hydroxyl radical production in Fenton reaction with ATP/ADP based on luminol chemiluminescence. J Chem Eng Jpn 34:956–959

Arcaro CA, Gutierres CA, Assis RP, Moreira TF, Costa PI, Baviera AM, Brunetti IL (2014) Piperine, a natural bioenhancer, nullifies the antidiabetic and antioxidant activities of curcumin in streptozotocin-diabetic rats. PLoS One 9, e113993

Arora V, Kuhad A, Tiwari V, Chopra K (2011) Curcumin ameliorates reserpine-induced pain-depression dyad: behavioural, biochemical, neurochemical and molecular evidences. Psychoneuroendocrinology 36:1570–1581

Aruoma OI, Halliwell B, Gajewski E, Dizdaroglu M (1991) Copper-ion-dependent damage to the bases in DNA in the presence of hydrogen peroxide. Biochem J 273:601–604

Asai A, Miyazawa T (2000) Occurrence of orally administered curcuminoid as glucuronide and glucuronide/sulfate conjugates in rat plasma. Life Sci 67:2785–2793

Atal CK, Dubey RK, Singh JJ (1985) Biochemical basis of enhanced drug bioavailability by piperine: evidence that piperine is a potent inhibitor of drug metabolism. J Pharmacol Exp Ther 232:258–262

Bae YS, Kang SW, Seo MS, Baines IC, Tekle E, Chock PB, Rhee SG (1997) Epidermal growth factor (EGF)-induced generation of hydrogen peroxide. Role in EGF receptor-mediated tyrosine phosphorylation. J Biol Chem 272:217–221

Balasubramanyam K, Varier RA, Altaf M, Swaminathan V, Siddappa NB, Ranga U, Kundu TK (2004) Curcumin, a novel p300/CREB-binding protein-specific inhibitor of acetyltransferase, represses the acetylation of histone/nonhistone proteins and histone acetyltransferase-dependent chromatin transcription. J Biol Chem 279:51163–51171

Balducci C, Mancini S, Minniti S, La Vitola P, Zotti M, Sancini G, Mauri M, Cagnotto A, Colombo L, Fiordaliso F, Grigoli E, Salmona M, Snellman A, Haaparanta-Solin M, Forloni G, Masserini M, Re F (2014) Multifunctional liposomes reduce brain β-amyloid burden and ameliorate memory impairment in Alzheimer's disease mouse models. J Neurosci 34:14022–14031

Barbusinski K (2009) Fenton reaction—controversy concerning the chemistry. Ecol Chem Eng 16:347–358

Barnham KJ, Masters CL, Bush AI (2004) Neurodegenerative diseases and oxidatives stress. Nat Rev Drug Discov 3:205–214

Barzegar A, Moosavi-Movahedi AA (2011) Intracellular ROS protection efficiency and free radical-scavenging activity of curcumin. PLoS One 6, e26012

Basuli D, Stevens RG, Torti FM, Torti SV (2014) Epidemiological associations between iron and cardiovascular disease and diabetes. Front Pharmacol 5:117

Begum AN, Jones MR, Lim GP, Morihara T, Kim P, Heath DD, Rock CL, Pruitt MA, Yang F, Hudspeth B, Hu S, Faull KF, Teter B, Cole GM, Frautschy SA (2008) Curcumin structure function, bioavailability, and efficacy in models of neuroinflammation and Alzheimer's disease. J Pharmacol Exp Ther 326:196–208

Bernabe-Pineda M, Ramirez-Silva MT, Romero-Romo MA, Gonzalez-Vergara E, Rojas-Hernandez A (2004) Spectrophotometric and electrochemical determination of the formation constants of the complexes Curcumin–Fe(III)–water and Curcumin– Fe(II)–water. Spectrochim Acta A Mol Biomol Spectrosc 60:1105–1113

Bertolasi V, Ferretti V, Gilli P, Yao X, Li CJ (2008) Substituent effects on keto-enol tautomerization of [small beta]-diketones from X-ray structural data and DFT calculations. New J Chem 32:694–704

Bisht S, Maitra A (2009) Systemic delivery of curcumin: 21st century solutions for an ancient conundrum. Curr Drug Discov Technol 6:192–199

Broom DC, Samad TA, Kohno T, Tegeder I, Geisslinger G, Woolf CJ (2004) Cyclooxygenase 2 expression in the spared nerve injury model of neuropathic pain. Neuroscience 124:891–900

Brouet I, Ohshima H (1995) Curcumin, an anti-tumour promoter and anti-inflammatory agent, inhibits induction of nitric oxide synthase in activated macrophages. Biochem Biophys Res Commun 206:533–540

Bush AI (2013) The metal theory of Alzheimer's disease. J Alzheimer's Dis 33(Suppl 1):S277–S281

Cameron HA, McKay RD (1999) Restoring production of hippocampal neurons in old age. Nat Neurosci 2:894–897

Chan JM, Zhang L, Yuet KP, Liao G, Rhee J-W, Langer R, Farokhzad OC (2009) PLGA–lecithin–PEG core–shell nanoparticles for controlled drug delivery. Biomaterials 30:1627–1634

Chen A, Xu J, Johnson AC (2006) Curcumin inhibits human colon cancer cell growth by suppressing gene expression of epidermal growth factor receptor through reducing the activity of the transcription factor Egr-1. Oncogene 25:278–287

Cheng AL, Hsu CH, Lin JK, Hsu MM, Ho YF, Shen TS, Ko JY, Lin JT, Lin BR (2001) Phase I clinical trial of curcumin, a chemopreventive agent, in patients with high-risk or pre-malignant lesions. Anticancer Res 21:2895–2900

Cheng KK, Yeung CF, Ho SW, Chow SF, Chow AH, Baum L (2013a) Highly stabilized curcumin nanoparticles tested in an in vitro blood-brain barrier model and in Alzheimer's disease Tg2576 mice. AAPS J 15:324–336

Cheng KW, Wong CC, Mattheolabakis G, Xie G, Huang L, Rigas B (2013b) Curcumin enhances the lung cancer chemopreventive efficacy of phospho-sulindac by improving its pharmacokinetics. Int J Oncol 43:895–902

Chin D, Huebbe P, Frank J, Rimbach G, Pallauf K (2014) Curcumin may impair iron status when fed to mice for six months. Redox Biol 2:563–569

Choi JH, Yoo KY, Lee CH, Park OK, Yan BC, Li H, Hwang IK, Park JH, Kim SK, Won MH (2010) Comparison of newly generated doublecortin-immunoreactive neuronal progenitors in the main olfactory bulb among variously aged gerbils. Neurochem Res 35:1599–1608

Chougala MB, Bhaskar JJ, Rajan MGR, Salimath PV (2012) Effect of curcumin and quercetin on lysosomal enzyme activities in streptozotocin-induced diabetic rats. Clin Nutr 31:749–755

Chu LF, Angst MS, Clark D (2008) Opioid-induced hyperalgesia in humans: molecular mechanisms and clinical considerations. Clin J Pain 24:479–496

Clark CA, McEachern MD, Shah SH, Rong Y, Rong X, Smelley CL, Caldito GC, Abreo FW, Nathan CO (2010) Curcumin inhibits carcinogen and nicotine-induced mammalian target of rapamycin pathway activation in head and neck squamous cell carcinoma. Cancer Prev Res (Phila) 3:1586–1595

Clevers H, Loh KM, Nusse R (2014) Stem cell signaling. An integral program for tissue renewal and regeneration: Wnt signaling and stem cell control. Science 346:1248012

Cohen-Sela E, Chorny M, Koroukhov N, Danenberg HD, Golomb G (2009) A new double emulsion solvent diffusion technique for encapsulating hydrophilic molecules in PLGA nanoparticles. J Control Release 133:90–95

Cornago P, Claramunt RM, Bouissane L, Alkorta I, Elguero J (2008) A study of the tautomerism of beta-dicarbonyl compounds with special emphasis on curcuminoids. Tetrahedron 64: 8089–8094

Cuadrado A, Moreno-Murciano P, Pedraza-Chaverri J (2009) The transcription factor Nrf2 as a new therapeutic target in Parkinson's disease. Expert Opin Ther Targets 13:319–329

Deng T, Sieglaff DH, Zhang A et al (2011) A peroxisome proliferator-activated receptor γ (PPARγ)/PPARγ coactivator 1β autoregulatory loop in adipocyte mitochondrial function. J Biol Chem 286:30723–30731

Dickinson BC, Peltier J, Stone D, Schaffer DV, Chang CJ (2011) Nox2 redox signaling maintains essential cell populations in the brain. Nat Chem Biol 7:106–112

Doggui S, Sahni JK, Arseneault M, Dao L, Ramassamy C (2012) Neuronal uptake and neuroprotective effect of curcumin-loaded PLGA nanoparticles on the human SK-N-SH cell line. J Alzheimers Dis 30:377–392

Dong Y, Benveniste EN (2001) Immune function of astrocytes. Glia 36:180–190

Duman RS (2004) Depression: a case of neuronal life and death? Biol Psychiatry 56:140–145

Duncan R, Gaspar R (2011) Nanomedicine (s) under the microscope. Mol Pharm 8:2101e41

El-Azab MF, Attia FM, El-Mowafy AM (2011) Novel role of curcumin combined with bone marrow transplantation in reversing experimental diabetes: effects on pancreatic islet regeneration, oxidative stress, and inflammatory cytokines. Eur J Pharmacol 658:41–48

El-Moselhy MA, Taye A, Sharkawi SS, El-Sisi SFI, Ahmed AF (2011) The antihyperglycemic effect of curcumin in high fat diet fed rats. Role of TNF-α and free fatty acids. Food Chem Toxicol 49:1129–1140

Emerit J, Edeas M, Bricaire F (2004) Neurodegenerative diseases and oxidative stress. Biomed Pharmacother 58:39–46

Farombi EO, Shrotriya S, Na HK, Kim SH, Surh YJ (2008) Curcumin attenuates dimethylnitrosamine-induced liver injury in rats through Nrf2-mediated induction of heme oxygenase-1. Food Chem Toxicol 46:1279–1287

Farooqui AA (2010) Neurochemical aspects of neurotraumatic and neurodegenerative diseases. Springer, New York

Farooqui AA (2013) Metabolic syndrome: an important risk factor for stroke, Alzheimer, and depression. Springer, New York

Farooqui AA (2014) Inflammation and oxidative stress in neurological disorders. Springer, New York

Farooqui AA (2015) High calorie diet and the human brain: consequences of long term consumption. Springer, Switzerland

Farooqui AA, Horrock LA (2007) Glycerophospholipids in brain. Springer, New York

Farooqui AA, Ong WY, Horrocks LA (2008) Neurochemical aspects of excitotoxicity. Springer, New York

Farooqui AA, Farooqui T, Panza F, Frisardi V (2012) Metabolic syndrome as a risk factor for neurological disorders. Cell Mol Life Sci 69:741–762

Frautschy SA, Hu W, Kim P, Miller SA, Chu T, Harris-White ME, Cole GM (2001) Phenolic anti-inflammatory antioxidant reversal of Abeta-induced cognitive deficits and neuropathology. Neurobiol Aging 22:993–1005

Fujiwara H, Hosokawa M, Zhou X, Fujimoto S, Fukuda K, Toyoda K, Nishi Y, Fujita Y, Yamada K, Yamada Y, Seino Y, Inagaki N (2008) Curcumin inhibits glucose production in isolated mice hepatocytes. Diabetes Res Clin Pract 80:185–191

Gaetke LM, Chow-Johnson HS, Chow CK (2014) Copper: Toxicological relevance and mechanisms. Arch Toxicol 88:1929–1938

Garcia-Garcia A, Rodriguez-Rocha H, Madayiputhiya N, Pappa A, Panayiotidis MI, Franco R (2012) Biomarkers of protein oxidation in human disease. Curr Mol Med 12:681–697

Garg AD, Kaczmarek A, Krysko O, Vandenabeele P, Krysko DV, Agostinis P (2012) ER stress-induced inflammation: does it aid or impede disease progression? Trends Mol Med 18:589–598

Genheden M, Kenney JW, Johnston HE, Manousopoulou A, Garbis SD, Proud CG (2015) BDNF stimulation of protein synthesis in cortical neurons requires the MAP kinase-interacting kinase MNK1. J Neurosci 35:972–984

Ghalandarlaki N, Alizadeh AM, Ashkani-Esfahani S (2014) Nanotechnology-applied curcumin for different diseases therapy. Biomed Res Int 2014:394264

Ghorbani Z, Hekmatdoost A, Mirmiran P (2014) Anti-hyperglycemic and insulin sensitizer effects of turmeric and its principle constituent curcumin. Int J Endocrinol Metab 12, e18081

Gispen WH, Biessels G-J (2000) Cognition and synaptic plasticity in diabetes mellitus. Trends Neurosci 23:542–549

Goel A, Kunnumakkara AB, Aggarwal BB (2008) Curcumin as "Curecumin": from kitchen to clinic. Biochem Pharmacol 75:787–809

Gota VS, Maru GB, Soni TG, Gandhi TR, Kochar N, Agarwal MG (2010) Safety and pharmacokinetics of a solid lipid curcumin particle formulation in osteosarcoma patients and healthy volunteers. J Agric Food Chem 58:2095–2099

Gozzelino R, Arosio P (2015) The importance of iron in pathophysiologic conditions. Front Pharmacol 6:26

Gururaj AE, Belakavadi M, Venkatesh DA, Marme D, Salimath BP (2002) Molecular mechanisms of anti-angiogenic effect of curcumin. Biochem Biophys Res Commun 297:934–942

Hasima N, Aggarwal BB (2012) Cancer-linked targets modulated by curcumin. Int J Biochem Mol Biol 3:328–351

He HJ, Wang GY, Gao Y, Ling WH, Yu ZW, Jin TR (2012) Curcumin attenuates Nrf2 signaling defect, oxidative stress in muscle and glucose intolerance in high fat diet-fed mice. World J Diabetes 3:94–104

Hentze MW, Muckenthaler MU, Andrews NC (2004) Balancing acts: molecular control of mammalian iron metabolism. Cell 117:285–297

Hoehle SI, Pfeiffer E, Metzler M (2007) Glucuronidation of curcuminoids by human microsomal and recombinant UDP-glucuronosyltransferases. Mol Nutr Food Res 51:932–938

Holland WL, Brozinick JT, Wang LP, Hawkins ED, Sargent KM, Liu Y, Narra K, Hoehn KL, Knotts TA, Siesky A, Nelson DH, Karathanasis SK, Fontenot GK, Birnbaum MJ, Summers SA (2007) Inhibition of ceramide synthesis ameliorates glucocorticoid-, saturated-fat- and obesity-induced insulin resistance. Cell Metab 5:167–179

Hu X, Huang F, Szymusiak M, Liu Y, Wang ZJ (2015) Curcumin attenuates opioid tolerance and dependence by inhibiting Ca2+/calmodulin-dependent protein kinase II α activity. J Pharmacol Exp Ther 352:420–428

Huang MT, Lou YR, Xie JG, Ma W, Lu YP, Yen P, Zhu BT, Newmark H, Ho CT (1998) Effect of dietary curcumin and dibenzoylmethane on formation of 7,12-dimethylbenz[a]anthracene-induced mammary tumors and lymphomas/leukemias in Sencar mice. Carcinogenesis 19:1697–1700

Hussein HK, Abu-Zinadah OA (2010) Antioxidant effect of curcumin extracts in induced diabetic wister rats. Inter J Zool Res 13:266–276

Itani SI, Ruderman NB, Schmieder F, Boden G (2002) Lipid-induced insulin resistance in human muscle is associated with changes in diacylglycerol, protein kinase C, and IkappaB-alpha. Diabetes 51:2005–2011

Jain SK, Rains J, Croad J, Larson B, Jones K (2009) Curcumin supplementation lowers TNF-α, IL-6, IL-8, and MCP-1 secretion in high glucose-treated cultured monocytes and blood levels of TNF-α, IL-6, MCP-1, glucose, and glycosylated hemoglobin in diabetic rats. Antioxid Redox Signal 11:241–249

Jenner P (1989) Clues to the mechanism underlying dopamine cell death in Parkinson's disease. J Neurol Neurosurg Psychiatry 52:22–28

Jeong GS, Oh GS, Pae HO, Jeong SO, Kim YC, Shin MK, Seo BY, Han SY, Lee HS, Jeong JG, Koh JS, Chung HT (2006) Comparative effects of curcuminoids on endothelial heme oxygenase-1 expression: ortho-methoxy groups are essential to enhance heme oxygenase activity and protection. Exp Mol Med 38:393–400

Jeong SO, Oh GS, Ha HY, Soon Koo B, Sung Kim H, Kim YC, Kim EC, Lee KM, Chung HT, Pae HO (2009) Dimethoxycurcumin, a synthetic curcumin analogue, induces heme oxygenase-1 expression through Nrf2 activation in RAW264.7 macrophages. J Clin Biochem Nutr 44:79–84

Jiao Y, Wilkinson J, Di X, Wang W, Hatcher H, Kock ND, D'Agostino R Jr (2009) Curcumin, a cancer chemopreventive and chemotherapeutic agent, is a biologically active iron chelator. Blood 113:462–469

Jones PA (2002) DNA methylation and cancer. Oncogene 21:5358–5360

Joshi G, Johnson JA (2012) The Nrf2-ARE pathway: a valuable therapeutic target for the treatment of neurodegenerative diseases. Recent Pat CNS Drug Discov 7:218–229

Jurenka JS (2009) Anti-inflammatory properties of curcumin, a major constituent of *Curcuma longa*: a review of preclinical and clinical research. Altern Med Rev 14:141–153

Kakkar V, Singh S, Singla D, Sahwney S, Chauhan AS, Singh G (2010) Pharmacokinetic applicability of a validated liquid chromatography tandem mass spectroscopy method for orally administered curcumin loaded solid lipid nanoparticles to rats. J Chromatogr B Analyt Technol Biomed Life Sci 878:3427–3431

Kanski J, Aksenova M, Stoyanova A, Butterfield DA (2002) Ferulic acid antioxidant protection against hydroxyl and peroxyl radical oxidation in synaptosomal and neuronal cell culture systems in vitro: structure-activity studies. J Nutr Biochem 13:273–281

Khalil NM, do Nascimento TC, Casa DM, Dalmolin LF, de Mattos AC, Hoss I, Romano MA, Mainardes RM (2013) Pharmacokinetics of curcumin-loaded PLGA and PLGA-PEG blend nanoparticles after oral administration in rats. Colloids Surf B Biointerfaces 101:353–360

Kim SJ, Son TG, Park HR, Park M, Kim MS, Kim HS, Chung HY, Mattson MP, Lee J (2008) Curcumin stimulates proliferation of embryonic neural progenitor cells and neurogenesis in the adult hippocampus. J Biol Chem 283:14497–14505

Kim T, Davis J, Zhang AJ, He X, Mathews ST (2009) Curcumin activates AMPK and suppresses gluconeogenic gene expression in hepatoma cells. Biochem Biophys Res Commun 388: 377–382

Kumar A, Naidu PS, Seghal N, Padi SS (2007a) Effect of curcumin on intracerebroventricular colchicine-induced cognitive impairment and oxidative stress in rats. J Med Food 10:486–494

Kumar P, Padi SS, Naidu PS, Kumar A (2007b) Possible neuroprotective mechanisms of curcumin in attenuating 3-nitropropionic acid-induced neurotoxicity. Methods Find Exp Clin Pharmacol 29:19–25

Kumar A, Dogra S, Prakash A (2009) Protective effect of curcumin (*Curcuma longa*), against aluminium toxicity: possible behavioral and biochemical alterations in rats. Behav Brain Res 205:384–390

Lammers T, Aime S, Hennink WE, Storm G, Kiessling F (2011) Theranostic nanomedicine. Acc Chem Res 44:1029e38

Lavoie S, Chen Y, Dalton TP, Gysin R, Cuénod M, Steullet P, DOKQ (2009) Curcumin, quercetin, and tBHQ modulate glutathione levels in astrocytes and neurons: importance of the glutamate cysteine ligase modifier subunit. J Neurochem 108:1410–1422

Letelier ME, Faúndez M, Jara-Sandoval J, Molina-Berríos A, Cortés-Troncoso J, Aracena-Parks P, Marín-Catalán R (2009) Mechanisms underlying the inhibition of the cytochrome P450 system by copper ions. J Appl Toxicol 29:695–702

Li Y, Li J, Li S, Li Y, Wang X, Liu B, Fu Q, Ma S (2015) Curcumin attenuates glutamate neurotoxicity in the hippocampus by suppression of ER stress-associated TXNIP/NLRP3 inflammasome activation in a manner dependent on AMPK. Toxicol Appl Pharmacol 286:53–63

Lichtenwalner RJ, Parent JM (2006) Adult neurogenesis and the ischemic forebrain. JCBFM 26:1–20

Lieu PT, Heiskala M, Peterson PA, Yang Y (2001) The roles of iron in health and disease. Mol Aspects Med 22:1–87

Lin YC, Chen HW, Kuo YC, Chang YF, Lee YJ, Hwang JJ (2010) Therapeutic efficacy evaluation of curcumin on human oral squamous cell carcinoma xenograft using multimodalities of molecular imaging. Am J Chin Med 38:343–358

Lin YL, Liu YK, Tsai NM, Hsieh JH, Chen CH, Lin CM (2012) A lipo-PEG-PEI complex for encapsulating curcumin that enhances its antitumor effects on curcumin-sensitive and curcumin-resistance cells. Nanomed Nanotechnol 8:318e27

Liochev SI, Fridovich I (2002) The Haber-Weiss cycle-70 years later: an alternative view. Redox Rep 7:55–57

Liu A, Lou H, Zhao L, Fan P (2006) Validated LC/MS/MS assay for curcumin and tetrahydrocurcumin in rat plasma and application to pharmacokinetic study of phospholipid complex of curcumin. J Pharm Biomed Anal 40:720–727

Maes M, Galecki P, Chang YS, Berk M (2011) A review on the oxidative and nitrosative stress (O&NS) pathways in major depression and their possible contribution to the (neuro)degenerative processes in that illness. Prog Neuropsychopharmacol Biol Psychiatry 35:676–692

Mahesh T, Sri Balasubashini MM, Menon VP (2004) Photo-irradiated curcumin supplementation in streptozotocin-induced diabetic rats: effect on lipid peroxidation. Therapie 59:639–644

Maines MD (2000) The heme oxygenase system and its functions in the brain. Cell Mol Biol 46:573–585

Marczylo TH, Verschoyle RD, Cooke DN, Morazzoni P, Steward WP, Gescher AJ (2007) Comparison of systemic availability of curcumin with that of curcumin formulated with phosphatidylcholine. Cancer Chemother Pharmacol 60:171–177

Martindale JL, Holbrook NJ (2002) Cellular response to oxidative stress: signaling for suicide and survival. J Cell Physiol 192:1–15

Mathew A, Fukuda T, Nagaoka Y, Hasumura T, Morimoto H, Yoshida Y, Maekawa T, Venugopal K, Kumar DS (2012) Curcumin loaded-PLGA nanoparticles conjugated with Tet-1 peptide for potential use in Alzheimer's disease. PLoS One 7, e32616

McClure R, Yanagisawa D, Stec D, Abdollahian D, Koktysh D, Xhillari D, Jaeger R, Stanwood G, Chekmenev E, Tooyama I, Gore JC, Pham W (2015) Inhalable curcumin: offering the potential for translation to imaging and treatment of Alzheimer's disease. J Alzheimers Dis 44:283–295

Melzack R, Casey KL (1968) Sensory, motivational and central control determinants of pain: a new conceptual model. In: Kenshalo DR (ed) The skin senses. Charles C Thomas, Springfield, pp 423–443

Menon VP, Sudheer AR (2007) Antioxidant and anti-inflammatory properties of curcumin. Adv Exp Med Biol 595:105–125

Minghetti L, Ajmone-Cat MA, De Berardinis MA, De Simone R (2005) Microglial activation in chronic neurodegenerative diseases: roles of apoptotic neurons and chronic stimulation. Brain Res Rev 48:251–256

Mittal N, Joshi R, Hota D, Chakrabarti A (2009) Evaluation of antihyperalgesic effect of curcumin on formalin-induced orofacial pain in rat. Phytother Res 23:507–512

Monari M, Foschi J, Calabrese C, Liguori G, Di Febo G, Rizzello F, Gionchetti P, Trinchero A, Serrazanetti GP (2009) Implications of antioxidant enzymes in human gastric neoplasms. Int J Mol Med 24:693–700

Murugan P, Pari L (2007) Influence of tetrahydrocurcumin on erythrocyte membrane bound enzymes and antioxidant status in experimental type 2 diabetic rats. J Ethnopharmacol 113:479–486

National Cancer Institute (1996) Clinical development plan: curcumin. J Cell Biochem Suppl. 26:72–85

Nishiyama T, Mae T, Kishida H, Tsukagawa M, Mimaki Y, Kuroda M, Sashida Y, Takahashi K, Kawada T, Nakagawa K, Kitahara M (2005) Curcuminoids and sesquiterpenoids in turmeric (*Curcuma longa* L.) Suppress an increase in blood glucose level in type 2 diabetic KK-Aγ mice. J Agric Food Chem 53:959–963

Olivera A, Moore TW, Hu F, Brown AP, Sun A, Liotta DC, Snyder JP, Yoon Y, Shim H, Marcus AI, Miller AH, Pace TW (2012) Inhibition of the NF-κB signaling pathway by the curcumin analog, 3,5-Bis(2-pyridinylmethylidene)-4-piperidone (EF31): anti-inflammatory and anti-cancer properties. Int Immunopharmacol 12:368–377

Pae HO, Kim EC, Chung HT (2008) Integrative survival response evoked by heme oxygenase-1 and heme metabolites. J Clin Biochem Nutr 42:197–203

Panchal HD, Vranizan K, Lee CY, Ho J, Ngai J, Timiras PS (2008) Early anti-oxidative and anti-proliferative curcumin effects on neuroglioma cells suggest therapeutic targets. Neurochem Res 33:1701–1710

Park HR, Lee J (2011) Neurogenic contributions made by dietary regulation to hippocampal neurogenesis. Ann N Y Acad Sci 1229:23–28

Perkins S, Verschoyle RD, Hill K, Parveen I, Threadgill MD, Sharma RA, Williams ML, Steward WP, Gescher AJ (2002) Chemopreventive efficacy and pharmacokinetics of curcumin in the min/+ mouse, a model of familial adenomatous polyposis. Cancer Epidemiol Biomarkers Prev 11:535–540

Phillis JW, Horrocks LA, Farooqui AA (2006) Cyclooxygenases, lipoxygenases, and epoxygenases in CNS: their role and involvement in neurological disorders. Brain Res Rev 52:201–243

Phillips J, Moore-Medlin T, Sonavane K, Ekshyyan O, McLarty J, Nathan CA (2013) Curcumin inhibits UV radiation-induced skin cancer in SKH-1 mice. Otolaryngol Head Neck Surg 148:797–803

Pluta R, Bogucka-Kocka A, Ułamek-Kozioł M, Furmaga-Jabłońska W, Januszewski S, Brzozowska J, Jabłoński M, Kocki J (2015) Neurogenesis and neuroprotection in postischemic brain neurodegeneration with Alzheimer phenotype: is there a role for curcumin? Folia Neuropathol 53:89–99

Poole KM, Nelson CE, Joshi RV, Martin JR, Gupta MK, Haws SC, Kavanaugh TE, Skala MC, Duvall CL (2015) ROS-responsive microspheres for on demand antioxidant therapy in a model of diabetic peripheral arterial disease. Biomaterials 41:166–175

Prasad S, Tyagi AK, Aggarwal BB (2014) Recent developments in delivery, bioavailability, absorption and metabolism of curcumin: the golden pigment from golden spice. Cancer Res Treat 46:2–18

Priyadarsini KI (2013) Chemical and structural features influencing the biological activity of curcumin. Curr Pharm Des 19:2093–2100

Prousek J (2007) Fenton chemistry in biology and medicine. Pure Appl Chem 79:2325–2338

Rajapurkar MM, Shah SV, Lele SS, Hegde UN, Lensing SY, Gohel K, Mukhopadhyay B, Gang S, Eigenbrodt ML (2012) Association of catalytic iron with cardiovascular disease. Am J Cardiol 109:438–442

Rajasekar N, Dwivedi S, Tota SK, Kamat PK, Hanif K, Nath C, Shukla R (2013) Neuroprotective effect of curcumin on okadaic acid induced memory impairment in mice. Eur J Pharmacol 715:381–394

Rajasekaran SA (2011) Therapeutic potential of curcumin in gastrointestinal diseases. World J Gastrointest Pathophysiol 2:1–14

Ravindran J, Prasad S, Aggarwal BB (2009) Curcumin and cancer cells: how many ways can curry kill tumor cells selectively? AAPS J 11:495–510

Ray B, Bisht S, Maitra A, Maitra A, Lahiri DK (2011) Neuroprotective and neurorescue effects of a novel polymeric nanoparticle formulation of curcumin (NanoCurc™) in the neuronal cell culture and animal model: implications for Alzheimer's disease. J Alzheimers Dis 23:61–77

Re F, Cambianica I, Zona C, Sesana S, Gregori M, Rigolio R (2011) Functionalization of liposomes with ApoE-derived peptides at different density affects cellular uptake and drug transport across a blood-brain barrier model. Nanomed Nanotechnol 7:551e9

Reagan LP (2007) Insulin signaling effects on memory and mood. Curr Opin Pharmacol 7:633–637

Reagan LP (2012) Diabetes as a chronic metabolic stressor: causes, consequences, and clinical complications. Exp Neurol 233:68–78

Reagan LP, Grillo CA, Piroli GG (2008) The As and Ds of stress: metabolic, morphological and behavioral consequences. Eur J Pharmacol 585:64–75

Reger MA, Craft S (2006) Intranasal insulin administration: a method for dissociating central and peripheral effects of insulin. Drugs Today (Barc) 42:729–739

Rhee SG, Bae YS, Lee SR, Kwon J (2000) Hydrogen peroxide: a key messenger that modulates protein phosphorylation through cysteine oxidation. Sci STKE 53:pe1

Richardson DR (2002) Therapeutic potential of iron chelators in cancer therapy. Adv Exp Med Biol 509:231–249

Rinwa P, Kumar A (2012) Piperine potentiates the protective effects of curcumin against chronic unpredictable stress-induced cognitive impairment and oxidative damage in mice. Brain Res 1488:38–50

Rojo AI, Medina-Campos ON, Rada P, Zúñiga-Toalá A, López-Gazcón A, Espada S, Pedraza-Chaverri J, Cuadrado A (2012) Signaling pathways activated by the phytochemical nordihydroguaiaretic acid contribute to a Keap1-independent regulation of Nrf2 stability: role of glycogen synthase kinase-3. Free Radic Biol Med 52:473–487

Roney C, Kulkarni P, Arora V, Antich P, Bonte F, Wu A, Mallikarjuana NN, Manohar S, Liang HF, Kulkarni AR, Sung HW, Sairam M, Aminabhavi TM (2005) Targeted nanoparticle delivery through blood–brain barrier for Alzheimer's disease. J Control Release 108:193–214

Rushworth SA, Ogborne RM, Charalambos CA, O'Connell MA (2006) Role of protein kinase C delta in curcumin-induced antioxidant response element-mediated gene expression in human monocytes. Biochem Biophys Res Commun 341:1007–1016

Ryan JL, Heckler CE, Ling M, Katz A, Williams JP, Pentland AP, Morrow GR (2013) Curcumin for radiation dermatitis: a randomized, double-blind, placebo-controlled clinical trial of thirty breast cancer patients. Radiat Res 180:34–43

Sahay A, Scobie KN, Hill AS, O'Carroll CM, Kheirbek MA, Burghardt NS, Fenton AA, Dranovsky A, Hen R (2011) Increasing adult hippocampal neurogenesis is sufficient to improve pattern separation. Nature 472:466–470

Schiborr C, Kocher A, Behnam D, Jandasek J, Toelstede S, Frank J (2014) The oral bioavailability of curcumin from micronized powder and liquid micelles is significantly increased in healthy humans and differs between sexes. Mol Nutr Food Res 58:516–527

Schultze SM, Hemmings BA, Niessen M, Tschopp O (2012) PI3K/AKT, MAPK and AMPK signalling: protein kinases in glucose homeostasis. Expert Rev Mol Med 14, e1

Sehgal A, Kumar M, Jain M, Dhawan DK (2012) Piperine as an adjuvant increases the efficacy of curcumin in mitigating benzo(a)pyrene toxicity. Hum Exp Toxicol 31:473–482

Sena LA, Li S, Jairaman A, Prakriya M, Ezponda T, Hildeman DA, Wang C-R, Schumacker PT, Licht JD, Perlman H, Bryce PJ, Chandel NS (2013) Mitochondria are required for antigen-specific T cell activation through reactive oxygen species signaling. Immunity 38:225–236

Seo K-I, Choi M-S, Jung UJ, Kim HJ, Yeo J, Jeon SM, Lee MK (2008) Effect of curcumin supplementation on blood glucose, plasma insulin, and glucose homeostasis related enzyme activities in diabetic db/db mice. Mol Nutr Food Res 52:995–1004

Serhan CN, Savill J (2005) Resolution of inflammation: the beginning programs the end. Nat Immunol 6:1191–1197

Sharma RA, Ireson CR, Verschoyle RD, Hill KA, Williams ML, Leuratti C, Manson MM, Marnett LJ, Steward WP, Gescher A (2001) Effects of dietary curcumin on glutathione S-transferase and malondialdehyde-DNA adducts in rat liver and colon mucosa: relationship with drug levels. Clin Cancer Res 7:1452–1458

Sharma S, Kulkarni SK, Agrewala JN, Chopra K (2006) Curcumin attenuates thermal hyperalgesia in a diabetic mouse model of neuropathic pain. Eur J Pharmacol 536:256–261

Shehzad A, Lee YS (2013) Molecular mechanisms of curcumin action: signal transduction. Biofactors 39:27–36

Shehzad A, Ha T, Subhan F, Lee YS (2011) New mechanisms and the anti-inflammatory role of curcumin in obesity and obesity-related metabolic diseases. Eur J Nutr 50:151–161

Shishodia S (2013) Molecular mechanisms of curcumin action: gene expression. Biofactors 39:37–55

Shishodia S, Sethi G, Aggarwal BB (2005) Curcumin: getting back to the roots. Ann N Y Acad Sci 1056:206–217

Shishodia S, Singh T, Chaturvedi MM (2007) Modulation of transcription factors by curcumin. Adv Exp Med Biol 595:127–148

Shoba G, Joy D, Joseph T, Majeed M, Rajendran R, Srinivas PS (1998) Influence of piperine on the pharmacokinetics of curcumin in animals and human volunteers. Planta Med 64:353–356

Singh S, Aggarwal BB (1995) Activation of transcription factor NF-kappa B is suppressed by curcumin (diferuloylmethane) [corrected]. J Biol Chem 270:24995–25000

Singh DV, Godbole MM, Misra K (2013) A plausible explanation for enhanced bioavailability of P-gp substrates in presence of piperine: simulation for next generation of P-gp inhibitors. J Mol Model 19:227–238

Soetikno V, Sari FR, Veeraveedu PT, Thandavarayan RA, Harima M, Sukumaran V, Lakshmanan AP, Suzuki K, Kawachi H, Watanabe K (2011) Curcumin ameliorates macrophage infiltration by inhibiting NF-B activation and proinflammatory cytokines in streptozotocin induced-diabetic nephropathy. Nutr Metab (Lond) 8:35

Srinivasan M, Sudheer AR, Menon VP (2007) Ferulic acid: therapeutic potential through its antioxidant property. J Clin Biochem Nutr 40:92–100

Stone SS, Teixeira CM, Devito LM, Zaslavsky K, Josselyn SA, Lozano AM, Frankland PW (2011) Stimulation of entorhinal cortex promotes adult neurogenesis and facilitates spatial memory. J Neurosci 31:13469–13484

Stoner GD, Wang L, Casto BC (2008) Laboratory and clinical studies of cancer chemoprevention by antioxidants in berries. Carcinogenesis 29:1665–1674

Stumvoll M, Goldstein BJ, van Haeften TW (2005) Type 2 diabetes: principles of pathogenesis and therapy. Lancet 365:1333–1346

Sun GY, Horrocks LA, Farooqui AA (2007) The roles of NADPH oxidase and phospholipases A_2 in oxidative and inflammatory responses in neurodegenerative diseases. J Neurochem 103:1–16

Suresh D, Srinivasan K (2010) Tissue distribution & elimination of capsaicin, piperine & curcumin following oral intake in rats. Indian J Med Res 131:682–691

Suryanarayana P, Satyanarayana A, Balakrishna N, Kumar PU, Bhanuprakash Reddy G (2007) Effect of turmeric and curcumin on oxidative stress and antioxidant enzymes in streptozotocin-induced diabetic rat. Med Sci Monit 13:BR286–BR292

Svenson S (2012) Clinical translation of nanomedicines. Curr Opin Solid State Mater 16:287e94

Svensson CI, Yaksh TL (2002) The spinal phospholipase-cyclooxygenase-prostanoid cascade in nociceptive processing. Annu Rev Pharmacol Toxicol 42:553–583

Tajik H, Tamaddonfard E, Hamzeh-Gooshchi N (2007) Interaction between curcumin and opioid system in the formalin test of rats. Pak J Biol Sci 10:2583–2586

Tajik H, Tamaddonfard E, Hamzeh-Gooshchi N (2008) The effect of curcumin (active substance of turmeric) on the acetic acid-induced visceral nociception in rats. Pak J Biol Sci 11:312–314

Teiten MH, Dicato M, Diederich M (2013) Curcumin as a regulator of epigenetic events. Mol Nutr Food Res 57:1619–1629

Thiyagarajan M, Sharma SS (2004) Neuroprotective effect of curcumin in middle cerebral artery occlusion induced focal cerebral ischemia in rats. Life Sci 74:969–985

Thomas SL, Zhong D, Zhou W, Malik S, Liotta D, Snyder JP, Hamel E, Giannakakou P (2008) EF24, a novel curcumin analog, disrupts the microtubule cytoskeleton and inhibits HIF-1. Cell Cycle 7:2409–2417

Tiwari SK, Agarwal S, Seth B, Yadav A, Nair S, Bhatnagar P, Karmakar M, Kumari M, Chauhan LK, Patel DK, Srivastava V, Singh D, Gupta SK, Tripathi A, Chaturvedi RK, Gupta KC (2014) Curcumin-loaded nanoparticles potently induce adult neurogenesis and reverse cognitive deficits in Alzheimer's disease model via canonical Wnt/β-catenin pathway. ACS Nano 8:76–103

Todd PH Jr (1991) Curcumin complexed on water-dispersible substrates. US Patent US4999205

Tomren MA, Masson M, Loftsson T, Tonnesen HH (2007) Studies on curcumin and curcuminoids XXXI.Symmetric and asymmetric curcuminoids. Stability activity and complexation with cyclodextrin. Int J Pharm 338:27–34

Tønnesen HH, Másson M, Loftsson T (2002) Studies of curcumin and curcuminoids. XXVII. Cyclodextrin complexation: solubility, chemical and photochemical stability. Int J Pharm 244:127–135

Torchilin V (2009) Multifunctional and stimuli-sensitive pharmaceutical nanocarriers. Eur J Pharm Biopharm 7:431e44

Torchilin V (2012) Tumor delivery of macromolecular drugs based on the EPR effect. Adv Drug Deliv Rev 63:131e5

Tuorkey MJ (2014) Curcumin a potent cancer preventive agent: mechanisms of cancer cell killing. Interv Med Appl Sci 6:139–146

Uauy R, Olivares M, Gonzalez M (1998) Essentiality of copper in humans. Am J Clin Nutr 67(5 Suppl):952S–959S

van Praag H, Schinder AF, Christie BR, Toni N, Palmer TD, Gage FH (2002) Functional neurogenesis in the adult hippocampus. Nature 415:1030–1034

Vogelstein B, Kinzler KW (2004) Cancer genes and the pathways they control. Nat Med 10:789–799

Wahlstrom B, Blennow G (1978) A study on the fate of curcumin in the rat. Acta Pharmacol Toxicol 43:86–92

Wang YJ, Pan MH, Cheng AL, Lin LI, Ho YS, Hsieh CY (1997) Stability of curcumin in buffer solutions and characterization of its degradation products. J Pharm Biomed Anal 15:1867–1876

Wang R, Li YB, Li YH, Xu Y, Wu HL, Li XJ (2008) Curcumin protects against glutamate excitotoxicity in rat cerebral cortical neurons by increasing brain-derived neurotrophic factor level and activating TrkB. Brain Res 1210:84–91

Wang R, Li YH, Xu Y, Li YB, Wu HL, Guo H, Zhang JZ, Zhang JJ, Pan XY, Li XJ (2010) Curcumin produces neuroprotective effects via activating brain-derived neurotrophic factor/TrkB-dependent MAPK and PI-3K cascades in rodent cortical neurons. Prog Neuropsychopharmacol Biol Psychiatry 34:147–153

Weisberg SP, Leibel R, Tortoriello DV (2008) Dietary curcumin significantly improves obesity-associated inflammation and diabetes in mouse models of diabesity. Endocrinology 149:3549–3558

Weiss N, Miller F, Cazaubon S, Couraud PO (2009) The blood–brain barrier in brain homeostasis and neurological diseases. Biochim Biophys Acta 1788:842–857

Wu A, Ying Z, Gomez-Pinilla F (2006) Dietary curcumin counteracts the outcome of traumatic brain injury on oxidative stress, synaptic plasticity, and cognition. Exp Neurol 197:309–317

Xu Y, Ku B, Cui L, Li X, Barish PA, Foster TC, Ogle WO (2007) Curcumin reverses impaired hippocampal neurogenesis and increases serotonin receptor 1A mRNA and brain-derived neurotrophic factor expression in chronically stressed rats. Brain Res 1162:9–18

Yadav VR, Prasad S, Kannappan R, Ravindran J, Chaturvedi MM, Vaahtera L, Parkkinen J, Aggarwal BB (2010) Cyclodextrin-complexed curcumin exhibits anti-inflammatory and antiproliferative activities superior to those of curcumin through higher cellular uptake. Biochem Pharmacol 80:1021–1032

Yanagisawa D, Amatsubo T, Morikawa S, Taguchi H, Urushitani M, Shirai N, Hirao K, Shiino A, Inubushi T, Tooyama I (2011) In vivo detection of amyloid beta deposition using (1)(9)F magnetic resonance imaging with a (1)(9)F-containing curcumin derivative in a mouse model of Alzheimer's disease. Neuroscience 184:120–127

Yanagisawa D, Ibrahim NF, Taguchi H, Morikawa S, Hirao K, Shirai N, Sogabe T, Tooyama I (2015) Curcumin derivative with the substitution at C-4 position, but not curcumin, is effective against amyloid pathology in APP/PS1 mice. Neurobiol Aging 36:201–210

Yang F, Lim GP, Begum AN, Ubeda OJ, Simmons MR, Ambegaokar SS, Chen PP, Kayed R, Glabe CG, Frautschy SA, Cole GM (2005) Curcumin inhibits formation of amyloid beta oligomers and fibrils, binds plaques, and reduces amyloid in vivo. J Biol Chem 280:5892–5901

Yang K-Y, Lin L-C, Tseng T-Y, Wang S-C, Tsai T-H (2007) Oral bioavailability of curcumin in rat and the herbal analysis from Curcuma longa by LC-MS/MS. J Chromatogr B Analyt Technol Biomed Life Sci 853:183–189

Yang R, Zhang S, Kong D, Gao X, Zhao Y, Wang Z (2012) Biodegradable polymer-curcumin conjugate micelles enhance the loading and delivery of low-potency curcumin. Pharm Res 29:3512–3525

Yang Y, Duan W, Lin Y, Yi W, Liang Z, Yan J, Wang N, Deng C, Zhang S, Li Y, Chen W, Yu S, Yi D, Jin Z (2013) SIRT1 activation by curcumin pretreatment attenuates mitochondrial oxidative damage induced by myocardial ischemia reperfusion injury. Free Radic Biol Med 65C: 667–679

Yarru LP, Settivari RS, Gowda NK, Antoniou E, Ledoux DR, Rottinghaus GE (2009) Effects of turmeric (*Curcuma longa*) on the expression of hepatic genes associated with biotransformation, antioxidant, and immune systems in broiler chicks fed aflatoxin. Poult Sci 88:2620–2627

Ye SF, Hou ZQ, Zhong LM, Zhang QQ (2007) Effect of curcumin on the induction of glutathione *S*-transferases and NADP(H):quinone oxidoreductase and its possible mechanism of action. Yao Xue Xue Bao 42:376–380

Yeon KY, Kim SA, Kim YH, Lee MK, Ahn DK, Kim HJ, Kim JS, Jung SJ, Oh SB (2010) Curcumin produces an antihyperalgesic effect via antagonism of TRPV1. J Dent Res 89:170–174

Yoysungnoen P, Wirachwong P, Bhattarakosol P, Niimi H, Patumraj S (2006) Effects of curcumin on tumor angiogenesis and biomarkers, COX-2 and VEGF, in hepatocellular carcinoma cell-implanted nude mice. Clin Hemorheol Microcirc 34:109–115

Yu SY, Zhang M, Luo J, Zhang L, Shao Y, Li G (2013a) Curcumin ameliorates memory deficits via neuronal nitric oxide synthase in aged mice. Prog Neuropsychopharmacol Biol Psychiatry 45C:47–53

Yu SY, Gao R, Zhang L, Luo J, Jiang H, Wang S (2013b) Curcumin ameliorates ethanol-induced memory deficits and enhanced brain nitric oxide synthase activity in mice. Prog Neuropsychopharmacol Biol Psychiatry 44C:210–216

Zhao C, Deng W, Gage FH (2008) Mechanisms and functional implications of adult neurogenesis. Cell 132:645–660

Zhao X, Xu Y, Zhao Q, Chen CR, Liu AM, Huang ZL (2012) Curcumin exerts antinociceptive effects in a mouse model of neuropathic pain: Descending monoamine system and opioid receptors are differentially involved. Neuropharmacology 62:843–854

Zhao X, Wang C, Zhang JF, Liu L, Liu AM, Ma Q, Zhou WH, Xu Y (2013) Chronic curcumin treatment normalizes depression-like behaviors in mice with mononeuropathy: involvement of supraspinal serotonergic system and GABA receptor. Psychopharmacology (Berl) 231:2171–2187

Zhou H, Beevers CS, Huang S (2011) The targets of curcumin. Curr Drug Targets 12:332–347

Zhu Q, Sun Y, Yun X, Ou Y, Zhang W, Li J-X (2014) Antinociceptive effects of curcumin in a rat model of postoperative pain. Sci Rep 4:4932

Chapter 4
Effects of Curcumin on Transcription Factors and Enzyme Activities in Visceral Organs and the Brain

4.1 Introduction

Curcumin (diferuloyl methane) is a yellow polyphenolic diketone with most potent anti-inflammatory and antioxidant activities. Curcumin resembles ubiquinols in its structure. It is insoluble in water, but is readily soluble in organic solvents such as dimethylsulfoxide, acetone and ethanol. It is stable at acidic pH but unstable at neutral and basic pH, under which conditions it is degraded to ferulic acid and feruloylmethane (Wang et al. 1997). Most curcumin (>90 %) is rapidly degraded within 30 min of placement in phosphate buffer systems of pH 7.2 (Wang et al. 1997). The antioxidant, antiinflammatory, anticancer, and cholesterol lowering properties of curcumin are mediated by the regulation of transcription factors and enzymic activities (Aggarwal and Harikumar 2009; Jurenka 2009; Aggarwal et al. 2013). Curcumin also protects from metabolic syndrome (Madkor et al. 2011) by decreasing insulin resistance, obesity, hypertriglyceridemia, and hypertension (Jitoe-Masuda et al. 2013). In addition, curcumin significantly decreases lipid peroxidation, increases intracellular antioxidant (GSH), regulates antioxidant enzymes, and scavenges hyperglycemia-induced ROS (Strasser et al. 2005). Curcumin stabilizes lysosomal membranes and causes uncoupling of oxidative phosphorylation. Curcumin not only reduces the oxidative damage and cognitive deficits associated with aging, but also enhances hippocampal neurogenesis supporting the view that curcumin may enhance neural plasticity and repair (Kim et al. 2008). At the molecular level curcumin not only modulates the activity of transcription factors, but also regulates activities of many enzymes (Table 4.1). Curcumin also binds directly to DNA and RNA through interactions between nucleic acid and β-diketone moiety of curcumin. The interaction other macromolecules with curcumin is mediated through the α, β-unsaturated β-diketone moiety, carbonyl and enolic groups of the β-diketone moiety, methoxy and phenolic hydroxyl groups, and phenyl rings (Gupta et al. 2011). As a result of these interactions, curcumin can inhibit tumor proliferation, growth, metastasis, invasion, and angiogenesis. Curcumin mediates tumor cell death mainly

© Springer International Publishing Switzerland 2016

A.A. Farooqui, *Therapeutic Potentials of Curcumin for Alzheimer Disease*,

DOI 10.1007/978-3-319-15889-1_4

Table 4.1 Target genes for NF-κB and STAT3 transcription factors

Target genes for NF-κB	Target genes for STAT3
cMyc	cMyc
MDM2	Pim1
Cyclin D1	Cyclin D1
Pim1	AKT
AKT	Jun B
P53	P53
IL-6 and IL-8	–
MCP-1	–
sPLA$_2$	–
COX-2	–
MMP-9	–
iNOS	–

through apoptosis, but in cells that are apoptosis resistant, curcumin has been reported to induce mitotic catastrophe (Wolanin et al. 2006) and autophagy (Shinojima et al. 2007). At the cellular level, curcumin induces a mild oxidative and lipid-metabolic stress resulting into an adaptive cellular stress response through the hormetic stimulation of these cellular antioxidant defense systems and lipid metabolic enzymes (Farooqui 2013). Thus, curcumin increases biogenic amines (e.g., dopamine, serotonin, and norepinephrine) in the cortex and hippocampus after stress and olfactory bulbectomy induced depression-like behavior (Kulkarni et al. 2008; Xu et al. 2005). Curcumin treatment not only up-regulates the expression of hippocampal BDNF, but also down-regulates the expression of mRNA and protein of pro-inflammatory cytokines (TNF-α and IL-1β) (Arora et al. 2011). Curcumin has also been reported to ameliorate the adverse effects of the mutagenic neurotoxin *N*-methyl *N*-nitrosourea in cerebrum and cerebellum of mice (Singula and Dhawan 2012). Collective evidence suggests that curcumin is a multifunction phytochemical, which interacts with multiple molecular targets, modulating cell growth, inflammation, and apoptosis signaling pathways (Kurapati et al. 2012). Curcumin-mediated decrease in oxidative and lipid-mediated stress may not only explain the beneficial effects of curcumin on inflammation, cardiovascular, and neurodegenerative diseases, but may also contribute to the increase in maximum life-span observed in animal models.

4.2   Effect of Curcumin on Transcription Factors

Transcription factors are proteins responsible for the coordinated expression of genes through specific binding to gene promoter and enhancer sites. They control the transfer of genetic information from DNA to mRNA. Once bound to DNA,

Fig. 4.1 Modulation of
transcription factors by
curcumin

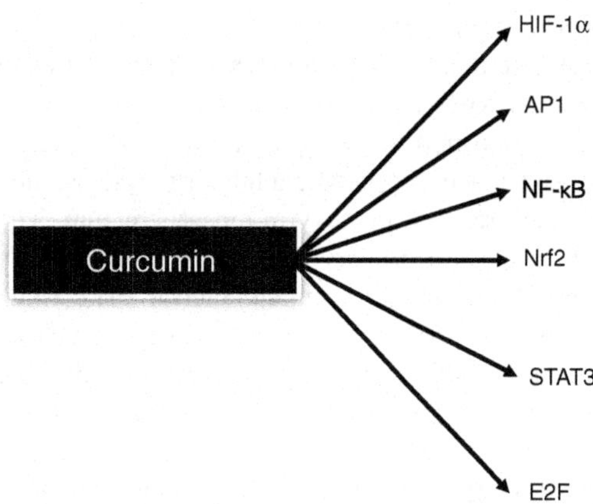

transcription factors can promote or block the enzyme that controls the "transcription," of genes. Thus, transcription factors can turn on or off the genes by binding to DNA and other proteins. Transcription factors are not only vital for the normal development, but are also involved in cellular functions and response to disease. As stated above, curcumin modulates the activity of several transcription factors including NF-κB, AP-1, STAT3, PPARγ; HIF-1α; Nrf2, β-catenin, FOXO1/3a, CREB, and E2F (Fig. 4.1), which not only modulate the expression of genes involved in free radicals scavenging (e.g., catalase, MnSOD, and heme oxygenase-1), but are also involved in maintaining lipid homeostasis.

4.2.1   Modulation of NF-κB by Curcumin

NF-κB is a critical regulator of immediate responses not only to pathogens and injury, but also plays an important role in regulating cell proliferation and survival (Atkinson et al. 2010; Hayden and Ghosh 2012). NF-κB is found in neuronal and glial cells, where it is involved in activation and modulation of a large number of genes in response to metabolic and traumatic injuries, immune responses, neuroinflammation, macrophage infiltration factors, cell adhesion molecules, cell survival, and other stressful situations requiring rapid reprogramming of gene expression (Hayden and Ghosh 2012). NF-κB family of transcription factor is composed of NF-κB1, NF-κB2, RelA, RelB, and c-Rel, which bind to κB enhancers as homo- or hetero-dimers. NF-κB signaling is mediated through the canonical and non-canonical pathways (Atkinson et al. 2010; Sun 2011; Razani et al. 2011). NF-κB1-mediated signaling is linked with the activation of distinct cell-surface receptors (e.g. NMDA receptors, toll like receptors, interleukin-1 receptor and tumor necrosis factor receptor-α) or cytoplasmic sensors (RIG-I-like receptors) induce the canonical response, which is mediated by the IκB kinase (IKK) complex (Sun 2011;

Razani et al. 2011). This complex is composed of three core proteins (IKK1/IKKα, IKK2/IKKβ, and NEMO/IKKγ). The first two core proteins are structurally related kinases whereas the third is a regulatory subunit exhibiting affinity for upstream activators modified by polyubiquitin chains. NF-κB proteins are normally held in the cytoplasm by specific inhibitory proteins, inhibitors if NF-κB (IκBs), characterized by the presence of ankyrin-repeat structure. The prototypical IκB member is IκBα, which plays a primary role in inducing the activation of the canonical (or classical) NF-κB pathway (Hayden and Ghosh 2012; Chen 2005). Degradation of IκB releases the NF-κB p50/p65 complex promoting translocation NF-κB to the nucleus, where it interacts with NF-κB-response element and initiates expression of many genes including genes for pro-inflammatory cytokines (TNF-α, IL-8, MCP-1 and IL-6), chemokines, adhesion molecules, secretary and cytosolic phospholipase A2 (PLA$_2$), cyclooxygenase-2 (COX-2), growth factors, metallo-proteinases (MMPs), and nitric oxide synthese (NOS) (Fig. 4.2 and Table 4.2) (Gilmore and Wolenski 2012; Amălinei et al. 2010). It should be noted that interactions of ROS with NF-κB also promote the translocation of NF-κB to the nucleus, where NFκB promotes the expression of above mentioned genes. Antioxidants prevent NF-κB translocation to the nucleus (Stephenson et al. 2000).

The noncanonical NF-κB signaling pathway involves the activation of p52/RelB NF-κB complex leading to the regulation of specific immunological processes. This NF-κB pathway relies on the inducible processing of NF-κB2 precursor protein, p100, as opposed to the degradation of IκBα in the canonical NF-κB pathway (Xiao et al. 2001). A central signaling component of the noncanonical NF-κB pathway is NF-κB-inducing kinase (NIK), which functions together with a downstream kinase, IKKα (inhibitor of NF-κB kinase α), to induce phosphorylation-dependent ubiquitination and processing of p100 (Xiao et al. 2001). In non-stimulated neural cells,

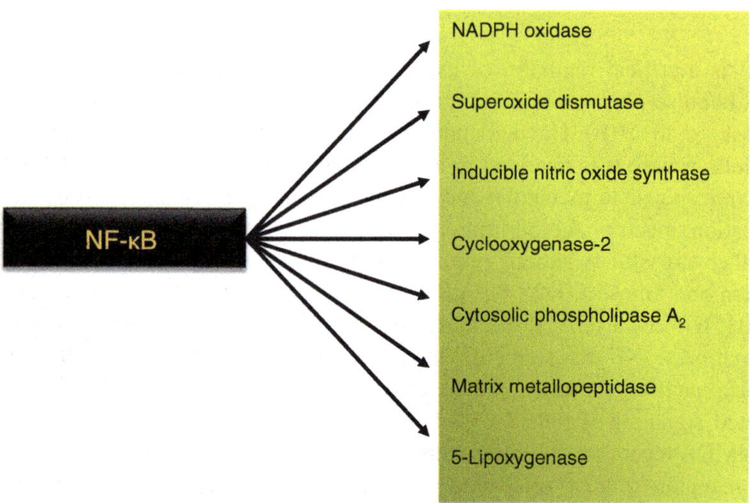

Fig. 4.2 Modulation of enzymes by NF-κB

Table 4.2 Nrf2 regulated genes in neural and nonneural tissues

Nrf2 regulated genes and enzymes	Reference
NAD(P)H:quinone reductase (NQO1)	Benson et al. (1980) and Tkachev et al. (2011)
Glutathione S-transferase (GST)	Mann et al. (2007) and Gao et al. (2014b)
γ-Glutamylcysteine synthetase	Mann et al. (2007) and Gao et al. (2014b)
Uridine diphosphate-glucuronosyltransferases	Mann et al. (2007) and Gao et al. (2014b)
Epoxide hydrolase	Mann et al. (2007) and Gao et al. (2014b)
Thioredoxin reductase	Mann et al. (2007) and Gao et al. (2014b)
Glucose 6-phosphate dehydrogenase	Mann et al. (2007) and Gao et al. (2014b)
6-Phosphogluconate dehydrogenase	Mann et al. (2007) and Gao et al. (2014b)
Malic enzyme	Mann et al. (2007) and Gao et al. (2014b)
Glutathione peroxidase	Mann et al. (2007) and Gao et al. (2014b)
Glutathione reductase	Harvey et al. (2009) and Gao et al. (2014b)
Ferritin	Li and Kong (2009)
Haptaglobin	Mann et al. (2007) and Gao et al. (2014b)
Biosynthetic enzymes of the glutathione conjugation pathway	Mann et al. (2007) and Gao et al. (2014b)
Biosynthetic enzymes of the glucuronidation conjugation pathway	Mann et al. (2007) and Gao et al. (2014b)
Heme oxygenase	Ryter and Choi (2010)
Aldoketo reductase	Barski et al. (2008)
ABC transporters	Wang et al. (2014)
Thioredoxin	Lee et al. (2003)

a TNF receptor-associated factor-3 (TRAF3)-dependent E3 ubiquitin ligase targets NIK for continuous degradation. In response to signals induced by a subset of TNF receptor superfamily members, NIK becomes stabilized as a result of TRAF3 degradation leading to the activation of noncanonical NF-κB (Sun 2012).

Curcumin inhibits the activation of constitutive and inducible NF-κB (Fig. 4.3). The molecular mechanism of curcumin induced inhibition of NF-κB activation is not clear. However, it is reported that curcumin has the ability to inhibit IKK activation, one of the major kinase involved in NF-κB activation pathway (Bharti et al. 2003). Curcumin completely blocks both oxidative stress-induced and TNF-α-mediated activation of NF-κB. In vitro studies have indicated that curcumin inhibits both serine/threonine protein kinase and protein tyrosine kinase (Reddy and Aggarwal 1994). It is also suggested that curcumin not only increases the glutathione levels, but also quenches superoxide and hydroxyl radicals. Collective evidence suggests that curcumin blocks inflammation and oxidative stress by inhibiting the activation of NF-κB and quenching ROS production.

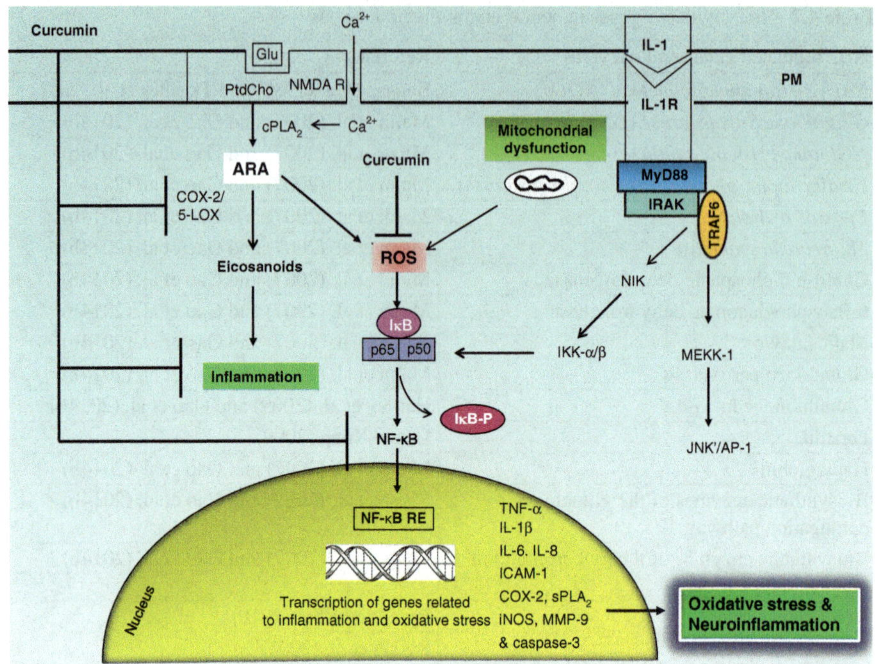

Fig. 4.3 Hypothetical diagram showing the effect of curcumin on oxidative stress, inflammation, and NF-κB-mediated gene transcription. *N*-Methyl-D-aspartate receptor (*NMDA-R*); Glutamate (*Glu*); Phosphatidylcholine (*PtdCho*); cytosolic phospholipase A₂ (*cPLA₂*); cyclooxygenase-2 (*COX-2*); 5-lipoxygenase (*5-LOX*); arachidonic acid (*ARA*); reactive oxygen species (*ROS*); nuclear factor-κB (*NF-κB*); nuclear factor-κB-response element (*NF-κB-RE*); inhibitory subunit of NF-κB (*I-κB*); tumor necrosis factor-α (*TNF-α*); interleukin-1β (*IL-1β*); interleukin-6 (*IL-6*); inducible nitric oxide synthase (*iNOS*); secretory phospholipase A₂ (*sPLA₂*); interleukin-1 (*IL-1*); interleukin-1 receptor (*IL-1R*); adaptor protein (*MyD88*); 5-lipoxygenase (*5-LOX*); 5-LOX activating protein (*FLAP*); tumor necrosis factor receptor-associated factor adaptor protein 6 (*TRAF6*); NF-κB-inducing kinase (*NIK*); and IκB kinase (*IKK*)

4.2.2   Modulation of AP1 by Curcumin

AP-1is a dimeric inducible transcription factor, which controls virtually all areas of eukaryotic cellular behavior, from cell cycle proliferation and development to stress response and apoptosis through its binding with AP1/TRE or CRE DNA motifs found in a wide variety of genes (Shaulian and Karin 2001; Shaulian and Karin 2002). AP-1 is composed of members of Fos (c-Fos, FosB, Fra-1 and Fra-2) and Jun (c-Jun, JunB and JunD) families (Shaulian and Karin 2001; Shaulian and Karin 2002; Ozanne et al. 2006; Ozanne et al. 2007). AP1 protein abundance and activity are controlled by intermingled transcriptional and post-transcriptional mechanisms, which are themselves regulated by intracellular signaling (Lopez-Bergami et al. 2010; Murphy and Blenis 2006). AP1 belongs to the class of basic leucine zipper (bZIP) transcription factors, which bind to promoters of its target genes in a sequence-specific manner, and transactivates or represses them. As stated above, AP1 proteins are involved in the modulation of a variety of cellular processes including proliferation and survival, differentiation, growth, apoptosis, cell migration, morphogenesis, and transformation. Depending upon the abundance of dimerization partners, dimer-composition, post-translational regulation, and interaction with accessory proteins, cells make the decision if AP1 transcription factor mediates positive or negative effect (Kaminska et al. 2000; Vesely et al. 2009). AP1 transcription factor is activated not only by growth factors, neurotransmitters, and cellular stress, but also by ionizing and ultraviolet irradiation, cytoskeletal rearrangements, and variety of cytokines (Wisdom 1999; Kaminska et al. 2000; Eferl and Wagner 2003). External stimuli mediate their effect on AP1 transcription factor mainly through mitogen-activated protein kinases (MAPKs) cascade (Pulverer et al. 1991; Chen et al. 1996) using extracellular signal regulated kinases (ERKs) 1 and 2, p38 kinases and c-Jun N-terminal kinases (JNKs). Among these kinases, ERK1 and ERK2 are widely expressed and involved in the regulation of critical cellular functions, including proliferation, differentiation, migration and apoptosis (Shaul and Seger 2007). Activation of upstream MAPK kinase results in translocation of ERK1 and ERK2 into the nucleus, where ERK1/2 further activates the transcription factors including AP1. The activated AP1 interacts with the 12-*O*-tetradecanoylphorbol-13-acetate (TPA) responsive element (5′-TGAG/CTCA-3′) in the gene promoter or enhancer and thereby modulates expression of target genes (*matrix metalloproteinases* (*MMPs*) and *CD44*) (Ozanne et al. 2007).

AP1 has been found constitutively active in many cancers including breast, ovarian, cervical, and lung. Many studies have shown that inhibition of AP1 may have a profound effect on the behavior of cancer cells and tumors supporting the view that AP1 may be a promising target for cancer therapy (Shaulian and Karin 2002). Curcumin suppresses constitutive AP1 activity not only in HL-60, Raji cells, but also in several prostate cancer cell lines (Huang et al. 1991; Hergenhahn et al. 2002; Mukhopadhyay et al. 2001). Curcumin also suppresses LPS-mediated COX-2 gene expression by inhibiting AP1 DNA binding in BV2 microglial cells (Kang et al. 2004). It is suggested that curcumin directly interacts with Jun-Fos dimer and inhibits its binding to DNA (AP1 site) (Park et al. 1998; Huang et al. 1991).

4.2.3 Modulation of STAT3 by Curcumin

Signal transducer and activator of transcription 3 (STAT3) is a member of STATs family of transcription factors, which are tightly regulated under physiologic conditions and become transiently activated in response to cytokine and growth factor receptor as well as nonreceptor kinase stimulation, all of which are found to be aberrantly active in glial tumors (Brantley and Benveniste 2008). STAT3 has been reported to regulate cell growth, survival, angiogenesis, metastasis, differentiation, motility, inflammation, and immune responses. STAT3 protein contains 5 domains, including oligomerization domain, coiled coil domain, DNA binding domain, phosphotyrosine binding domain, and transcriptional activation domain. The DNA-binding domain is necessary for the recognition of specific binding sequences. Seven members of STAT family have been identified and characterized. Among STAT family members, STAT3 is over-expressed or activated by various carcinogenic agents. STAT3 modulates cell proliferation and differentiation and anti-apoptosis through the activation of the target genes, including STAT3, c-Myc and p53 (Kathiria et al. 2012). STAT3 protein exists in a latent or inactive form in the cytoplasm. Transient phosphorylation of STAT3 by JAK kinases (JAKs) at various phosphorylation sites (particularly at Tyr-705 and Ser-727) results in its dimerization and translocation to the nucleus, where it binds to specific promoter sequences and regulates gene expression (Fig. 4.4) (Darnell 2002). STAT3 can be also phosphorylated by other tyrosine kinases, such as the Src family. However, such Src-mediated STAT3 phosphorylation does not always leads to the STAT3 activation (Michels et al. 2013). In addition, serine phosphorylation at residue 727 of STAT3 also results in the up-regulation of the transcriptional activity. STAT3 phosphorylation at Ser-727 is mediated by MAPK, P38 and c-Jun N-terminal kinase (JNK) pathways, and is associated with the transcriptional regulation of the target genes of STAT3 (Tworkoski et al. 2011). Ser-727 mutant STAT3 knock-in mice display impaired development and survival process (Shen et al. 2004). Recent studies have also indicated that STAT3 may also migrate into mitochondria, where it regulates mitochondrial respiration and homeostatic mechanism coupling cell cycle progression to the intracellular redox potential. These activities of STAT3 are independent of its

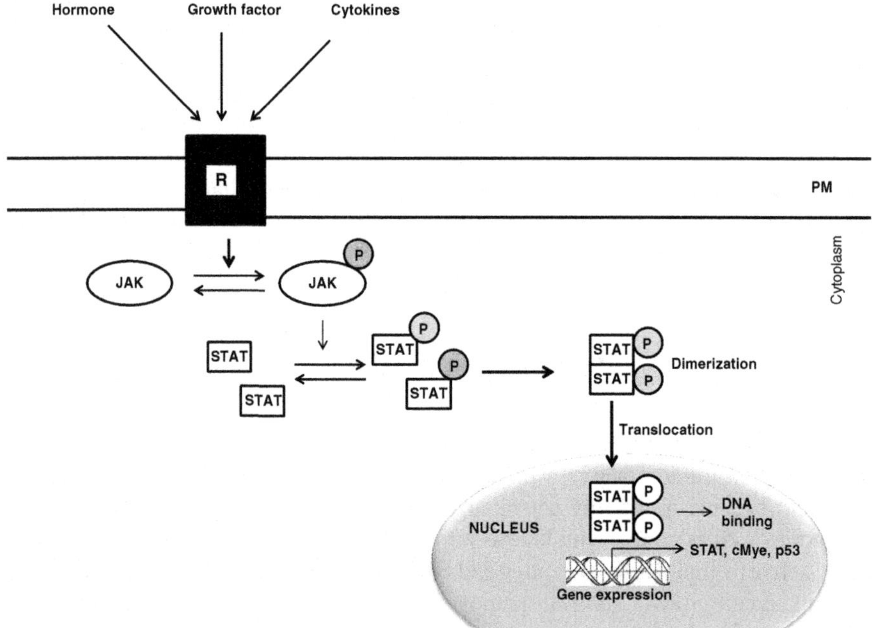

Fig. 4.4 Hypothetical diagram showing JAK/STAT signaling. Plasma membrane (*PM*); Receptor (*R*); Janus kinase (*JAK*); signal transducer and activator of transcription (*STAT*); and phosphate (*P*)

transcriptional regulatory activity (Shaw 2010). Presence of phosphorylated STAT3 (p-STAT3) has been reported to occur in nearly 70 % of human cancers (Yu et al. 2009).

Inappropriate activation of JAK/STAT signaling occurs with high frequency in human cancers and is associated with cancer cell survival and proliferation. JAK/ STAT pathway plays a critical role in tumor growth including glioblastoma. Microglial cells, the resident macrophages of the brain, have been known to play a critical role in the progression of brain glioblastoma. Curcumin inhibits STAT3 activity, cell growth, cell cycle progression, and invasiveness of 3 different murine glioma cell lines. Daily consumption of curcumin inhibits tumor growth and produces significant tumor-free survival in immunocompetent mice with orthotopically implanted gliomas. The molecular mechanisms in curcumin-mediated effects are not fully understood. However, it is suggested that curcumin mediates its effects through the suppression of Janus kinase (JAK)-STAT signaling pathway. Several JAK isoforms (JAK1, JAK2, JAK3 and TYK2) have been reported to occur in mammalian tissues including brain. Curcumin inhibits the phosphorylation of JAK1 and JAK2 via the increased phosphorylation of SHP-2 and its association with JAK1/2, thus attenuating inflammatory response (Kim et al. 2003).

4.2.4   Modulation of HIF-1α by Curcumin

HIF-1 is a basic helix-loop-helix transcription factor, which transactivates genes encoding proteins that modulate and participate in homeostatic responses to hypoxia. HIF-1 is composed of a constitutively expressed HIF-1β subunit and one of three α subunits (HIF-1α, HIF-2α or HIF-3α). Both subunits are part of the basic Helix-Loop-Helix PER-ARNT-SIM (bHLH-PAS) family of transcription factors and these domains are important for DNA binding and dimerization (Wang et al. 1995). The HIF-α subunits contains several domains including an oxygen-dependent degradation domain (ODDD) and two transactivation domains. Both the *N*-terminal transactivation domain (N-TAD) and the *C*-terminal transactivation domain (C-TAD) are essential for the regulation of HIF-dependent gene expression (Pugh et al. 1997; Jiang et al. 1997). This transcription factor regulates the expression of multiple genes involved in oxygen transportation, angiogenesis, proliferation, and metabolism, enabling a cell to counteract a hypoxic environment (Semenza 2011).

All cells require specific levels of O_2 for respiration as well as for maintaining optimal cellular homeostasis. Low oxygen tension is a unique environmental stress, which produces global changes in a complex regulatory network of transcription factors and signaling proteins to coordinate cellular adaptations in metabolism, proliferation, DNA repair, and apoptosis. Cells respond to conditions of low levels of O_2 by altering gene expression patterns, which affects levels of several hundred proteins involved in cell survival as well as microRNAs (miRNAs), which are short noncoding RNAs (about 22 nucleotides) that regulate gene expression through posttranscriptional mechanisms, as key elements in this response to hypoxia. The biogenesis of miRNA is a highly-orchestrated process which essentially requires the co-ordination of ribonucleases, RNA-binding proteins and the miRNA gene itself (Slezak-Prochazka et al. 2010). Low levels of O_2 induce a distinct shift in the expression of a specific group of miRNAs, termed hypoxamirs. Among these miR-NAs, miR-210 is the master hypoxamiR and regulates a variety of cellular events in non-lymphoid tissues (Chan et al. 2012). Emerging evidence indicates that low levels of O_2 regulate several facets of hypoxamir transcription, maturation, and function (Nallamshetty et al. 2013). These changes are initiated, regulated, and maintained by HIF-1. This transcription factor anchors in cytoplasm. Under adequate levels of oxygen a rapid degradation of HIFα subunits by prolyl hydroxylase domain-containing enzymes (PHD) keep HIF-1α in an inactive state. Hypoxia leads to inhibition of prolyl hydroxylation and, as a consequence, to HIF stabilization. Subsequently, HIF-α migrates to the nucleus where it forms a heterodimer with HIF-β and this complex interacts with the hypoxia-responsive elements (HREs), which is composed of the core sequence 5′-(A/G)CGTG-3′ of the HIF-responsive genes (Fig. 4.5) (Wang and Semenza 1993). HIF then activates a highly complex transcription program, comprising in excess of one hundred genes that regulate

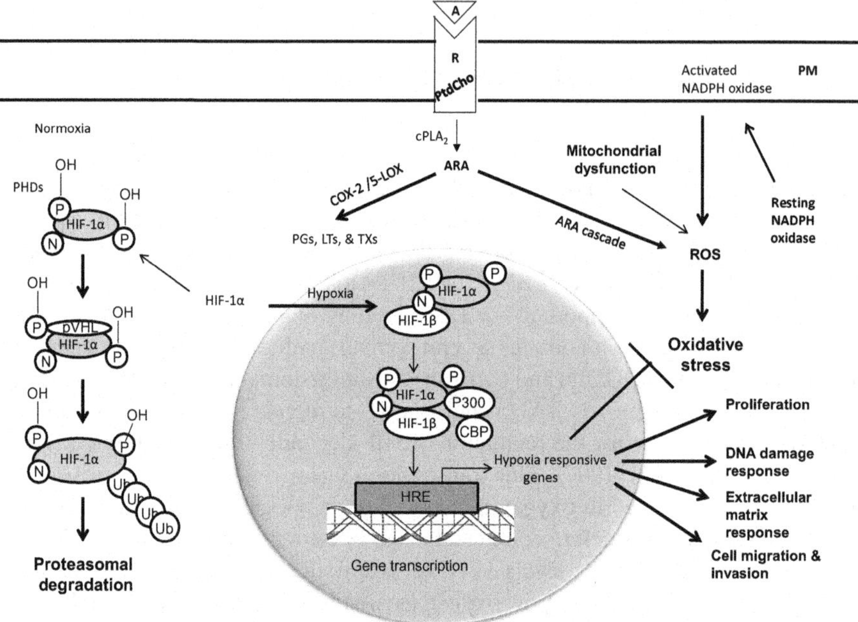

Fig. 4.5 Hypothetical diagram showing transcription factors hypoxia-inducible factors-1α (HIF-1α). Plasma membrane (*PM*); Receptor (*R*); agonist (*A*); Phosphatidylcholine (*PtdCho*); cytosolic phospholipase A₂ (*cPLA₂*); cyclooxygenase-2 (*COX-2*); 5-lipoxygenase (*5-LOX*); prostaglandins (*PGs*); leukotrienes (*LTs*); tromboxanes (*TXs*); reactive oxygen species (*ROS*); transcription factors hypoxia-inducible factors-1α (*HIF-1α*); transcription factors hypoxia-inducible factors-1β (*HIF-1β*); hypoxia-responsive elements (*HRE*); prolyl hydroxylase domain proteins (*PHDs*); protein Von Hippel Lindau (*pVHL*); co-factors p300; Creb-binding protein (*CBP*); and ubiquitin (*Ub*)

processes such as angiogenesis, glucose metabolism, migration, survival and death (Semenza 2003; Pouysségur et al. 2006; Semenza 2007). These processes are supported by the transcription of many genes including heme-oxygenase-1 (HO-1), erythropoietin (EPO), and numerous molecules involved in vascular reactivity such as nitric oxide synthase (NOS). Hypoxia is also accompanied by the stimulation of glycogenolysis resulting in increased availability of glucose. The upregulation of glucose transporter (GLUT) augments glucose uptake. This overexpression of GLUTs is mediated by the activation of AMP kinase (AMPK) and p38 mitogen-activated kinase. Stimulation of AMPK results from a decreased cytoplasmic ATP/AMP ratio together with altered cellular redox status. HIF-1α decreases mitochondrial oxygen consumption and induces the expression of pyruvate dehydrogenase kinase, the main inhibitor of pyruvate dehydrogenase and of the entry of acetylCoA into mitochondria. Collective evidence suggests that HIF-1 not only controls glucose metabolism, but also modulates cell proliferation and vascularization along with regulation of mammalian development. HIF-1α also regulates many important steps of the metastatic processes, especially epithelial-mesenchymal transition (EMT) that is one of the crucial mechanisms to cause early stage of tumor metastasis (Nauta et al. 2014). The activation/overexpression of the alpha subunit(s) of HIF in cancer is a very common occurrence and is suspected to account at least in part for the well-established tumor-associated properties of deregulated glycolysis and angiogenesis (Gordan and Simon 2007).

It is well known that the vascular endothelial growth factor (VEGF) family consists of several members including VEGF-A-D and PlGF. Each member has distinct affinity for VEGF receptors 1–3 (VEGFR1–3) and neuropilins. VEGFR1/Flt1 is a high-affinity receptor for VEGF-A, -B and PlGF versus VEGFR2/Flk1, which is a low-affinity receptor for VEGF-A, -C and –D (Ferrara et al. 2003). VEGF inhibitor treatment decreases fasting blood glucose levels and improves glucose tolerance in mice and humans through unclear mechanisms (Kamba et al. 2006) and specific VEGF-B inhibition improves glucose tolerance through enhanced peripheral glucose uptake (Hagberg et al. 2012). Studies on the effect of curcumin on HIF-1α have indicated that curcumin modulates the production of HIF1α and vascular endothelial growth factor A (VEGFA), two key components involved in tumor neovascularization through angiogenesis. Curcumin dose-dependently inhibits basal VEGFA secretion in corticotroph AtT20 mouse and lactosomatotroph GH3 rat pituitary tumor cells as well as in all human pituitary adenoma cell cultures. Under hypoxic conditions in AtT20 and GH3 cells as well as in all human pituitary adenoma cell cultures, curcumin not only suppresses the induction of mRNA synthesis, but also down-regulates the production of HIF1α protein. Furthermore, curcumin also blocks hypoxia-induced mRNA synthesis and secretion of VEGFA in all cell cultures. These observations support the view that curcumin inhibits pituitary adenoma progression not only through anti-proliferative and pro-apoptotic actions but also by its suppressive effects on pituitary tumor neovascularization (Shan et al. 2012). It is also reported that both curcumin and its analog (EF24) decrease intracellular levels of both HIF-1α and HIF-1β proteins This in turn results in down-regulation of HIF's transcriptional activity in various human epithelial cancer cell lines (Thomas et al. 2008). Although, the molecular mechanism associated with curcumin-mediated inhibition of HIF protein is not fully understood, but based on detailed investigations it is suggested that curcumin dose-dependently blocks hypoxia-stimulated angiogenesis in vitro and down-regulated HIF-1α and VEGF expression in vascular endothelial cells. These findings suggest that curcumin may play pivotal roles in tumor suppression via the inhibition of HIF-1α-mediated angiogenesis.

4.2.5   *Modulation of Nrf2 by Curcumin*

Nuclear factor E2-related factor 2 (Nrf2) is a redox-sensitive transcription factor, which controls and mediates adaptive responses to intrinsic and extrinsic cellular stresses leading to the expression of antioxidant, detoxification, and cell defense genes (Kaspar et al. 2012). Nrf2 in combination with PPARγ and PGC-1α also plays an important role in biogenesis of mitochondria (Itoh et al. 2015). Nrf2 is composed of six functional domains known as Nrf2-ECH homologies (Neh) and designated as Neh1-6, respectively. Each Neh domain serves its own function and the details have been well characterized (Baird and Dinkova-Kostova 2011). Under physiological conditions, Nrf2 anchors in the cytoplasm where its activity is repressed through binding with a 624 amino acids protein called Kelch like ECH-associated protein 1 (Keap1), a protein, which has multiple electrophile-reactive cysteine residues (Fig. 4.6) (Itoh et al. 2010; Kobayashi and Yamamoto 2005; Kobayashi et al. 2009). Nrf2 is constantly ubiquitinated through Keap1 in the cytoplasm and degraded in the proteasome. Because the Keap1-Nrf2 pathway and particularly the thiols of Keap1 are sensitive to oxidative modification and adduction by electrophiles, this pathway represents an important detoxification and cell protection mechanism. Nrf2 deficient mice are more susceptible to vascular damage due to diminished glutathione levels and an impaired compensatory induction of GSH synthesis, highlighting a fundamental role for Nrf2 in antioxidant defenses against oxidative stress (Chan and Kwong 2000).

The exposure cells to oxidative or electrophilic stress (15-deoxy-PGJ$_2$, NO$_2$-fatty acids, isothiocyanates, and 4-HNE) promotes the dissociation of Nrf2 from Keap1complex. Free Nrf2 migrates to the nucleus, where in combination with other transcription factors (e.g., sMaf, ATF4, JunD, PMF-1) binds to the antioxidant response elements (ARE) located in the regulatory regions of cellular defense enzyme genes (5′-NTGAG/CNNNGC-3′) (Itoh et al. 2010; Kobayashi et al. 2009) and facilitates expression of cytoprotective genes, numerous protective enzymes and scavengers (Table 4.2). A number of other processes have been reported to stimulate the migration of Nrf2 to the nucleus. These include caloric restriction, moderate exercise, hyperbaric therapy, ozone therapy, and consumption of polyphenol supplements (Bocci and Valacchi 2015). Repression of Nrf2 signaling has been demonstrated. It is reported that either Nrf2 breaks down in the nucleus (Kaspar et al. 2012) or Keap1 is imported into the nucleus, where it binds to Nrf2, which is exported of out of the nucleus as a Nrf2-Keap1 complex and then degraded in the

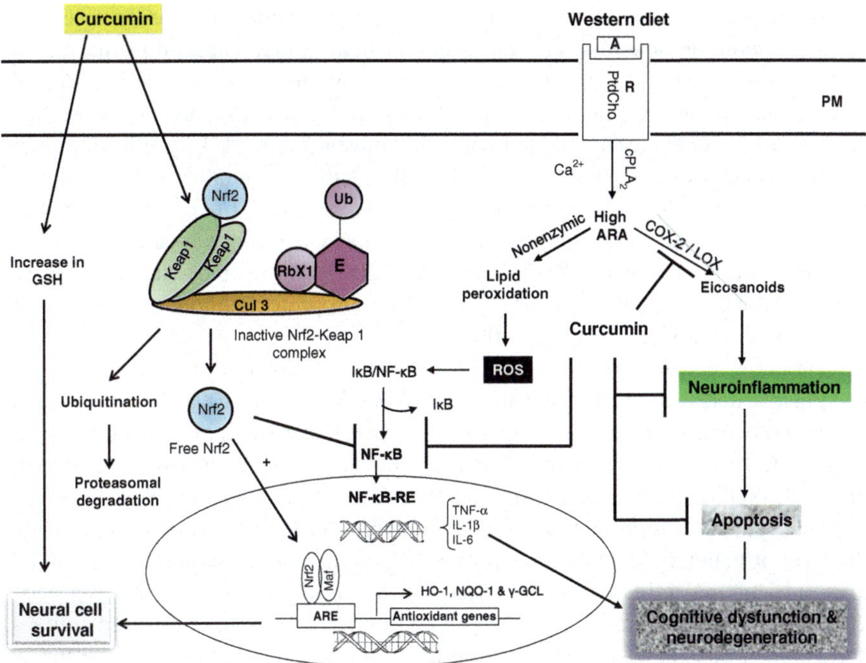

Fig. 4.6 Hypothetical diagram showing beneficial effects of curcumin on long term consumption of western diet. Receptor (*R*); agonist (*A*); Phosphatidylcholine (*PtdCho*); cytosolic phospholipase A₂ (*cPLA₂*); cyclooxygenase-2 (*COX-2*); lipoxygenase (*LOX*); arachidonic acid (*ARA*); reactive oxygen species (*ROS*); nuclear factor-κB (*NF-κB*); nuclear factor-βB-response element (*NF-κB-RE*); inhibitory subunit of NF-κB (*I-κB*); tumor necrosis factor-α (*TNF-α*); interleukin-1β (*IL-1β*); interleukin-6 (*IL-6*); inducible nitric oxide synthase (*iNOS*); secretory phospholipase A₂ (*sPLA₂*); nuclear factor E2-related factor 2 (*Nrf2*); kelch-like ECH-associated protein 1 (*Keap1*); antioxidant response element (*ARE*); KEAP1 Cullin-3 E3-ubiquitin ligase complex (*KEAP1-CUL3-RBX1 E3*); small leucine zipper proteins (*Maf*); heme oxygenase (*HO-1*); NADPH quinine oxidoreductase (*NQO-1*); γ-glutamate cystein ligase (*γ-GCL*)

cytosol (Sun et al. 2007). Both processes constitute a means of turning off Nrf2 signaling and preventing permanent induction of Nrf2-regulated genes. Once in the cytoplasm, the Nrf2-Keap1 complex associates with the Cul3-Rbx1 core ubiquitin machinery, leading to the degradation of Nrf2 and termination of the Nrf2/ARE signaling pathway.

In the brain, antioxidant properties of Nrf2-overexpressing glial cells are more pronounced than neurons. Thus, the activation of ARE-regulated genes has been reported to occur in astrocytes, which consequently have more efficient detoxification and antioxidant defences than neurons. Astrocytes closely interact with neurons to provide structural, metabolic and trophic support, as well as actively participating in the modulation of neuronal excitability and neurotransmission (Vargas and Johnson 2009). Therefore, functional alterations in astrocytes metabolism may influence metabolic and functional activities of surrounding cells, such as neurons and microglia. Activation of Nrf2 in astrocytes protects neurons from a wide array of insults in different in vitro and in vivo conditions including neurodegenerative diseases, cerebral ischemia, and intracerebral hemorrhage supporting the role of astrocytes in determining the vulnerability of neurons to noxious stimuli (Gao et al. 2014a; van Muiswinkel and Kuiperij 2005; Shih et al. 2005). In astrocytes, Nrf2 also modulates enzymes associated with GSH biosynthesis (xCT cystine antiporter, gamma-glutamylcysteine synthetase; and GSH synthase), glutathione S-transferase and glutathione reductase, and multidrug resistance protein 1 overexpression. This regulation may lead to an increase in intracellular GSH contents and decrease in oxidative stress (Armstrong 1997; Hayes et al. 2005). Collective evidence suggests that Nrf2/ARE activation in astrocytes confers neuroprotection to neighboring neurons (Shih et al. 2003; Kraft et al. 2004).

Curcumin increases survival of cortical neurons following exposure to oxygen and glucose deprivation/reoxygenation (OGD/R) (Yang et al. 2009; Wu et al. 2013). At the molecular level, curcumin disrupts the Nrf2-Keap1 complex, leading to elevated Nrf2 binding to ARE and subsequent increase in the expression and activity of HO-1, NQO1 in neural and non-neural cells via activation of p38 MAP kinase and UDP-glucuronosyltransferase (UGT) isozymes. In addition, curcumin decreases the level of malondialdehyde, increases SOD activity and the expression of NQO1 after intraluminal middle cerebral artery occlusion (Yang et al. 2009; Wu et al. 2013). It is also demonstrated that after 24 h reoxygenation with curcumin cultured cortical neurons increase the transcription level of NQO1 and the addition of LY294002 (signaling inhibitors for PtdIns 3 K) reduces the NQO1 expression indicating that up-regulation NQO1 expression in neurons by curcumin requires activation of the PtdIns 3-K/Akt pathway in vitro (Pugazhenthi et al. 2007). Finally, curcumin may also provide neuroprotection by increasing levels of reduced glutathione (Stridh et al. 2010).

4.2.6 *Modulation of E2F by Curcumin*

E2F, a family of transcription factor, plays an important role in controlling development, proliferation, differentiation, and survival. The E2F proteins consist of 8 family members (E2F1, E2F2, E2F3, E2F4, E2F5, E2F6, E2F7, and E2F8), which function as transcription factors. They are divided into groups: activators and repressors. E2F1-6 all contain one E2F-TDP domain, and E2F1, E2F2, and E2F3 are generally considered as the 'active E2Fs' on the basis of their ability to potently activate transcription. E2F4 and E2F5 are called as the 'repressive E2Fs', which bind their targets coincident with their repression in G0/G1, and only modestly activate transcription (Iaquinta and Lees 2007; Polager and Ginsberg 2008). E2F1-5, along with DP family proteins, can bind with the 'pocket protein' family protein (RB1, RBL1, and RBL2). DP family proteins, which contain one E2F-TDP domain and one DP domain, are dimerization partners of E2Fs. E2F6 also possesses one E2F-TDP domain but no RB binding domain, and it cannot bind to RB family proteins (Trimarchi et al. 1998). E2F, a family of transcription factor plays an important role in controlling development, proliferation, differentiation, and survival. These roles of E2F are mediated through the involvement of cyclin-dependent kinase (CDK) and its inhibitors of the INK4 family (Muller et al. 2001; DeGregori and Johnson 2006; Garneau et al. 2009). The E2Fs regulate the timely expression of a series of genes whose products are essential for cell proliferation (Helin 1998; Muller et al. 2001; Polager et al. 2002).

Curcumin-induced ROS down-regulation of E2F4 expression and modulation of E2F4 target genes which finally lead to the apoptotic cell death in HCT116 colon cancer cells, suggesting that E2F4 appears to be a novel determinant of curcumin-induced cytotoxicity.

Treatment of HCT116 colon cancer cells with curcumin results not only in the expression of E2F4, but also several E2F4 downstream genes, E2F1, c-myc, CDK2, cyclin A, cyclin D1, p21, and p27, which are involved in cell cycle control and related to the inhibition of apoptotic cell death. This inhibition can be restored by pretreatment of N-acetylcystein and doxycyclin-induced E2F4 expression. Moreover, E2F4 overexpression partially restores curcumin-mediated growth inhibition, confirming the role of E2F4 in the cell proliferation supporting the view that E2F4 seems to be a novel determinant of curcumin-induced cytotoxicity (Kim and Lee 2010).

4.2.7   Modulation of FOXO by Curcumin

Forkhead box class O (FOXO) is a family of transcription factors, which plays an important role in proliferation, apoptosis, differentiation and metabolic processes. In mammals, members of the FOXO family include FOXO1, FOXO3A, FOXO4, and FOXO6 (Monsalve and Olmos 2011). These proteins remain transcriptionally active in the nucleus in the absence of environmental and growth factors (Huang and Tindall 2007). Modification of FOXO leads to its translocation to the cytoplasm and/or its degradation, resulting in the suppression of transcriptional activity. The FOXO genes are homologues of daf-16, a key regulator of the insulin-IGF1 signaling pathway and a modulator of lifespan in Caenorhabditis elegans. Members of FOXO family share the characteristic of being regulated by the insulin/PI3K/Akt signaling pathway (Dillin et al. 2002). Targeting of FOXO factors has also been proposed for the treatment of metabolic dysfunctions such as diabetes mellitus, immunological disorders and neurodegenerative diseases (Farooqui 2013). Activation of FOXO proteins also upregulates expression of genes involved in cell cycle arrest, apoptosis and DNA repairs, implying that these proteins function as tumor suppressors (Huang and Tindall 2011). FOXO proteins contribute to tumor suppression by transcriptionally regulating expression of genes involved in cell cycle arrest, apoptosis, DNA repair and oxidative stress resistance.

Nothing is known about the effect of curcumin and its metabolite on FOXO. However, there is one report on the effect of tetrahydrocurcumin (a metabolite of curcumin) in Drosophila melanogaster have indicated that tetrahydrocurcumin-mediated modulation of FOXO is closely associated with anti-oxidative stress response and extension of life span (Xiang et al. 2011).

4.3   Effect of Curcumin on Enzyme Activities

It is well known that curcumin produces its effects by modulating activities of many enzymes including protein kinases (Protein kinase C, mitogen-activated protein kinases, and Akt), arachidonic acid metabolizing enzymes (such as COX-2 and 5-LOX), nitric oxide producing enzymes (iNOS) (Zhou et al. 2011), apoptosis promoting enzymes (caspases), and chromosomal DNA protecting and repairing enzyme (telomerase) (Table 4.3). The molecular mechanism by which curcumin activates protein kinases (ERK and p38 MAP kinases) is unknown. However, it is proposed that curcumin influences structurally unrelated membrane proteins across several signaling pathways (Bilmen et al. 2001). For a small minority of these proteins, specific binding of curcumin to the protein has been detected with a binding constant typically in the nanomolar range. However, for the majority of the proteins whose activity is modulated by curcumin, a curcumin binding site has not been

Table 4.3 Effect of curcumin on enzyme activities in neural and non-neural tissues

Enzyme	Effect	Reference
COX-2	Inhibition	Anand et al. (2008) and Hatcher et al. (2008)
5-LOX	Inhibition	Anand et al. (2008) and Hatcher et al. (2008)
iNOS	Inhibition	Chan et al. (1998)
MMP9	Inhibition	Anand et al. (2008) and Hatcher et al. (2008)
IKK	Inhibition	Anand et al. (2008) and Hatcher et al. (2008)
Akt	Inhibition	Anand et al. (2008) and Hatcher et al. (2008)
JAK2	Inhibition	Kim et al. (2003)
PKA	Inhibition	Anand et al. (2008) and Hatcher et al. (2008)
PKC	Inhibition	Anand et al. (2008) and Hatcher et al. (2008)
PLD	Inhibition	Yamamoto et al. (1997)
Histone acetyltransferase	Inhibition	Yamamoto et al. (1997) and Balasubramanyam et al. (2004)
Acidic sphingomyelinase	Inhibition	Cheng et al. (2007)
Ornithine decarboxylase	Inhibition	Oyagbemi et al. (2009)
Glyoxalase	Inhibition	Santel et al. (2008)
Phosphorylase kinase	Inhibition	Sun et al. (2013)
GST	Activation	Anand et al. (2008) and Hatcher et al. (2008)
NQO1	Activation	Shih et al. (2003) and Vargas and Johnson (2009)
Heme oxygenase	Activation	Shih et al. (2003) and Vargas and Johnson (2009)
CAT	Activation	Shih et al. (2003) and Vargas and Johnson (2009)
SOD	Activation	Shih et al. (2003) and Vargas and Johnson (2009)
UGT	Activation	Shih et al. (2003) and Vargas and Johnson (2009)
Epoxide hydrolase	Activation	Anand et al. (2008) and Hatcher et al. (2008)

Cycooxygenase-2 (*COX-2*); 5-lipoxygenase (*5-LOX*); inducible nitric oxide synthase (*iNOS*); matrix metallopeptidase9; IκB kinase (*I-κK*); a serine/threonine protein kinase (*Akt or Protein kinase B*); NADPH:quinone oxidoreductase (*NQO1*); Janus kinase 2 (*JAK2*); protein kinase A (*PKA*); protein kinase C (*PKC*); phospholipase D (*PLD*); glutathione S-transferase (*GST*); catalase (*CAT*); superoxide dismutase (*SOD*); and UDP-glucuronosyl transferase (*UGT*) identified and characterized. Furthermore, curcumin modulates activity of these proteins at approximately similar concentrations (micromolar range) despite the lack of a consensus binding motif (Ingolfsson et al. 2007). Since curcumin modulates many unrelated membrane proteins at approximately similar concentrations, it is proposed that curcumin regulates the action of membrane proteins indirectly by changing the physical properties of the membrane rather than by the direct binding of curcumin to the protein (Ingolfsson et al. 2007). Furthermore, curcumin either directly binds with the kinases or phosphatases upstream of the MAP kinases or the activation of ERK and p38 MAP kinases by curcumin may represent an adaptive response of the cell to stress (Kim et al. 2008).

4.3.1 Effect of Curcumin on Cyclooxygenases and Lipoxygenases

COXs and LOXs are enzymes that catalyze the conversion of arachidonic acid into prostaglandins (PGs), and leukotrienes (LTs). These metabolites not only contribute to inflammation, but also are the key mediators of exaggerated pain sensation (Farooqui 2011). Therefore, non steroidal anti-inflammatory drugs (NSAIDs), which are strong inhibitors of COXs have been used in the management of chronic inflammation and postoperative pains (Rainsford 2006). Treatment of HT-29 human colon cancer cells with various concentrations of curcumin results in inhibition of the cell growth of HT-29 cells in a concentration- and time-dependent manner. Detailed investigations have indicated that curcumin markedly inhibits the mRNA and protein expression of COX-2, but not cyclooxygenase-1 (COX-1). This inhibition of COX-2 reduces levels of PGs leading to decrease in the intensity of inflammation (Goel et al. 2001).

4.3.2 Effect of Curcumin on Kinases

Curcumin modulates activities of many protein kinases, such as protein kinase C (PKC), mammalian target of rapamycin (mTOR), and EGFR tyrosine kinase in time and dose-dependent manner. PKC is a Ca^{2+}- and phospholipid-dependent enzyme that catalyzes the phosphorylation of serine and threonine residues in a wide variety of cellular proteins. PKC comprises a family of enzymes that significantly differ in structure, cofactor requirements and cellular compartmentalization. The PKC family has been classified into three classes: conventional isoforms (α, β_I, β_{II} and γ) that require Ca^{2+} and diacylglycerol (DAG) for activation; novel isoforms (δ, ε, η, θ and μ) that require only DAG and atypical isoforms (ζ, ι and λ) that require neither Ca^{2+} nor DAG (Newton 2001). Curcumin inhibits PKC activity at low Ca^{2+} concentrations, whereas activation is observed at saturating Ca^{2+} concentrations. The binding of curcumin with PKC depends on the presence of membranes. Furthermore, the addition of PtdSer-containing membrane vesicles results in strong activation of PKC by curcumin. The Ca^{2+} and membrane effects can be explained in the light of previous investigations on the role of these cofactors in PKC activation (Mahmmoud 2007).

mTOR is a 289 kDa serine/threonine (Ser/Thr) protein kinase, which lies downstream of insulin-like growth factor I receptor (IGF-IR) and PtdIns 3' kinase (PtdIns 3 K) and acts as a master regulator of a large array of cellular processes, including cell proliferation, growth, differentiation, survival, and motility (Guertin and Sabatini 2007). Curcumin inhibits proliferation and induced apoptosis of RMS cells in a concentration-dependent manner. Curcumin inhibition of cell proliferation by curcumin is related to the arrest of cells in the G_1/G_0 phase of the cell cycle (Beevers et al. 2006). Curcumin also blocks the basal and IGF-1-stimulated cell motility of these RMS cells (Beevers et al. 2006; Beevers et al. 2009).

Receptor tyrosine kinases (RTKs) are key regulators of normal cellular processes. They play a critical role in the development and progression of many types of cancer by binding either with polypeptide growth factors or cytokines or hormones (Zwick et al. 2001). It is becoming increasingly evident that protein kinases (PKC, mTOR, and EGFR tyrosine kinase) are the major upstream molecular targets for curcumin intervention, whereas the nuclear oncogenes such as c-jun, c-fos, c-myc, CDKs, FAS, and iNOS act as downstream molecular targets for curcumin actions (Lin 2007). TPA activates PKC by reacting with zinc thiolates present within the regulatory domain. In contrast, curcumin inactivates PKC by oxidizing the vicinal thiols present within the catalytic domain (Lin 2004). Due to its ability to inhibit Src, JNK, and Smad3 phosphorylations, curcumin abrogates the TGFβ1-induced tissues overgrowth (Yang et al. 2013). Curcumin directly induces a tumor-suppressive miR-203-mediated regulation of the Src-Akt axis in bladder cancer (Saini et al. 2011). Another potential mechanism by which curcumin inhibits phosphorylation of Src and stat3 is partly through regenerating liver-3 (PRL-3) downregulation (Wang et al. 2009). Curcumin may produce its anticancer actions partly via suppressing PtdIns 3 K/Akt signal transduction pathway in several tumor models (Jiang et al. 2014). Furthermore, curcumin significantly inhibits NF-κB and attenuates the effect of irradiation-mediated prosurvival signaling through the PtdIns 3 K/Akt/mTOR and NF-κB pathways in these gut-specific endothelial cells (Rafiee et al. 2010).

Curcumin activates ERK and p38 MAP kinases in a concentration and dose-dependent manner. The molecular mechanisms by which curcumin activates these enzymes remain unknown. One possibility is that curcumin directly interacts with the kinases themselves, or kinases or phosphatases upstream of the MAP kinases. This has not been clearly established, but it is becoming increasingly evident that there is direct interaction of polyphenols with signal transduction molecules. Earlier studies suggest that electrophilic curcumin can activate the Nrf2-antioxidant response element pathway by interacting with and dissociating the Keap-Nrf2 complex (Balogun et al. 2003). Curcumin may also cause the activation of ERK and p38 MAP kinases through the adaptive mechanism. Thus, high doses of curcumin (micromolar) can cause oxidative stress leading to apoptotic cell death in cancer cells (Salvioli et al. 2007). However, subtoxic doses of curcumin may cause a mild adaptive stress response that involve the activation of ERK and p38 MAP kinases leading to an increase in proliferation and survival of neural stem cells. The latter mechanism is consistent with the hormesis hypothesis for the beneficial actions phytochemicals on neurons (Mattson and Cheng 2006).

4.3.3   Effect of Curcumin on Matrix Metalloproteinase

Matrix metalloproteinases (MMPs) are zinc-dependent endopeptidases, which hydrolyze components of the extracellular matrix (ECM) (Visse and Nagase 2003). MMPs family consists of more than 26 endopeptidases, which share homologous protein sequences, with conserved domain structures and specific domains related to substrate specificity and recognition of other proteins (Visse and Nagase 2003; Fields 2014). Timely degradation of ECM is an important feature of development, morphogenesis, tissue repair and remodeling. MMPs hydrolyze a peptide bond before a residue with a hydrophobic side chain, such as Leu, Ile, Met, Phe, or Tyr. The hydrophobic residues fit into the S1′ specificity pocket, whose size and shape differ considerably among MMPs (Bode et al. 1999). In addition to the S1′ pocket, other substrate contact sites (subsites) also participate in the substrate specificity of MMPs (Visse and Nagase 2003). These enzymes play a central role in many biological processes, such as embryogenesis, normal tissue remodeling, wound healing, and angiogenesis, and in diseases such as atheroma, arthritis, cancer, and tissue ulceration. MMPs are important regulators of tumor growth both at the primary site and in distant metastases suggesting that MMP can be considered as important targets for cancer therapy.

In indomethacin-induced model of gastric damage, treatment of Curcumin not only protects gastric mucosal cell damage and oxidative stress, but it also regulates expression and activities of MMPs during protection and healing. Detailed investigations have revealed that the antiulcer activity of curcumin may be due to the inhibition of MMP-9 (Swarnakar et al. 2005).

Treatment with curcumin is not only known to suppress NF-κB-regulated gene products involved in inflammation (COX-2, MMP-3, MMP-9, vascular endothelial growth factor), but also inhibits apoptosis-mediated cell death and prevents the activation of caspase-3 (Shakibaei et al. 2007).

4.3.4   Effect of Curcumin on Caspases

Caspases or cysteine-aspartic acid proteases contribute to the apoptotic cell death. By hydrolyzing critical proteins, caspases contribute to morphological and biochemical changes, which result in apoptosis, such as chromatin condensation, loss of cell adhesion, cell shrinkage, membrane blebbing, DNA fragmentation, and finally formation of apoptotic bodies, which stimulate their own engulfment by phagocytes (Taylor et al. 2008). Caspases are expressed as single chain polypeptides composed of three domains: N-terminal pro-peptide, a large subunit and a small sub-unit. The mammalian caspase family contains at least 14 members that are categorized according to substrate specificity, domain composition or their intracellular role (Nicholson 1999; Chowdhury et al. 2008). Activation of caspases involves cleavage of their proforms into active forms. Based on protein structure and activation mechanism, caspases can be subdivided into initiator caspases and executioner caspases. Initiator caspases include Caspase-8 and Caspase-10, Caspase-9, and Caspase-2, whereas executioner caspases consists of Caspase-3, -6, and -7 (Taylor et al. 2008). In addition to apoptosis, caspases play an important role in other cellular processes. Thus, caspase-1 and caspase-11 are not only involved in inflammation, but also contribute to inflammatory cell death by pyroptosis (Shalini et al. 2015). Similarly, caspase-8 has dual role in cell death, mediating both receptor-mediated apoptosis and in its absence, necroptosis (Shalini et al. 2015). Caspase-8 also functions in the maintenance and homeostasis of the adult T-cell population. Caspase-3 is involved in tissue differentiation, regeneration and neural development in ways that are distinct and do not involve any apoptotic activity. Caspase-2 has emerged as a unique caspase with potential roles in maintaining genomic stability, metabolism, autophagy and aging (Shalini et al. 2015).

Curcumin acts by producing DNA damage and endoplasmic reticulum (ER) stress and mitochondrial-dependent-induced apoptosis through the activation of caspase-3 (Ravindran et al. 2009). Curcumin-mediated apoptotic cell death is accompanied with upregulation of the protein expression of Bax and downregulation of the protein levels of Bcl-2, resulting in dysfunction of mitochondria and subsequently led to cytochrome c release and sequential activation of caspase-9 and caspase-3 in various types of cancer cells in a time-dependent manner. These findings support the view that mitochondria, AIF caspase-3- dependent pathways play a vital role in curcumin-induced G2/M phase arrest and apoptosis of cancer cells in vitro (Ravindran et al. 2009).

4.3.5 *Effect of Curcumin on Glutathione S-Transferase*

The glutathione S-transferase (GST) is the detoxification enzyme, which plays an important role in inactivating endogenous and exogenous toxic products under oxidative stress. Humans contain three distinct classes of GST's: cytosolic, mitochondrial and microsomal. Cytosolic GST is the predominant enzyme in the brain (Hayes et al. 2005). At least 7 forms of cytosolic GST (named alpha, mu, pi, sigma, theta, omega, and zeta) have been reported to occur in the brain (Higgins and Hayes 2011). Each GST molecule composed of two characteristic domains: I and II. Domain I consist of the N-terminal residues (1–80) of the protein whose structure contains a series of beta pleated sheets and alpha helices. Domain II comprises the remaining residues ($81–209 \pm 11$) and is also referred to as the hydrophobic site (H site) (Dirr et al. 1994). The GSH binding domain is found in Domain I and is structurally conserved in each of the isoforms (Armstrong 1997). Among isoforms of

GST, Domain II contains structural differences among the isoforms. The variable residues in this domain contribute to the array of substrate specificity found among the GSTs (Babbitt 2000).

Curcumin protects neural and nonneural cells by increasing not only the expression of GST, but also several phase II detoxification enzymes, such as γ-glutamyl cysteine ligase (γ-GCL), heme oxygenase-1 (HO-1) (Dinkova-Kostova and Talalay 2008; Reyes-Fermin et al. 2012). These proteins reduce electrophiles and free radicals to less toxic intermediates whilst increasing the ability of the cell to repair any subsequent damage.

4.3.6 Effect of Curcumin on Inducible Nitric Oxide Synthase

NO, a short-lived gaseous lipophilic molecule synthesized in almost all tissues and organs, is formed from its precursor L-arginine by a family of enzymes called nitric oxide synthases (NOSs) (Alderton et al. 2001). NOS system consists of three distinct isoforms including neuronal nitric oxide synthase (nNOS), inducible nitric oxide synthase (iNOS), and endothelial nitric oxide synthase (eNOS), which differ from each other in their expression, subcellular localizations and mechanistic features and are responsible for their unique pathophysiological roles. NO is a free radical synthetized in immune cells and other cell types such as epithelial cells in order to get rid of pathogens (Eckmann et al. 2000). Enhanced expression of the iNOS with the sustained overproduction of NO is closely associated with the pathogenesis of chronic inflammatory diseases of brain and visceral organs. Demethoxycurcumin, active constituents of *Curcuma longa* L shows higher potency in suppressing lipopolysaccharide (LPS)-mediated NO production, iNOS, COX-2, and NF-κB activities in a RAW 264.7 macrophage cell line (Guo et al. 2008). Similarly, demethoxy curcumin exhibits the strongest inhibitory activity on NO production and TNF-α expression in LPS-activated microglia compared to curcumin (CUR) and BDMC (Zhang et al. 2008).

4.3.7 Effect of Curcumin on Telomerase

Telomeres are nucleoprotein structures that protect the ends of chromosomes from DNA repair and degradation. In vertebrates, telomeres consist of TTAGGG repeats, which are bound by a multiprotein complex known as "shelterin" or "the telosome". This complex has fundamental roles in the regulation of telomere length and protection (de Lange and Shelterin 2005). Telomerase is a reverse transcriptase associated with the elongation of telomeres *de novo* during cell division (Collins and Mitchell 2000; Collins 2008). Human telomerase is composed of three molecules: (a) human telomerase reverse transcriptase (TERT), (b) telomerase RNA (TR or TERC), and (c) dyskerin (DKC1) (Cohen et al. 2007). TR serves as a template for telomeric

DNA synthesis and the reverse transcriptase activity of hTERT adds a template region of the hTR RNA onto chromosomal ends as the telomeric DNA sequences (de Lange and Shelterin 2005). After birth, in most of somatic cells telomeres are silenced and progressively shorten with aging (Canela et al. 2007). Some cell types, such as cells of the hematopoietic lineage, stem cells, and germ cells, have the ability to activate telomerase, but this is not sufficient to inhibit telomere shortening with aging (Flores et al. 2008). Critically short telomeres cannot be repaired by any of the known DNA repair mechanisms and consequently trigger a persistent DNA damage response (DDR) leading to cellular senescence and/or apoptosis (Collado et al. 2007), eventually compromising tissue regenerative capacity and function, and contributing to organismal aging (Blasco 2007).

Curcumin inhibits expression of telomerase mRNA in several types of neural and nonneural cell lines in dose and time-dependent manner (Ramachandran et al. 2002) especially when its concentration is increased. Exposure time with curcumin also plays a key role in the inhibition of expression levels (a time-and dose-dependent manner similar to that of the cell growth inhibition). Inhibitory effect of curcumin on telomerse activity in breast cancer (MCF-7) cell line has also been observed (Ramachandran et al. 2002; Nasiri et al. 2013). Detailed investigations on glioblastoma and medulloblastoma cells have indicated that curcumin treatment results in higher cytotoxicity in the cells that express telomerase enzyme, highlighting its potential as an anticancer agent. Gene and protein expression analyses indicate that curcumin acts by down-regulating CCNE1, E2F1 and CDK2 and up-regulating the expression of PTEN genes, resulting in growth arrest at G2/M phase (Khaw et al. 2013). Curcumin-mediated apoptotic cell death not only involves an increase in caspase-3/7 activity and overexpression of Bax, but also inhibition of telomerase activity and down-regulation of hTERT mRNA expression leading to telomere shortening (Sundin and Hentosh 2012; Khaw et al. 2013). In addition, down-regulation of Bcl2 and survivin is also observed in curcumin-treated cells.

4.4 Conclusion

Curcumin, a dietary polyphenol, is the active constituent from the *Curcuma longa* plant, commonly known as turmeric. Turmeric exhibits anti-tumor and anti-inflammatory activities with low toxicity. The most important feature of curcumin is that it has very little or no side effects despite being a therapeutic agent with multiple beneficial functions. The antioxidant mechanism of curcumin is due to its specific conjugated structure of two methoxylated phenols and an enol form of β-diketone. This structure is responsible for free radical trapping ability as a chain breaking antioxidant. It can significantly inhibit the generation of ROS both in vitro and in vivo. Due to its polyphenolic structure, curcumin prevents carcinogen-induced cancer in rodents and inhibits the growth of tumors in humans and animals. These actions of curcumin are mediated through the modulation of transcription factors, such as NF-κB, AP-1, STAT3, HIF-1α; Nrf2, FOXO, and E2F factors and enzyme activities. Curcumin also exerts its effects by blocking cell cycle progression and triggering apoptotic cell death. All three stages of carcinogenesis including initiation, promotion and progression are suppressed by curcumin in dose and time dependent manner. Suppression of NF-κB, AP-1, STAT3, HIF-1α; Nrf2, FOXO, and E2F plays a central role in regulating the expression of various genes involved in cell survival, apoptosis, carcinogenesis and inflammation. These properties make curcumin a powerful therapeutic agent for the treatment of cancer and other chronic diseases. Furthermore curcumin may also act by downregulating cyclins and cyclin-dependent kinases (cdk), upregulating cdk inhibitors, and inhibiting the synthesis of DNA synthesis. However, most physiological responses are triggered by curcumin depend on the cell type, the concentration of curcumin and the time of treatment. Despite its demonstrated ability and efficacy, the limited bioavailability of curcumin continues to be an insurmountable barrier and obstacle. Discovery of new structural analogs of curcumin may be a better way of targeting cells and tissues affected by chronic pathological conditions.

References

Aggarwal BB, Harikumar KB (2009) Potential therapeutic effects of curcumin, the anti-inflammatory agent, against neurodegenerative, cardiovascular, pulmonary, metabolic, autoimmune and neoplastic diseases. Int J Biochem Cell Biol 41:40–59

Aggarwal BB, Yuan W, Li S, Gupta SC (2013) Curcumin-free turmeric exhibits anti-inflammatory and anticancer activities: identification of novel components of turmeric. Mol Nutr Food Res 57:1529–1542

Alderton WK, Cooper CE, Knowles RG (2001) Nitric oxide synthases: structure, function and inhibition. Biochem J 357:593–615

Amălinei C, Căruntu ID, Giuşcă SE, Bălan RA (2010) Matrix metalloproteinases involvement in pathologic conditions. Rom J Morphol Embryol 51:215–228

Anand P, Thomas SG, Kunnumakkara AB, Sundaram C, Harikumar KB, Sung B, Tharakan ST, Misra K, Priyadarsini IK, Rajasekharan KN, Aggarwal BB (2008) Biological activities of curcumin and its analogues (Congeners) made by man and Mother Nature. Biochem Pharmacol 76:1590–1611

Armstrong RN (1997) Structure, catalytic mechanism, and evolution of the glutathione transferases. Chem Res Toxicol 10:2–18

Arora V, Kuhad A, Tiwari V, Chopra K (2011) Curcumin ameliorates reserpine-induced pain-depression dyad: behavioural, biochemical, neurochemical and molecular evidences. Psychoneuroendocrinology 36:1570–1581

Atkinson GP, Nozell SE, Benveniste ET (2010) NF-κB and STAT3 signaling in glioma: targets for future therapies. Expert Rev Neurother 10:575–586

Babbitt PC (2000) Reengineering the glutathione S-transferase scaffold: a rational design strategy pays off. Proc Natl Acad Sci U S A 97:10298–10300

Baird L, Dinkova-Kostova AT (2011) The cytoprotective role of the Keap1-Nrf2 pathway. Arch Toxicol 85:241–272

Balasubramanyam K, Varier RA, Altaf M, Swaminathan V, Siddappa NB, Ranga U, Kundu TK (2004) Curcumin, a novel p300/CREB binding protein-specific inhibitor of acetyltransferase, represses the acetylation of histone/nonhistone proteins and histone acetyltransferase-dependent chromatin transcription. J Biol Chem 279:51163–51171

Balogun E, Hoque M, Gong P, Killeen E, Green CJ, Foresti R, Alam J, Motterlini R (2003) Curcumin activates the haem oxygenase-1 gene via regulation of Nrf2 and the antioxidant-responsive element. Biochem J 371:887–895

Barski OA, Tipparaju SM, Bhatnagar A (2008) The aldo-keto reductase superfamily and its role in drug metabolism and detoxification. Drug Metab Rev 40:553–624

Beevers CS, Li F, Liu L, Huang S (2006) Curcumin inhibits the mammalian target of rapamycin-mediated signaling pathways in cancer cells. Int J Cancer 119:757–764

Beevers CS, Chen L, Liu L, Luo Y, Webster NJ, Huang S (2009) Curcumin disrupts the Mammalian target of rapamycin-raptor complex. Cancer Res 69:1000–1008

Benson AM, Hunkeler MJ, Talalay P (1980) Increase of NAD(P)H: quinone reductase by dietary antioxidants: possible role in protection against carcinogenesis and toxicity. Proc Natl Acad Sci U S A 77:5216–5220

Bharti AC, Donato N, Singh S, Aggarwal BB (2003) Curcumin (diferuloylmethane) down-regulates the constitutive activation of nuclear factor-kappa B and IkappaBalpha kinase in human multiple myeloma cells, leading to suppression of proliferation and induction of apoptosis. Blood 101:1053–1062

Bilmen JG, Khan SZ, Javed MUH, Michelangeli F (2001) Inhibition of the SERCA Ca2+ pumps by curcumin. Curcumin putatively stabilizes the interaction between the nucleotide-binding and phosphorylation domains in the absence of ATP. Eur J Biochem 268:6318–6327

Blasco MA (2007) Telomere length, stem cells and aging. Nat Chem Biol 3:640–649

Bocci V, Valacchi G (2015) Nrf2 activation as target to implement therapeutic treatments. Front Chem 3:4

Bode W, Fernandez-Catalan C, Tschesche H, Grams F, Nagase H, Maskos K (1999) Structural properties of matrix metalloproteinases. Cell Mol Life Sci 55:639–652

Brantley EC, Benveniste EN (2008) Signal transducer and activator of transcription-3: a molecular hub for signaling pathways in gliomas. Mol Cancer Res 6:675–684

Canela A, Vera E, Klatt P, Blasco MA (2007) High-throughput telomere length quantification by FISH and its application to human population studies. Proc Natl Acad Sci U S A 104:5300–5305

Chan JY, Kwong M (2000) Impaired expression of glutathione synthetic enzyme genes in mice with targeted deletion of the Nrf2 basic-leucine zipper protein. Biochim Biophys Acta 1517:19–26

Chan MM, Huang HI, Fenton MR, Fong D (1998) In vivo inhibition of nitric oxide synthase gene expression by curcumin, a cancer preventive natural product with anti-inflammatory properties. Biochem Pharmacol 55:1955–1962

Chan YC, Banerjee J, Choi SY, Sen CK (2012) miR-210: the master hypoxamir. Microcirculation 19:215–223

Chen ZJ (2005) Ubiquitin signalling in the NF-kappaB pathway. Nat Cell Biol 7:758–765

Chen RH, Juo PC, Curran T, Blenis J (1996) Phosphorylation of c-Fos at the C-terminus enhances its transforming activity. Oncogene 12:1493–1502

Cheng Y, Kozubek A, Ohlsson L, Sternby B, Duan RD (2007) Curcumin decreases acid sphingo-myelinase activity in colon cancer Caco-2 cells. Planta Med 73:725–730

Chowdhury I, Tharakan B, Bhat GK (2008) Caspases – an update. Comp Biochem Physiol B Biochem Mol Biol 151:10–27

Cohen SB, Graham ME, Lovrecz GO, Bache N, Robinson PJ, Reddel RR (2007) Protein composition of catalytically active human telomerase from immortal cells. Science 315:1850–1853

Collado M, Blasco MA, Serrano M (2007) Cellular senescence in cancer and aging. Cell 130:223–233

Collins K, Mitchell JR (2000) Telomerase in the human organism. Oncogene 21:564–579

Collins K (2008) Physiological assembly and activity of human telomerase complexes. Mech Ageing Dev 129:91–98

Darnell JE Jr (2002) Transcription factors as targets for cancer therapy. Nat Rev Cancer 2:740–749

de Lange T, Shelterin F (2005) The protein complex that shapes and safeguards human telomeres. Genes Dev 19:2100–2110

DeGregori J, Johnson DG (2006) Distinct and overlapping roles for E2F family members in transcription, proliferation and apoptosis. Curr Mol Med 6:739–748

Dillin A, Crawford DK, Kenyon C (2002) Timing requirements for insulin/IGF-1 signaling in C. elegans. Science 298:830–834

Dinkova-Kostova AT, Talalay P (2008) Direct and indirect antioxidant properties of inducers of cytoprotective proteins. Mol Nutr Food Res 52(Suppl 1):S128–S138

Dirr H, Reinemer P, Huber R (1994) X-ray crystal structures of cytosolic glutathione S-transferases. Implications for protein architecture, substrate recognition and catalytic function. Eur J Biochem 220:645–661

Eckmann L, Laurent F, Langford TD, Hetsko ML, Smith RJ, Kagnoff MF, Gillin FD (2000) Nitric oxide production by human intestinal epithelial cells and competition for arginine as potential determinants of host defense against the lumen-dwelling pathogen *Giardia lamblia*. J Immunol 164:1478–1487

Eferl R, Wagner EF (2003) AP1: a double-edged sword in tumorigenesis. Nat Rev Cancer 3:859–868

Farooqui AA (2011) Metabolism of lipid mediators in the brain. Springer, New York

Farooqui AA (2013) Metabolic syndrome: an important risk factor for stroke, Alzheimer disease, and depression. Springer, New York

Ferrara N, Gerber HP, LeCouter J (2003) The biology of VEGF and its receptors. Nat Med 9:669–676

Fields GB (2014) Biophysical studies of matrix metalloproteinase/triple-helix complexes. Adv Protein Chem Struct Biol 97:37–48

Flores I, Canela A, Vera E, Tejera A, Cotsarelis G, Blasco MA (2008) The longest telomeres: a general signature of adult stem cell compartments. Genes Dev 22:654–667

Gao Y, Chu SF, Li JP, Zuo W, Wen ZL, He WB, Yan JQ, Chen NH (2014a) Do glial cells play an anti-oxidative role in Huntington's disease? Free Radic Res 48:1135–1144

Gao B, Doan A, Hybertson BM (2014b) The clinical potential of influencing Nrf2 signaling in degenerative and immunological disorders. Clin Pharmacol 6:19–34

Garneau H, Paquin MC, Carrier JC, Rivard N (2009) E2F4 expression is required for cell cycle progression of normal intestinal crypt cells and colorectal cancer cells. J Cell Physiol 221:350–358

Gilmore TD, Wolenski FS (2012) NF-κB: where did it come from and why? Immunol Rev 246:14–35

Goel A, Boland CR, Chauhan DP (2001) Specific inhibition of cyclooxygenase-2 (COX-2) expression by dietary curcumin in HT-29 human colon cancer cells. Cancer Lett 172:111–118

Gordan JD, Simon MC (2007) Hypoxia-inducible factors: central regulators of the tumor phenotype. Curr Opin Genet Dev 17:71–77

Guertin DA, Sabatini DM (2007) Defining the role of mTOR in cancer. Cancer Cell 12:9–22

Guo L, Cai X, Lee J, Kang SS, Shin EM, Zhou HY, Jung JW, Kim YS (2008) Comparison of suppressive effects of demethoxycurcumin and bisdemethoxycurcumin on expressions of inflammatory mediators in vitro and in vivo. Arch Pharm Res 31:490–496

Gupta SC, Prasad S, Kim JH, Patchva S, Webb LJ, Priyadarsini IK, Aggarwal BB (2011) Multitargeting by curcumin as revealed by molecular interaction studies. Nat Prod Rep 28: 1937–1955

Hagberg CE, Mehlem A, Falkevall A, Muhl L, Fam BC, Ortsäter H, Scotney P, Nyqvist D, Samén E, Lu L, Stone-Elander S, Proietto J, Andrikopoulos S, Sjöholm A, Nash A, Eriksson U (2012) Targeting VEGF-B as a novel treatment for insulin resistance and type 2 diabetes. Nature 490:426–430

Harvey CJ, Thimmulappa RK, Singh A, Blake DJ, Ling G, Wakabayashi N, Fujii J, Myers A, Biswal S (2009) Nrf2-regulated glutathione recycling independent of biosynthesis is critical for cell survival during oxidative stress. Free Radic Biol Med 46:443–453

Hatcher H, Planalp R, Cho J, Torti FM, Torti SV (2008) Curcumin: from ancient medicine to current clinical trials. Cell Mol Life Sci 65:1631–1652

Hayden MS, Ghosh S (2012) NF-kappaB, the first quarter-century: remarkable progress and outstanding questions. Genes Dev 26:203–234

Hayes JD, Flanagan JU, Jowsey IR (2005) Glutathione transferases. Annu Rev Pharmacol Toxicol 45:51–88

Helin K (1998) Regulation of cell proliferation by the E2F transcription factors. Curr Opin Genet Dev 8:28–35

Hergenhahn M, Soto U, Weninger A, Polack A, Hsu CH, Cheng AL, Rösl F (2002) The chemopreventive compound curcumin is an efficient inhibitor of Epstein-Barr virus BZLF1 transcription in Raji DR-LUC cells. Mol Carcinog 33:137–145

Higgins LG, Hayes JD (2011) Mechanisms of induction of cytosolic and microsomal glutathione transferase (GST) genes by xenobiotics and pro-inflammatory agents. Drug Metab Rev 43:92–137

Huang H, Tindall DJ (2007) Dynamic FoxO transcription factors. J Cell Sci 120:2479–2487

Huang H, Tindall DJ (2011) Regulation of FOXO protein stability via ubiquitination and proteasome degradation. Biochim Biophys Acta 1813:1961–1964

Huang T-S, Lee SC, Lin J-K (1991) Suppression of c-Jun/AP-1 activation by an inhibitor of tumor promotion in mouse fibroblast cells. Proc Natl Acad Sci U S A 88:5292–5296

Iaquinta PJ, Lees JA (2007) Life and death decisions by the E2F transcription factors. Curr Opin Cell Biol 19(6):649–657

Ingolfsson HI, Koeppe RE, Andersen OS (2007) Curcumin is a modulator of bilayer material properties. Biochemistry 46:10384–10391

Itoh K, Mimura J, Yamamoto M (2010) Discovery of the negative regulator of Nrf2, Keap1: a historical overview. Antioxid Redox Signal 13:1665–1678

Itoh K, Ye P, Matsumiya T, Tanji K, Ozaki T (2015) Emerging functional cross-talk between the Keap1-Nrf2 system and mitochondria. J Clin Biochem Nutr 56:91–97

Jiang B-H, Zheng JZ, Leung SW, Roe R, Semenza GL (1997) Transactivation and inhibitory domains of hypoxia-inducible factor 1alpha. Modulation of transcriptional activity by oxygen tension. J Biol Chem 272:19253–19260

Jiang QG, Li TY, Liu DN, Zhang HT (2014) PI3K/Akt pathway involving into apoptosis and invasion in human colon cancer cells LoVo. Mol Biol Rep 41:3359–3367

Jitoe-Masuda A, Fujimoto A, Masuda T (2013) Curcumin: from chemistry to chemistry-based functions. Curr Pharm Des 19:2084–2092

Jurenka JS (2009) Anti-inflammatory properties of curcumin, a major constituent of Curcuma longa: a review of preclinical and clinical research. Altern Med Rev 14:141–153

Kamba T, Tam BY, Hashizume H, Haskell A, Sennino B, Mancuso MR, Norberg SM, O'Brien SM, Davis RB, Gowen LC, Anderson KD, Thurston G, Joho S, Springer ML, Kuo CJ, McDonald DM (2006) VEGF-dependent plasticity of fenestrated capillaries in the normal adult microvasculature. Am J Physiol Heart Circ Physiol 290:H560–H576

Kaminska B, Pyrzynska B, Ciechomska I, Wisniewska M (2000) Modulation of the composition of AP-1 complex and its impact on transcriptional activity. Acta Neurobiol Exp (Wars) 60:395–402

Kang G, Kong PJ, Yuh YJ, Lim SY, Yim SV, Chun W, Kim SS (2004) Curcumin suppresses lipopolysaccharide-induced cyclooxygenase-2 expression by inhibiting activator protein 1 and nuclear factor kappab bindings in BV2 microglial cells. J Pharmacol Sci 94:325–328

Kaspar JW, Niture SK, Jaiswal AK (2012) Antioxidant-induced INrf2 (Keap1) tyrosine 85 phosphorylation controls the nuclear export and degradation of the INrf2-Cul3-Rbx1 complex to allow normal Nrf2 activation and repression. J Cell Sci 125:1027–1038

Kathiria AS, Neumann WL, Rhees J, Hotchkiss E, Cheng Y, Genta RM, Meltzer SJ, Souza RF, Theiss AL (2012) Prohibitin attenuates colitis-associated tumorigenesis in mice by modulating p53 and STAT3 apoptotic responses. Cancer Res 72:5778–5789

Khaw AK, Hande MP, Kalthur G, Hande MP (2013) Curcumin inhibits telomerase and induces telomere shortening and apoptosis in brain tumour cells. J Cell Biochem 114:1257–1270

Kim K-C, Lee L (2010) Curcumin induces downregulation of E2F4 Expression and apoptotic cell death in HCT116 human colon cancer cells; involvement of Reactive Oxygen Species. Korean J Physiol Pharmacol 14:391–397

Kim HY, Park EJ, Joe EH, Jou I (2003) Curcumin suppresses Janus kinase-STAT inflammatory signaling through activation of Src homology 2 domain-containing tyrosine phosphatase 2 in brain microglia. J Immunol 17:6072–6079

Kim SJ, Son TG, Park HR, Park M, Kim MS, Kim HS, Chung HY, Mattson MP, Lee J (2008) Curcumin stimulates proliferation of embryonic neural progenitor cells and neurogenesis in the adult hippocampus. J Biol Chem 283:14497–14505

Kobayashi M, Yamamoto M (2005) Molecular mechanisms activating the Nrf2-Keap1 pathway of antioxidant gene regulation. Antioxid Redox Signal 7:385–394

Kobayashi M, Li L, Iwamoto N, Nakajima-Takagi Y, Kaneko H, Nakayama Y, Eguchi M, Wada Y, Kumagai Y, Yamamoto M (2009) The antioxidant defense system Keap1-Nrf2 comprises a multiple sensing mechanism for responding to a wide range of chemical compounds. Mol Cell Biol 29:493–502

Kraft AD, Johnson DA, Johnson J (2004) Nuclear factor E2-related factor 2-dependent antioxidant response element activation by tert-butylhydroquinone and sulforaphane occurring preferentially in astrocytes conditions neurons against oxidative insult. J Neurosci 24:1101–1112

Kulkarni SK, Bhutani MK, Bishnoi M (2008) Antidepressant activity of curcumin: involvement of serotonin and dopamine system. Psychopharmacology (Berl) 201:435–442

Kurapati KR, Samikkannu T, Kadiyala DB, Zainulabedin SM, Gandhi N, Sathaye SS, Indap MA, Boukli N, Rodriguez JW, Nair MP (2012) Combinatorial cytotoxic effects of Curcuma longa and Zingiber officinale on the PC-3M prostate cancer cell line. J Basic Clin Physiol Pharmacol 23:139–146

Lee J-M, Calkins MJ, Chan K, Kan YW, Johnson JA (2003) Identification of the NF-E2-related factor-2-dependent genes conferring protection against oxidative stress in primary cortical astrocytes using oligonucleotide microarray analysis. J Biol Chem 278:12029–12038

Li W, Kong AN (2009) Molecular mechanisms of Nrf2-mediated antioxidant response. Mol Carcinog 48:91–104

Lin JK (2004) Suppression of protein kinase C and nuclear oncogene expression as possible action mechanisms of cancer chemoprevention by Curcumin. Arch Pharm Res 27:683–692

Lin JK (2007) Molecular targets of curcumin. Adv Exp Med Biol 595:227–243

Lopez-Bergami P, Lau E, Ronai Z (2010) Emerging roles of ATF2 and the dynamic AP1 network in cancer. Nat Rev Cancer 10:65–76

Madkor HR, Mansour SW, Ramadan G (2011) Modulatory effects of garlic, ginger, turmeric and their mixture on hyperglycaemia, dyslipidaemia and oxidative stress in streptozotocin-nicotinamide diabetic rats. Br J Nutr 105:1210–1217

Mahmmoud YA (2007) Modulation of protein kinase C by curcumin; inhibition and activation switched by calcium ions. Br J Pharmacol 150:200–208

Mann GE, Niehueser-Saran J, Watson A, Gao L, Ishii T, de Winter P, Siow RC (2007) Nrf2/ARE regulated antioxidant gene expression in endothelial and smooth muscle cells in oxidative stress: implications for atherosclerosis and preeclampsia. Acta Physiol Sinica 59:117–127

Mattson MP, Cheng A (2006) Neurohormetic phytochemicals: low-dose toxins that induce adaptive neuronal stress responses. Trends Neurosci 29:632–639

Michels S, Trautmann M, Sievers E, Kindler D, Huss S, Renner M, Friedrichs N, Kirfel J, Steiner S, Endl E, Wurst P, Heukamp L, Penzel R, Larsson O, Kawai A, Tanaka S, Sonobe H, Schirmacher P, Mechtersheimer G, Wardelmann E, Büttner R, Hartmann W (2013) SRC signaling is crucial in the growth of synovial sarcoma cells. Cancer Res 73:2518–2528

Monsalve M, Olmos Y (2011) The complex biology of FOXO. Curr Drug Targets 12:1322–1350

Mukhopadhyay A, Bueso-Ramos C, Chatterjee D, Pantazis P, Aggarwal BB (2001) Curcumin downregulates cell survival mechanisms in human prostate cancer cell lines. Oncogene 20:7597–7609

Muller H, Bracken AP, Vernell R, Moroni MC, Christians F, Grassilli E, Prosperini E, Vigo E, Oliner JD, Helin K (2001) E2Fs regulate the expression of genes involved in differentiation, development, proliferation, and apoptosis. Genes Dev 15:267–285

Murphy LO, Blenis J (2006) MAPK signal specificity: the right place at the right time. Trends Biochem Sci 31:268–275

Nallamshetty S, Chan SY, Loscalzo J (2013) Hypoxia: a master regulator of microRNA biogenesis and activity. Free Radic Biol Med 64:20–30

Nasiri M, Zarghami N, Koshki KN, Mollazadeh M, Moghaddam MP, Yamchi MR, Esfahlan RJ, Barkhordari A, Alibakhshi A (2013) Curcumin and silibinin inhibit telomerase expression in T47D human breast cancer cells. Asian Pac J Cancer Prev 14:3449–3453

Nauta TD, van Hinsbergh VW, Koolwijk P (2014) Hypoxic signaling during tissue repair and regenerative medicine. Int J Mol Sci 15:19791–19815

Newton AC (2001) Protein kinase C: structural and spatial regulation by phosphorylation, cofactors, and macromolecular interactions. Chem Rev 101:2353–2364

Nicholson DW (1999) Caspase structure, proteolytic substrates, and function during apoptotic cell death. Cell Death Differ 6:1028–1042

Oyagbemi AA, Saba AB, Ibraheem AO (2009) Curcumin: from food spice to cancer prevention. Asian Pac J Cancer Prev 10:963–967

Ozanne BW, Spence HJ, McGarry LC, Hennigan RF (2006) Invasion is a genetic program regulated by transcription factors. Curr Opin Genet Dev 16:65–70

Ozanne BW, Spence HJ, McGarry LC, Hennigan RF (2007) Transcription factors control invasion: AP-1 the first among equals. Oncogene 26:1–10

Park S, Lee DK, Yang CH (1998) Inhibition of fos-jun-DNA complex formation by dihydroguaiaretic acid and in vitro cytotoxic effects on cancer cells. Cancer Lett 127:23–28

Polager S, Ginsberg D (2008) E2F-at the crossroads of life and death. Trends Cell Biol 18:528–535

Polager S, Kalma Y, Berkovich E, Ginsberg D (2002) E2Fs up-regulate expression of genes involved in DNA replication, DNA repair and mitosis. Oncogene 21:437–446

Pouysségur J, Dayan F, Mazure NM (2006) Hypoxia signalling in cancer and approaches to enforce tumour regression. Nature 441:437–443

Pugazhenthi S, Akhov L, Selvaraj G, Wang M, Alam J (2007) Regulation of heme oxygenase-1 expression by demethoxy curcuminoids through Nrf2 by a PI3-kinase/Akt-mediated pathway in mouse beta-cells. Am J Physiol Endocrinol Metab 293:E645–E655

Pugh CW, Rourke JFO, Nagao M, Gleadle JM, Ratcliffe PJ (1997) Activation of hypoxia-inducible factor-1; definition of regulatory domains within the alpha subunit. J Biol Chem 272:11205–11214

Pulverer BJ, Kyriakis JM, Avruch J, Nikolakaki E, Woodgett JR (1991) Phosphorylation of c-jun mediated by MAP kinases. Nature 353:670–674

Rafiee P, Binion DG, Wellner M, Behmaram B, Floer M, Mitton E, Nie L, Zhang Z, Otterson MF (2010) Modulatory effect of curcumin on survival of irradiated human intestinal microvascular endothelial cells: role of Akt/mTOR and NF-{kappa}B. Am J Physiol Gastrointest Liver Physiol 298:G865–G877

Rainsford KD (2006) Current status of the therapeutic uses and actions of the preferential cyclooxygenase-2 NSAID, nimesulide. Inflammopharmacology 14:120–137

Ramachandran C, Fonseca HB, Jhabvala P, Escalon EA, Melnick SJ (2002) Curcumin inhibits telomerase activity through human telomerase reverse transcritpase in MCF-7 breast cancer cell line. Cancer Lett 184:1–6

Ravindran J, Prasad S, Aggarwal BB (2009) Curcumin and cancer cells: how many ways can curry kill tumor cells selectively? AAPS J 11:495–510

Razani B, Reichardt AD, Cheng G (2011) Non-canonical NF-kappaB signaling activation and regulation: principles and perspectives. Immunol Rev 244:44–54

Reddy S, Aggarwal BB (1994) Curcumin is a non-competitive and selective inhibitor of phosphorylase kinase. FEBS Lett 341:19–22

Reyes-Fermin LM, Gonzalez-Reyes S, Tarco-Alvarez NG, Hernández-Nava M, Orozco-Ibarra M, Pedraza-Chaverri J (2012) Neuroprotective effect of alpha-mangostin and curcumin against iodoacetate-induced cell death. Nutr Neurosci 15:34–41

Ryter SW, Choi AM (2010) Heme oxygenase-1/carbon monoxide: novel therapeutic strategies in critical care medicine. Curr Drug Targets 11:1485–1494

Saini S, Arora S, Majid S, Shahryari V, Chen Y, Deng G, Yamamura S, Ueno K, Dahiya R (2011) Curcumin modulates microRNA-203-mediated regulation of the Src-Akt axis in bladder cancer. Cancer Prev Res (Phila) 4:1698–1709

Salvioli S, Sikora E, Cooper EL, Franceschi C (2007) Curcumin in cell death processes: a challenge for CAM of age-related pathologies. Evid Based Complement Alternat Med 4:181–190

Santel T, Pflug G, Hemdan NY, Schäfer A, Hollenbach M, Buchold M, Hintersdorf A, Lindner I, Otto A, Bigl M, Oerlecke I, Hutschenreuther A, Sack U, Huse K, Groth M, Birkemeyer C, Schellenberger W, Gebhardt R, Platzer M, Weiss T, Vijayalakshmi MA, Krüger M, Birkenmeier G (2008) Curcumin inhibits glyoxalase 1: a possible link to its anti-inflammatory and antitumor activity. PLoS One 3, e3508

Semenza GL (2003) Targeting HIF-1 for cancer therapy. Nat Rev Cancer 3:721–732

Semenza GL (2007) Evaluation of HIF-1 inhibitors as anticancer agents. Drug Discov Today 12:853–859

Semenza GL (2011) Oxygen sensing, homeostasis, and disease. N Engl J Med 365:537–547

Shakibaei M, John T, Schulze-Tanzil G, Lehmann I, Mobasheri A (2007) Suppression of NF-kappaB activation by curcumin leads to inhibition of expression of cyclo-oxygenase-2 and matrix metalloproteinase-9 in human articular chondrocytes: implications for the treatment of osteoarthritis. Biochem Pharmacol 73:1434–1445

Shalini S, Dorstyn L, Dawar S, Kumar S (2015) Old, new and emerging functions of caspases. Cell Death Differ 22:526–539

Shan B, Schaaf C, Schmidt A, Lucia K, Buchfelder M, Losa M, Kuhlen D, Kreutzer J, Perone MJ, Arzt E, Stalla GK, Renner U (2012) Curcumin suppresses HIF1A synthesis and VEGFA release in pituitary adenomas. J Endocrinol 214:389–398

Shaul YD, Seger R (2007) The MEK/ERK cascade: from signaling specificity to diverse functions. Biochim Biophys Acta 1773:1213–1226

Shaulian E, Karin M (2001) AP-1 in cell proliferation and survival. Oncogene 20:2390–2400

Shaulian E, Karin M (2002) AP-1 as a regulator of cell life and death. Nat Cell Biol 4:E131–E136

Shaw PE (2010) Could STAT3 provide a link between respiration and cell cycle progression? Cell Cycle 9:4294–4296

Shen Y, Schlessinger K, Zhu X, Meffre E, Quimby F, Levy DE, Darnell JE (2004) Essential role of STAT3 in postnatal survival and growth revealed by mice lacking STAT3 serine 727 phosphorylation. Mol Cell Biol 24:407–419

Shih AY, Johnson DA, Wong G, Kraft AD, Jiang L, Erb H, Johnson JA, Murphy TH (2003) Coordinate regulation of glutathione biosynthesis and release by Nrf2-expressing glia potently protects neurons from oxidative stress. J Neurosci 23:3394–3406

Shih AY, Li P, Murphy TH (2005) A small-molecule-inducible Nrf2-mediated antioxidant response provides effective prophylaxis against cerebral ischemia in vivo. J Neurosci 25:10321–10335

Shinojima N, Yokoyama T, Kondo Y, Kondo S (2007) Roles of the Akt/mTOR/p70S6K and ERK1/2 signaling pathways in curcumin-induced autophagy. Autophagy 3:635–637

Singula N, Dhawan DK (2012) N-methyl N-nitrosourea induced functional and structural alterations in mice brain-role of curcumin. Neurotox Res 22:115–126

Slezak-Prochazka I, Durmus S, Kroesen BJ, van den Berg A (2010) MicroRNAs, macrocontrol: regulation of miRNA processing. RNA 16:1087–1095

Stephenson D, Yin T, Smalstig EB, Hsu MA, Panetta J, Little S, Clemens J (2000) Transcription factor nuclear factor-kappa B is activated in neurons after focal cerebral ischemia. J Cereb Blood Flow Metab 20:592–603

Strasser EM, Wessner B, Manhart N, Roth E (2005) The relationship between the anti-inflammatory effects of curcumin and cellular glutathione content in myelomonocytic cells. Biochem Pharmacol 70:552–559

Stridh MH, Correa F, Nodin C, Weber SG, Blomstrand F, Nilsson M, Sandberg M (2010) Enhanced glutathione efflux from astrocytes in culture by low extracellular Ca^{2+} and curcumin. Neurochem Res 35:1231–1238

Sun SC (2011) Non-canonical NF-kappaB signaling pathway. Cell Res 21:71–85

Sun SC (2012) The noncanonical NF-κB pathway. Immunol Rev 246:125–140

Sun Z, Zhang S, Chan JY, Zhang DD (2007) Keap1 controls postinduction repression of the Nrf2-mediated antioxidant response by escorting nuclear export of Nrf2. Mol Cell Biol 27:6334–6349

Sun J, Zhao Y, Hu J (2013) Curcumin inhibits imiquimod-induced psoriasis-like inflammation by inhibiting IL-1beta and IL-6 production in mice. PLoS One 8, e67078

Sundin T, Hentosh P (2012) InTERTesting association between telomerase, mTOR and phytochemicals. Expert Rev Mol Med 14, e8

Swarnakar S, Ganguly K, Kundu P, Banerjee A, Maity P, Sharma AV (2005) Curcumin regulates expression and activity of matrix metalloproteinases 9 and 2 during prevention and healing of indomethacin-induced gastric ulcer. J Biol Chem 280:9409–9415

Taylor RC, Cullen SP, Martin SJ (2008) Apoptosis: controlled demolition at the cellular level. Nat Rev Mol Cell Biol 9:231–241

Thomas SL, Zhong D, Zhou W, Malik S, Liotta D, Snyder JP, Hamel E, Giannakakou P (2008) EF24, a novel curcumin analog, disrupts the microtubule cytoskeleton and inhibits HIF-1. Cell Cycle 7:2409–2417

Tkachev V, Menshchikova E, Zenkov N (2011) Mechanism of the Nrf2/Keap1/ARE signaling system. Biochemistry (Moscow) 76:407–422

Trimarchi JM, Fairchild B, Verona R, Moberg K, Andon N, Lees JA (1998) E2F-6, a member of the E2F family that can behave as a transcriptional repressor. Proc Natl Acad Sci U S A 95:2850–2855

Tworkoski K, Singhal G, Szpakowski S, Zito CI, Bacchiocchi A, Muthusamy V, Bosenberg M, Krauthammer M, Halaban R, Stern DF (2011) Phosphoproteomic screen identifies potential therapeutic targets in melanoma. Mol Cancer Res 9:801–812

van Muiswinkel FL, Kuiperij HB (2005) The Nrf2-ARE signalling pathway: promising drug target to combat oxidative stress in neurodegenerative disorders. Curr Drug Targets CNS Neurol Disord 4:267–281

Vargas MR, Johnson JA (2009) The Nrf2-ARE cytoprotective pathway in astrocytes. Expert Rev Mol Med 11, e17

Vesely PW, Staber PB, Hoefler G, Kenner L (2009) Translational regulation mechanisms of AP-1 proteins. Mutat Res 682:7–12

Visse R, Nagase H (2003) Matrix metalloproteinases and tissue inhibitors of metalloproteinases structure, function, and biochemistry. Circ Res 92:827–839

Wang GL, Semenza GL (1993) Characterization of hypoxia-inducible factor 1 and regulation of DNA binding activity by hypoxia. J Biol Chem 268:21513–21518

Wang GL, Jiang BH, Rue EA, Semenza GL (1995) Hypoxia-inducible factor 1 is a basic-helix-loop-helix-PAS heterodimer regulated by cellular O2 tension. Proc Natl Acad Sci U S A 92:5510–5514

Wang YJ, Pan MH, Cheng AL, Lin LI, Ho YS, Hsieh CY (1997) Stability of curcumin in buffer solutions and characterization of its degradation products. J Pharm Biomed Anal 15:1867–1876

Wang L, Shen Y, Song R, Sun Y, Xu J, Xu Q (2009) An anticancer effect of curcumin mediated by down-regulating phosphatase of regenerating liver-3 expression on highly metastatic melanoma cells. Mol Pharmacol 76:1238–1245

Wang X, Campos CR, Peart JC, Smith LK, Boni JL, Cannon RE, Miller DS (2014) Nrf2 upregulates ATP binding cassette transporter expression and activity at the blood-brain and blood-spinal cord barriers. J Neurosci 34:8585–8593

Wisdom R (1999) AP1: One switch for many signals. Exp Cell Res 253:180–185

Wolanin K, Magalska A, Mosieniak G, Klinger R, McKenna S, Vejda S, Sikora E, Piwocka K (2006) Curcumin affects components of the chromosomal passenger complex and induces mitotic catastrophe in apoptosis-resistant Bcr-Abl-expressing cells. Mol Cancer Res 4:457–469

Wu J, Li Q, Wang X, Yu S, Li L, Wu X, Chen Y, Zhao J, Zhao Y (2013) Neuroprotection by curcumin in ischemic brain injury involves the Akt/Nrf2 pathway. PLoS One 8, e59843

Xiang L, Nakamura Y, Lim YM, Yamasaki Y, Kurokawa-Nose Y, Maruyama W, Osawa T, Matsuura A, Motoyama N, Tsuda L (2011) Tetrahydrocurcumin extends life span and inhibits the oxidative stress response by regulating the FOXO forkhead transcription factor. Aging (Albany NY) 3:1098–1109

Xiao G, Harhaj EW, Sun SC (2001) NF-kappaB-inducing kinase regulates the processing of NF-kappaB2 p100. Mol Cell 7:401–409

Xu Y, Ku BS, Yao HY, Lin YH, Ma X, Zhang YH, Li XJ (2005) Antidepressant effects of curcumin in the forced swim test and olfactory bulbectomy models of depression in rats. Pharmacol Biochem Behav 82:200–206

Yamamoto H, Hanada K, Kawasaki K, Nishijima M (1997) Inhibitory effect on curcumin on mammalian phospholipase D activity. FEBS Lett 417:196–198

Yang C, Zhang X, Fan H, Liu Y (2009) Curcumin upregulates transcription factor Nrf2, HO-1 expression and protects rat brains against focal ischemia. Brain Res 1282:133–141

Yang WH, Kuo MY, Liu CM, Deng YT, Chang HH, Chang JZ (2013) Curcumin inhibits TGFβ1-induced CCN2 via Src, JNK, and Smad3 in gingiva. J Dent Res 92:629–634

Yu H, Pardoll D, Jove R (2009) STATs in cancer inflammation and immunity: a leading role for STAT3. Nat Rev Cancer 9:798–809
Zhang LJ, Wu CF, Meng XL, Yuan D, Cai XD, Wang QL, Yang JY (2008) Comparison of inhibitory potency of three different curcuminoid pigments on nitric oxide and tumor necrosis factor production of rat primary microglia induced by lipopolysaccharide. Neurosci Lett 447:48–53
Zhou H, Beevers CS, Huang S (2011) The targets of curcumin. Curr Drug Targets 12:332–347
Zwick E, Bange J, Ullrich A (2001) Receptor tyrosine kinase signalling as a target for cancer intervention strategies. Endocr Relat Cancer 8:161–173

Chapter 5
Effect of Curcumin on Growth Factors and Their Receptors, Ion Channels, and Transporters in the Visceral Organs and the Brain

5.1 Introduction

Curcumin, the main component of turmeric, is a natural polyphenols that has been used for the treatment of many diseases since ancient times (Goel et al. 2008). Curcumin is insoluble in water and is eliminated mostly unchanged and partly after alterations in the gut. In preclinical and clinical studies, oral administration of curcumin results in low plasma and tissue concentrations. This may be due to low absorption, rapid metabolism, elimination, and limited systemic bioavailability (Anand et al. 2007). As stated in Chap. 3, tumeric contains three natural analogs collectively known as curcuminoids. Each curcuminoid contains the basic aromatic ring native to parent compound, but is differentiated by a unique methoxy group substitution in place of the aryl moieties. Curcumin is the most abundant, comprising 77 %, demethoxycurcumin (17 %), and finally bismethoxycurcumin (3 %) (Aggarwal and Sung 2009). Despite being structurally distinct, curcuminoids have similar biological properties.

Curcumin has ability to interact with a multitude of different molecules not only due to the presence of a central beta diketone moiety, but also in part due to the presence of phenyl rings in its structure. Curcumin produces its beneficial physiological and pharmacological effects (anti-proliferative, anti-oxidative, anti-inflammatory, anti-tumor, and anti-infectious effects) through the modulation of cytokines, growth factors, nuclear transcription factors, and intercellular adhesion molecules (Meng et al. 2013; Aggarwal and Sung 2009; Bar-Sela et al. 2010; Sikora et al. 2010). Because of above mentioned physiological and pharmacological effects, curcumin has been used as anticancer, anti-inflammatory, antioxidant, antithrombotic, antiatherosclerotic, antirheumatic, anti-infectious, and cardioprotective agent. Collective evidence suggests that the effect of curcumin is in part due to its ability to interfere with multiple signaling cascades such as cell cycle regulators, apoptotic proteins, pro-inflammatory cytokines, proliferative regulators and transcription factors such as nuclear factor-kappa B (NF-κB) and STAT3 (Shishodia et al. 2007). It suppresses

© Springer International Publishing Switzerland 2016
A.A. Farooqui, *Therapeutic Potentials of Curcumin for Alzheimer Disease*,
DOI 10.1007/978-3-319-15889-1_5

cancer cell and tumor growth, suppresses proliferation through the involvement and modulation of growth factors, which block angiogenesis and inflammation. The pleiotropic effects of curcumin are mediated by the regulation of various signaling pathways and genes have been reported in different cancer cell lines (Shehzad et al. 2013).

Curcumin also produces neuroprotective, antiparkinsonism, and anticonvulsant effects in animal models of neurological disorders (Jeong et al. 2007; Levi et al. 2012; Libby 2012). Because of its favorably affecting all leading components of metabolic syndrome including insulin resistance, obesity, hypertriglyceridemia, decreased HDL-C and hypertension, curcumin has been used for the treatment of obesity, diabetes, metabolic syndrome and Alzheimer disease (Meydani and Hasan 2010; Maradana et al. 2013a, b: Sahebkar 2013: Fiala et al. 2007; Garcia-Alloza et al. 2007; Brondino et al. 2014; Huang et al. 2013a, b). Curcumin also has ability to directly interact with a number of enzymes such as COX-2, histone acetyltransferases, Src and glycogen synthase kinase 3β (GSK3β), and a variety of other molecules including RNA and DNA. Curcumin also activates 5' AMP-activated protein kinase (AMPK) and down-regulates acetyl-CoA carboxylase (ACC) activity through phosphorylation of this enzyme, which in turn down-regulates the flow of acetyl CoA to malonyl CoA, leading to up-regulation of carnitine palmitoyltransferase-1 (CPT-1) that transfers cytosolic long-chain fatty acyl CoA into the mitochondria for oxidation (Ruderman et al. 2003). In addition, through activation of AMPK, curcumin down-regulated synthesis of glycerol lipids by inhibiting glycerol-3-phosphate acyl transferase-1 (GPAT-1) activity, which esterifies fatty acids to glycerol to form triglycerides for storage (Ejaz et al. 2009).

5.2   Effect of Curcumin on Growth Factors in Visceral Organs

Growth factors are naturally occurring substances, which play a critical role in the normal and tumor cell proliferation and growth. Curcumin suppresses angiogenesis (formation of new blood vessels) and restricts the growth of tumors through the down-regulation of several growth factors including vascular endothelial growth factor (VEGF), basic fibroblast growth factor (bFGF) and epidermal growth factor (EGF), as well as angiopoietin and hypoxia-inducible factors (HIF)-1α (Bae et al. 2006; Gururaj et al. 2002; Xia et al. 2006). The molecular mechanisms associated with suppression of angiogenesis and restricts the growth of tumors are not fully understood. However, it is well known that curcumin mediates these effects through the inhibition of EGF-stimulated phosphorylation of EGFR in MDA-MB-468 cells (Squires et al. 2003). Other studies have indicated that in 253JB-V and KU7 bladder cancer cells, curcumin can mediate its effects by inducing apoptotic cell death through the decrease in expression of the proapoptotic protein survivin and VEGF and VEGFR1 (Chadalapaka et al. 2008).

5.2.1   Effect of Curcumin on Vascular Endothelial Growth Factor (VEGF)

Vascular endothelial growth factor (VEGF) is an important growth factor associated with the regulation of angiogenesis, a highly regulated process (Fig. 5.1). Angiogenesis plays a critical role not only in various biological processes such as normal development, wound healing, proliferation, migration, new vessel formation, embryological development, the menstrual cycle, inflammation, but also in pathogenesis of various diseases such as cancer, diabetic retinopathy, and rheumatoid arthritis. Angiogenesis also contributes to recovery from ischemic disease, and organ regeneration. Angiogenesis is regulated by VEGF. Interactions between VEGF isoforms (VEGFR1 (FLT1), VEGFR2 (KDR), and NRP1) and their receptors result in facilitation of angiogenesis through activation of a kinase cascade that includes RAS and MAPK. VEGF plays an important role in normal developmental physiology and also contribute to tumor development (Ferrara 2004; Ferrara et al. 2005). Phosphorylation of VEGFR enhances endothelial cell protection and activation of protein kinase C and RAS/RAF/MAPK (ERK) pathway (Millauer et al. 1993). At least 50-fold over-regulation of VEGF is observed in glioma cells compared normal brain tissues (Plate et al. 1992). VEGF signaling is associated with PtdIns 3 K/AKT pathway during hypoxia conditions and RAS/RAF/MAPK (ERK) signaling during normoxia and the activation of EGFR in non-hypoxic conditions. In addition, as VEGF promoter contains NF-κB putative response element, its expression is therefore enhanced by NFκB. Indeed, it has been demonstrated that the inhibition of NFκB in glioma cell lines shows a decrease in VEGF expression, resulting in suppressed angiogenesis in vivo and in vitro. Through the involvement of above mechanisms, VEGF activates a broad spectrum of biological responses in endothelial cells, including cell proliferation, migration, survival, differentiation and permeability to macromolecules (Suarez and Ballmer-Hofer 2001a, b). In brain, VEGF contributes to neurogenesis, angiogenesis, improvement in cognitive function, and neuronal survival (Fig. 5.2). Although VEGFR-1 and VEGFR-2 are structurally highly similar, their mechanisms of angiogenic signal transduction are different. Targeted deletion of *VEGFR-2* has indicated that VEGFR-2 has essential requirement in embryogenesis. Homozygous *VEGFR-2* null mice exhibit early embryonic lethality due to the absence of endothelial cells (Shalaby et al. 1995a, b). In contrast, targeted deletion of *VEGFR-1* causes in early embryonic lethality due to overgrowth of endothelial cells, suggesting a negative role for VEGFR-1 in angiogenesis (Kearney et al. 2002). The molecular mechanisms responsible for these differences are not fully understood.

Dysregulation of VEGFR-2 signaling mediates neoangiogenesis thereby promoting tumor development and metastasis (Hicklin and Ellis 2005). VEGF acts as a vasodilator and inhibition of VEGF results in vasoconstriction leading to hypertension. The exact pathophysiology of VEGF inhibition-mediated hypertension is not entirely understood. However, proposed mechanisms not only include reduction in generation of nitric oxide (NO) and increase in the reaction to vasoconstricting

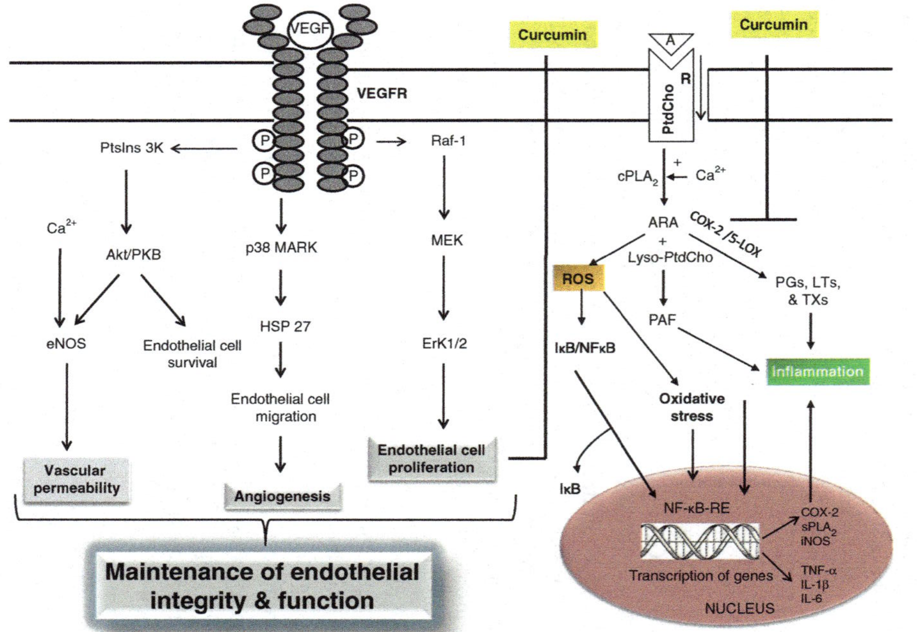

Fig. 5.1 Hypothetical diagram showing VEGF receptor signaling. Vascular endothelial growth factor (*VEGF*); VEGF receptor (*VEGFR*); phosphophate (*P*); phosphatidylinositol 3-kinase (*PtdIns 3 K*); serine-threonine kinase Akt (also known as *PKB*); endothelial nitric oxide synthase (*eNOS*); p38 mitogen-activated protein kinase (*p38 MARK*); heat shock protein 27 (*HSP27*); RAF proto-oncogene serine/threonine-protein kinase (*Raf-1*); Ras/Raf/Mitogen-activated protein kinase/ERK kinase (*MEK*)/extracellular-signal-regulated kinase1/2 (*ERK1/2*); agonist (*A*); receptor (*R*); phosphatidylcholine (*PtdCho*); Lyso-phosphatidylcholine (*Lyso-PtdCho*); cytosolic phospholipase A₂ (*cPLA₂*); cyclooxygenase (*COX*); lipoxygenase (*LOX*); arachidonic acid (*ARA*); platelet-activating factor (*PAF*); reactive oxygen species (*ROS*); nuclear factor-κB (*NF-κB*); nuclear factor-κB-response element (*NF-κB-RE*); inhibitory subunit of NF-κB (*I-κB*); tumor necrosis factor-α (*TNF-α*); interleukin-1β (*IL-1β*); interleukin-6 (*IL-6*); inducible nitric oxide synthase (*iNOS*); secretory phospholipase A₂ (*sPLA₂*)

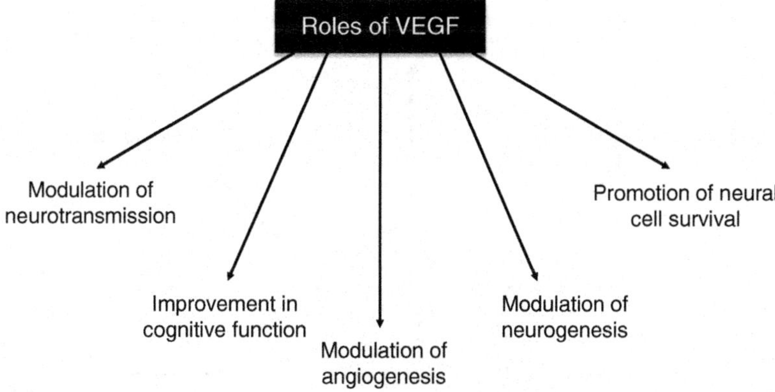

Fig. 5.2 Roles of VEGF in the brain

stimuli, but also a reduction in compliance and distensibility of the vascular wall, and microvascular rarefaction (Kamba and McDonald 2007; Verheul and Pinedo 2007). Because microvessels (arterioles and capillaries) are a major contributor to total peripheral vascular resistance, functional rarefaction (a decrease in perfused microvessels) or anatomic rarefaction (a reduction in capillary density) may play an important role in the development of hypertension.

Curcumin and tetrahydrocurcumin, a major metabolite of curcumin with phenolic and β-diketo moiety exert their effects by interacting with VEGF receptors. These interactions do not result in inhibition of tumor cell proliferation and suppression of tumor angiogenesis. Although the precise mechanisms that lead to tumor angiogenesis are not fully understood, several studies have indicated that tumor angiogenesis requires the expressions of cyclooxygenase-2 (COX-2), 5-lipooxygenase (5-LOX), and matrix metalloproteinase-9 (MMP-9) (Aggarwal et al. 2006). Curcumin not only acts by down-regulating the expression of COX-2, 5-LOX, PtdIns 3 K and MMP-9 (Huang et al. 2013a, b), but also by inhibiting MAPK activity (Fig. 5.1).

5.2.2   *Effect of Curcumin on Basic Fibroblast Growth Factor (bFGF)*

Fibroblast growth factors, or FGFs, are a family of growth factors, which is expressed in both in brain and visceral tissues. FGFs are synthesized as high molecular weight (>20 kDa, Hi-) or low molecular weight (18 kDa, Lo-) isoforms from a single mRNA, translated, respectively, from CUG or AUG start sites (Yu et al. 2007). In visceral organs FGF members are key players in the processes of proliferation and differentiation of wide variety of cells and tissues. They are closely associated with angiogenesis, wound healing, embryonic development and various

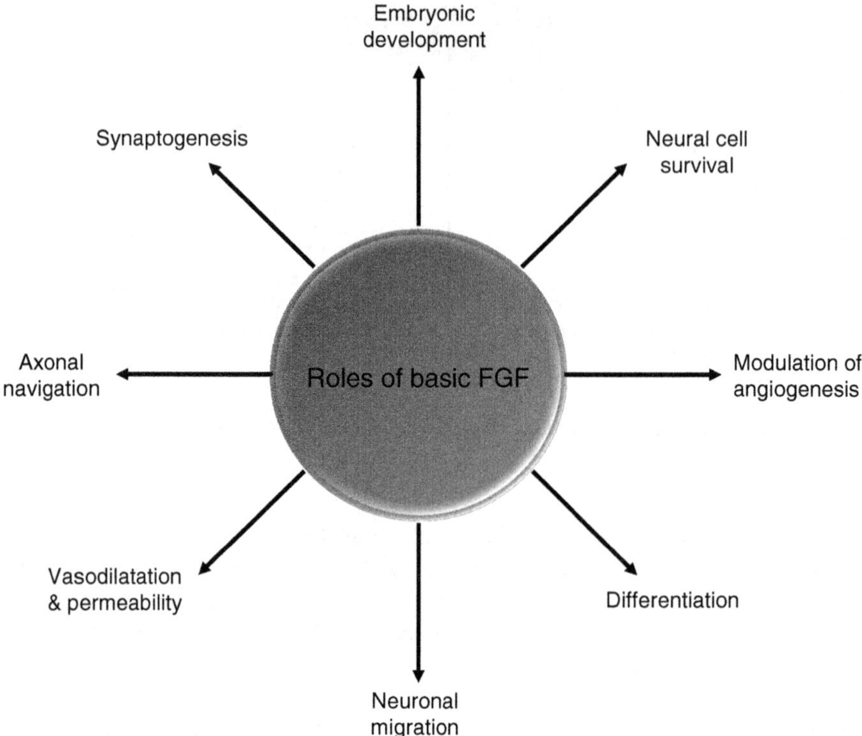

Fig. 5.3 Roles of bFGF in the brain

endocrine signaling pathways. The major role of FGF in the brain is the regulation of neuronal and glial cell proliferation, migration, differentiation and survival (Fig. 5.3) (Vergaño-Vera et al. 2009). Abnormal expression of FGF may lead to depression (Turner et al. 2012). Among various members of FGF family, acidic FGF (FGF-1) and basic FGF (FGF-2) play important role in endothelial cell proliferation and the physical organization into tube-like structures promoting angiogenesis in cancer. FGF exerts its biologic effect through interaction with cell surface FGF receptors (FGFR). The mammalian FGFR family has four members, FGFR-1, FGFR-2, FGFR-3, and FGFR-4. FGF-1 and FGF-2 are ligands of all different FGFR. The binding between FGF ligands and receptors are regulated by proteoglycan cofactors (heparin) and extracellular binding proteins. The activation of FGF receptors recruits particular target proteins such as Ras, a MAPK kinase kinase (MAPKKK) that triggers the mitogen activated protein kinase (MAPK) pathway (Powers et al. 2000). MARK pathway includes extracellular signal-regulated kinases (ERK1/2), the p38 MAPKs and c-Jun N-terminal kinases (JNK1/2/3). Many studies have indicated that activation of JNK signaling is critical for the pathogenesis of apoptotic cell death that occurs in I/R injury (Bogoyevitch et al. 1996).

Activated FGFRs phosphorylate specific tyrosine residues that promote interaction with cytosolic adaptor proteins and the RAS-MAPK, PtdIns 3 K-Akt, PLCγ, and STAT intracellular signaling pathways (Schultz 2002). FGFs also interact with and regulate the family of voltage gated sodium channels. Members of the FGF family not only function in the earliest stages of embryonic development, but also modulate growth, differentiation, survival, and patterning. FGFs also play an important role in metabolic functions, tissue repair, and regeneration through the generation of variety of mediators. It is proposed that aberrant FGF signaling may be involved in the developmental defects that disrupt organogenesis, and impair the response to injury leading in metabolic disorders, and cancer. Curcumin inhibits bFGF-mediated endothelial cell proliferation in vitro. Its antitumor activity is mediated by the suppression of bFGF-mediated angiogenesis (Arbiser et al. 1998).

5.2.3   *Effect of Curcumin on Epidermal Growth Factor (EGF)*

Epidermal growth factor receptor (EGFR) is a 170-kDa transmembrane protein, which belongs to ErbB family of receptor tyrosine kinases. EGFR is a transmembrane receptor that consists of an extracellular domain with conserved two cysteine-rich domains necessary for ligand binding; a single membrane-spanning domain, which has a passive role in signaling and functions as an "anchor" of receptor in plasma membrane, and a cytoplasmic protein tyrosine kinase domain, where six tyrosine autophosphorylation sites are located (Fig. 5.4) (Kashles et al. 1988). EGFR and its family members are stimulated by several distinct ligands, including epidermal growth factor (EGF), TGF-α, amphiregulin, betacellulin, epigen, epiregulin, and heparin binding EGF-like growth factor (Scaltriti and Baselga 2006; Burgess et al. 2003). The binding of ligands with EGFR results in dimerization (homodimers or heterodimers). This stimulates the auto- and/or cross-phosphorylation of key tyrosine residues in the C-terminal tails of the receptor, which can function to initiate phosphorylation/signaling cascades via interaction with SH2- and phosphotyrosine-binding domain containing proteins (Carpenter and Cohen 1990; Scaltriti and Baselga 2006; Blaikie et al. 1994). It is shown that the EGFR migrates to the nucleus, where it acts as a transcription factor for cyclin D1 (Scaltriti and Baselga 2006; Lin et al. 2001) and as a co-activator for STAT3 (Lo et al. 2005) and E2F1 (Hanada et al. 2006). These pathways play important roles in regulating cell proliferation, differentiation, migration, invasion, adhesion, angiogenesis and apoptosis (Yarden and Sliwkowski 2001). In addition to tumorigenesis, EGF signaling is critically involved in renal electrolyte homeostasis (Melenhorst et al. 2008). Dysregulated EGFR signaling has been implicated as a major contributing factor to many types of cancers (Ahmed et al. 2006), such as breast, lung, colorectal, and head/neck cancer. Collective evidence suggests that the EGFR pathway plays critical roles in cancer cell proliferation, migration, survival, angiogenesis, and invasion (Dancey and Freidlin 2003).

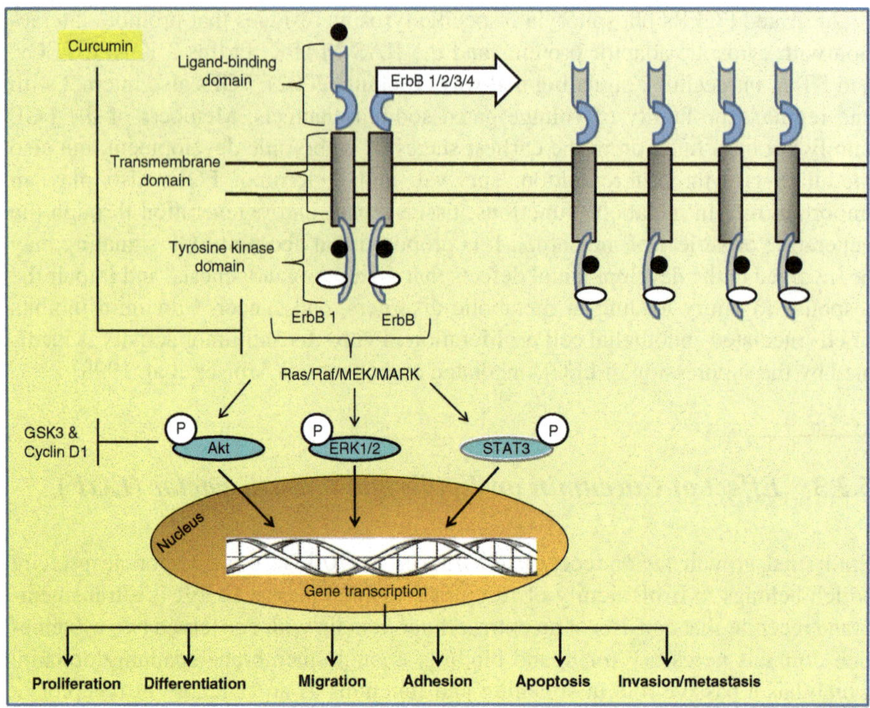

Fig. 5.4 Hypothetical diagram showing EGF signaling. Epidermal growth factor (EGF) or ErbB receptors (ErbB1, ErbB2, ErbB3, and ErbB4); serine-threonine kinase Akt (also known as PKB); *extracellular-signal-regulated kinase*1/2 (ERK1/2); and signal transducer and activator of transcription 3 (STAT3)

EGF is also involved in expansion of renal cysts. Cyst growth is associated with the proliferation of incompletely differentiated epithelial cells and accumulation of fluid within the cysts. Epithelial cells from cysts from patients with both Autosomal dominant (ADPKD) and autosomal recessive (ARPKD) polycystic kidney diseases are unusually susceptible to the proliferative stimulus of EGF (Wilson 2004). Moreover, cyst fluids from these patients contain mitogenic concentrations of EGF, and this EGF is secreted into the lumens of cysts in amounts that can induce cell proliferation (Du and Wilson 1995).

Curcumin acts not only by blocking EGFR signaling, but also by preventing EGFR tyrosine phosphorylation leading to the suppression of EGFR gene expression, which is modulated by the activation of PPAR-γ (Chen and Xu 2005). Curcumin also inhibits SCC-25 cells proliferation and induce G2/M phase arrest in a dose-dependent manner. Curcumin not only inhibits SCC-25 cells invasion, but also down-regulates MMP-2, MMP-9, uPA and uPAR expression (Zhen et al. 2014). Moreover, curcumin regulates the p-EGFR and EGFR downstream signaling molecules including Akt, ERK1/2 and STAT3. Based on these results, it is proposed that curcumin reduces the EGF-induced phosphorylation of EGFR and suppresses

EGF-mediated SCC-25 cells invasion. Collective evidence suggests that curcumin decreases SCC-25 cells proliferation and invasion by inhibiting the phosphorylation of EGFR and EGFR downstream signaling molecules Akt, ERK1/2 and STAT3 (Zhen et al. 2014).

Curcumin has been used for the treatment of kidney diseases. Thus, curcumin has shown renal protective properties against gentamicin- and cisplatin- induced renal toxicities (Ueki et al. 2013) as well as diabetic nephropathy (Huang et al. 2013a, b). It is demonstrated that in cisplatin-induced nephropathy, curcumin inhibits the mitochondrial LPO level, decreases protein carbonyl levels and increases antioxidant enzymes (Waseem and Parvez 2013). In cisplatin-induced nephropathy, curcumin also promotes nephroprotection due to the inhibition of TNF-α and ICAM-1 (Ueki et al. 2013).

5.2.4   Effect of Curcumin on Hypoxia-Inducible Factor (HIF)-α

Hypoxia-inducible factors (HIFs) are DNA-binding transcription factors that transactivate a series of hypoxia-associated genes under hypoxic conditions to adapt to the decreased oxygen tension. As stated in Chap. 4, HIF is a heterodimer composed of an α subunit and a β subunit (Wang et al. 1995). HIF-β is a constitutive subunit expressed in the nucleus, and its activity is not affected by hypoxia, whereas HIF-α is a functional subunit, and its protein stability, subcellular localization, and transcriptional potency are affected by oxygen levels (Li et al. 1996). Three members of HIF family (HIF-1, HIF-2, and HIF-3) are known to occur. These members have the same β subunit but have different α subunits (HIF-1α, HIF-2α, and HIF-3α). The HIF-α subunits contains several domains including an oxygen-dependent degradation domain (ODDD) and two transactivation domains. Both the *N*-terminal transactivation domain (N-TAD) and the *C*-terminal transactivation domain (C-TAD) are essential for the regulation of HIF-dependent gene expression (Pugh et al. 1997; Jiang et al. 1997). Under normal conditions (normoxia), HIF-α is inactivated through the process of hydroxylation by prolyl hydroxylase domain proteins (PHDs). In its inactive form, it is recognized by the ubiquitin E3 ligase, which facilitates its degradation through the proteasomes. Under hypoxic condition, the catalytic activity of PHDs is inhibited and HIF-α translocates to the nucleus where it forms a heterodimer with HIF-β and this complex interacts with the hypoxia-responsive elements (HREs) (Wang and Semenza 1993). Interactions between heteromer and HREs promote the transcription of hypoxia-response genes (Ema et al. 1999). Approximately 100 target genes of HIFs have been identified, consisting of erythropoietin (EPO), vascular endothelial growth factor, heme oxygenase-1 (HO-1), inducible nitric oxide synthase (iNOS), glucose transporter protein 1 (Glut-1), insulin-like growth factor 2 (IGF-2), endothelin 1, transferrin, and others. HIF target genes are particularly relevant to angiogenesis, glucose metabolism and glycolytic enzymes, migration, survival and death (Semenza 2003, 2007; Pouysségur et al. 2006).

Curcumin not only inhibits hypoxia-induced reactive oxygen species (ROS) upregulation, but significantly decreases the mRNA and protein expression levels of hypoxia-inducible factor-1α (HIF-1α) in K1 cells (Tan et al. 2014). Curcumin also decreases the DNA binding ability of HIF-1α to hypoxia response element (HRE). Furthermore, curcumin enhances E-cadherin expression, inhibits metalloprotein-ase-9 (MMP-9) enzyme activity, and weakens K1 cells migration under hypoxic conditions (Tan et al. 2014) supporting the view that curcumin possesses a potent anti-metastatic effect and may be an effective tumoristatic agent for the treatment of aggressive papillary thyroid cancers.

The angiogenic effects of HIF1 are mediated through the stimulation of different angiogenic factors, among which VEGFA is the most important. However, VEGF is also involved in other physiological functions such as maintenance the existing blood vessel system and regulation of vessel permeability through the stimulation of capillary fenestration (Ferrara 2004). It is well known that the presence of a dense intrapituitary vascular system (Viacava et al. 2003) and highly permeable capillary endothelial cells is necessary for the rapid regulation and release of hormones within the anterior pituitary (Lafont et al. 2010). Curcumin acts by attenuating the above-mentioned physiological functions of VEGF through the suppression of VEGF secretion in folliculostellate (FS) cells in the normal pituitary (Schaaf et al. 2010; Shan et al. 2012). Curcumin and its analog (EF24) decrease intracellular levels of both HIF-1α and HIF-1β proteins This in turn results in down-regulation of HIF's transcriptional activity in various human epithelial cancer cell lines (Thomas et al. 2008). Although, the molecular mechanism associated with curcumin-mediated inhibition of HIF protein is not fully understood, but based on detailed investigations it is suggested that curcumin dose-dependently blocks hypoxia-stimulated angiogenesis in vitro and down-regulates HIF-1α and VEGF expression in vascular endothelial cells. These findings suggest that curcumin may play pivotal roles in tumor suppression via the inhibition of HIF-1α-mediated angiogenesis.

5.2.5   *Effect of Curcumin on Peroxisome Proliferator-Associated Receptor γ*

PPARs comprise a family of nuclear receptors that function as transcription factors to regulate gene expression (Belvisi et al. 2006). PPARs regulate fatty acid storage and glucose metabolism. There are three PPAR isoforms (PPAR α, γ, and δ), which are differentially expressed in brain and spinal cord. Among these isoforms, PPAR-γ is the most widely studied form. It not only regulates cell differentiation, apoptosis, but also modulates lipid metabolism and inflammation. The natural ligands for PPAR receptor family are fatty acids and lipid metabolites (J class prostaglandin PGJ$_2$ (15d-PGJ$_2$), and each PPAR family member displays a distinct pattern of ligand specificity. Upon ligand binding, PPAR-γ forms heterodimers with the reti-noid X receptor. This dimer then migrates to the nucleus, where it interacts with a peroxisome proliferation response element (PPRE) leading to the regulation of gene transcription (Forman et al. 1996). In addition of lipid metabolites, PPARγ is also activated by nonsteroidal anti-inflammatory drugs (NSAIDs) and thiazolidinedione. Activation of PPARγ downregulates proinflammatory cytokine expression by antagonizing the activity of the transcription factors NF-κB, AP-1, and STAT proteins in microglial cells, astrocytes, and macrophages (Lemberger et al. 1996; Ricote et al. 1998; Jiang et al. 1998). PPARγ agonists are shown to inhibit the Aβ-stimulated expression of the cytokine genes interleukin-6 and tumor necrosis factor alpha. Furthermore, PPARγ agonists inhibit the expression of cyclooxygenase-2. These data provide direct evidence that PPARγ plays a critical role in regulating the inflammatory responses of microglia and monocytes to Aβ. It is proposed that the efficacy of NSAIDs in the treatment of AD may be a consequence of their actions on PPARγ rather than on their canonical targets the cyclooxygenases. Importantly, the efficacy of these agents in inhibiting a broad range of inflammatory responses suggests PPARγ agonists may provide a novel therapeutic approach to AD (Combs et al. 2000). Curcumin acts by activating PPAR-γ and inhibiting neuroinflammation and oxidative stress caused by advanced glycation end-products (AGEs) in the brain and liver (Liu et al. 2013a; Lin et al. 2012).

5.2.6 Effect of Curcumin on Signal Transducer and Activator of Transcription

Signal Transducer and Activator of Transcription (STATs) are a family of transcription factors consisting of seven members: STAT1, STAT2, STAT3, STAT4, STAT5a, STAT5b, and STAT6. They range in size from 750 to 850 amino acids (Nicolas et al. 2013). These transcription factors are linked with immune regulation, development, metabolism, cell death, and tumorigenesis, amongst other cellular roles. Among other STAT family proteins, STAT3 has received considerable attention during the last decade since it is a convergent point for a number of cellular activities. STAT3 in its inactive form docks in the cytosol. The transcriptional activity of STAT3 is regulated by Janus kinase (JAK)-mediated phosphorylation, resulting in dimerization allowing STAT3 to migrate to the nucleus where it binds to the promotor region of genes containing gamma-activated sequences. In the adult nervous system, this pathway is mostly dormant (Raible et al. 2014). However, under pathological conditions (traumatic insult), the activation of STAT3 by Janus kinase (JAK) occurs in astrocytes, microglia, endothelial cells, and neurons in response to a variety of cytokines and growth factors (Choi et al. 2003; Justicia et al. 2000) leading not only to neuroinflammation, but also to regulation of the transcription of numerous genes. JAK2-STAT3 pathway plays an important role in the regulation of synaptic transmission. The induction of long-term depression (LTD) via NMDA receptors depends on STAT3 activation by JAK2 that can be localized specifically to postsynaptic structures (Hofmann and Kirsch 2012). It is suggested that in brain dysregulation of the JAK-STAT pathway is mainly related to neuroinflammation and neuronal/glial survival. It is also reported that STAT3 modulates the migration of reactive astrocytes to the injury site (Okada et al. 2006).

Anti-inflammatory effects of curcumin are mediated through the modulation of Janus kinase (JAK)-STAT signaling. In both rat primary microglia and murine BV2 microglial cells, curcumin inhibits cytokine-mediated induction of cyclooxygenase-2 and inducible NO synthase. These enzymes mediate and support inflammatory processes. Curcumin inhibits the phosphorylation of JAK1 and JAK2 via the increased phosphorylation of SHP-2 and its association with JAK1/2, thus attenuating inflammatory response supporting the view that curcumin acts via a novel anti-inflammatory mechanism and is also a negative regulator of the JAK-STAT pathway by the activation of SHP-2 (Kim et al. 2003).

5.3   Effects of Curcumin on Ion Channels and Transporters in Visceral Organs and Brain

Biomembrane are composed of phospholipids, sphingolipids, cholesterol, and proteins and are penetrated by types of receptors and ion channels. Ion channels are pore-forming membrane proteins, which are involved in establishing a resting membrane potential, shaping action potentials along with maintenance of electrical signals by gating the flow of ions across the hydrophobic cell membrane. Ion channels also control the flow of ions across hydrophobic biomembranes, and regulate cell volume. Ion channel proteins are important for normal cellular metabolism. Similarly, transport are membrane or cytoplasmic protein associated with the movement of ions (sodium-potassium pump, sodium-calcium exchanger, and sodium-glucose transport proteins etc), small molecules, or macromolecules, such as another protein, across a biological membrane. Transporters are crucial for normal cellular metabolism. Ion channels and transporters regulate the fluxes of ions, nutrients and other molecules across the membranes of all brain and visceral cells. The activities of ion channels and transporters are necessary for diverse physiological processes such as brain electrical activity, muscle contraction, water and solute transport in the kidney, hormone secretion and the immune response. Curcumin influences membrane structure in a manner analogous to lipophillic drugs, which are inserted deeply into the membrane in a transbilayer orientation and anchored by hydrogen bonding to the phosphate group of lipids (Barry et al. 2009). Curcumin binding sites of majority of membrane proteins, which may be components of ion channels, have not been characterized. However, it is becoming increasingly evident that curcumin modulates activity of many transporters proteins at approximately similar concentrations (micromolar range) despite the lack of a consensus binding motif (Ingolfsson et al. 2007). Since curcumin modulates many unrelated ion channels transporter proteins at approximately similar concentrations (Figs. 5.5 and 5.6); it is proposed

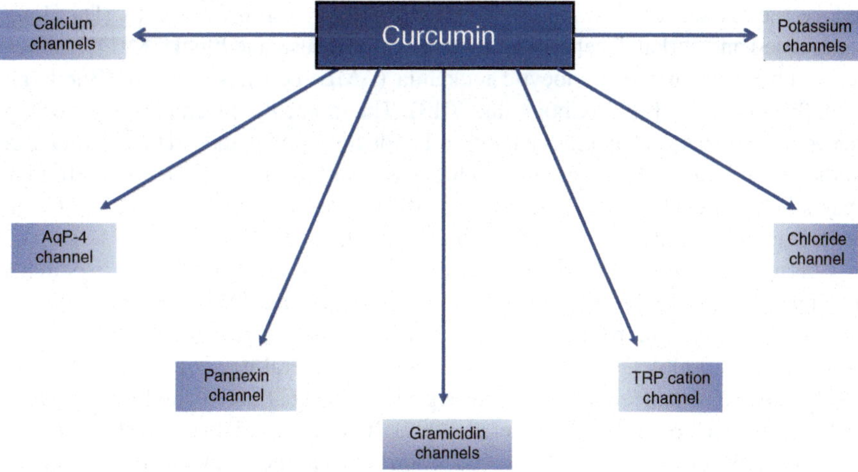

Fig. 5.5 Modulation of ion channels by curcumin

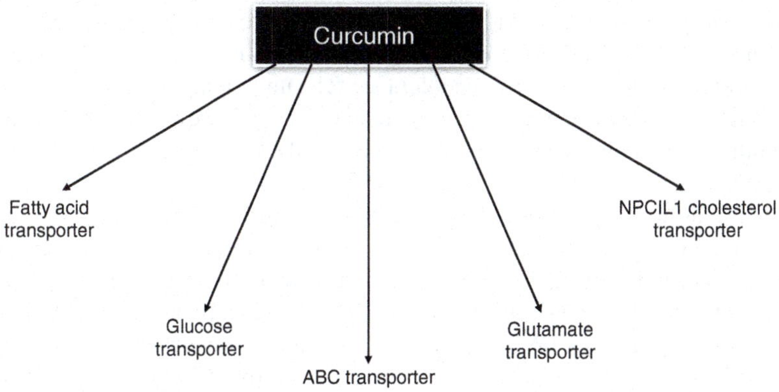

Fig. 5.6 Modulation of transporters by curcumin

that curcumin regulates the action of membrane proteins indirectly by changing the physical properties of the membrane rather than by direct binding of curcumin to the protein (Ingolfsson et al. 2007).

5.3.1   *Modulation of Ion Channels by Curcumin*

Curcumin modulates a number of ion channels in time and dose-dependent manner (Fig. 5.5) (Zhang et al. 2014). It is well known that Kv channels contribute to the regulation of resting membrane potential. Curcumin has not only been reported to reduce the Kv current in rabbit coronary arterial smooth muscle cells (Da Hong et al. 2013) and Jurkat T cells (Shin et al. 2011), but also and blocks Kv11.1 potassium current in human monocytic leukemia (AML) cell lines THP-1 (Banderali et al. 2011) and HEK293 (Choi et al. 2013). The molecular mechanisms by which curcumin reduces Kv channel activation is not fully understood. However, recent studies have indicated that curcumin blocks Kv current via direct action on the Kv channels (Choi et al. 2013; Da Hong et al. 2013; Lian et al. 2013), possibly through its interaction with the pore blocker binding site (Choi et al. 2013).

Curcumin also has antinociceptive activity. It reduces pain and fatigue in patients undergoing laparoscopic cholecystectomy (Agarwal et al. 2011). Systemic administration of curcumin results in significant decrease in pain in animal models (Sharma et al. 2006) such as acetic acid-induced visceral nociception (Tajik et al. 2008), capsaicin-induced thermal hyperalgesia (Yeon et al. 2010), and the formalin-induced orofacial pain test (Mittal et al. 2009). It is suggested that activation of K_{ATP} channels, possibly by direct stimulation, contributes to the antinociceptive effect of curcumin (De Paz-Campos et al. 2012). Additional potential mechanisms of the antinociceptive effect of curcumin include activating $G_{i/o}$ proteins, stimulating the particular form of guanylyl cyclase or acting through the hydrogen sulfide-KATP channel pathway. Collective evidence suggests that curcumin modulates activity of potassium channels. Chronic dose of large amount of curcumin (100 mg/kg) has been reported to attenuate morphine tolerance. The underlying mechanism involved in attenuation of morphine tolerance remains unclear. However, it is reported that curcumin acts by inhibiting morphine-induced CaMKIIα activation in the brain (Hu et al. 2015)

Calcium influx plays an important role in neurodegeneration through the activation of calcium-dependent enzymes, such as phospholipases A_2, calcium-dependent proteases, and nitric oxide synthases. Calcium channel blockers been reported to mediate neuroprotective effects (Farooqui 2010). In rat hippocampal neurons curcumin reversibly inhibits HVGCC currents via a novel protein kinase C-θ -dependent pathway which is associated with the neuroprotective effects (Liu et al. 2013b). Furthermore, curcumin also inhibits glutamate release from rat prefrontocortical synaptosomes by suppressing presynaptic voltage-gated calcium channels Cav2.2 and Cav2.1 (Lin et al. 2011). It is proposed that these effects of curcumin may be related to the mechanisms underlying the antidepressant effect of curcumin. Curcumin has also been observed to inhibit Ca^{2+} release-activated Ca^{2+} (CRAC) channels (Shin et al. 2011, 2012).

In contrast to cation (potassium and calcium) channels, which are inhibited by curcumin, chloride channels are activated by this polyphenol. Thus, cystic fibrosis transmembrane regulator (CFTR) chloride channel is activated by curcumin (0.5–10 μM) both wild-type and ΔF508-CFTR in excised membrane patches (Berger et al. 2005; Wang et al. 2007). It has been pointed out that the structure of curcumin (two aromatic rings separated by a hydrocarbon spacer) is similar to that of 5-nitro-2-(3 phenylpropylamino)benzoic acid-AM (NPPB-AM), which is an uncharged form of the well-known chloride channel blocker NPPB. This compound acts as a CFTR agonist by increasing the channel opening rate.

Curcumin down-regulates AQP-4, a predominant isoform of water channels in the brain, which regulates fast water transport within the perivascular end feet of astrocytes (Nicchia et al. 2004). It plays an important role in fluid generation, transfer, and absorption in brain. The down-regulation of AQP-4 by curcumin protects from hypoxic–ischemic brain damage in the hippocampus by the attenuation of cerebral edema and inhibition of nitric oxide synthase (NOS) activity (Yu et al. 2012). Curcumin also decreases the expression of IL-1β and preferentially attenuates the activation of NFκB, a transcriptional regulator of IL-1β and downstream mediator of IL-1β-induced signaling (Moynagh 2005), within pericontusional astrocytes. Curcumin also retards IL-1β -mediated AQP-4 expression in the cultured astrocytes and further reduce glial activation and cerebral edema following neurotrauma (Laird et al. 2010).

In addition to potassium and calcium channels, curcumin also modulates volume-regulated anion channel (VRAC), chloride channel (CFTR), aquaporin-4 (AQP-4), and Transient receptor potential (TRP) cation channel (Yu et al. 2011; Zhang et al. 2014).

5.3.2   *Modulation of Transporters by Curcumin*

As stated above, curcumin modulates activities of many unrelated transporter proteins at approximately similar concentrations (Fig. 5.6), it is proposed that curcumin regulates the action of membrane proteins indirectly by changing the physical properties of the membrane rather than by the direct binding of curcumin to the protein (Ingolfsson et al. 2007). Studies on the effect of tetrahydrocurcumin (THC) on three ABC drug transporter proteins, P-glycoprotein (P-gp or ABCB1), mitoxantrone resistance protein (MXR or ABCG2) and multidrug resistance protein 1 (MRP1 or ABCC1) have indicated that THC inhibits the function of P-gp, MXR and MRP1 in time and dose-dependent manner (Zhang et al. 2014). In the multidrug-resistant human cervical carcinoma cell line KB-V1, curcumin significantly lowers the Pgp expression and reduces the function of Pgp. Curcumin is not a substrate for Pgp, but it binds directly with drug binding site of the transporter (Anuchapreeda et al. 2002; Limtrakul et al. 2007). Treatment of KB-V1 cells with curcumin increases their sensitivity to vinblastine (Anuchapreeda et al. 2002; Limtrakul et al. 2007).

5.4   Effect of Curcumin on Obesity

Obesity is a multifactorial (interaction of social, behavioral, psychological, metabolic, cellular, and molecular factors) condition mediated by an imbalance in energy intake and energy expenditure leading to the pathological growth of adipocytes. Obesity is modulated by lifestyle (energy dense diet, lack of exercise, and sleep), behavioral, environmental as well as genetic factors (Taheri et al. 2004; Kavanagh et al. 2007). At the cellular level obesity is accompanied by the hypertrophy of adipocytes along with increase in numbers of adipocytes (hyperplasia) (Sikaris 2004). Adipose tissue is an important energy storage organ, which contributes to endocrine function by secreting pro- and anti-inflammatory adipokines (Kalupahana et al. 2012). These adipokines play critical roles to overall energy balance and homeostasis. Increase in production and secretion of pro-inflammatory adipokines, chemokines, and cytokines is accompanied by systemic inflammation, insulin resistance and obesity-related metabolic disorders arise (Kalupahana et al. 2012). In addition to adipocytes, stroma vascular cells—preadipocytes, adipose stem cells—and immune cells within adipose tissue such as macrophages and lymphocytes are rich sources of adipokines and contribute to obesity-associated inflammation (Kalupahana et al. 2012). Thus, obesity is a chronic low-grade inflammatory condition, which contributes to the pathogenesis of heart disease, diabetes, and common forms of cancer (Després et al. 2001). Metabolic syndrome, a pathological condition mediated by clustering of obesity, insulin resistance, hypertension, and dyslipidemia, with high triglycerides and low HDL cholesterol, has emerged as important risk factors not only for atherosclerotic disease, but also for stroke, Alzheimer disease, and depression (Farooqui et al. 2012; Farooqui 2013). Consumption of Western diet (high fat, high protein diet, which is also enriched in sugar and refined flour) induces metabolic dysfunction capable of disrupting brain homeostasis and is likely to contribute to the development of brain dysfunction associated with the pathogenesis of above mentioned neurological disorders (Fig. 5.7) (Farooqui 2013). The molecular mechanisms associated with harmful effects of western diet are not fully understood. However, it is becoming increasingly evident that western diet-mediated neurochemical changes in the brain not only generate elevated levels of proinflammatory eicosanoids, platelet activating factor (PAF), and proinflammatory cytokines (TNF-α, and IL-1β), but also upregulate of gp91(phox) subunit of NADPH oxidase and downregulate superoxide dismutase (SOD) isoforms, glutathione peroxidase (GPX), and heme oxygenase-2 (HO-2) in brain and other body tissues (Roberts et al. 2006). In addition, consumption of western diet results in reduction in levels of BDNF in the hippocampus leading to impairment in synaptic function, metabolic perturbation, and atrophy of dendritic spines (Farooqui et al. 2012; Farooqui 2013). BDNF is widely expressed throughout the brain and by most cell types, including neurons, astrocytes, microglia, and vascular endothelial cells. The classical induction mechanism involves calcium influx and phosphorylation of the transcription factor CREB. Under pathological conditions, the phosphorylation of CREB is associated with induction of 7 neurotrophic and growth factor genes and 5 receptor genes including FGF-2, ILG-1, VEGF A, NGFb, and BDNF—and epidermal growth factor (Qu et al. 2007) (see below).

In visceral tissues, curcumin produces significant effect on adiposity and lipid metabolism through several mechanisms including modulation of energy metabolism, inflammation, and suppression of angiogenesis. Thus, curcumin prevents adipogenesis through suppression of angiogenesis into the adipose tissue (Cao 2007; Lijnen 2008; Voros et al. 2005; Rupnick et al. 2002). Suppression of angiogenesis is promoted through the involvement of VEGF, bFGF, EGF, angiopoietin, and

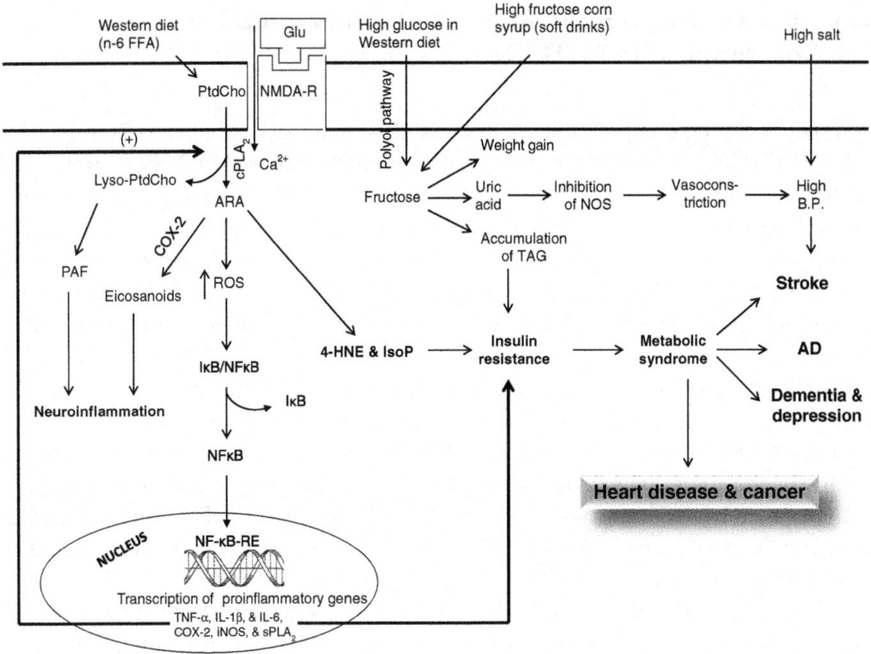

Fig. 5.7 Hypothetical diagram showing metabolic consequences of western diet consumption. Glutamate (*Glu*); NMDA receptor (*NMDA-R*); Phosphatidylcholine (*PtdCho*); Lyso-phosphatidylcholine (*Lyso-PtdCho*); cytosolic phospholipase A_2 (*cPLA₂*); cyclooxygenase-2 (*COX-2*); arachidonic acid (*ARA*); platelet-activating factor (*PAF*); reactive oxygen species (*ROS*); 4-hydroxynonenal (*4-HNE*); isopropstane (*IsoP*); nuclear factor-κB (*NF-κB*); nuclear factor-κB-response element (*NF-κB-RE*); inhibitory subunit of NF-κB (*I-κB*); tumor necrosis factor-α (*TNF-α*); interleukin-1β (*IL-1β*); interleukin-6 (*IL-6*); inducible nitric oxide synthase (*iNOS*); secretory phospholipase A_2 (*sPLA₂*); triacylglycerol (*TAG*); blood pressure (*B.P.*); Alzheimer disease (*AD*)

HIF-1α. In adipose tissue, angiogenesis is mediated by adipose tissue secretion of adipokines including leptin, adiponectin, resistin, visfatin, tumor necrosis factor (TNF)-α, interleukin-6 (IL)-6, IL-1, and VEGF (Tilg and Moschen 2006). Therefore, the inhibition of angiogenesis in adipose tissue can be used as a strategy to prevent the growth of adipose tissue and thus reducing obesity. It is demonstrated that supplementation of a high fat diet of C57BL/6 J mice with curcumin reduces microvessel density as an indication of suppression of angiogenesis in adipose tissue (Ejaz et al. 2009). Furthermore, curcumin also activates AMPK through the down-regulation of ACC activity via phosphorylation of this enzyme. This process decreases the flow of acetyl CoA to malonyl CoA leading to up-regulation of carnitine palmitoyltransferase-1 (CPT-1), which transfers cytosolic long-chain fatty acyl CoA into the mitochondria for oxidation (Ruderman et al. 2003). In addition, through activation of AMPK, curcumin down-regulated synthesis of glycerol lipids by inhibiting glycerol-3-phosphate acyl transferase-1 (GPAT-1) activity, which esterifies fatty acids to glycerol to form triglycerides for storage (Ejaz et al. 2009; Anand et al. 2007; Agarwal et al. 2011).

5.5 Effect of Curcumin and Mammalian Target of Rapamycin (mTOR)

Another important target of curcumin is the mammalian target of rapamycin (mTOR), which is a serine/threonine protein kinase composed of 2549 amino acids with molecular mass of 289 kDa (Dennis et al. 2001). This kinase plays an important role in regulating many fundamental molecules mediating cell growth and cell cycle progression in response to cellular signals in eukaryotes (Liu et al. 2009). The mTOR signaling pathway is involved in cellular processes such as cell survival, cell growth and proliferation, and cell death. The mTOR activity is regulated by many growth factors, such as EGF, VEGF, insulin, insulin-like growth factor 1 (IGF- 1), transforming growth factor (TGF), vascular endothelial growth factor (VEGF), hepatocyte growth factor (HGF) and platelet-derived growth factor (PDGF) (Gomez-Pinillos and Ferrari 2012). mTOR functions in vivo as the catalytic subunit of one or more supramolecular protein complexes that control the intrinsic kinase activity of mTOR (Kim et al. 2002). So far two targets, mTORC1 and mTORC2 have been identified (Beevers et al. 2013). mTORC1 is composed of mTOR, regulatory associated protein of mTOR (Raptor), GβL, proline-rich Akt1 substrate of 40 kDa (PRAS40) and DEPTOR (Kim et al. 2002; Johnston 2006; Beevers et al. 2009; Rafiee et al. 2010). mTORC1 signals to two primary downstream targets, p70 S6 kinase 1 (S6K1) and eukaryotic translation initiation factor 4E (eIF4E)-binding protein 1 (4E-BP1) (Hay and Sonenberg 2004). In contrast, mTORC2 is composed of mTOR, rapamycin insensitive companion of mTOR (Rictor), G-protein β-subunit like protein (GβL), mammalian Sin1 (mSin1), protein observed with Rictor-1 and -2 (protor-1 and protor-2), and DEP domain containing mTOR-interacting protein (DEPTOR) (Pearce et al. 2007). mTOR has been recognized as a key therapeutic target for the prevention and/or treatment of cancer and many other diseases because of the fact that it functions as a central hub for the regulation of numerous key cellular processes that are not only critical for normal cellular function but are also absolutely necessary for tumorigenesis (Guertin and Sabatini 2007).

Curcumin promotes G_2/M cell cycle arrest and autophagy in two human malignant glioma cell lines (U87-MG and U373-MG) (Aoki et al. 2007). These effects are associated with curcumin-mediated inhibition of mTORC2 (as indicated by the phosphorylation of Akt) and activation of the ERK pathway, as this inhibition can prevent curcumin-mediated autophagy (Aoki et al. 2007). Curcumin treatment also retards the growth of these tumor cell lines in vivo by inducing autophagy (Aoki et al. 2007). Detailed investigations support the view that curcumin acts by inhibiting the phosphorylation of S6K1 and 4E-BP1 in the cells pretreated with PP2A inhibitor (okadaic acid) or AMPK inhibitor (compound C), or in the cells expressing dominant-negative (dn) PP2A, shRNA to PP2A-A subunit, or dn-AMPKα. Curcumin does not alter the TSC1/2 interaction. Knockout of TSC2 does not affect curcumin inhibition of mTOR signaling. It is also reported that curcumin promote the dissociation of raptor from mTOR, leading to inhibition of mTORC1 activity (Beevers et al. 2009).

5.6 Conclusion

Extensive research over the last decade has revealed that curcumin produces its effects by interacting with several growth factors (VEGF, bFGF, EGF, and HIF), ion channels, and transporters. These growth factors not only regulate inflammatory responses and control apoptotic cell death, but also modulate angiogenesis. Main molecular mechanisms of curcumin action not only include the inhibition of proinflammatory enzymes, such as COX-2, LOX, and iNOS, and the inhibition of PtdIns 3 K, tyrosine kinases, and NF-κB, but also include the activation of MAPK, PKC, and the modulation of several cell survival/cell-cycle genes. Collective evidence suggests that by regulating the activation of above mentioned growth factors, ion channels, and transporters, curcumin inhibits cancer cell proliferation, induces apoptosis and autophagy. Curcumin also blocks cell invasion and migration and suppresses inflammatory responses.

References

Agarwal KA, Tripathi CD, Agarwal BB, Saluja S (2011) Efficacy of turmeric (curcumin) in pain and postoperative fatigue after laparoscopic cholecystectomy: a double-blind, randomized placebo-controlled study. Surg Endosc 25:3805–3810

Aggarwal BB, Sung B (2009) Pharmacological basis for the role of curcumin in chronic diseases: an age-old spice with modern targets. Trends Pharmacol Sci 30:85–94

Aggarwal S, Ichikawa H, Takada Y, Sandur SK, Shishodia S, Aggarwal BB (2006) Curcumin (diferuloylmethane) down-regulates expression of cell proliferation and antiapoptotic and metastatic gene products through suppression of IkappaBalpha kinase and Akt activation. Mol Pharmacol 69:195–206

Ahmed K, Cao N, Li J (2006) HER-2 and NF-kappaB as targets for therapy-resistant breast cancers. Anticancer Res 26:4235–4243

Anand P, Kunnumakkara AB, Newman RA, Aggarwal BB (2007) Bioavailability of curcumin: problems and promises. Mol Pharm 4:807–818

Anuchapreeda S, Leechanachai P, Smith MM, Ambudkar SV, Limtrakul PN (2002) Modulation of P-glycoprotein expression and function by curcumin in multidrug-resistant human KB cells. Biochem Pharmacol 64:573–582

Aoki H, Takada Y, Kondo S, Sawaya R, Aggarwal BB, Kondo Y (2007) Evidence that curcumin suppresses the growth of malignant gliomas in vitro and in vivo through induction of autophagy: role of Akt and extracellular signal-regulated kinase signaling pathways. Mol Pharmacol 72:29–39

Arbiser JL, Klauber N, Rohan R, van Leeuwen R, Huang MT, Fisher C, Flynn E, Byers HR (1998) Curcumin is an in vivo inhibitor of angiogenesis. Mol Med 4:376–383

Bae MK, Kim SH, Jeong JW, Lee YM, Kim HS, Kim SR, Yun I, Bae SK, Kim KW (2006) Curcumin inhibits hypoxia-induced angiogenesis via down-regulation of HIF-1. Oncol Rep 15:1557–1562

Banderali U, Belke D, Singh A, Jayanthan A, Giles WR, Narendran A (2011) Curcumin blocks Kv11.1 (erg) potassium current and slows proliferation in the infant acute monocytic leukemia cell line THP-1. Cell Physiol Biochem 28:1169–1180

Barry J, Fritz M, Brender JR, Smith PE, Lee DK, Ramamoorthy A (2009) Determining the effects of lipophilic drugs on membrane structure by solid-state NMR spectroscopy: the case of the antioxidant curcumin. J Am Chem Soc 131:4490–4498

Bar-Sela G, Epelbaum R, Schaffer M (2010) Curcumin as an anti-cancer agent: review of the gap between basic and clinical applications. Curr Med Chem 17:190–197

Beevers CS, Chen L, Liu L, Luo Y, Webster NJ, Huang S (2009) Curcumin disrupts the Mammalian target of rapamycin-raptor complex. Cancer Res 69:1000–1008

Beevers CS, Zhou H, Huang S (2013) Hitting the golden TORget: curcumin's effects on mTOR signaling. Anticancer Agents Med Chem 13:988–994

Belvisi MG, Hele DJ, Birrell MA (2006) Peroxisome proliferator-activated receptor gamma agonists as therapy for chronic airway inflammation. Eur J Pharmacol 533:101–109

Berger AL, Randak CO, Ostedgaard LS, Karp PH, Vermeer DW, Welsh MJ (2005) Curcumin stimulates cystic fibrosis transmembrane conductance regulator Cl⁻ channel activity. J Biol Chem 280:5221–5226

Blaikie P, Immanuel D, Wu J, Li N, Yajnik V, Margolis B (1994) A region in Shc distinct from the SH2 domain can bind tyrosine-phosphorylated growth factor receptors. J Biol Chem 269:32031–32034

Bogoyevitch MA, Gillespie-Brown J, Ketterman AJ, Fuller SJ, Ben-Levy R, Ashworth A, Marshall CJ, Sugden PH (1996) Stimulation of the stress-activated mitogen-activated protein kinase subfamilies in perfused heart. p38/RK mitogen-activated protein kinases and c-Jun N-terminal kinases are activated by ischemia/reperfusion. Circ Res 79:162–173

Brondino N, Re S, Boldrini A, Cuccomarino A, Lanati N, Barale F, Politi P (2014) Curcumin as a therapeutic agent in dementia: a mini systematic review of human studies. ScientificWorldJournal 2014:174282

Burgess A, Cho H, Eigenbrot C, Ferguson KM, Garrett TP, Leahy DJ, Lemmon MA, Sliwkowski MX, Ward CW, Yokoyama S (2003) An open-and-shut case? Recent insights into the activation of EGF/ErbB receptors. Mol Cell 12:541–552

Cao Y (2007) Angiogenesis modulates adipogenesis and obesity. J Clin Invest 117:2362–2368

Carpenter G, Cohen S (1990) Epidermal growth factor. J Biol Chem 265:7709–7712

Chadalapaka G, Jutooru I, Chintharlapalli S, Papineni S, Smith R 3rd, Li X, Safe S (2008) Curcumin decreases specificity protein expression in bladder cancer cells. Cancer Res 68:5345–5354

Chen A, Xu J (2005) Activation of PPAR{gamma} by curcumin inhibits Moser cell growth and mediates suppression of gene expression of cyclin D1 and EGFR. Am J Physiol Gastrointest Liver Physiol 288:G447–G456

Choi JS, Kim SY, Cha JH, Choi YS, Sung KW, Oh ST, Kim ON, Chung JW, Chun MH, Lee SB, Lee MY (2003) Upregulation of gp130 and STAT3 activation in the rat hippocampus following transient forebrain ischemia. Glia 41:237–246

Choi SW, Kim KS, Shin DH, Yoo HY, Choe H, Ko TH, Youm JB, Kim WK, Zhang YH, Kim SJ (2013) Class 3 inhibition of hERG K(+) channel by caffeic acid phenethyl ester (CAPE) and curcumin. Pflugers Arch 465:1121–1134

Combs CK, Johnson DE, Karlo JC, Cannady SB, Landreth GE (2000) Inflammatory mechanisms in Alzheimer's disease: inhibition of beta-amyloid-stimulated proinflammatory responses and neurotoxicity by PPARgamma agonists. J Neurosci 20:558–567

Da Hong H, Son YK, Choi IW, Park WS (2013) The inhibitory effect of curcumin on voltage-dependent K(+) channels in rabbit coronary arterial smooth muscle cells. Biochem Biophys Res Commun 430:307–312

Dancey J, Freidlin B (2003) Targeting epidermal growth factor receptor--are we missing the mark? Lancet 362:62–64

De Paz-Campos MA, Chavez-Pina AE, Ortiz MI, Castaneda-Hernandez G (2012) Evidence for the participation of ATP-sensitive potassium channels in the antinociceptive effect of curcumin. Korean J Pain 25:221–227

Dennis PB, Jaeschke A, Saitoh M, Fowler B, Kozma SC, Thomas G (2001) Mammalian TOR: a homeostatic ATP sensor. Science 294:1102–1105

Després JP, Lemieux I, Prud'homme D (2001) Treatment of obesity: need to focus on high risk abdominally obese patients. BMJ 322:716–720

Du J, Wilson PD (1995) Abnormal polarization of EGF receptors and autocrine stimulation of cyst epithelial growth in human ADPKD. Am J Physiol 269:C487–C495

Ejaz A, Wu D, Kwan P, Meydani M (2009) Curcumin inhibits adipogenesis in 3T3-L1 adipocytes and angiogenesis and obesity in C57/BL mice. J Nutr 139:919–925

Ema M, Hirota K, Mimura J, Abe H, Yodoi J, Sogawa K, Poellinger L, Fujii-Kuriyama Y (1999) Molecular mechanisms of transcription activation by HLF and HIF1α in response to hypoxia: their stabilization and redox signal-induced interaction with CBP/p300. EMBO J J18: 1905–1914

Farooqui AA (2010) Neurochemical aspects of neurotraumatic and neurodegenerative diseases. Springer, New York

Farooqui AA (2013) Metabolic syndrome: an important risk factor for stroke, Alzheimer disease, and depression. Springer, New York

Farooqui AA, Farooqui T, Panza F, Frisardi V (2012) Metabolic syndrome as a risk factor for neurological disorders. Cell Mol Life Sci 69:741–762

Ferrara N (2004) Vascular endothelial growth factor: basic science and clinical progress. Endocr Rev 25:581–611

Ferrara N, Hillan KJ, Novotny W (2005) Bevacizumab (Avastin), a humanized anti-VEGF monoclonal antibody for cancer therapy. Biochem Biophys Res Commun 333:328–335

Fiala M, Liu PT, Espinosa-Jeffrey A, Rosenthal MJ, Bernard G, Ringman JM, Sayre J, Zhang L, Zaghi J, Dejbakhsh S, Chiang B, Hui J, Mahanian M, Baghaee A, Hong P, Cashman J (2007) Innate immunity and transcription of MGAT-III and Toll-like receptors in Alzheimer's disease patients are improved by bisdemethoxycurcumin. Proc Natl Acad Sci U S A 104: 12849–12854

Forman BM, Chen J, Evans RM (1996) The peroxisome proliferator-activated receptors: ligands and activators. Ann N Y Acad Sci 804:266–275

Garcia-Alloza M, Borrelli LA, Rozkalne A, Hyman BT, Bacskai BJ (2007) Curcumin labels amyloid pathology in vivo, disrupts existing plaques, and partially restores distorted neurites in an Alzheimer mouse model. J Neurochem 102:1095–1104

Goel A, Kunnumakkara AB, Aggarwal BB (2008) Curcumin as "Curecumin": from kitchen to clinic. Biochem Pharmacol 75:787–809

Gomez-Pinillos A, Ferrari AC (2012) mTOR signalling pathway and mTOR inhibitors in cancer therapy. Hematol Oncol Clin North Am 26:483–505

Guertin DA, Sabatini DM (2007) Defining the role of mTOR in cancer. Cancer Cell 12:9–22

Gururaj AE, Belakavadi M, Venkatesh DA, Marme D, Salimath BP (2002) Molecular mechanisms of anti-angiogenic effect of curcumin. Biochem Biophys Res Commun 297:934–942

Hanada N, Lo H, Day C, Pan Y, Nakajima Y, Hung MC (2006) Co-regulation of B-Myb expression by E2F1 and EGF receptor. Mol Carcinog 45:10–17

Hay N, Sonenberg N (2004) Upstream and downstream of mTOR. Genes Dev 18:1926–1945

Hicklin DJ, Ellis LM (2005) Role of the vascular endothelial growth factor pathway in tumor growth and angiogenesis. J Clin Oncol 23:1011–1027

Hofmann HD, Kirsch M (2012) JAK2-STAT3 signaling: a novel function and a novel mechanism. JAKSTAT 1:191–193

Hu X, Huang F, Szymusiak M, Liu Y, Wang ZJ (2015) Curcumin attenuates opioid tolerance and dependence by inhibiting Ca2+/calmodulin-dependent protein kinase II α activity. J Pharmacol Exp Ther 352:420–428

Huang CZ, Huang WZ, Zhang G, Tang DL (2013a) In vivo study on the effects of curcumin on the expression profiles of anti-tumour genes (VEGF, CyclinD1 and CDK4) in liver of rats injected with DEN. Mol Biol Rep 40:5825–5831

Huang J, Huang K, Lan T, Xie X, Shen X, Liu P, Huang H (2013b) Curcumin ameliorates diabetic nephropathy by inhibiting the activation of the SphK1-S1P signaling pathway. Mol Cell Endocrinol 365:231–240

Ingolfsson HI, Koeppe RE, Andersen OS (2007) Curcumin is a modulator of bilayer material properties. Biochemistry 46:10384–10391

Jeong JM, Choi CH, Kang SK, Lee IH, Lee JY, Jung H (2007) Antioxidant and chemosensitizing effects of flavonoids with hydroxy and/or methoxy groups and structure-activity relationship. J Pharm Pharm Sci 10:537–546

Jiang B-H, Zheng JZ, Leung SW, Roe R, Semenza GL (1997) Transactivation and inhibitory domains of hypoxia-inducible factor 1alpha. Modulation of transcriptional activity by oxygen tension. J Biol Chem 272:19253–19260

Jiang C, Ting AT, Seed B (1998) PPAR-gamma agonists inhibit production of monocyte inflammatory cytokines. Nature 391:82–86

Johnston SR (2006) Targeting downstream effectors of epidermal growth factor receptor/HER2 in breast cancer with either farnesyltransferase inhibitors or mTOR antagonists. Int J Gynecol Cancer 16(Suppl 2):543–548

Justicia C, Gabriel C, Planas AM (2000) Activation of the JAK/STAT pathway following transient focal cerebral ischemia: signaling through Jak1 and Stat3 in astrocytes. Glia 30:253–270

Kalupahana NS, Moustaid-Moussa N, Claycombe KJ (2012) Immunity as a link between obesity and insulin resistance. Mol Aspects Med 33:26–34

Kamba T, McDonald DM (2007) Mechanisms of adverse effects of anti-VEGF therapy for cancer. Br J Cancer 96:1788–1795

Kashles O, Szapary D, Bellot F, Ullrich A, Schlessinger J, Schmidt A (1988) Ligand-induced stimulation of epidermal growth factor receptor mutants with altered transmembrane regions. Proc Natl Acad Sci U S A 85:9567–9571

Kavanagh K, Jones KL, Sawyer J, Kelley K, Carr JJ, Wagner JD, Rudel LL (2007) Trans fat diet induces abdominal obesity and changes in insulin sensitivity in monkeys. Obesity (Silver Spring) 15:1675–1684

Kearney JB, Ambler CA, Monaco KA, Johnson N, Rapoport RG, Bautch VL (2002) Vascular endothelial growth factor receptor Flt-1 negatively regulates developmental blood vessel formation by modulating endothelial cell division. Blood 99:2397–2407

Kim DH, Sarbassov DD, Ali SM, King JE, Latek RR, Erdjument-Bromage H, Tempst P, Sabatini DM (2002) mTOR interacts with raptor to form a nutrient-sensitive complex that signals to the cell growth machinery. Cell 110:163–175

Kim HY, Park EJ, Joe EH, Jou I (2003) Curcumin suppresses Janus kinase-STAT inflammatory signaling through activation of Src homology 2 domain-containing tyrosine phosphatase 2 in brain microglia. J Immunol 17:6072–6079

Lafont C, Desarmenien MG, Cassou M, Molino F, Lecoq J, Hodson D, Lacampagne A, Mennessier G, El Yandouzi T, Carmignac D, Fontanaud P, Christian H, Coutry N, Fernandez-Fuente M, Charpak S, Le Tissier P, Robinson IC, Mollard P (2010) Cellular in vivo imaging reveals coordinated regulation of pituitary microcirculation and GH cell network function. Proc Natl Acad Sci U S A 107:4465–4470

Laird MD, Sukumari-Ramesh S, Swift AE, Meiler SE, Vender JR, Dhandapani KM (2010) Curcumin attenuates cerebral edema following traumatic brain injury in mice: a possible role for aquaporin-4? J Neurochem 113:637–648

Lemberger T, Desvergne B, Wahli W (1996) Peroxisome proliferator-activated receptors: a nuclear receptor signaling pathway in lipid physiology. Annu Rev Cell Dev Biol 12:335–363

Levi M, van der Poll T, Schultz M (2012) Infection and inflammation as risk factors for thrombosis and atherosclerosis. Semin Thromb Hemost 38:506–514

Li H, Ko HP, Whitlock JP Jr (1996) Induction of phosphoglycerate kinase 1 gene expression by hypoxia. Roles of Arnt and HIF1α. J Biol Chem 271:21262–21267

Lian YT, Yang XF, Wang ZH, Yang Y, Shu YW, Cheng LX, Kiu K (2013) Curcumin serves as a human Kv1.3 blocker to inhibit effector memory T Lymphocyte activities. Phytother Res 27:1321–1327

Libby P (2012) Inflammation in atherosclerosis. Arterioscler Thromb Vasc Biol 32:2045–2051

Lijnen HR (2008) Angiogenesis and obesity. Cardiovasc Res 78:286–293

Limtrakul P, Chearwae W, Shukla S, Phisalphong C, Ambudkar SV (2007) Modulation of function of three ABC drug transporters, P-glycoprotein (ABCB1), mitoxantrone resistance protein

(ABCG2) and multidrug resistance protein 1 (ABCC1) by tetrahydrocurcumin, a major metabolite of curcumin. Mol. Cell. Biochem. 296:85–95

Lin S, Makino K, Xia W, Matin A, Wen Y, Kwong KY, Bourguignon L, Hung MC (2001) Nuclear localization of EGF receptor and its potential new role as a transcription factor. Nat Cell Biol 3:802–808

Lin TY, Lu CW, Wang C-C, Wang Y-C, Wang S-J (2011) Curcumin inhibits glutamate release in nerve terminals from rat prefrontal cortex: possible relevance to its antidepressant mechanism. Prog Neuropsychopharmacol Biol Psychiatry 35:1785–1793

Lin J, Tang Y, Kang Q, Feng Y, Chen A (2012) Curcumin inhibits gene expression of receptor for advanced glycation end-products (RAGE) in hepatic stellate cells *in vitro* by elevating PPARgamma activity and attenuating oxidative stress. Br J Pharmacol 166:2212–2227

Liu Q, Thoreen C, Wang J, Sabatini D, Gray NS (2009) mTOR mediated anti-cancer drug discovery. Drug Discov Today Ther Strateg 6:47–55

Liu ZJ, Liu W, Liu L, Xiao C, Wang Y, Jiao JS (2013a) Curcumin protects neuron against cerebral ischemia-induced inflammation through improving PPAR-gamma function. Evid Based Complement Alternat Med 2013:470975

Liu K, Gui B, Sun Y, Shi N, Gu Z, Zhang T, Sun X (2013b) Inhibition of L-type Ca^{2+} channels by curcumin requires a novel protein kinase-theta isoform in rat hippocampal neurons. Cell Calcium 53:195–203

Lo H, Hsu S, Ali-Seyed M, Gunduz M, Xia W, Wei Y, Bartholomeusz G, Shih JY, Hung MC (2005) Nuclear interaction of EGFR and STAT3 in the activation of the iNOS/NO pathway. Cancer Cell 7:575–589

Maradana MR, Thomas R, O'Sullivan BJ (2013a) Targeted delivery of curcumin for treating type 2 diabetes. Mol Nutr Food Res 57:1550–1556

Maradana MR, Thomas R, O'Sullivan BJ (2013b) Targeted delivery of curcumin for treating type 2 diabetes. Mol Nutr Food Res 57:1550–1556

Melenhorst WB, Mulder GM, Xi Q, Hoenderop JG, Kimura K, Eguchi S, van GH (2008) Epidermal growth factor receptor signaling in the kidney: key roles in physiology and disease. Hypertension 52:987–993

Meng Z, Yan C, Deng Q, Gao DF, Niu XL (2013) Curcumin inhibits LPS-induced inflammation in rat vascular smooth muscle cells *in vitro* via ROS-relative TLR4-MAPK/NF-κB pathways. Acta Pharmacol Sin 34:901–911

Meydani M, Hasan ST (2010) Dietary polyphenols and obesity. Nutrients 2:737–751

Millauer B, Wizigmann-Voos S, Schnurch H, Martinez R, Moller NP, Risau W, Ullrich A (1993) High affinity VEGF binding and development expression suggest Flk-1 as a major regulator of vasculogenesis and angiogenesis. Cell 72:835–846

Mittal N, Joshi R, Hota D, Chakrabarti A (2009) Evaluation of antihyperalgesic effect of curcumin on formalin-induced orofacial pain in rat. Phytother Res 23:507–512

Moynagh PN (2005) The interleukin-1 signalling pathway in astrocytes: a key contributor to inflammation in the brain. J Anat 207:265–269

Nicchia GP, Nico B, Camassa LM, Mola MG, Loh N, Dermietzel R, Spray DC, Svelto M, Frigeri A (2004) The role of aquaporin-4 in the blood-brain barrier development and integrity: studies in animal and cell culture models. Neuroscience 129:935–945

Nicolas CS, Amici M, Bortolotto ZA, Doherty A, Csaba Z, Fafouri A, Dournaud P, Gressens P, Collingridge GL, Peineau S (2013) The role of JAK-STAT signaling within the CNS. JAKSTAT 2, e22925

Okada S, Nakamura M, Katoh H, Miyao T, Shimazaki T, Ishii K, Yamane J, Yoshimura A, Iwamoto Y, Toyama Y, Okano H (2006) Conditional ablation of Stat3 or Socs3 discloses a dual role for reactive astrocytes after spinal cord injury. Nat Med 12:829–834

Pearce LR, Huang X, Boudeau J, Pawlowski R, Wullschleger S, Deak M, Ibrahim AF, Gourlay R, Magnuson MA, Alessi DR (2007) Identification of Protor as a novel Rictor-binding component of mTOR complex-2. Biochem J 405:513–522

Plate KH, Breier G, Weich HA, Risau W (1992) Vascular endothelial growth factor is a potential tumour angiogenesis factor in human gliomas in vivo. Nature 359:845–847

Pouysségur J, Dayan F, Mazure NM (2006) Hypoxia signalling in cancer and approaches to enforce tumour regression. Nature 441:437–443

Powers CJ, McLeskey SW, Wellstein A (2000) Fibroblast growth factors, their receptors and signaling. Endocr Relat Cancer 7:165–197

Pugh CW, Rourke JFO, Nagao M, Gleadle JM, Ratcliffe PJ (1997) Activation of hypoxia-inducible factor-1; definition of regulatory domains within the alpha subunit. J Biol Chem 272:11205–11214

Qu R, Li Y, Gao Q, Shen L, Zhang J, Liu Z, Chen X, Chopp M (2007) Neurotrophic and growth factor gene expression profiling of mouse bone marrow stromal cells induced by ischemic brain extracts. Neuropathology 27:355–363

Rafiee P, Binion DG, Wellner M, Behmaram B, Floer M, Mitton E, Nie L, Zhang Z, Otterson MF (2010) Modulatory effect of curcumin on survival of irradiated human intestinal microvascular endothelial cells: role of Akt/mTOR and NF-κB. Am J Physiol Gastrointest Liver Physiol 298:G865–G877

Raible DJ, Frey LC, Brooks-Kayal AR (2014) Effects of JAK2-STAT3 signaling after cerebral insults. JAKSTAT 3, e29510

Ricote M, Li AC, Willson TM, Kelly CJ, Glass CK (1998) The peroxisome proliferator-activated receptor-gamma is a negative regulator of macrophage activation. Nature 391:79–82

Roberts CK, Barnard RJ, Sindhu RK, Jurczak M, Ehdaie A, Vaziri ND (2006) Oxidative stress and dysregulation of NAD(P)H oxidase and antioxidant enzymes in diet-induced metabolic syndrome. Metabolism 55:928–934

Ruderman NB, Park H, Kaushik VK, Dean D, Constant S, Prentki M, Saha AK (2003) AMPK as a metabolic switch in rat muscle, liver and adipose tissue after exercise. Acta Physiol Scand 178:435–442

Rupnick MA, Panigrahy D, Zhang CY, Dallabrida SM, Lowell BB, Langer R, Folkman MJ (2002) Adipose tissue mass can be regulated through the vasculature. Proc Natl Acad Sci U S A 99:10730–10735

Sahebkar A (2013) Why it is necessary to translate curcumin into clinical practice for the prevention and treatment of metabolic syndrome? Biofactors 39:197–208

Scaltriti M, Baselga J (2006) The epidermal growth factor receptor pathway: a model for targeted therapy. Clin Cancer Res 12:5268–5272

Schaaf C, Shan B, Onofri C, Stalla GK, Arzt E, Schilling T, Perone MJ, Renner U (2010) Curcumin inhibits the growth, induces apoptosis and modulates the function of pituitary folliculostellate cells. Neuroendocrinology 91:200–210

Schultz JE (2002) FGF and the cardiovascular system in 45-66*in*. In: Cuevas P (ed) Fibroblast growth factors in the cardiovascular system, vol I. Holzapfel Publishers, Munich, pp 45–66

Semenza GL (2003) Targeting HIF-1 for cancer therapy. Nat Rev Cancer 3:721–732

Semenza GL (2007) Evaluation of HIF-1 inhibitors as anticancer agents. Drug Discov Today 12:853–859

Shalaby F, Rossant J, Yamaguchi TP, Gertsenstein M, Wu XF, Breitman ML, Schuh AC (1995a) Failure of blood-island formation and vasculogenesis in Flk-1-deficient mice. Nature 376:62–66

Shalaby F, Rossant J, Yamaguchi TP, Gertsenstein M, Wu XF, Breitman ML, Schuh AC (1995b) Failure of blood island formation and vasculogenesis in Flk-1-deficient mice. Nature 376:62–66

Shan B, Schaaf C, Schmidt A, Lucia K, Buchfelder M, Losa M, Kuhlen D, Kreutzer J, Perone MJ, Arzt E, Stalla GK, Renner U (2012) Curcumin suppresses HIF1A synthesis and VEGFA release in pituitary adenomas. J Endocrinol 214:389–398

Sharma S, Kulkarni SK, Agrewala JN, Chopra K (2006) Curcumin attenuates thermal hyperalgesia in a diabetic mouse model of neuropathic pain. Eur J Pharmacol 536:256–261

Shehzad A, Lee J, Lee YS (2013) Curcumin in various cancers. Biofactors 39:56–68

Shin DH, Seo EY, Pang B, Nam JH, Kim HS, Kim WK, Kim SJ (2011) Inhibition of Ca(2+)-release-activated Ca^{2+} channel (CRAC) and K(+) channels by curcumin in Jurkat-T cells. J Pharmacol Sci 115:144–154

Shin DH, Nam JH, Lee ES, Zhang Y, Kim SJ (2012) Inhibition of Ca^{2+} release-activated Ca(2+) channel (CRAC) by curcumin and caffeic acid phenethyl ester (CAPE) via electrophilic addition to a cysteine residue of Orai1. Biochem Biophys Res Commun 428:56–61

Shishodia S, Singh T, Chaturvedi MM (2007) Modulation of transcription factors by curcumin. Adv Exp Med Biol 595:127–148

Sikaris KA (2004) The clinical biochemistry of obesity. Clin Biochem Rev 25:165–181

Sikora E, Scapagnini G, Barbagallo M (2010) Curcumin, inflammation, ageing and age-related diseases. Immun Ageing 7:1

Squires MS, Hudson EA, Howells L, Sale S, Houghton CE, Jones JL, Fox LH, Dickens M, Prigent SA, Manson MM (2003) Relevance of mitogen activated protein kinase (MAPK) and phosphotidylinositol-3-kinase/protein kinase B (PI3K/PKB) pathways to induction of apoptosis by curcumin in breast cells. Biochem Pharmacol 65:361–376

Suarez S, Ballmer-Hofer K (2001a) VEGF transiently disrupts gap junctional communication in endothelial cells. J Cell Sci 114:1229–1235

Suarez S, Ballmer-Hofer K (2001b) VEGF transiently disrupts gap junctional communication in endothelial cells. J Cell Sci 114:1229–1235

Taheri S, Lin L, Austin D, Young T, Mignot E (2004) Short sleep duration is associated with reduced leptin, elevated ghrelin, and increased body mass index. PLoS Med 1:e62

Tajik H, Tamaddonfard E, Hamzeh-Gooshchi N (2008) The effect of curcumin (active substance of turmeric) on the acetic acid-induced visceral nociception in rats. Pak J Biol Sci 11: 312–314

Tan C, Zhang L, Cheng X, Lin XF, Lu RR, Bao JD, Yu HX (2014) Exp Biol Med (Maywood). pii: 1535370214555665 [Epub ahead of print]

Thomas SL, Zhong D, Zhou W, Malik S, Liotta D, Snyder JP, Hamel E, Giannakakou P (2008) EF24, a novel curcumin analog, disrupts the microtubule cytoskeleton and inhibits HIF-1. Cell Cycle 7:2409–2417

Tilg H, Moschen AR (2006) Adipocytokines: mediators linking adipose tissue, inflammation and immunity. Nat Rev Immunol 6:772–783

Turner CA, Watson SJ, Akil H (2012) The fibroblast growth factor family: neuromodulation of affective behavior. Neuron 76:160–174

Ueki M, Ueno M, Morishita J, Maekawa N (2013) Curcumin ameliorates cisplatin-induced nephrotoxicity by inhibiting renal inflammation in mice. J Biosci Bioeng 115:547–551

Vergaño-Vera E, Méndez-Gómez HR, Hurtado-Chong A, Cigudosa JC, Vicario-Abejón C (2009) Fibroblast growth factor-2 increases the expression of neurogenic genes and promotes the migration and differentiation of neurons derived from transplanted neural stem/progenitor cells. Neuroscience 162:39–54

Verheul HM, Pinedo HM (2007) Possible molecular mechanisms involved in the toxicity of angiogenesis inhibition. Nat Rev Cancer 7:475–485

Viacava P, Gasperi M, Acerbi G, Manetti L, Cecconi E, Bonadio AG, Naccarato AG, Acerbi F, Parenti G, Lupi I, Genovesi M, Martino E (2003) Microvascular density and vascular endothelial growth factor expression in normal pituitary tissue and pituitary adenomas. J Endocrinol Invest 26:23–28

Voros G, Maquoi E, Demeulemeester D, Clerx N, Collen D, Lijnen HR (2005) Modulation of angiogenesis during adipose tissue development in murine models of obesity. Endocrinology 146:4545–4554

Wang GL, Semenza GL (1993) Characterization of hypoxia-inducible factor 1 and regulation of DNA binding activity by hypoxia. J Biol Chem 268:21513–21518

Wang GL, Jiang BH, Rue EA, Semenza GL (1995) Hypoxia-inducible factor 1 is a basic-helix-loop-helix-PAS heterodimer regulated by cellular O2 tension. Proc Natl Acad Sci U S A 92:5510–5514

Wang W, Bernard K, Li G, Kirk KL (2007) Curcumin opens cystic fibrosis transmembrane conductance regulator channels by a novel mechanism that requires neither ATP binding nor dimerization of the nucleotide-binding domains. J Biol Chem 282:4533–4544

Waseem M, Parvez SS (2013) Mitochondrial dysfunction mediated cisplatin induced toxicity: modulatory role of curcumin. Food Chem Toxicol 53:334–342

Wilson PD (2004) Polycystic kidney disease. N Engl J Med 350:151–164

Xia G, Kumar SR, Hawes D, Cai J, Hassanieh L, Groshen S, Zhu S, Masood R, Quinn DI, Broek D, Stein JP, Gill PS (2006) Expression and significance of vascular endothelial growth factor receptor 2 in bladder cancer. J Urol 175:1245–1252

Yarden Y, Sliwkowski MX (2001) Untangling the ErbB signalling network. Nat Rev Mol Cell Biol 2:127–137

Yeon KY, Kim SA, Kim YH, Lee MK, Ahn DK, Kim HJ, Kim JS, Jung SJ, Oh SB (2010) Curcumin produces an antihyperalgesic effect via antagonism of TRPV1. J Dent Res 89:170–174

Yu PJ, Ferrari G, Galloway AC, Mignatti P, Pintucci G (2007) Basic fibroblast growth factor (FGF-2): the high molecular weight forms come of age. J Cell Biochem 100:1100–1108

Yu YC, Miki H, Nakamura Y, Hanyuda A, Matsuzaki Y, Abe Y, Yasui M, Tanaka K, Hwang TC, Bompadre SG, Sohma Y (2011) Curcumin and genistein additively potentiate G551D-CFTR. J Cyst Fibros 10:243–252

Yu L, Yi J, Ye G, Zheng Y, Song Z, Yang Y, Song Y, Wang Z, Bao Q (2012) Effects of curcumin on levels of nitric oxide synthase and AQP-4 in a rat model of hypoxia-ischemic brain damage. Brain Res 1475:88–95

Zhang X, Chen Q, Wang Y, Peng W, Cai H (2014) Effects of curcumin on ion channels and transporters. Front Physiol 5:94

Zhen L, Fan D, Yi X, Cao X, Chen D, Wang L (2014) Curcumin inhibits oral squamous cell carcinoma proliferation and invasion via EGFR signaling pathways. Int J Clin Exp Pathol 7: 6438–6446

Chapter 6
Effects of Curcumin on Oxidative Stress in Animal Models and Patients with Alzheimer Disease

6.1 Introduction

Curcumin (molecular weight: 368.37) is a polyphenol, which contains a conjugated double bond heptadienone linker containing a bis-α,β-unsaturated β-diketone moiety (Sharma et al. 2005). This linker joins together the molecule's two methoxyphenol rings (Sharma et al. 2005). The diketone moiety undergoes tautomerization in a pH-dependent manner (Sharma et al. 2005). The bis-keto tautomer displays potent Brønsted acid activity, while the enol tautomer functions as a powerful Lewis Base (Jovanovic et al. 2001). As stated earlier, curcumin is insoluble in water, but is soluble in organic solvents such as acetone, ethanol, and dimethylsulfoxide (DMSO). In neutral/alkaline solutions curcumin displays a dark red color, while in acidic solutions it adopts a vibrant yellow color (Jovanovic et al. 2001). Curcumin is extremely unstable in phosphate-based buffers, normal cell culture media, and most alkaline solutions, while it is extremely stable in acidic solutions (Jovanovic et al. 2001; Sharma et al. 2005).

Antioxidants are a structural heterogeneous group of molecules, which have the ability to scavenge reactive oxygen and nitrogen species (free radicals) and contribute to the first defense mechanism against the toxic effects of free radicals. Curcumin is a powerful antioxidant. The antioxidant activity of curcumin is 10 times greater than vitamin E (Jayaprakasha et al. 2006; Ak and Gulcin 2008). The antioxidant activity of curcumin is due to its chemical structure. In curcumin molecule the keto-enol-enolate equilibrium of the heptadienone moiety of curcumin dictates its antioxidant properties. The central CH2 group adjacent to the highly activated carbon atom in the heptadienone link is the antioxidant site, in which the delocalization of unpaired electron on the adjacent oxygen atoms contributes to antioxidant effects of curcumin. In addition, phenolic OH is the most preferable group for the loss of proton from the one-electron oxidized species. As the resultant phenoxyl radical is stabilized by the delocalization of electrons, the ability for curcumin to scavenge the oxidizing free radicals is greatly increased. The resonance stabilized radicals can undergo further

© Springer International Publishing Switzerland 2016
A.A. Farooqui, *Therapeutic Potentials of Curcumin for Alzheimer Disease*,
DOI 10.1007/978-3-319-15889-1_6

loss of the second hydrogen atom from the second phenolic OH group, generating a diradical. This diradical is transformed into stable products like quinones or undergo cleavage to produce smaller phenols like ferulic acid (*trans*-4-hydroxy-3-methoxycinnamic acid) (Priyadarsini et al. 2003; Ghosh et al. 2015). In curcumin molecule, the C7 linker and its carbonyl functions are important for anti-inflammatory activity and the conjugated enones have been shown to act as Michael acceptors for curcumin's anti-cancer activity (Anand et al. 2008; Mosley et al. 2007). Many curcumin analogs have been synthesized, a better known analog is diphenyl difluoroketone (EF24), a fluorinated compound that does not have a phenolic hydroxyl group and possesses only one enone. EF24 has been reported to have greater biological activity and bioavailability than curcumin without increased toxicity and it also seems to possess a distinctive mechanism of action from curcumin (Thomas et al. 2008). EF24 induces more potent anti-cancer activity than curcumin. Its therapeutic potential in other diseases remains to be determined. Recent studies have demonstrated that 5-chlorocurcumin superior antioxidant compound compared to both curcumin and ascorbic acid (Al-Amiery et al. 2013). Curcumin inhibits lipid peroxidation by acting as a chain-breaking antioxidant at the 3′ position, resulting in an intramolecular Diels-Alder reaction and neutralization of the lipid radicals (Masuda et al. 2001). In addition to inhibiting lipid peroxidation, curcumin also possesses free radical-scavenging activity. It has been shown to scavenge various ROS produced by macrophages (including superoxide anions, hydrogen peroxide and nitrite radicals) both in vitro as well as in vivo using rat peritoneal macrophages as a model (Joe and Lokesh 1994; Joe et al. 2004). Curcumin also produces anti-proliferative effects on microglia. A minimal dose of curcumin affects neuroglial proliferation and differentiation. Studies on the effect of curcumin on microglial proliferation and differentiation have indicated that curcumin stops the proliferation of neuroglial cells in a dose-dependent manner (Ambegaokar et al. 2003). Curcumin has been shown to decrease the activity of glutamine synthetase, which is a marker enzyme for astrocytes and increases 2′3′- cyclic Nucleotide 3′-phosphohydrolase (CNP), an enzyme that is a marker for oligodendrocytes. Based on several studies, it is proposed that overall effect of curcumin on neuroglial cells involves decrease in astrocytes proliferation, improvement in myelogenesis and increase in activity and differentiation of oligodendrocytes (Ambegaokar et al. 2003). In vitro studies on the effect of curcumin on Aβ have indicated that curcumin retards the aggregation of amyloid β (Aβ) peptide (Hamaguchi et al. 2010; Ono et al. 2004), a process which is closely associated with the start of AD. Curcumin disrupts the formation of the long straight Aβ fibrils (Yang et al. 2005) and reduces Aβ toxicity. Moreover, it also disrupts the preformed Aβ fibrils (Ono et al. 2004). The reduction in Aβ toxicity is possibly related to above mentioned effects of curcumin. Alternative modes of action of curcumin have also been proposed. Thus, curcumin inhibits the process that generates Aβ from the amyloid precursor protein (Narlawar et al. 2008). In addition, curcumin reduces the generation and concentration of ROS (Rathore et al. 2008), a process, which is closely associated with neurodegeneration (Farooqui 2012a). Whatever the mechanism of action of curcumin may be, information about the disaggregation of Aβ may be valuable in understanding the role of curcumin in blocking the pathogenesis of

AD. Curcumin not only increases the expression of synapse-related proteins PSD95 and Shank1 in APP/PS1 double transgenic mice, but also improves structure and plasticity of synapse in APP/PS1 double transgenic mice leading to enhancement in learning and memory abilities (Wei et al. 2012).

6.2   Oxidative Stress, Nitrosative Stress, and Redox Systems in the Brain

Oxidative stress is defined as an imbalance between the production of reactive oxygen species (ROS) on one side and endogenous antioxidant and repair capacity on the other, in favor of the former. The high vulnerability of the brain, spinal cord, and peripheral nerves to ROS is due to its elevated bioenergetics and oxygen requirements. In fact neurons with axons and multiple synapses not only have high ATP demand, but both organelles have large demand for the massive consumption of oxygen in the respiratory chain. This coupled with the high content of polyunsaturated fatty acids and easily mobilizable iron from several areas of the brain can stimulate the production of ROS (Fenton and/or Haber-Weiss reactions) (Andorn et al. 1990). In neural cells, ROS generation occurs in two ways through enzymic and non-enzymic reactions. Enzymic reactions generating ROS occur in the mitochondrial respiratory chain, phagocytosis, prostaglandin synthesis and the cytochrome P450 system (Halliwell 2007). Non-enzymic reactions contributing to ROS formation are associated with non-enzymic reactions of oxygen with organic compounds as well as those initiated by ionizing radiations. The non-enzymic process can also occur during oxidative phosphorylation (i.e. aerobic respiration) in the mitochondria (Dröge and Schipper 2007). It is becoming increasingly evident that low levels of ROS may not only trigger adaptive cellular machinery, but also increase the organism's stress resistance (Ristow and Zarse 2010). The adaptive cellular machinery include antioxidant and heat shock responses, cell cycle regulation and apoptosis, DNA repair, fatty acid deacylation-reacylation, unfolded protein responses and autophagy stimulation (Farooqui et al. 2000). Elevation in ROS results in damage to macromolecules leading to the generation of (a) cytotoxic proteins that aggregate, (b) lipid oxidation that produces cytotoxic levels of lipid mediators and final oxidation products that can damage proteins and DNA and induce cell death, and (c) modifications of DNA (mitochondrial and nuclear) leading to mutagenesis that produce gain/loss of function. Moreover, many oncogenes and tumor suppressor genes are also altered by the oxidative stress. Converging evidence suggests that oxidative stress induces disease through the oxidative modification of biomolecules and the alteration of signaling pathways resulting in dysregulation of cell cycles, interference with cellular metabolism, genetic instability, epigenetic change and mutation (Farooqui 2014). Low levels of ROS play a pivotal role in cell division and survival, cell signaling, inflammation and immune functions, autophagy, and stress response. The role of oxygen in cell survival is linked to its high redox potential, which makes it an excellent oxidizing agent capable of accepting electrons easily from reduced substrates. At low level, ROS may contribute to neurogenesis, since neuronal precursor cells exhibit about four times higher ROS levels than other cell types. The concentration of ROS, which is dependent on the density of precursor cells is associated with the rate of proliferation (Limoli et al. 2004). Furthermore, at low levels, ROS may act as messengers and activate, up to a certain concentration, the self-renewal multipotent neural progenitors and neurogenesis via PtdIns 3K/Akt signaling (Le Belle et al. 2011). High levels of ROS are major disruptor of neural cell homeostasis. Brain represents only ~2 % of the total body mass and yet accounts for more than 25 % glucose and 20 % of the total consumption of oxygen (Schiavone et al. 2012). In addition to high oxygen utilization, brain contains high levels of unsaturated fatty acids and the large amount of iron and copper, which may interact with the diffusible hydrogen peroxide causing the generation of extremely reactive and toxic hydroxyl radicals. Hydroxyl radicals have been reported to cause damage to proteins, lipids and DNA (Halliwell 2001). This makes brain particularly vulnerable to ROS-mediated damage. The major sources of ROS in brain are uncontrolled ARA cascade, mitochondrial respiratory chain, xanthine/xanthine oxidase, myeloperoxidase, and activation of NADPH oxidase. ROS attack both glial cells and neurons, which are post-mitotic cells and therefore, they are particularly sensitive to ROS (free radicals), leading to neuronal damage (Gilgun-Sherki et al. 2001; Farooqui 2014).

ROS include superoxide anions ($O_2^{\cdot-}$), hydroxyl ($^{\cdot}OH$), alkoxyl ($RO^{\cdot}$), and peroxyl radicals ($ROO^{\cdot}$), and hydrogen peroxide (H_2O_2), which are generated as by-product of oxidative metabolism, in which energy activation and electron reduction are involved. Mitochondria are major intracellular source of ROS. Of total mitochondrial O_2 consumed, 1–2 % is diverted to the formation of ROS, mainly at the level of complex I and complex III of the respiratory chain (Fig. 6.1) (Turrens 2003). However, certain enzymic components are loosely associated with the inner mitochondrial membrane and, under conditions of cellular stress, can be released or inactivated, greatly diminishing the reducing capacity of the electron transport chain. The electrons are subsequently monoelectronically donated to oxygen (O_2), yielding increased production of $O_2^{\cdot-}$. Mitochondria-derived $O_2^{\cdot-}$ is dismutated to H_2O_2 by manganese superoxide dismutase, and, in the presence of metal ions. The chemical reactivity of ROS varies from the very toxic $^{\cdot}OH$ with very short half-life in tissues (10^{-9} s) to the less reactive $O_2^{\cdot-}$ anion (Pryor 1986). $^{\cdot}OH$ oxidizes virtually any cell component that it interacts with and is the primary cause of toxicity due to oxidative damage. Due to this reactivity and lack of specificity, the $^{\cdot}OH$ is not thought to have a signaling role in cells. $O_2^{\cdot-}$ is generated from the addition of a single electron to molecular oxygen. $O_2^{\cdot-}$ is able to cross mitochondrial and endoplasmic reticulum walls into the cytoplasm through the voltage-dependent anion channel (VDAC) (Han et al. 2003). $O_2^{\cdot-}$ has a high affinity for iron–sulfur clusters in proteins and reacts with them at rates limited only by diffusion. This can release free iron and promote structural changes to alter protein activity. At low levels, $O_2^{\cdot-}$ are not toxic in vivo because neural cells have mechanisms to minimize their accumulation. $O_2^{\cdot-}$ is rapidly neutralized by high concentrations of scavenging enzymes called superoxide dismutases (SOD) with distinct isoenzymes located in the mitochondria, cytoplasm, and extracellular compartments. Action of SODs yields H_2O_2 and O_2, which can be reused to generate superoxide radical. H_2O_2 is the most stable ROS, with an estimated half-life in cells of approximately 1 millisecond.

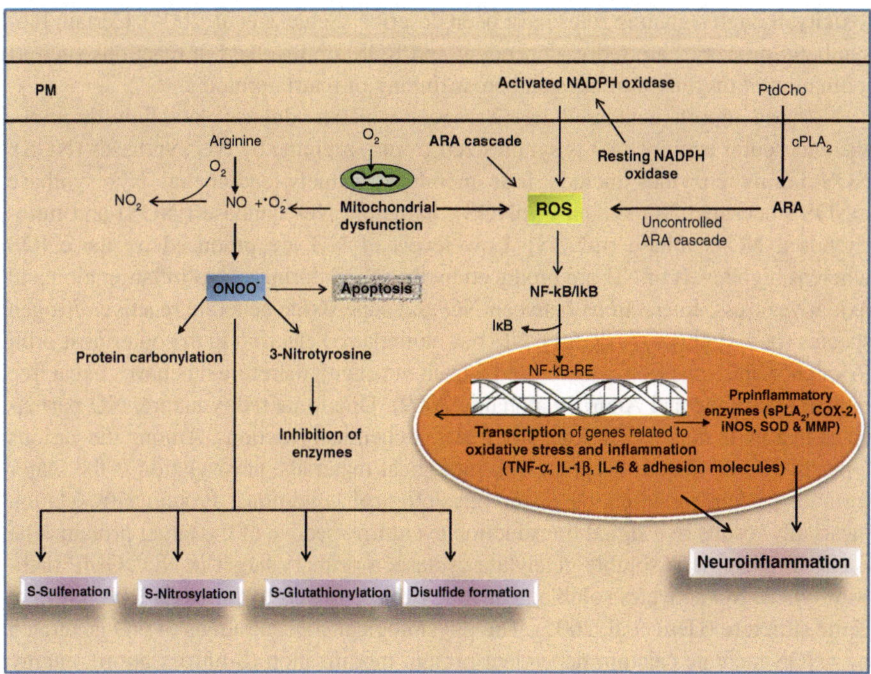

Fig. 6.1 Generation of ROS and NRS from various sources and harmful effects of ROS and RNS in the brain. Plasma membrane (*PM*); phosphatidylcholine (*PtdCho*); cytosolic phospholipase A_2 (*cPLA$_2$*); arachidonic acid (*ARA*); reactive oxygen species (*ROS*); nuclear factor-kappa B (*NF-κB*); nuclear factor-kappa B response element (*NF-κB-RE*); tumor necrosis factor-alpha (*TNF-α*); interleukin-1beta (*IL-1β*); interleukin-6 (*IL-6*); nitric oxide (*NO*); peroxynitrite (*ONOO⁻*); inducible nitric oxide synthase (*iNOS*); secretory phospholipase A_2 (*sPLA$_2$*); cycloxygenase-2 (*COX-2*); superoxide anions ($O_2^{\cdot-}$); superoxide dismutase (*SOD*); and matrix metalloproteinases (*MMP*)

As a nonpolar molecule, H_2O_2 can diffuse through neural membranes. Membrane transport of H_2O_2 is further facilitated by aquaporin channels (Bienert et al. 2007). Thus, chemical considerations point to H_2O_2 as the major ROS involved in redox signaling in both brain and visceral tissues. The generation of ROS in the brain is also linked with the participation of redox-active metals such as iron and copper. As a general principle, the chemical origin of the majority of ROS is the direct interactions between redox-active metals and oxygen species via reactions such as the Fenton and Haber-weiss reaction. In the presence of reduced transition metals (iron and copper), H_2O_2 can also be transformed into the highly reactive ·OH. In physiological signaling, ROS can not only modify redox-sensitive amino acids in a variety of proteins (phosphatases and kinases), but also produce changes in activities of ion channels, and transcription factors (Winterbourn and Hampton 2008). Calcium ions are not only known to stimulate the TCA cycle and enhance electron flow into the respiratory chain, but also cause stimulation of the nitric oxide synthase and subsequently promote nitric oxide (NO) generation leading to the inhibition of respiration at complex IV. The more highly reactive species such as singlet oxygen may be more involved in toxicity, though signaling roles have been describe (Schieke et al. 2004). Certain biosynthetic processes are redox-dependent and ROS are involved in reactions such as iodination of thyroid hormones and cross linking of matrix proteins.

NO is an important intercellular messenger, which modulates blood flow, thrombosis, and neural activity. NO is synthesized from L-arginine by NO synthases (NOS). NOS family enzymes include four members, namely endothelial NO synthase (eNOS), neuronal NO synthase (nNOS), inducible NO synthase (iNOS) and mitochondrial NO synthase (mNOS). Low levels of NO are produced by the eNOS whereas high levels of NO are produced by the iNOS during neuro-inflammation and oxidative stress. Interactions between NO and superoxide generate reactive nitrogen species (peroxynitrite) (Fig. 6.1). NO has an unpaired electron in the outermost orbit (6 valence electrons from oxygen and 5 from nitrogen), therefore it behaves like a free radical (Martínez and Andriantsitohaina 2009). Due to its redox nature, NO participates in a wide range of biologically relevant chemical reactions. Among the various types of NO-mediated reactions with biological materials, nitrosylation is the major form of protein modification under physiological conditions. Protein nitrosylation invariably results in a signal transduction event irrespective of the target protein. The classical example is soluble guanylate cyclase, a primary target in NO/cGMP pathway, where NO activates soluble guanylate cyclase in part by nitrosylation on iron in heme structure (Hess et al. 2005). The physiological concentrations of NO generated by nNOS mediate calcium dependent protein modification (S-nitrosylation), energy metabolism (through cytochrome c oxidase), synaptic plasticity, and neuroprotection. NO, which is generated by eNOS contributing to calcium dependent cyclic guanosine monophosphate (cGMP) mediated vasodilation to maintain vascular tone of cerebral blood vessels (Butler et al. 1998). The NO is produced by iNOS and plays an important role in immune response or killing pathogens by generating free radicals. However, excessive amount of NO produced by iNOS-mediated mechanism is harmful to the host cells in many ways (Fig. 6.2) (Thippeswamy et al. 2006).

RNS include NO and $ONOO^-$. Low levels of both ROS and RNS are continuously produced in mammalian cells and play important physiological roles in processes, such as gene expression, cell proliferation and survival, pathogen clearance by the immune system, and blood vessel permeability (Allen and Tresini 2000; Kamata and Hirata 1999). Oxidative stress-mediated cell protecting proteins include many enzymes (catalase, superoxide dismutase, hemoxygenase-1, ferritin, glutathione peroxidase, glutathione reductase, quinone reductase, thioredoxin, thioredoxin reductase, and cyclooxygenase-2), proteins (chaperons, superoxide dismutase, heme oxygenase-1), and transcription factors (AP-1, ATF/CREB, ETS, C/EBP, NF-κB) (Turpaev 2002). In contrast, high levels of ROS and RNS contribute to ROS and RNS-mediated neurodegeneration (Jung et al. 2006). Like hydrogen peroxide, $ONOO^-$ can modify proteins (Rubbo et al. 2002; Rubbo et al. 2009) and induce the formation of lipid modifications (Rubbo et al. 2009). $ONOO^-$-mediated DNA damage leads to activation of PARP, a nuclear enzyme which causes transfer of poly-ADP ribose units to DNA by utilizing NADH energy pool. Depletion of NADH leads to bioenergetic crisis and thus drives the cell towards necrosis (Szabó 2003). $ONOO^-$ can also modify mitochondrial DNA (mtDNA) (Misiaszek et al. 2004) contributing to neuronal injury.

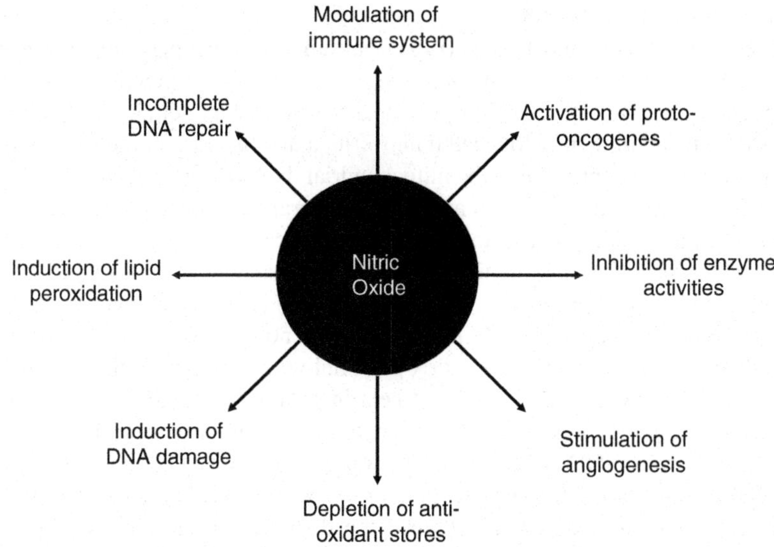

Fig. 6.2 Effects of high levels of NO on metabolic processes in the brain and visceral tissues

$ONOO^-$ contributes to the regulation of cellular reactions. The existence of a mitochondrial nitric oxide synthase (NOS) has been hypothesized. NOS generates NO within the organelle, in which generation of $O_2^{\cdot-}$ occurs. This leads to in situ generation of peroxynitrite. In addition, the effect of NO on mtDNA integrity is a key point for redox regulation of a mitochondrial enzyme called telomerase. The mitochondrial isoform of telomerase is critical for protecting mtDNA against damage, and is more effective than its nuclear counterpart in protecting cells against apoptosis, by limiting mitochondrial ROS formation (Maccarrone and Brüne 2009). High levels of ROS and RNS are closely linked to the pathogenesis of a variety of neurotraumatic, neurodegenerative, and neuropsychiatric diseases, including age-associated disorders (Jellinger 2009; Farooqui 2010).

Due to reactivity and short half lives, the direct detection of ROS and RNS is difficult, and hence, the amounts of ROS and RNS are often judged from the alteration of antioxidant status or the accumulation of relatively stable products of lipid, protein and DNA interactions. Levels of oxidative and nitrosative damage, besides the concentration and reactivity of ROS and RNS, are also influenced and modulated by the activity of the repair systems. Therefore, levels of oxidative and nitrosative modification of lipids, proteins and DNA are generally used as markers of oxidative and nitrosative damage, which is not only increased in normal aging, but also in neurotraumatic, neurodegenerative, and neuropsychiatric diseases (Farooqui 2010).

Neurodegeneration and synaptic loss in AD are primarily induced by defective mitochondrial biogenesis and axonal transport of mitochondria (Reddy et al. 2012). Normal mitochondrial dynamics is an essential function in maintaining cell viability. Metabolic changes in mitochondrial membrane metabolism are caused by imbalance between fusion and fission of mitochondria trigger serious cellular perturbations leading to apoptotic cell death. Thus, alterations in levels of mitochondrial fusion (MNF-1/2 and OPA1) and fission (FIS1) proteins in AD hippocampus not only decreases fusion and increases fission processes. Levels of mitochondrial fission protein DLP1 are also decreased in hippocampal neurons (Wang et al. 2009). In addition, Aβ-mediated mitochondrial calcium overload and increase in the generation of ROS and RNS also contribute to the mitochondrial dysfunction in AD.

The detrimental effects of ROS and RNS in neural cells are accompanied by the opening of mitochondrial permeability transition pore (mtPTP) leading to decline of the mitochondrial inner membrane potential ($\Delta\Psi_m$) and release of cytochrome c, ultimately resulting in apoptotic cell death (Fig. 6.3) (Bouchier-Hayes et al. 2005; Farooqui 2014). Neural cells contain several endogenous enzymic and non-enzymic antioxidant systems to control the redox potential within each subcellular compartment to retard ROS and RNS-mediated neuronal damage (Table 6.1) (Thannickal and Fanburg 2003). As mentioned earlier, enzymic antioxidant systems include enzymes, including SOD, which dismutes superoxide anion ($O_2^{\cdot-}$) into H_2O_2; catalase, which converts H_2O_2 to water; glutathione peroxidase, heme oxygenase, peroxiredoxins, and thioredoxin that reduce the disulfide bridges of various target proteins. The nonenzymic systems include GSH/GSSG, Cys/CySS, ascarbate/dehydroascarbate, and α-tocopherol systems neutralize oxidative stress.

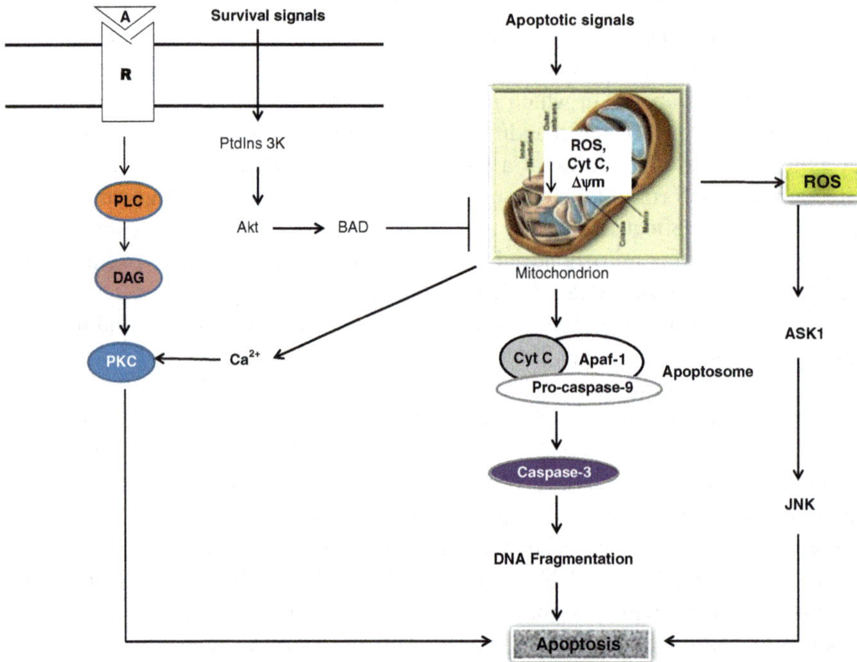

Fig. 6.3 A schematic diagram showing the onset of ROS-mediated cell death. ROS production decreases Δψm which can induce cytochrome *c* release from mitochondria promoting the formation of apoptosome and activation of cascase-9. Activation of caspase-9 initiates caspase cascade and activation of caspase-3 and onset of apoptotic cell death

Table 6.1 Antioxidants and antioxidant/pro-oxidant balance system in the brain

Antioxidant systems	Antioxidant/pro-oxidant balance
Glutathione	GSH/GSSG ratio
Ascorbic acid (vitamin C)	Cysteine redox state
α-tocopheral (vitamin E)	Thiol/disulfide state
Carnosine	Free-radical scavenger and metal chelating properties
Metatonin	Chronobiotic properties
Uric acid	Antioxidant properties
Lipoic acid	GSH/GSSG ratio

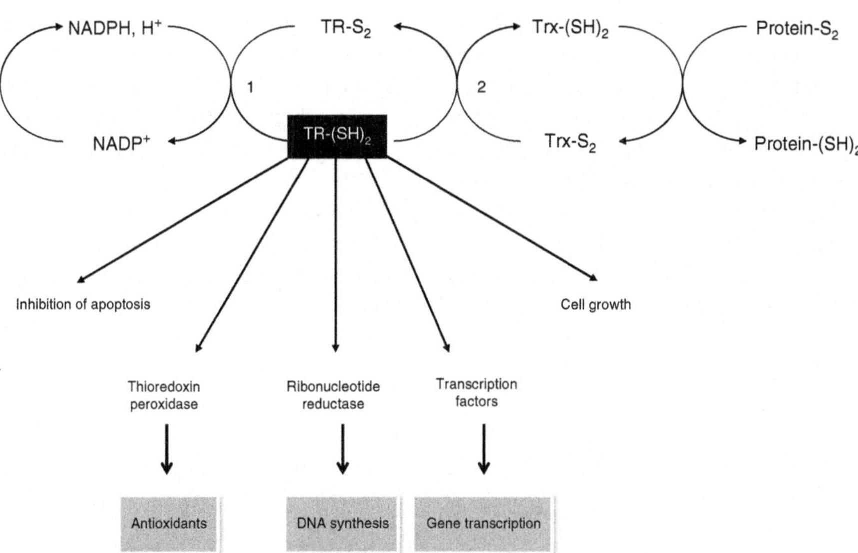

Fig. 6.4 Reactions involved in thioredoxin system and involvement of thioredoxin system in cell growth, DNA synthesis and transcription factors. Thioredoxin is a redox-regulating protein with a redox-active disulfide/dithiol within the conserved active site sequence -Cys-Gly-pro-Cys-. Thioredoxin reductase, a 55 kDa flavoprotein that catalyzes the NADPH-dependent reduction of thioredoxin (1) and thioredoxin oxidase (2), a flavin-dependent sulfhydryl oxidase that catalyzes the oxidative protein folding with the generation of disulfides

Thioredoxins are low molecular weight dithiol motif containing proteins, which support a range of biologic functions. Thioredoxin system is associated with induction and maintenance of strong reducing conditions in the nucleus and in the cytoplasm (Fig. 6.4) (Thannickal and Fanburg 2003; Holmgren et al. 2005), and presence of cysteines in intracellular proteins are typically maintained in the reduced state. The TRX system composed of TRX, TRX reductase (TrxR), TRX peroxidase/Prdx, and NADPH. It is a key antioxidant system in defense against ROS through its disulfide reductase activity regulating protein dithiol/disulfide balance. The TRX system provides the electrons to thiol-dependent peroxidases (peroxiredoxins) to capture ROS and RNS with a fast reaction rate. TRX system contributes to redox regulation through catalysis of thiol-disulfide interchanges that lead to modification of enzyme activity or provide thiol-dependent reductases with redox power (Thom et al. 2012). The TRX family contains two catalytically conserved active cysteine residues that can reduce the disulfide bonds of target proteins in the thiol proteome. These proteins are apoptosis signaling kinase-1 (ASK-1), TRX1-interacting protein (Txnip), various transcription factors, and actin (Saitoh et al. 1998). It is well known that TNF-α, which activates ASK1 also stimulates ROS production. Furthermore, ASK1 regulates the induction of downstream effectors such as the c-Jun N-terminal kinase (JNK) and the p38 MAPK pathway required for cell death. The link between ROS production and the subsequent activation of ASK1-dependent signaling appears to involve a redox-dependent interaction between ASK1 and TRX. In this scenario, TRX acts less as a classical antioxidant, and more as a sensor of intracellular oxidants and a regulator of redox signaling. In addition to its catalytic activity, TRXs interacts with target proteins, further controlling their cellular activity and biological function including DNA and protein repair by reducing ribonucleotide reductase, methionine sulfoxide reductases, and regulating the activity of many redox-sensitive transcription factors (Lu and Holmgren 2014).

Neural cells also contain several non-enzymic systems such as glutathione (L-γ-glutamyl-L-cysteinyl-glycine), ascorbic acid (vitamin C), and α-tocopherol (vitamin E) (Fig. 6.5) to neutralize oxidative stress. Two forms of glutathione namely reduced

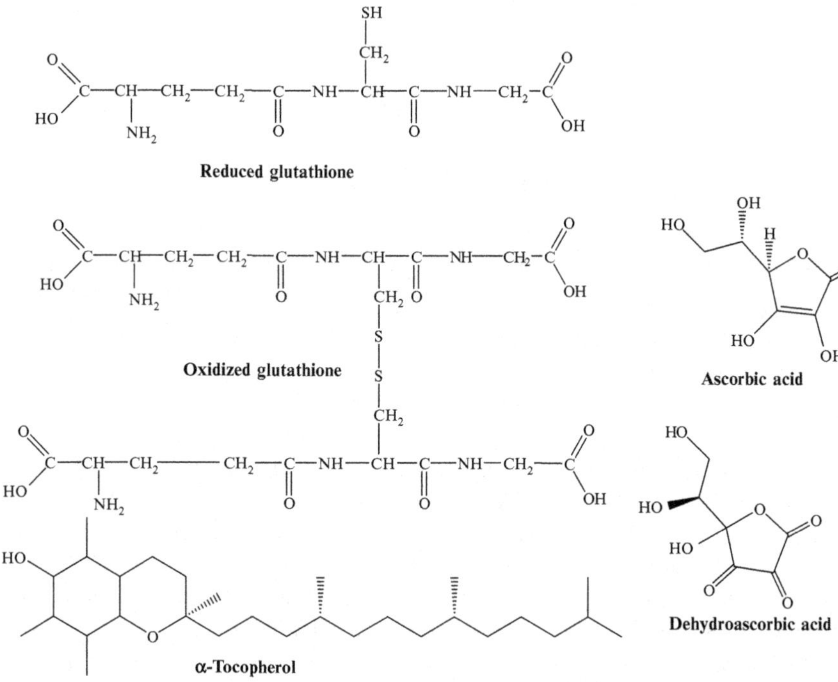

Fig. 6.5 Chemical structures of glutathione, α-tocopherol, and ascorbic acid

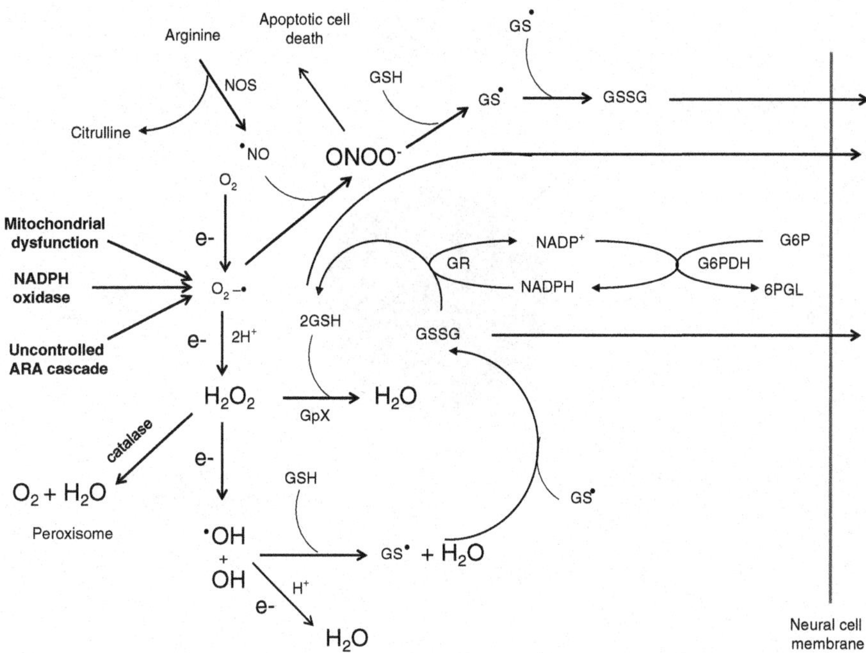

Fig. 6.6 Sources of ROS and RNS generation and glutathione-mediated elimination of ROS and RNS. Hydroxyl radical are generated by the decomposition of H_2O_2. Nitric oxide is formed from arginine. Superoxides are derived from mitochondrial dysfunction, activation of NADPH, and uncontrolled ARA cascade. Superoxide reacts with nitric oxide to generate peroxynitrite, which interacts directly with GSH leading to GSSG formation. Hydrogen peroxide may be removed by catalase or by glutathione peroxidase. The latter requires GSH to reduce peroxide. Glutathione reductase (*GR*); nitric oxide (*NO*); peroxynitrite (*ONOO⁻*); reduced glutathione (*GSH*); oxidized glutathione (*GSSG*); glutathione peroxidase (*GPx*); hydroxyl radical (*HO•*); thiyl radicals (*GS•*); glucose-6-P (*G6P*); glucose-6-phosphate dehydrogenase (*G6PDH*); and 6-phospho-D-glucono-1,5-lactone (*6PGL*)

form (GSH) and oxidized form (GSSG) are found in vivo (Fig. 6.6). Glutathione has many cellular functions (Fig. 6.7). The synthesis of GSH occurs in the cytoplasm. GSH is distributed in intracellular organelles, including the endoplasmic reticulum (ER), nucleus, and mitochondria. In endoplasmic reticulum glutathione is present in GSSG form and is the main source of oxidizing equivalents to provide the adequate environment necessary for favoring disulfide bond formation and the proper folding of nascent proteins (Hwang et al. 1992). In mitochondria, GSH is found mainly in reduced form and represents a minor fraction of the total GSH pool (10–15 %). It plays an important role in protecting mitochondria from xenobiotic- and ROS-induced toxicity. The antioxidant function of GSH is performed largely by GSH peroxidase (GPx)-catalyzed reactions, which reduce hydrogen peroxide and lipid peroxide as GSH, which is oxidized to GSSG. GSSG is reduced back to GSH by GSSG reductase at the expense of NADPH, forming a redox cycle (Lu 2009). Catalase, which is present in peroxisomes, catalyzes the decomposition of hydrogen peroxide in to water and oxygen. Levels of GSH in mitochondria make them important in defending against

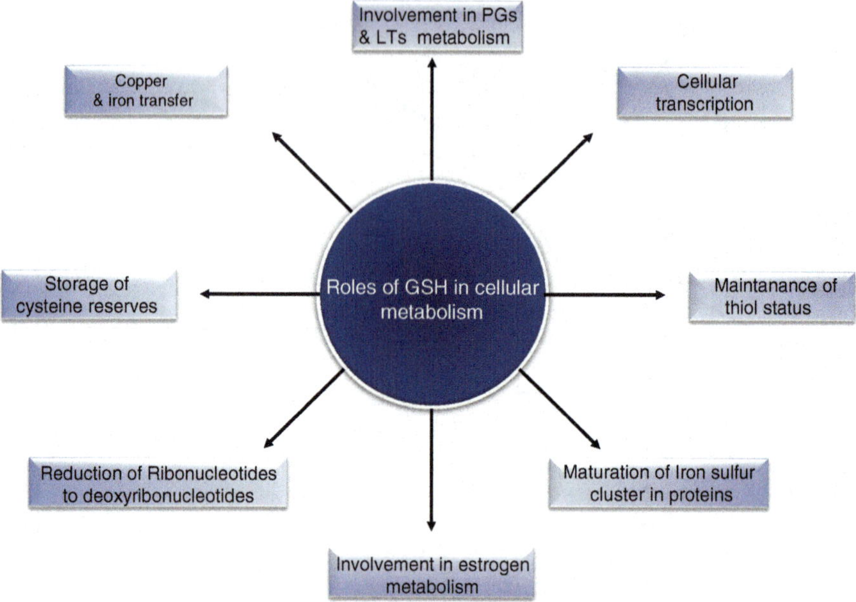

Fig. 6.7 Roles of glutathione in cellular metabolism

pathologically generated high levels of ROS (Garcia-Ruiz and Fernández-Checa 2006). The ratio between GSH to GSSG not only determines the intracellular redox potential (Forman et al. 2009), but also prevents a major shift in the redox equilibrium when ROS overcomes the ability of the cell to reduce GSSG to GSH. The GSSG is actively exported out of the cell or reacts with a protein sulfhydryl group leading to the formation of a mixed disulfide. This may be one of the endogenous mechanisms associated with the regulation of the mitochondrial permeability transition pore complex and, in consequence, thiol oxidation can be a causal factor in the mitochondrion-mediated mechanism that may contribute to the apoptotic cell death. Nevertheless)GSH depletion is a common feature not only of apoptosis but also of other types of cell death. Indeed rates of GSH synthesis and fluxes regulate its levels in cellular compartments, and potentially influence switching among different mechanisms of death. In addition, GSH is also involved in the detoxification of exogenous and endogenous substances such as xenobiotics, ionizing radiation, organic peroxides and heavy metals (Ortega et al. 2011; Solecki et al. 2013). In the brain, cortex contains highest levels of GSH followed by the cerebellum, hippocampus, and striatum, and it is lowest in the substantia nigra (Kang et al. 1999). Among the neural cells, astrocytes are most resistant to ROS attack than neurons because they contain higher glutathione content than other neural cells. Astrocytes also have the ability to secrete GSH into the extracellular space (Dringen 2000). GSH and GSSG act in concert with other redox-active metabolites (e.g., NADPH) to regulate and maintain cellular redox status (Jones et al. 2011). GSH is also oxidized $ONOO^-$. This metabolite is synthesized from the reaction between NO and $O_2^{\cdot-}$ and is reduced by GSH leading to the formation of highly reactive, peroxynitrous acid (ONOOH). This metabolite has the reactivity of nitrogen dioxide (NO_2), a very toxic free radical component in smog and cigarette smoke, and OH (Forman et al. 2009). Like ONOO-, GSH also reacts with NO and produces S-nitrosoglutathione, a modulator of cellular redox (Singh et al. 1996). S-Nitrosoglutathione is an efficient nitrosylating agent, which functions as a signaling pathway and plays a major role in regulating several physiological and pathological processes (Foster et al. 2009) (7, 8). Under physiological conditions, GSNO and S-nitrosothiols are present in blood and brain (Bryan et al. 2004).

Like glutathione, L-ascorbate also occurs in two forms in the diet namely, L-ascorbate and L-dehydroascorbate. Both ascorbate and dehydroascorbate are absorbed along the entire length of the human intestine. The reduced form, L-ascorbate is imported by an active mechanism, requiring two sodium-dependent vitamin C transporters (SVCT1 and SVCT2). The transport of the oxidized form, dehydroascorbate is mediated by glucose transporters GLUT1, GLUT3 and possibly GLUT4. Ascorbate plays a crucial role in various hydroxylation reactions. There are several ascorbate-dependent mono- and dioxygenations in various neurotransmitter and hormone formation processes and ascorbate is also required for the hydroxylation of carnitine. Furthermore, ascorbate not only contributes to the synthesis of collagen, catecholamine, and norepinephrine, but also plays an important role in protection against ROS-mediated cellular damage (Bürzle and Hediger 2012).

α-Tocopherol (vitamin E) is one of the most important lipid-soluble antioxidants (Fig. 6.5). It is involved in the protection against lipid peroxidation in biological membranes. Four homologues (α-, β-, γ-, δ-tocopherols and -tocotrienols) have been described. Among these homologues, α form has the greatest activities against ROS. α-Tocopherols has ability to scavenge the lipid peroxy radical from the target lipids. In this process tocopherols loses a hydrogen atom and is transformed into tocopheroxyl radical. The tocopheroxyl radical is reduced by ascorbate leading in the generation of dehydroascorbate, which is re-reduced by glutathione to ascorbate. The degradation products recycle into the carbohydrate metabolism (Braun et al. 1997) (Table 6.2).

α-Tocopherol appears to be especially critical for CNS function. Vitamin E regulates multiple signal transduction enzymes whose activities consequentialy affect gene expression (Zingg 2007). For example, α-tocopherol inhibits the activation of the protein kinase C (PKC) (Brigelius-Flohe 2009; Mahoney and Azzi 1988) by retarding its phosphorylation as well as its translocation to the membrane (Boscoboinik et al. 1991; Ricciarelli et al. 1998). Moreover α-tocopherol also increases the protein phosphatase 2A (PP2A) activity, an enzyme, which may be involved in pathophysiology of AD (Martin et al. 2013). Other effects of vitamin E include modulation of cell proliferation (Azzi et al. 1998) inflammation (Reiter et al. 2007) and cellular adhesion (Naito et al. 2005; Brigelius-Flohe 2009). Microarray data from rodent studies (Brigelius-Flohe 2009) indicate that vitamin E also regulates the expression of specific genes related to oxidative stress, muscles structure, cholesterol metabolism, amongst others (Brigelius-Flohe 2009). In rats, Vitamin E deprivation results in the expression of a number of genes linked to the onset and progression of AD in the hippocampus (Rota

Table 6.2 Transcription factors contributing to the pathogenesis of Alzheimer disease

Transcription factor	Pathway involved/processes	References
NF-κB	ROS, Iκ-B kinase, E3RS ligase,	Chen et al. (2012), Chami et al. (2012), and Niu et al. (2011)
AP-1	p38-MAPK, JNK,	Akhter et al. (2015)
Sp-1	p53, cytokine and chemokine expression	Marinho et al. (2014)
HIF-1α	PHD hydroxylation, ubiquitin-proteasome pathway, MEK/ERK pathway	Schubert et al. (2009) and Carvalho et al. (2009)
PPARγ	AP-1, STAT, cytokine and chemokine expression	Denner et al. (2012) and Jahrling et al. (2014)
Nrf2	Phase II enzymes (heme oxygenase, NADPH quinine oxidoreductase; and γ-glutamate cysteine ligase).	Joshi et al. (2015)

Nuclear factor kappa-light-chain-enhancer of activated B cells (*NF-κB*); specificity protein-1 (*Sp1*); activator protein 1 (*AP-1*); hypoxia-inducible factor 1-alpha (*HIF-1α*); Peroxisome proliferator-activated receptor gamma (*PPAR-γ*); and NF-E2-Related Factor 2 (*Nrf2*)

et al. 2005). These genes not only control apoptosis, and growth factor responses, but also contribute to neurotransmission and amyloid-beta metabolism (Rota et al. 2005). The hippocampus of rats deficient in vitamin E shows a reduction in the expression of the APP binding protein 1 (Rota et al. 2005), whose activity is to bind and stabilize APP, the precursor of the Aβ fragments, which contributes to the pathogenesis of AD (Nishida et al. 2006). Collective evidence suggests that vitamin E deficiency presents primarily as neurological and neuromuscular disorders, specifically spinocerebellar ataxia (Gohil et al. 2010; Muller 2010). Beside above mentioned antioxidant systems, brain also contains histidine-related compounds (carnosine, homocarnosine, and anserine), melatonin, uric acid, and lipoic acid (Watson 1993). These systems also play an important role in minimizing oxidative stress. However, cells can synthesize only a limited number of these molecules (e.g., GSH and carnosine). The majority of low-molecular-weight antioxidants are derived from dietary sources. These low molecular weight antioxidants are not only capable of chelating metal ions and reducing their catalytic activity to form ROS, but also able to scavenge metal ions. Thus, the main function of non-enzymic antioxidant systems is to liquidate the uncontrolled production of ROS, which is closely linked with the pathogenesis of cardiovascular disease, malignancy, type 2 diabetes, and neurological disorders (Farooqui 2010, 2013). Collective evidence suggests that changes in oxidation/reduction (redox) state of thiol/disulfide is linked with protein conformation, enzyme activity, transporter activity, ligand binding to receptors, protein-protein interactions, protein-DNA interactions, protein trafficking and protein degradation (Jordan and Gibbins 2006; Yang et al. 2006). The main function of above mentioned enzymic and non-enzymic mechanisms (redox regulators) is to protect cellular organelles and maintain appropriate environment for physiological functions (Hsieh and Yang 2013).

The maintenance of redox homeostasis is critical for cell function and survival, and alterations in redox homeostasis are involved in the pathophysiology of many human diseases, such as cardiovascular diseases, diabetes, neurological disorders, and cancer (Dhalla et al. 2000; Sayre et al. 2001; Valko et al. 2007; Farooqui 2010). The detoxification of ROS is a major prerequisite for the survival of aerobic life. The detoxification is achieved through the involvement of several enzymic and non-enzymic antioxidant mechanisms, which are available to the cells in different cellular compartments (Droge 2002; Farooqui 2012a). Buffering mechanisms that restore and repair altered small molecular antioxidants are also needed for the regeneration of proteins damaged by ROS for facilitating neural cell survival (Droge 2002; Poli et al. 2004). These mechanisms are not only dependent on the intermediary metabolism for ATP and NADPH, but also on the diet to maintain pro-oxidant reactions and cellular damage at a minimum level under physiological conditions (Poli et al. 2004).

6.3   Sources Contributing to ROS Formation in Brains from Normal Subjects and Patients with Alzheimer Disease

As stated in Chaps. 1 and 2, Alzheimer disease (AD) is an age-related neurodegenerative disease, which is neuropathologically characterized by the accumulation of β-amyloid (Aβ) in senile plaques and neurofibrillary tangles, which are composed of hyperphosphorylated tau protein. Another pathological hallmark of AD is brain atrophy with swollen axons, neuritis, glia cells, and loss of synaptic connections within selective brain regions along with gradual loss of neurons mainly in hippocampus, frontal cortex and limbic areas along with onset of chronic inflammation. Clinically, AD is characterized by a gradual decline in cognition along with progressive loss of memory, and changes in behavior and personality including difficulty in reasoning, disorientation, and language problems. Aging is the most important and prevalent risk factor for 90 % AD cases (Price et al. 2009) possibly because of the decline of cellular protein quality control processing (Powers et al. 2009). Other risk factors for AD include heart diseases related factors, long term consumption of high calorie diet, type II diabetes, incidence of head injury, stroke, alcohol consumption, smoking, sedentary lifestyle, elevated plasma levels of homocysteine, obesity, and severe adverse stress (Fig. 6.8) (Mejía et al. 2003; Farooqui 2010, 2012a, b). Among these factors ischemic stroke and AD, despite being distinct disease entities, share numerous pathophysiological mechanisms. The complex interrelations between AD and ischemic stroke are enhanced by atherosclerosis and inflammatory cascade. These processes not only increase the risk of immune exhaustion (Brod 2000), but are also magnified by diabetes, obesity, and hypertension (Pinti et al. 2014; Purkayastha and Cai 2013; Farooqui 2013). Oxidative stress and mitochondrial dysfunction also contribute to the neurodegeneration in AD

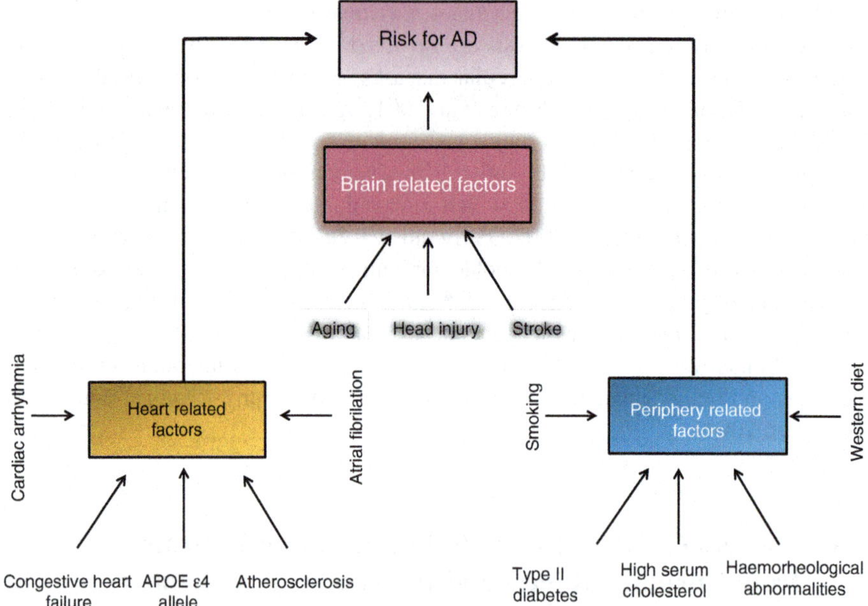

Fig. 6.8 Risk factors contributing to Alzheimer disease

(Farooqui 2010). Many studies have indicated that the oligomeric form of Aβ is highly toxic and associated with oxidative stress (Walsh et al. 2001). Aβ(1–42) acts as a free radical inducer. Its interactions with lipid bilayer result in abstraction of allylic hydrogen-atom from the unsaturated acyl chains of phospholipid molecules within the lipid bilayer initiating the process of lipid peroxidation (Lauderback et al. 2001), which generates highly reactive products, such as (4-hydroxy-2-nonenal, acrolein, malondialdehyde, isoprostanes, isoketals, and isofurans) from arachidonic acid through enzymic and non-enzymic pathways (Farooqui 2011). These metabolites can further react with proteins and enzymes, effectively amplifying the effects of Aβ(1–42) to induce more free radical formation or ROS production (Butterfield et al. 2006; Farooqui 2011; Pocernich and Butterfield 2012).

As stated above, oxidative stress in AD is accompanied by elevation in the generation of ROS and RNS and oxidative damage to various biological macromolecules, including lipids, proteins, and nucleic acids. Certain antioxidants including glutathione, α-tocopherol (vitamin E), carotenoids, ascorbic acid, antioxidant enzymes such as catalase and glutathione peroxidases are able to detoxify H_2O_2 by converting it to O_2 and H_2O under physiological conditions (Yu 1994). However, when ROS levels exceed the removal capacity of antioxidant system in AD or normal aging the onset of oxidative stress takes place inducing biological dysfunction (Yu 1994). Levels of vitamin E and glutathione are significantly decreased not only in brain samples from AD patients, but also in CSF samples from AD patients (Nishida et al. 2009; La Fata et al. 2014). Additionally, in brain tissue, ROS and

RNS are generated by neurons, microglia, and astrocytes. They not only modulate synaptic and nonsynaptic communication between neurons and glia, but also contribute to neuroinflammation and cell death, triggering neurodegeneration and memory loss (Popa-Wagner et al. 2013).

6.3.1   Contribution of Phospholipids in the Induction of Oxidative Stress

It is well known that in neural membrane phospholipids ARA is located at the *sn*-2 position of glycerol moiety. It is released by the action of $cPLA_2$ (Farooqui and Horrocks 2007). The free ARA is oxidized by cyclooxygenases (COXs), lipoxygenases (LOXs), and epoxygenases (EPOXs) leading to the generation of prostaglandins (PGs), leukotriene (LTs), lipoxins (LXs), and thromboxanes (TXs). These metabolites are collectively known as eicosanoids. Eicosanoids mediate their action by interacting with their receptors and modulating several important processes, such as neuroinflammation, vasodilation, vasoconstriction, apoptosis and immune responses (Phillis et al. 2006; Farooqui 2011). In addition to generating above mentioned metabolites, non-enzymic peroxidation of ARA generates 4-hydroxynonenals (4-HNE), isoprostanes (IsoP), isoketal (IsoK) isofuran (IsoF), acrolein (Ac), and malonaldehyde (MDA). The first step in the production of ARA-derived lipid mediators is the formation of hydroperoxide (the primary product). The breakdown of hydroperoxide results in the formation of secondary products, such as 4-HNE, IsoP, IsoK, IsoF, Ac, and MDA (Farooqui 2011). These metabolites react with proteins and deoxyribonucleic acid DNA to modify their structures leading to neural cell dysfunction and death in AD. Thus, 4-HNE modifies proteins, resulting in many harmful effects, including inhibition of neuronal glucose and glutamate transporters, inhibition of Na^+-K^+-ATPases, activation of kinases and dysregulation of intracellular calcium signaling, that ultimately induce an apoptotic cascade mechanism (Farooqui 2011).

6.3.2   Contribution of Carbohydrates in the Induction of Oxidative Stress

Protein synthesis, calcium signaling, lipid biosynthesis, and protein folding and maturation occur in the endoplasmic reticulum. All these processes are supported and driven by glucose metabolism. A variety of physiological conditions, such as perturbations in calcium homeostasis, glucose/energy deprivation, and redox changes can produce endoplasmic reticulum perturbations leading to changes in endoplasmic reticulum homeostasis disrupt protein folding. These processes may contribute to the accumulation of unfolded proteins and protein aggregates that can be injurious to the neural cells (Farooqui 2010). The consumption of carbohydrate-enriched diet produces insulin resistance and reduces insulin transport across the BBB, resulting in lowering insulin levels and activity in brain. Accumulation of Aβ oligomer in AD brain induces neuronal insulin resistance not only by inhibiting the insulin network by targeting the insulin/Akt pathway (Townsend et al. 2007), but also by removing insulin receptors after binding at the dendrites of synaptic sites (Zhao et al. 2008). Impaired insulin signaling cannot efficiently block GSK3β and therefore, the activated GSK3β not only triggers APP γ-secretase activity, but also increases tau phosphorylation (Hooper et al. 2008).

Advanced glycation end products (AGEs) are a complex, heterogeneous, and diverse class of post-translational macroprotein derivatives, which are formed by non-enzymic reaction between reducing sugars and amino groups of proteins, lipids and nucleic acids. Glycation of proteins starts as a nonenzymic process with the spontaneous condensation of ketone or aldehyde groups of sugars with the free amino groups of a protein or amino acid specifically lysine, arginine, and histidine (Harrington and Colaco 1994). Non-enzymic glycation and cross-linking of proteins induce alterations in their structural integrity and functions. Glycated proteins are also resistant to the degradation by lysosomal enzymes. For example, glycated haemoglobin is used as marker for type II diabetes and glycated LDL are poorly recognized by lipoprotein receptors and scavenger

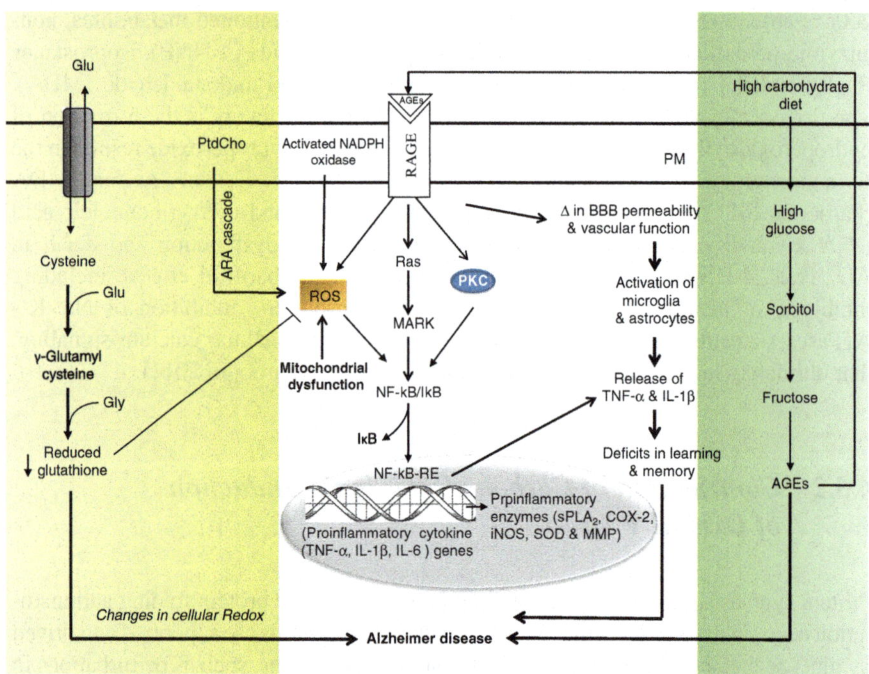

Fig. 6.9 Generation of AGE from carbohydrate enriched diet and interactions of AGEs with RAGE and RAGE-mediated signal transduction processes in the brain. Advanced glycation end-product (*AGE*); receptors for advanced glycation end-product (*RAGE*); plasma membrane (*PM*); phosphatidylcholine (*PtdCho*); arachidonic acid (*ARA*); reactive oxygen species (*ROS*); mitogen-activated protein kinase (*MARK*); nuclear factor-kappa B (*NF-κB*); nuclear factor-kappa B response element (*NF-κB-RE*); tumor necrosis factor-alpha (*TNF-α*); interleukin-1beta (*IL-1β*); interleukin-6 (*IL-6*); and glutamate (*glu*)

receptors (Zimmermann et al. 2001). AGE-modified proteins accumulate in all cells and tissues in normal ageing and their levels correlate with the glucose concentration in the blood. During AGE formation, oxygen radicals are also generated, which, are involved in neuronal cell damage by oxygen stress and apoptotic processes (Yamagishi et al. 2002). Collective evidence suggests that formation of AGE and glycated proteins can damage visceral and neural cells not only by several general mechanisms: (a) AGE-induced modification of intracellular proteins leads to alterations in their functions, (b) interactions of AGE with RAGE results in the production of ROS, which in turn activates the pleiotropic transcription factor, nuclear factor kappa B (NF-κB) causing multiple pathological changes in gene expression (Fig. 6.9) (Goldin et al. 2006).

AGEs bind to their receptors (RAGE), which are multiligand receptors of the immunoglobulin superfamily. RAGE are composed of three immunoglobulin-like domains: a "V"-type domain involved in ligand binding, and two "C"-type domains (a short transmembrane domain, and cytoplasmic tail involved in intracellular signaling) (Fig. 6.10) (Bierhaus et al. 2005). RAGE may contribute to the pathogenesis of AD through interactions with Aβ and AGEs (Schmidt et al. 2009). RAGE is

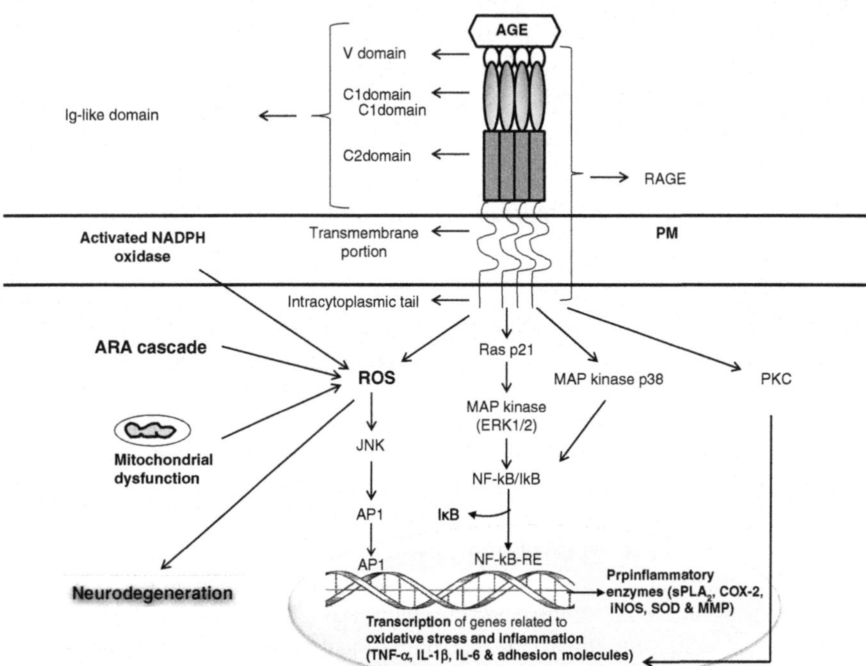

Fig. 6.10 Hypothetical diagram showing various domains of RAGE receptor. Plasma membrane (*PM*); advanced glycation end products (*AGEs*); reactive oxygen species (*ROS*); nuclear factor-kappa B (*NF-κB*); nuclear factor-kappa B response element (*NF-κB-RE*); tumor necrosis factor-alpha (*TNF-α*); interleukin-1beta (*IL-1β*); interleukin-6 (*IL-6*); RAS p21 protein activator 1 or RasGAP (Ras GTPase activating protein (*Ras p21*); MAP kinases, p38 kinases (*MAP kinase p38*); and Protein kinase C (*PKC*)

not only upregulated in the brain in AD, colocalizes with plaques, and induces oxidative stress, but also contributes to neuroinflammation and neurodegeneration (Fang et al. 2010). RAGE has been reported to play an important role in the transport of Aβ across the blood–brain barrier (Deane et al. 2003). Some studies have shown that in AD, AGEs can bind with Aβ and Tau protein and promote the formation of β-sheet fibrils (Yan et al. 1994; González et al. 1998). These)findings have led to the proposal that AGEs may act to promote increased β-amyloid and plaque deposition within the AD brain (Münch et al. 2012). Inhibition of RAGE activation may retard the pathogenic events in AD (Galasko et al. 2014). TransTech Pharma, Inc has discovered PF-04494700, a drug, which inhibits in vitro interactions of RAGE with Aβ42 at nanomolar concentration. The drug is orally bioavailable and crosses the blood–brain barrier. Chronic oral dosing of PF-04494700 in AD transgenic mice results in reduction of amyloid load in the brain, improvement in performance on tests of spatial memory, and normalization of electrophysiologic recordings from hippocampal slices (data on file, Trans Tech Pharma, unpublished) (Sabbagh et al. 2011). A phase I clinical human trial has indicated that PF-04494700 at 20 mg/d is associated with increased adverse events and cognitive decline. At 5 mg/d, PF-04494700 had a good safety profile. A potential benefit for this low dose on the ADAS-cog is not conclusive, because of high dropout and discontinuation rates subsequent to the interim analyses. Thus, more studies are needed on the use of PF-04494700 in human patients.

6.3.3 Contribution of Proteins in the Induction of Oxidative Stress

Effect of oxidative stress in brain causes both irreversible and reversible protein oxidative modifications in the proteins. Irreversible modifications mainly include protein carbonylation, tyrosine nitration, s-sulfenation, s-nitrosylation, s-glutathionylation, and disulfide formation (Beckman et al. 1999; Prokai et al. 2007; Rao and Moller 2011; Cai and Yan 2013) which involve oxidative damage to proteins and have been used as biomarkers for assessment of oxidative stress in aging and age-related diseases (Yan and Sohal 1998; Stadtman 2001; Prokai et al. 2007). In general three types of amino acid oxidative modifications have been reported to occur that can give rise to protein carbonyls: (a) direct attack by reactive oxygen species on certain amino acid side chains (Glu, Thr, Asp, Lys, Arg, and Pro) (Morgan et al. 2002; Stadtman 2006); (b) modification of histidine, cysteine, and lysine residues by lipid peroxidation products such as malondialdehyde and 4-HNE (Uchida and Stadtman 1992; Szweda et al. 1993); and (c) reaction with reducing sugars, forming advanced glycation end products adducts (Stadtman 2001; Picklo et al. 2002). The existence of all three mechanisms of protein carbonylation has been well established during normal aging and in age-related degenerative diseases, such as AD (Stadtman 2001; Picklo et al. 2002). The direct protein carbonylation take place through a variety of reactions. Thus, oxidation of proline or arginine side chain with metals and H_2O_2 forms glutamic semialdehyde, while lysine oxidation results in aminoadipic semialdehyde, introducing carbonyl groups $(RR'C=O)$ in the protein structure. The hydroxyl group of threonine side chain can also be oxidized to form the carbonyl group (Stadtman and Berlett 1991). Protein carbonylation is monitored via the formation of hydrazone derivative. 2,4-Dinitrophenylhydrazine (DNPH) reacts with the carbonyl group of aldehyde or ketone and forms a hydrazone derivative (DNP) in the following reaction:

$$RR'C = O + C_6H_3\left(NO_2\right)_2 NHNH_2 \rightarrow C_6H_3\left(NO_2\right)_2 NHNCRR' + H_2O$$

This hydrazone derivative can be detected spectrophotometrically. Protein carbonylation produces mitochondrial dysfunction not only by producing alterations in activities of phosphate carrier protein, NADH dehydrogenase 1α subcomplexes 2 and 3, but also in promoting the translocation of inner mitochondrial membrane 50, and valyl-tRNA synthetase (Curtis et al. 2012). Elevations in protein carbonylation is also accompanied by a reduction in activity of complex I, impaired respiration, increase in superoxide production, and a reduction in membrane potential without changes in mitochondrial number, area, or density indicating that protein carbonylation contribute to mitochondrial dysfunction (Fig. 6.11). It is now recognized that protein carbonylation plays a positive role in many cellular functions. This gradual realization of the beneficial roles of protein carbonylation may be attributed to accumulating evidence that ROS and RNS are indispensible for cell survival and regeneration. Apparently, the adaptation to the increased oxidant or reductant status is

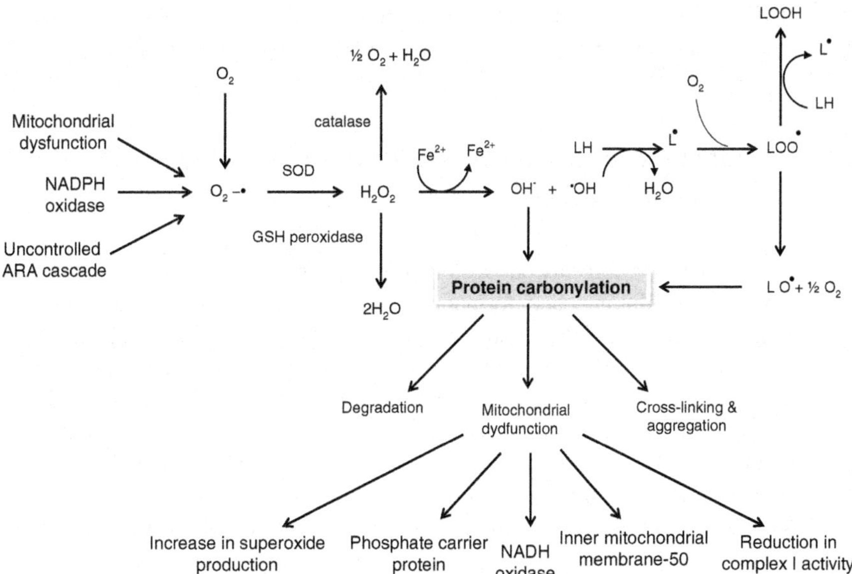

Fig. 6.11 Protein carboxylation and its effects on mitochondrial dysfunction

important to cell survival and the cellular redox homeostasis is essential to maintain a normal long life (Cai and Yan 2013). Accumulation of protein carbonyls has been implicated, in the pathophysiology of AD (Aksenov et al. 2001).

3-Nitrotyrosine is formed by a reaction between reactive nitrogen species and protein's tyrosine residue (Feeney and Schoneich 2012). This modification is a highly selective process as not all proteins or all tyrosine residues on a target protein can get nitrated (Radi 2013). Elevation in NO production is accompanied by tyrosine nitration and onset of chronic inflammatory diseases (Salvemini et al. 2006; Sugiura and Ichinose 2011). However, recent studies have indicated that the formation of 3-nitrotyrosine also occurs under normal physiological conditions during healthy pregnancy (Horvath et al. 2009). Nitrated proteins are usually removed by the proteasomes. However, inhibition of the proteasome or decreasing proteasomal activity may result in the accumulation of abnormal proteins (Jung et al. 2009). The overload of abnormal protein in the proteasome may stimulates generation of ROS via the activation of microglia cells and astrocytes.

Reversible protein oxidative modifications include protein cysteine modifications (Cai and Yan 2013). This is particularly true for reversible cysteine oxidation, which not only reflects changes in cellular redox state, but can also protect the target proteins from further damage. Additionally, reversible cysteine oxidation is also involved in redox signaling cascades (Finkel 2011; Chung et al. 2013). These days there is a lot of interest in cysteine modifications including s-sulfenation, s-nitrosylation, s-glutathionylation, and disulfide formation that are all reversible (Reddie and Carroll 2008). Protein oxidative modifications have deleterious effects in health and disease. The degradation of oxidized protein occurs in the proteasomes. However, alterations in proteasomal function have been reported to occur during aging process (Farout and Friguet 2006). This age-related impairment in proteasomal function may also contribute to the accumulation of oxidized proteins during aging (Farout and Friguet 2006; Grune et al. 2004).

6.3.4 Contribution of Nucleic Acids in the Induction of Oxidative Stress

The attack of hydroxyl radicals on DNA results in strand breaking, DNA–DNA and DNA–protein cross-linking, and formation of at least 20 modified bases adducts (Barciszewski et al. 1999; Lovell and Markesbery 2007). Among various DNA bases (5-hydroxyuracil (5-OHU), thymine glycol, 8-oxo-7,8-dihydroguanine (8-oxoG), 2,6-diamino-4-hydroxy-5-formamidopyrimidine (Fapy-G) and 4,6-diamino-5-formamidopyrimidine (Fapy-A)), the guanine residues are most sensitive to hydroxyl radicals ($^\bullet$OH) and singlet oxygen ($^1O_2$) attack (Fig. 6.12) (Steenken and Jovanovic 1997). This results in formation of 8-Oxo-7,8-dihydroguanine (8-oxoG), a product, which has been implicated in mutagenesis (Lu and Liu 2010; Damsma and Cramer 2009). 8-Oxoguanine DNA glycosylase (OGG1) plays an important role in the removal of oxidized bases, such as 8-oxo-G, 2,6-diamino-4-hydroxy-5-formamidopyrimidine (FaPyG) and 7,8-dihydro-8-oxoadenine (8-oxo-A) from the DNA

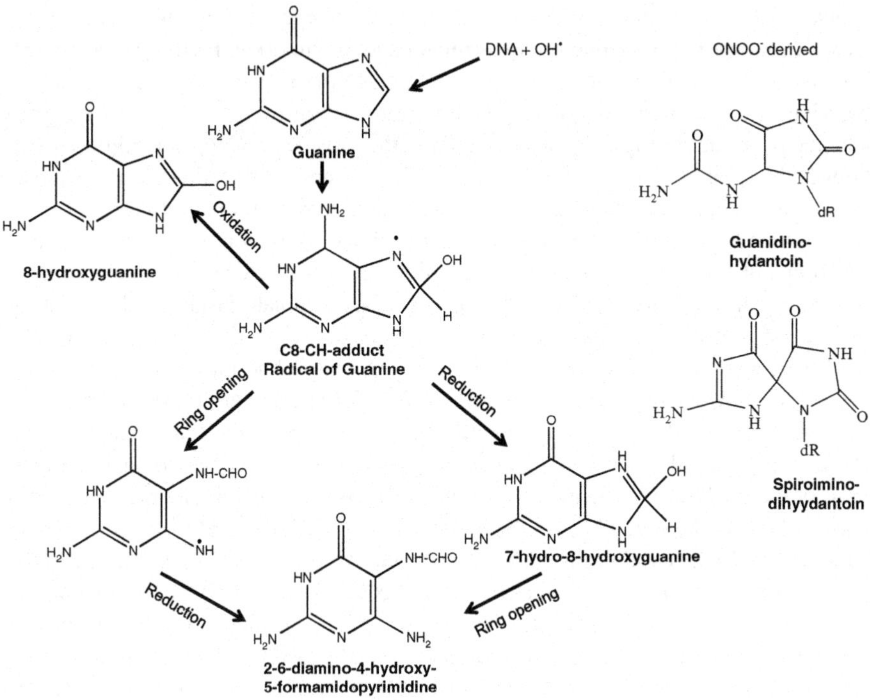

Fig. 6.12 Reactions showing attack of hydroxyl radical on guanine base of DNA

(Morales-Ruiz et al. 2003). This enzyme is specific for incising 8-oxoG to avoid the transversion of GC → TA, or DNA damage–induced apoptosis. It is recently shown that OGG1 can rescue neurons subjected to ischemic conditions (Liu et al. 2010).

Single strand DNA breaks, which arise from the disintegration of the sugar phosphate backbone of DNA following ROS-mediated attacks, which are of physiological relevance to neural cells (Brochier and Langley 2013). In contrast, double stranded DNA breaks are mostly associated with genotoxic DNA lesions. These lesions are closely associated with neural cell death. Both single strand and double strand DNA breaks are typically detected in brains from patients with age-associated neurodegenerative diseases including AD, Parkinson disease (PD) and amyotrophic lateral sclerosis (ALS). ROS-mediated generation of 8-oxodG has used as an indicator of oxidative DNA damage in relation to oxidative stress-driven diseases (Kawanishi et al. 2001; Ohshima et al. 2003). There is growing clinical interest in the measurement of urinary 8-oxodG as a mean in determining the role of oxidative stress in neurodegenerative diseases and to evaluate intervention strategies (Olinski et al. 2006; Cooke et al. 2008). Damage to RNA is caused by HO• rising from $O_2^{•-}$ and H_2O_2 by the Fenton and Haber-Weiss reactions. The reaction of ROS with free nucleobases, nucleosides, nucleotides, or oligonucleotides produces numerous distinct modifications in nucleic acids. The generation of 8-hydroxyguanosine occurs through the attack of HO• on guanine followed by oxidation or by reaction of guanine with singlet oxygen followed by its reduction (Wurtmann and Wolin 2009; Fleming et al. 2015). This oxidative modification is also the most deleterious since 8-hydroxyguanosine contributes to the alterations in the genetic information by incorrectly pairing with adenine at similar or higher efficiency than with cytosine in RNA and induces mutations at the level of transcription (Neeley and Essigmann 2006). Alterations of ribose, base excision, and strand break can represent other modifications caused by ROS on RNA (Rhee et al. 1995). Oxidative damage to RNA not only alters its structure and function, but also interferes with the interaction between RNA and other cellular molecules. As an example, oxidative damage to RNA template produces the block of reverse transcription (Rhee et al. 1995). Furthermore, oxidative damage to mRNA may also result in a decrease in translation efficiency and abnormal protein production and cause ribosome dysfunction (Shan and Lin 2006; Ding et al. 2005). Oxidative damage to RNA occurs more frequently than DNA, because RNA molecules are mostly single stranded and its bases are less protected by hydrogen bonding. Moreover, most of the mRNAs are not associated with chromatin and are distributed in the cytoplasm, closer to the site, where ROS generation occurs (Radak and Boldogh 2010). In addition, the high susceptibility of RNA to oxidative damage may also be due to the iron-binding properties of certain classes of RNAs. As stated above, iron catalyzes Haber-Weiss and Fenton reactions that produce ROS. Purified ribosomes from AD patients have elevated levels of associated redox-active iron. Moreover, in vitro studies have shown that rRNAs have higher iron binding than tRNA or mRNA. Accordingly, iron-rich rRNA had a 13-fold greater formation of 8-oxo-7,8-dihydroguanosine in oxidation experiments as compared to the iron-poor tRNA (Honda et al. 2005). There is evidence that oxidized mRNA causes errors in translation, eventually leading to the production of abnormal proteins (Tanaka et al. 2007), such as Aβ in AD, α-synuclein in PD, and mutated huntingtin in HD (Nunomura et al. 2009; Farooqui 2010). Furthermore, increase in levels of oxidatively damaged nucleic acids has been observed in numerous diseases including AD (Gackowski et al. 2008), PD (Nakabeppu et al. 2007), autoimmune diseases (Bashir et al. 1993) and cardiovascular diseases (Lee and Blair 2001). It is suggested that such damage may play an important role in the etiology of these diseases (Cooke et al. 2003, 2006).

Accumulating evidence suggests that lipid peroxidation, protein carbonylation, and generation of nucleic acid oxidation products are important markers of oxidative stress. Recent progress in redox proteomics and mass spectrometry has indicated high levels of non-enzymic lipid mediators of phospholipid metabolism, carbonylated proteins, and oxidative products of nucleic acid metabolism in neurotraumatic, neurodegenerative, and neuropsychiatric diseases (Farooqui 2010, 2011). Lipid peroxidation and protein carbonylation also mediate redox signaling processes, which may contribute to the pathogenesis of chronic diseases in a positive or negative way. Thus, the development of agents, which can control cellular lipid peroxidation, protein carbonylation status may contribute to the development of therapeutic strategies against various chronic age-related neurological diseases.

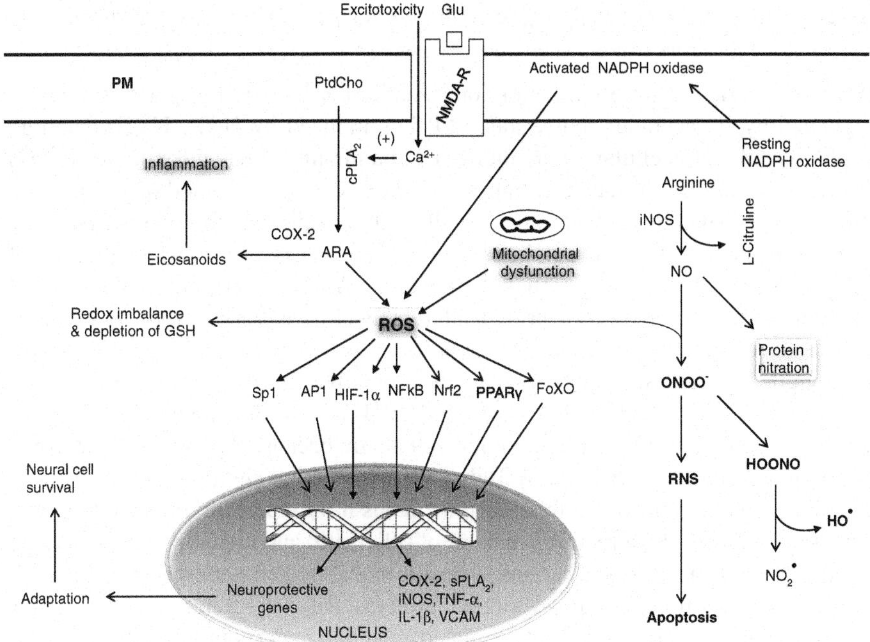

Fig. 6.13 Hypothetical model showing generation of ROS from various sources and effect of ROS of various transcription factors. Plasma membrane (*PM*); *N*-methyl-D-aspartate receptor (*NMDA-R*); glutamate (*Glu*); phosphatidylcholine (*PtdCho*); lyso-phosphatidylcholine (*lyso-PtdCho*); cytosolic phospholipase A₂ (*cPLA₂*); secretory phospholipase A₂ (*sPLA₂*); cyclooxygenase (*COX-2*); arachidonic acid (*ARA*); reactive oxygen species (*ROS*); nuclear factor kappaB (*NF-κB*); Heat shock transcription factor-1α (*HIF-1α*); nuclear factor-kappaB (*NF-κB*); inducible nitric oxide synthase (*iNOS*); platelet activating factor (*PAF*); vascular cell adhesion molecule-1 (*VCAM-1*); activator protein 1 (*AP-1*); hypoxia-inducible transcription factor-1α (*HIF-1α*); NF-E2-related factor 2 (*Nrf2*); peroxisome proliferator-activated receptor gamma (*PPAR-γ*); specificity protein 1(*sp1*) and hydroxyl radical (*OH*)

6.4   Contribution of Transcription Factors in Oxidative Stress Associated with Alzheimer Disease

Accumulation of Aβ peptide produces ROS-mediated neurodegeneration in brains from AD patients. ROS interact with a number of redox sensitive transcription factors including activator protein-1(AP-1), nuclear factor kappa-light-chain-enhancer of activated B (NF-κB), cAMP responsive element binding protein (CREB), hypoxia-inducible factors (HIF), and tumor protein p53 (p53), play a crucial role in cell death signaling pathways (Fig. 6.13).

6.4.1 AP1 Activity and Oxidative Stress in Alzheimer Disease

AP-1, a dimeric transcription factor composed of the Jun and Fos gene families, is regulated by intracellular redox state. AP-1 is induced by H_2O_2, UV irradiation, depletion of intracellular glutathione by buthionine-(SR)-sulfoximine (BSO) (Meyer et al. 1993). It activates transcription of genes, which modulate both pro-inflammatory and antioxidant pathways through phosphorylation of JNK pathway (Yao et al. 2005). AD is accompanied by the activation of the c-Jun N-terminal kinase (JNK) pathway along with induction of the AP-1 transcription factor. Targets of JNK/c-Jun in Aβ-induced neuron death remain elusive. Earlier studies have indicated that Aβ significantly reduces expression of antiapoptotic Bcl-w and Bcl-x(L), mildly affects expression of bim, Bcl-2, and bax, but does not alter expression of bak, bad, bik, bid, or BNIP3 (Yao et al. 2005). Aβ-induced downregulation of Bcl-w has been reported to link with apoptotic cell death, because Aβ-mediated neuronal death is significantly increased by Bcl-w suppression but significantly reduced by Bcl-w overexpression. In contrast, recent studies have indicated that pro-apoptotic proteins, Bim (Bcl-2 interacting mediator of cell death) and Puma (p53 up-regulated modulator of apoptosis) are targets of c-Jun in Aβ-treated neurons (Akhter et al. 2015). Thus, two important pro-apoptotic proteins, Bim (Bcl-2 interacting mediator of cell death) and Puma (p53 up-regulated modulator of apoptosis) are targets of c-Jun in Aβ-treated neurons. There is a functional co-operation of both JNK and p53 pathway in the regulation of Puma under Aβ toxicity. Detailed investigations have identified a novel AP1-binding site on rat puma gene, which is necessary for direct binding of c-Jun with Puma promoter. Thus, converging evidence suggests that both Bim and Puma are target of c-Jun and elucidate the intricate regulation of Puma expression by JNK/c-Jun and p53 pathways in neurons upon Aβ toxicity (Akhter et al. 2015). Curcumin not only inhibits AP-1 activity by blocking the activity of c-Jun N-terminal kinase, protein tyrosine kinases and protein serine/threonine kinases, but also decreases the low-density lipoprotein oxidation and the free radicals generation that cause the deterioration of neurons, not only in AD but also in other neuron degenerative disorders such as Parkinson's and Huntington's disease (Aggarwal et al. 2003; Kim et al. 2005).

6.4.2 NF-κB Activity, Oxidative Stress in Alzheimer Disease

As stated in Chap. 3, NFκB, a redox-sensitive transcription factor, plays an important role in both pro-inflammatory and protective pathways including transcription of cytokine genes and antioxidant genes (Memet 2006). The NF-κB transcription factor family consists of five proteins: RelA (p65), RelB, cRel, p105/p50 (NF-κB1) and p100/p52 (NF-κB2). They can form different homo- and heterodimers that are able to induce or repress gene expression of specific target genes (Farooqui 2010). Normally, NF-κB resides in the neuronal cytoplasm as an inactive form bound with

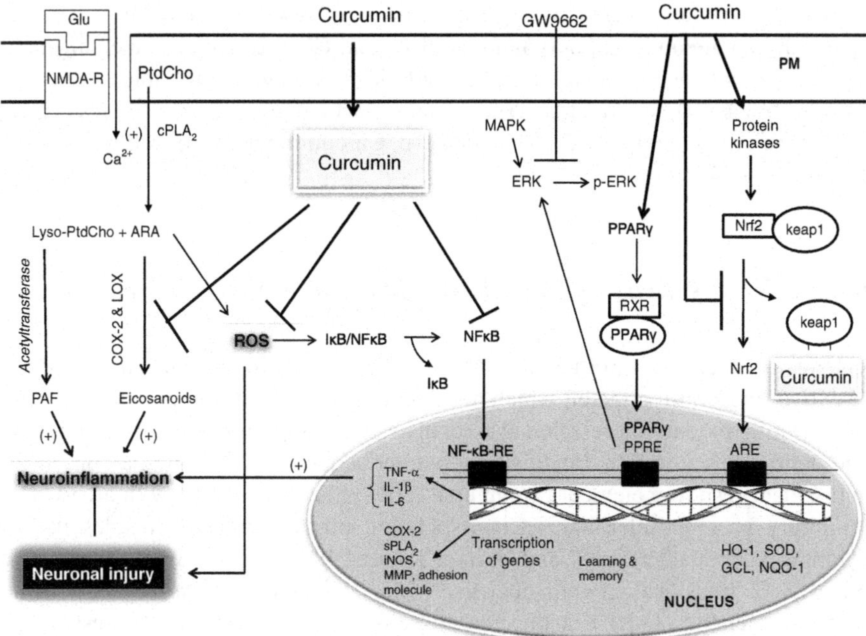

Fig. 6.14 Hypothetical diagram showing the effect transcription factors on oxidative stress and neuroinflammation along with the effect of curcumin. N-Methyl-D-asparetate receptor (*NMDA-R*); phosphatidylcholine (*PtdCho*); cytosolic phospholipase A₂ (*cPLA₂*); secretory phospholipase A₂ (*sPLA₂*); arachidonic acid (*ARA*); lysophosphatidylcholine (*lyso-PtdCho*); platelet activating factor (*PAF*); cyclooxygenase (*COX*); lipoxygenase (*LOX*); prostaglandins (*PGs*); leukotrienes (*LTs*); thromboxanes (*TXs*); nuclear factor κB (*NF-κB*); nuclear factor κB-response element (*NFκB-RE*); inhibitory subunit of NFκB (*IκB*); tumor necrosis factor-α (*TNF-α*); interleukin-1β (*IL-1β*); interleukin-6 (*IL-6*); inducible nitric oxide synthase (*iNOS*); matrix metalloproteinases (*MMPs*); NF-E2 related factor 2 (*Nrf2*); kelch-like erythroid Cap'n'Collar homologue-associated protein 1 (*Keap1*); antioxidant response-element (*ARE*); heme oxygenase (*HO-1*); superoxide Dismutase (*SOD*); NADPH quinine oxidoreductase (*NQO-1*); γ-glutamate cysteine ligase (*γ-GCL*); The peroxisome-proliferator activator receptors (*PPARs*); extracellular signal-regulated protein kinase (*ERK*); and PPARγ antagonist (*GW9662*)

the inhibitory subunit I-κB. During oxidative stress NF-κB/I-κB complex dissociates and free NK-κB migrates to the nucleus, where it binds to promoter regions of its target genes and modulates the transcription of genes for proinflammatory cytokines and chemokines, genes for proinflammatory enzymes, genes encoding anti-apoptotic proteins such as Bcl-2 and the antioxidant enzyme Mn-superoxide dismutase (Fig. 6.14) (Mattson and Meffert 2006). I-κB is degraded by ubiquitination in proteasomes. In contrast, activation of NF-κB in glial and immune cells mediates pathological processes through the induction of inflammatory cytokines release and production of ROS. In glial and immune cells, oxidative stress and inflammatory response are typically accompanied by activation of PLA₂, induction of arachidonic acid cascade, generation of ARA-derived enzymic and non-enzymic lipid mediators, and increased expression of cytokines and chemokines (Granic et al. 2009). Curcumin not only inhibits PLA_2 and COX-2 activities, the enzymes, which transform neural membrane phospholipids proinflammatory lipid mediators, but also blocks NF-κB, the transcription factor, which promotes the expression of proinflammatory cytokines TNF-α and IL-1β. Curcumin retards oxidative stress by chelating metal ions like Cu^{2+} and Fe^{3+}.

6.4.3 Nrf2 Activity and Oxidative Stress in Alzheimer Disease

The nuclear factor erythroid 2 related factor (Nrf2) is a basic leucine zipper redox-sensitive transcription factor, which not only plays a pivotal role in redox homeostasis during oxidative stress, but also regulates the expression of many antioxidant and detoxifying genes by binding to promoter sequences containing a consensus antioxidant response element (Singh et al. 2010). The Nrf2 along with a Cys-rich inhibitory protein Kelch-like ECH-associating protein 1(Keap1) resides in the cytoplasm. Oxidative stress and protein kinases liberate Nrf2 from Keap1 and allow Nrf2 translocation into nucleus, where it binds to stress or antioxidant response elements (ARE) in the presence of Maf facilitating the expression of cytoprotective genes and numerous neuroprotective enzymes, such as superoxide dismutase (SOD), catalase (CAT), sulfaredoxin (Srx), thioredoxin (Trx), peroxiredoxin (Prdx) system, glutathione synthesis and metabolism (glutathione peroxidase (Gpx), glutathione reductase (GR), γ-glutamine cysteine ligase (GCL) and synthase (GCS)), quinone recycling (NAD(P)H quinone oxidoreducase (NQO1)) and heme oxygenase 1 (HO-1) (Fig. 6.14) (Venugopal and Jaiswal 1998; Joshi and Johnson 2012). These observations support the view that Nrf2 may be the major transcription factor necessary for ARE activation and thus essential for the induction of phase II detoxification enzymes. It is recently reported that Nrf2/Hmox activation may enhance cell proliferation and survival in the subventricular zone (SVZ) of aged brains by reverting microglial phenotype into the proneurogenic phenotype (Pérez-de-Puig et al. 2013). Collective evidence suggests that Nrf2 contributes to the modulation of both antioxidant and anti-inflammatory signaling. Very little information is available on the status of Nrf2 in AD (Joshi and Johnson 2012). Recent studies have indicated that significant changes are observed in APP/PS1 mice after genetic ablation of Nrf2 not only in APP/Aβ processing and/or aggregation, but also in autophagic dysfunction in APP/PS1 mice (Joshi et al. 2015). There is a significant increase in inflammatory response in APP/PS1 mice lacking Nrf2. This is accompanied by increase in intracellular levels of APP, Aβ (1-42), and Aβ (1-40), without a change total full-length APP (Joshi et al. 2015). Detailed investigation has indicated that there is a shift of APP and Aβ into the insoluble fraction, as well as increase in poly-ubiquitin conjugated proteins in mice lacking Nrf2. APP/PS1-mediated autophagic dysfunction is also elevated in Nrf2-deficient mice. Finally, neurons in the APP/PS1/Nrf2-/- mice show an increase in the accumulation of multivesicular bodies, endosomes, and lysosomes. Collective evidence suggests that changes in Nrf2 status may protect mice from neurochemical changes in AD (Joshi et al. 2015).

Curcumin is a natural antioxidant. It induces heme oxygenase, a protein, which contributes to an efficient cytoprotection against various forms of oxidative stress. By promoting the activation and migration of Nrf2, curcumin induces the expression of heme oxygenase activity. The incubation of astrocytes with curcumin at a concentration that promotes heme oxygenase activity and results in an early increase in reduced glutathione, followed by a significant elevation in oxidized glutathione content (Calabrese et al. 2003; Jeong et al. 2006). As stated above, glutathione is an important water-phase antioxidant and essential cofactor for antioxidant enzymes, which protect mitochondria against endogenous oxygen radicals. Its level reflects the free radical scavenging capacity of neural cells. GSH depletion in the brain may lead to brain damage due to lipid peroxidation.

6.4.4 Hypoxia-Inducible Factor Activity and Oxidative Stress in Alzheimer Disease

Cellular responses to hypoxia are mainly regulated by hypoxia-inducible factors (HIFs), group of heterodimeric transcription factors, which is central to oxygen homeostasis. It is composed of two subunits. An oxygen regulated subunit (HIF-1α) and the other subunit, which is constitutively expressed is known as HIF-1β subunits (Prabhakar and Semenza 2012). Active HIFs are heterodimers (HIF-α/β) that regulate a number of genes, which provide compensation for hypoxia, metabolic changes, and oxidative stress (Siddiq et al. 2007). Hypoxic insult contributes to the pathophysiology of AD (Carvalho et al. 2009). Acute hypoxic injury results in increased expression BACE1 by upregulating the level of BACE1 mRNA. Increased expression of BACE1 results in significant increase in the APP C-terminal fragment-β (βCTF) and Aβ (Zhang et al. 2007). In AD, the accumulation of Aβ peptide activates astrocyte causing not only a long-term decrease in hypoxia-inducible factor (HIF)-1α expression, but also a reduction in the rate of glycolysis (Schubert et al. 2009). Glial activation and the glycolytic alterations are reversed by the maintenance of HIF-1α levels with conditions that prevent the proteolysis of HIF-1α. It is reported that Aβ-mediated long-term ROS production along with activation of NADPH oxidase reduces the amount of HIF-1α via the activation of proteasomes. Collectively, these studies not only suggest the importance of HIF-1α-mediated transcription in maintaining the metabolic integrity of the AD brain but also identify the probable cause of lower energy metabolism in afflicted areas (Schubert et al. 2009; Carvalho et al. 2009). Curcumin significantly not only modulates HIF-1α protein level in HepG2 cells, but also induces angiogenesis by improving the expression of angiogenesis modulating genes (Duan et al. 2014).

6.4.5   Peroxisome-Proliferator Activator Receptors and Oxidative Stress in Alzheimer Disease

The peroxisome-proliferator activator receptors (PPARs) form a family of ligand-activated nuclear receptor transcription factors that regulates the function and expression of complex gene networks, especially involved in energy homeostasis and inflammation (Varga et al. 2011) Several PPAR family comprises three known members: PPARα, PPARγ, and PPAR β/δ. Among PPAR family members, PPARγ shares a protein structure common to the other PPARs and to most of the nuclear receptors characterized by 4 functional domains (called from the N terminal to the C terminal, A/B, C, D, and E/F) (Desvergne and Wahli 1999). PPARγ dimerizes with the retinoid X receptor (RXR) (Desvergne and Wahli 1999; Varga et al. 2011) and migrates to the nucleus, where it binds to PPAR response elements (PPREs) located in the promoter region of target genes (Fig. 6.14). PPARγ activity not only depends on the presence of coactivators and corepressors, expression of other PPARs, and availability of RXR, but also on the status of promoter of the target genes, and presence of endogenous ligands (Galli et al. 1998). The activity of PPARγ is modulated by posttranslational modifications, such as phosphorylations, independently from ligand binding (Burns and Vanden Heuvel 2007; Ceni et al. 2006).

Insulin resistance has been reported to contribute to the pathogenesis of AD (van Himbergen et al. 2012) and PPARγ is a valuable target for the treatment of diabetes and AD. The activation of PPARγ reduces neuropathological changes in different AD mice models. For instance, PPARγ activation by some thiazolidinediones (TZDs) drugs reduces amyloid deposition and reverses cognitive and memory decline in some AD transgenic mice models (Nicolakakis et al. 2008). Thus, PPARγ agonism enhances cognition through the involvement of extracellular signal-regulated protein kinase (ERK), a member of MAPK family of protein kinases (Escribano et al. 2009; Rodriguez-Rivera et al. 2011; Hort et al. 2007; Hoefer et al. 2008; Denner et al. 2012). Based on requirement of ERK2 activity in hippocampus-dependent learning and memory in rodents (Atkins et al. 1998; Selcher et al. 2001), it is shown that rosiglitazone (RSG), a PPARγ agonist enhances cognitive function through the involvement of PPARγ-mediated activation and phosphorylation of ERK (pERK) in a multiprotein complex during memory consolidation for a hippocampus-dependent cognitive task. Acute inhibition of hippocampal PPARγ by GW9662 not only blocks this type of memory consolidation, but also prevents the increase in recruitment of PPARγ to pERK, supporting the view that formation of this protein complex is requisite for memory formation (Jahrling et al. 2014). In addition, the stimulation of ERKs is also triggered by inflammatory cytokines including the TNF family. These mechanisms of ERK activation play important roles in inflammation and innate immunity (Kyriakis and Avruch 2012). Collectively, these studies indicate that acute pharmacological antagonism of PPARγ with GW9662 not only blocks hippocampal memory consolidation in RSG-treated Tg2576 (Denner et al. 2012) via prevention of hippocampal PPARγ association with pERK, but also modulates neuroinflammatory processes in the hippocampus. Curcumin enhances PPAR-γ expression. It also retards the expression of genes, such as B-cell lymphoma (Bcl)-2, cyclooxygenase-2, matrix metalloproteinase-9, cyclin D1 and the adhesion molecules. Increased activity of these proteins contributes to neurodegeneration in the brain.

6.4.6   *Signal Transducer and Activator of Transcription 3 and Oxidative Stress in Alzheimer Disease*

The signal transducer and activator of transcription 3 (STAT3) is a member of the Janus kinase (JAK)-signal transducer and activator of transcription (STAT) signaling family. The JAK-STAT pathway is involved in cell growth, survival, development, differentiation, and gene regulation (Oyinbo 2011; Nicolas et al. 2012, 2013). In neural tissues JAK-STAT pathway is involved in the neurogenesis and maintenance of neuroplasticity. The JAK-STAT pathway involves the activation of cell membrane receptors by polypeptides, such as growth factors, hormones, or cytokines, which induce the activation of JAK in cell membranes (Nicolas et al. 2013). Protein tyrosine phosphorylation is an important mechanism, by which growth factors or cytokines regulate cellular processes. Initially, the activation of JAKs involves tyrosine phosphorylation with cell membrane receptor binding with subsequent activation of STATs in the cytoplasm through tyrosine phosphorylation, leading to the dimerization of STATs (Stahl et al. 1994). The STAT dimers migrate to the nucleus, where they interact with specific cis-elements, followed by the transcription of various target genes (Darnell et al. 1994). The activation of the JAK-STAT pathway has been reported to occur in the ischemic brain injury (Choi et al. 2003; Justicia et al. 2000) and spinal cord injury (Oyinbo 2011). Several studies have indicated that the activation of JAK/STAT3 pathway in reactive astrocytes of two transgenic mouse models of AD has indicated that SOCS3 (suppressor of cytokine signaling 3) not only significantly inhibits JAK/STAT3 pathway in astrocytes, but also prevents astrocyte reactivity and decreases microglial activation in animal models of AD (Ben Haim et al. 2015). Similar observation have been made in animal models of Huntington disease (HD), where is reported that inhibition of the JAK/STAT3 pathway within reactive astrocytes also increases the number of huntingtin aggregates, a neuropathological hallmark of HD, but does not influence neuronal death. Collective evidence suggests that the JAK/STAT3 pathway is a common mediator of astrocyte reactivity that is highly conserved in AD (Ben Haim et al. 2015).

6.5   Effects of Curcumin on Downregulation of Oxidative Stress in Alzheimer Disease

Curcumin reduces oxidative damage and reverses the amyloid pathology in an AD transgenic mouse (Yang et al. 2005; Garcia-Alloza et al. 2007). Direct injections of curcumin into the brains of the mice with AD not only hampere further development

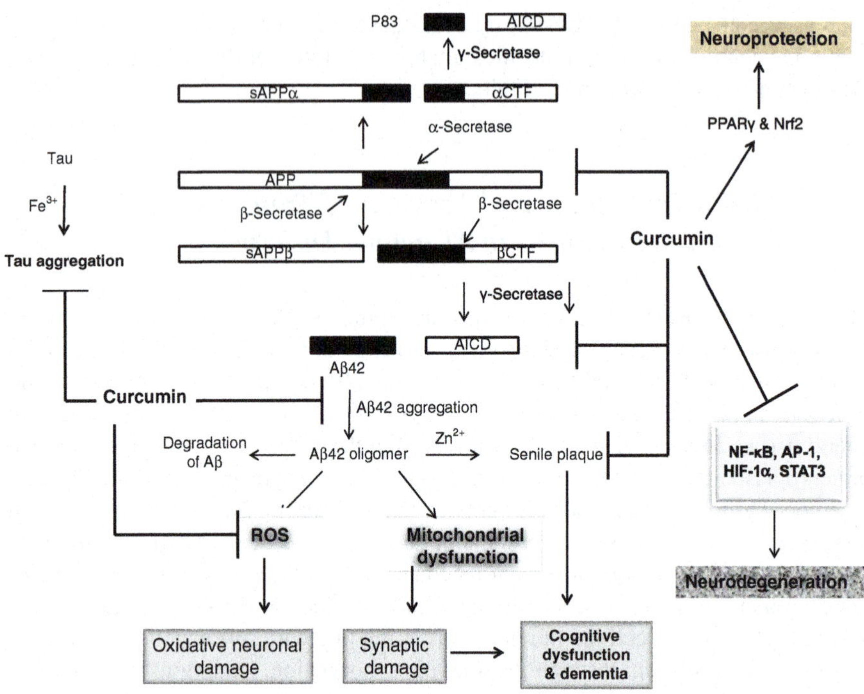

Fig. 6.15 Effect of curcumin on APP processing, Aβ aggregation, and transcription factors associated with oxidative stress. Amyloid precursor protein (*APP*); substrate inhibitory domain (*ASID*); and reactive oxygen species (*ROS*)

of plaque but also reduce the levels of senile plaque (Fig. 6.15). In APPswe/PS1dE9 double transgenic mice, 3 month curcumin treatment not only reduces Aβ40, Aβ42, and aggregation of Aβ-derived diffusible ligands in the mouse hippocampal CA1 area, but also reduces the expression of the γ-secretase component presenilin-2 and increases the expression of β-amyloid-degrading enzymes, including insulin-degrading enzymes and neprilysin supporting the view that curcumin can reduce β-amyloid pathological aggregation, possibly through mechanisms that prevent its production by inhibiting presenilin-2 and/or by accelerating its clearance by increasing degrading enzymes such as insulin-degrading enzyme and neprilysin (Murakami et al. 2011; Wang et al. 2014). Similarly, in other animal models of AD, curcumin or curcumin phospholipid conjugate reduces Aβ-mediated increase in the ROS level (Lim et al. 2001; Park and Kim 2002). Curcumin can cross blood brain barrier (Mukhopadhyay et al. 2002; Mishra and Palanivelu 2008; Tsai et al. 2011). It not only retards plaque formation, but also disrupts existing plaques in Alzheimer transgenic mouse models either after oral uptake (Yang et al. 2005; Begum et al. 2008) or intravenous (i.v.) injection (Garcia-Alloza et al. 2007). It shows dose-dependent inhibitory effect on the formation of Aβ fibrils with an EC$_{50}$ of 0.09–0.63 μM (Ono et al. 2004). In addition, the destabilization of preformed Aβ fibrils by curcumin was also dose-dependent at 0.1–1.0 μM (Ono et al. 2004). Curcumin, demethoxycurcumin, bisdemethoxycurcumin display better neuroprotective effect against $A\beta_{1-42}$ insults with ED_{50} values of 3.0–7.1 µg/mL compared to α-tocopherol (Kim et al. 2001). The mechanism through which curcumin disrupts Aβ aggregation and plaque formation is not fully understood. However, curcumin may also act by: (a) increasing Aβ uptake by macrophages (Zhang et al. 2006), (b) controlling the amyloid protein precursor maturation (Zhang et al. 2010), (c) modulating APP processing enzymes (Xiong et al. 2011), (d) protecting neurons from Aβ induced toxicity (Yang et al. 2005), and (e) influencing the expression of genes associated with apoptosis and inflammation (Ahmed and Gilani 2011) through the regulation of transcription factors, such as NF-κB, AP-1, Sp1, STAT3, PPARγ; HIF-1α; and Nrf2 (Orlando et al. 2012). Curcumin increases the expression of thioredoxin, an antioxidant protein in the Nrf2 pathway, and protects neurons from death caused by oxygen-glucose deprivation in an in vitro model of ischemia/reperfusion (Yang et al. 2009; Wu et al. 2015). Curcumin also induces the expression of heme oxygenase (HO-1) (Pugazhenthi et al. 2007) and aldoketo reductase (Kang et al. 2007) in vitro through a PtdIns 3K/Akt-mediated signaling pathway involving the transcription factor Nrf2. It is well known that multiple signaling pathways mediate the induction of Nrf2-driven phase II enzymes including PtdIns 3K/Akt, MEK/ERK, p38MAPK, JNK, and protein kinase C (Lee and Surh 2005). In addition, curcumin upregulates the expression of Bcl2, an integral mitochondrial membrane protein and acts as anti apoptotic and antioxidant factor (Jang and Surh 2003). Bcl-2 reduces the generation of ROS through binding to cytochrome c or prevents its entry to the cytosol (Shimizu et al. 1995). Curcumin also increases the expression of synapse-related proteins (PSD95 and Shank1) in APP/PS1 double transgenic mice, improving structure and plasticity of synapse in APP/PS1 double transgenic mice leading to enhancement in the learning and memory abilities (Wei et al. 2012). Converging evidence suggests that the activation of Nrf2 target genes and particularly HO-1 and increased expression of synaptic proteins in astrocytes and neurons produce strongly neuroprotective effects against oxidative stress and apoptotic cell death. In addition, curcumin has been reported to promote induction of heat shock proteins and reduce protein misfolding and aggregation (Dickey et al. 2009; Ma et al. 2013).

Curcumin also has ability to scavenge metal ions (Banerjee 2014). As stated above, curcumin has two o-methoxy phenolic OH groups attached to the β-diketone (heptadiene-dione) moiety that chelates with copper, iron, and other transition metals (Banerjee 2014). Although the metal chelation may occur through both the o-methoxy phenol and the β-diketo group, in most of the cases, the complexation of curcumin with metal ions involves the diketo group (Baum and Ng 2004; Anand et al. 2008). In physiological buffer solution, the mononuclear Cu^{2+} and Zn^{2+} complexes of curcumin have similar binding affinity to the monomeric $A\beta$ peptide compared to curcumin. However, the Cu^{2+} and Zn^{2+} curcumin complexes cause better inhibition in the secondary structural transition of the peptide from oligomers to protofibrils than the free curcumin. In addition, these complexes also have improved efficacy to inhibit the fibrillization and aggregation of the peptide (Banerjee 2014). Chelation of metal ions by curcumin may also reduce the generation of ROS and decrease oxidative stress in the brain of animal model of AD. These results may have implications in understanding the molecular mechanism of action of antioxidant-metal complexes in protein misfolding diseases. It is also reported that curcumin may inhibit Aβ-mediated neurodegeneration by maintaining telomeres length through enhancing human telomerase reverse transcriptase activity, the catalytic subunit of telomerase, which stabilizes telomere length during chromosomal replication (Xiao et al. 2014; Kim et al. 2003).

Little is known about the effects of curcumin on Tau protein. Curcumin not only reduces levels of total and p-Tau dimers, also restores dysregulated excitatory synaptic proteins supporting the view that Tau dimers represent a significant synaptotoxic species. Tau protein has been reported to induce excessive microtubule bundling producing deficits in organelle transport deficits and dendritic spine loss (Thies and Mandelkow 2007). Tau also binds and bundles actin (He et al. 2009), raising the possibility that Tau dimers disrupt dendritic spine actin dynamics, critical for normal learning and memory (Hotulainen and Hoogenraad 2010). Soluble Tau species have been implicated in synaptic deficits in the rTg4510 model (Ma et al. 2013). In control mice Tau is predominantly axonal, but in mutant mice it redistributes to dendritic spines where it is bound to actin (Hoover et al. 2010). It is well known that activated JNK is present in neurons and dystrophic neurites of both mice and brain AD models, which coincides with phosphorylation of protein insulin receptor substrate-1 (IRS-1) and tau, deposition of amyloid plaques and synapthophysin loss (Shen and Liu 2006). The treatment of 3xTg-AD mice with high-fat diet of DHA-rich fish oil or curcumin or a combination of both for 4 months suppresses the activities of JNK and phosphorylation of both IRS-1 and tau (Ma et al. 2009). Administration of either curcumin or tetrahydrocurcumin to aged Tg2576 APPsw mice or lipopolysaccharide-injected wild-type mice reduces JNK activity and soluble Aβ accumulation (Begum et al. 2008). Meanwhile, curcumin but not tetrahydrocurcumin shows significant effect in the prevention of Aβ aggregation, which indirectly indicates that the presence of dienone bridge in curcumin is needed to reduce senile plaque deposition (Begum et al. 2008). In *Drosophila* model of AD, curcumin improves lifespan and locomotor activity (Caesar et al. 2012; Alavez et al. 2011). However, in this model no changes are observed in Aβ deposition. The authors have proposed that curcumin accelerates conversion of pre-fibrilllar to $A\beta_{1-42}$ in treated flies leading to reduction in Aβ-mediated neurotoxicity (Caesar et al. 2012; Alavez et al. 2011). It is also reported that in animal models of AD, curcumin also corrects behavioral, synaptic, and heat shock proteins (HSPs) deficits (Ma et al. 2013). Detailed investigations have indicated that curcumin differentially impacts HSP90 client kinases, reducing Fyn without reducing Akt. Thus, converging evidence suggests that curcumin not only blocks aggregation of Aβ peptide, but also suppresses soluble Tau dimer and elevates HSPs. HSPs are involved in Tau clearance, supporting the view that curcumin may act after tangle formation and Tau-dependent behavioral and synaptic deficits can be corrected (Ma et al. 2013). Thus, the consumption of curcumin from young to the old age may not only facilitate good health by inhibiting oxidative stress and neuroinflammation, but also by promoting disaggregation of existing amyloid deposits, preventing aggregation of new amyloid deposits, and facilitating reduction of amyloid deposits in animal models of AD. Finally, studies on the effects of curcumin on the distribution of astrocytes and in hippocampus have indicated that curcumin inhibits astrocyte activation in the hippocampus and downregulates GFAP expression in the Aβ induced AD model in rats (Wang et al. 2013). These observations support the view that curcumin has real therapeutic potential for the treatment of AD. Despite of promising research mentioned above, the clinical use of orally administered curcumin is severely limited because of low bioavailability, which is a direct consequence of their poor solubility in aqueous solutions and their rapid metabolic conversion (Hamaguchi et al. 2010). Another important problem related to curcumin bioavailability is its rapid metabolism firstly in the gastrointestinal tract, and then in the liver even after its absorption, limiting the concentration of curcumin in plasma or in target organs. To improve systemic availability, several curcumin formulations have been developed (Naksuriya et al. 2014). These formulations have been described in detail earlier in Chap. 3.

So far, large and systematic clinical trials with curcumin on AD patients have not been performed. Seven clinical trials on the safety, tolerance, pharmacokinetics, and treatment effects of curcumin in AD (NIH registry) are underway. Observations from many of these trials are still pending. Amongst these, two are still recruiting, one has unknown status, one has been terminated due to unstated reasons, and two have been completed. The first one was a 6-month randomized, double blind, placebo-controlled study supplementing 1 or 4 g/d of oral curcumin in 34 patients with AD (Baum et al. 2008). In this trial patients received simultaneous supplementation with 120 mg/day ginkgo leaf extract. No significant differences have been observed reported in serum Aβ40 and F2-isoprostane levels, as well as MMSE scores after 6 month supplementation (Baum and Ng 2004). In this study biomarkers are only measured in blood, which may not be necessarily correlated with the levels in the brain or CSF (Baum et al. 2008). The second study was a 24 week randomized, double blind, placebo-controlled study with 2 or 4 g/d oral supplementation of curcumin in 36 patients with mild-to-moderate AD. Similarly, this study was unable to demonstrate any clinical or biochemical evidence of the neuroprotective efficacy of curcumin (Ringman et al. 2012). One small study conducted in Japan has indicated that three patients with AD whose behavioral symptoms improved remarkably as a result of curcumin treatment (Hishikawa et al. 2012). After 12 weeks of treatment, the total score of the Neuropsychiatric Inventory-brief questionnaire is decreased significantly in both acuity of symptoms and burden on caregivers. In one case, the minimental state examination (MMSE) score was up by five points, from 12/30 to 17/30. In the other two cases, no significant change is observed in the MMSE; however, patient came to recognize their family within 1 year of treatment. In all cases that have been consuming curcumin for more than 1 year reexacerbation of behavioral and psychological symptoms of dementia (BPSD) have not observed. Though it is a small sample size, three AD cases treated with curcumin suggest a significant improvement of the cognitive and behavioral symptoms, suggesting a probable benefit in the use of curcumin in individuals with AD for BPSD. Limitations to these trials include small number of AD patients and shorter trial duration. Importantly, determination of effective dose and optimal bioavailability will be the fundamental requirement for successful curcumin trials in normal human subjects and AD patients. This is tempting to speculate that more studies are required on pharmacokinetics of curcumin in the brains of human subjects before successful trials can be performed in AD patients.

6.6 Conclusion

Mitochondrial dysfunction and increase in oxidative stress (ROS production) is closely associated with the pathogenesis of AD. The molecular mechanisms through which ROS mediate neurodegeneration in AD are not fully understood. However, considerable evidence supports the view that low levels of ROS not only activate a stress response, and adaptation, but also protect cells from neurodegeneration by maintaining the redox balance. These adaptive responses often require upregulation of endogenous antioxidant enzymes, and their expression levels can be regulated by several transcription factors, such as Nrf2 and PPARγ. High levels of ROS supported by the accumulation of Aβ perturb cellular Ca^{2+} homeostasis linked with energy metabolism. These processes may contribute to neurodegeneration in AD. Furthermore, High levels of ROS may not only impair the function of ion-motive ATPases, glucose, glutamate transporters, and GTP-binding proteins, but also cause damage to proteins due to covalent modification of the proteins by 4-hydroxynonenal and oxidative damage to the nucleic acids. By disrupting cellular ion homeostasis and energy metabolism, relatively high levels of membrane-associated oxidative stress can render neurons vulnerable to excitotoxicity and apotosis. Increase in intracellular Ca^{2+} may also cause elevation in RNS by activating neuronal NO synthases. NO activates the epsilon isoform of PKC, which in turn activates the Src family of protein tyrosine kinases and transcription factors leading to upregulation in expression of many genes and proteins. Here it must be mentioned here that oxidative stress and neuroinflammation are closely intertwined processes, which generally function in parallel, particularly in the brain, an organ especially prone to oxidative stress. However, the specificity of various regions to oxidative stress response in AD is very poorly understood.

Treatment with curcumin produces beneficial effects in animal models of AD probably due to its antioxidant and anti-inflammatory properties. The molecular mechanisms associated with beneficial effects of curcumin in animal models of AD are not fully understood. However, it is suggested that mediates its beneficial effects through multiple activities including (a) direct binding and inhibition of Aβ aggregate formation, (b) reduction in expression of the β-secretase enzyme BACE1 (c) inhibition of pro-inflammatory cytokine expression, (d) suppression of the peroxidation product 4-hydroxynonenal, (e) chelation of metal ions, such as Cu^{2+}, Zn^{2+}, Al^{3+}, Fe^{3+} and Mn^{2+}, and (f) lowering of cholesterol levels. Metal–curcumin complexes can mediate protective effects against stress, ROS, and neuronal damage. Converging evidence suggests that curcumin also inhibits the AGEs-mediated increase in NF-κB and AP-1 activities. Curcumin treatment can not only lowers the elevation of oxidized proteins and decrease brain Aβ levels and Aβ plaque burden, but also produces positive effects on cognitive function in animal models of AD. Very little is known about the redox changes that occur in the extracellular and intracellular compartments of various types of neural cells with respect to curcumin treatment, neuroinflammation and oxidative stress in age-related disease like AD.

References

Aggarwal BB, Kumar A, Bharti AC (2003) Anti-cancer potential of curcumin: preclinical and clinical studies. Anticancer Res 23:363–398

Ahmed T, Gilani AH (2011) A comparative study of curcuminoids to measure their effect on inflammatory and apoptotic gene expression in an Aβ plus ibotenic acid-infused rat model of Alzheimer's disease. Brain Res 1400:1–18

Ak T, Gulcin I (2008) Antioxidant and radical scavenging properties of curcumin. Chem Biol Interact 174:27–37

Akhter R, Sanphui P, Das H, Saha P, Biswas SC (2015) The regulation of p53 up-regulated modulator of apoptosis by JNK/c-Jun pathway in β-amyloid-induced neuron death. J Neurochem 134:1091–1103. doi:10.1111/jnc.13128

Aksenov MY, Aksenova MV, Butterfield DA, Geddes JW, Markesbery WR (2001) Protein oxidation in the brain in Alzheimer's disease. Neuroscience 103:373–383

Al-Amiery AA, Kadhum AA, Obayes HR, Mohamad AB (2013) Synthesis and antioxidant activities of novel 5-chlorocurcumin, complemented by semiempirical calculations. Bioinorg Chem Appl 2013:354982

Alavez S, Vantipalli MC, Zucker DJS, Klang IM, Lithgow GJ (2011) Amyloid-binding compounds maintain protein homeostasis during ageing and extend lifespan. Nature 472:226–229

Allen RG, Tresini M (2000) Oxidative stress and gene regulation. Free Radic Biol Med 28:463–499

Ambegaokar SS, Wu L, Alamshahi K, Lau J, Jazayeri L, Chan S, Khanna P, Hsieh E, Timiras PS (2003) Curcumin inhibits dose-dependently and time-dependently neuroglial proliferation and growth. Neuro Endocrinol Lett 24:469–473

Anand P, Thomas SG, Kunnumakkara AB, Sundaram C, Harikumar KB, Sung B, Tharakan ST, Misra K, Priyadarsini IK, Rajasekharan KN, Aggarwal BB (2008) Biological activities of curcumin and its analogues (Congeners) made by man and Mother Nature. Biochem Pharmacol 76:1590–1611

Andorn AC, Britton RS, Bacon BR (1990) Evidence that lipid peroxidation and total iron are increased in Alzheimer's brain. Neurobiol Aging 11:316–320

Atkins CM, Selcher JC, Petraitis JJ, Trzaskos JM, Sweatt JD (1998) The MAPK cascade is required for mammalian associative learning. Nat Neurosci 1:602–609

Azzi A, Aratri E, Boscoboinik D, Clément S, Özer NK, Ricciarelli R, Spycher S (1998) Molecular basis of α-tocopherol control of smooth muscle cell proliferation. Biofactors 7:3–14

Banerjee R (2014) Inhibitory effect of curcumin-Cu(II) and curcumin-Zn(II) complexes on amyloid-beta peptide fibrillation. Bioinorg Chem Appl 2014:325873

Barciszewski J, Barciszewska MZ, Siboska G, Rattan SIS, Clark BFC (1999) Some unusual nucleic acid bases are products of hydroxyl radical oxidation of DNA and RNA. Mol Biol Rep 26:231–238

Bashir S, Harris G, Denman MA, Blake DR, Winyard PG (1993) Oxidative DNA damage and cellular sensitivity to oxidative stress in human autoimmune diseases. Ann Rheum Dis 52:659–666

Baum L, Ng A (2004) Curcumin interaction with copper and iron suggests one possible mechanism of action in Alzheimer's disease animal models. J Alzheimers Dis 6:367–377

Baum L, Lam CW, Cheung SK, Kwok T, Lui V, Tsoh J, Lam L, Leung V, Hui E, Ng C, Woo J, Chiu HF, Goggins WB, Zee BC, Cheng KF, Fong CY, Wong A, Mok H, Chow MS, Ho PC, Ip

SP, Ho CS, Yu XW, Lai CY, Chan MH, Szeto S, Chan IH, Mok V (2008) Six-month randomized, placebo-controlled, double-blind, pilot clinical trial of curcumin in patients with Alzheimer disease. J Clin Psychopharmacol 28:110–113

Beckman JS (1996) Oxidative damage and tyrosine nitration from peroxynitrite. Chem Res Toxicol 9:836–844

Begum AN, Jones MR, Lim GP, Morihara T, Kim P, Heath DD, Rock CL, Pruitt MA, Yang F, Hudspeth B, Hu S, Faull KF, Teter B, Cole GM, Frautschy SA (2008) Curcumin structure-function, bioavailability, and efficacy in models of neuroinflammation and Alzheimer's disease. J Pharmacol Exp Ther 326:196–208

Ben Haim L, Ceyzériat K, Carrillo-de Sauvage MA, Aubry F, Auregan G, Guillermier M, Ruiz M, Petit F, Houitte D, Faivre E, Vandesquille M, Aron-Badin R, Dhenain M, Déglon N, Hantraye P, Brouillet E, Bonvento G, Escartin C (2015) The JAK/STAT3 pathway is a common inducer of astrocyte reactivity in Alzheimer's and Huntington's diseases. J Neurosci 35:2817–2829

Bienert GP, Møller AL, Kristiansen KA, Schulz A, Møller IM, Schjoerring JK (2007) Specific aquaporins facilitate the diffusion of hydrogen peroxide across membranes. J Biol Chem 282:1183–1192

Bierhaus A, Humpert PM, Morcos M, Wendt T, Chavakis T, Arnold B, Stern DM, Nawroth PP (2005) Understanding RAGE, the receptor for advanced glycation end products. J Mol Med 83:876–886

Boscoboinik D, Szewczyk A, Hensey C, Azzi A (1991) Inhibition of cell proliferation by alpha-tocopherol. Role of protein kinase C. J Biol Chem 266:6188–6194

Bouchier-Hayes L, Lartigue L, Newmeyer DD (2005) Mitochondria: pharmacological manipulation of cell death. J Clin Invest 115:2640–2647

Braun L, Puskás F, Csala M, Mészáros G, Mandl J, Bánhegyi G (1997) Ascorbate as a substrate for glycolysis or gluconeogenesis: evidence for an interorgan ascorbate cycle. Free Radic Biol Med 23:804–808

Brigelius-Flohe R (2009) Vitamin E: The shrew waiting to be tamed. Free Radic Biol Med 46:543–554

Brochier C, Langley B (2013) Chromatin modifications associated with DNA double-strand breaks repair as potential targets for neurological diseases. Neurotherapeutics 10:817–830

Brod SA (2000) Unregulated inflammation shortens human functional longevity. Inflamm Res 49:561–570

Bryan NS, Rassaf T, Maloney RE, Rodriguez CM, Saijo F, Rodriguez JR, Feelisch M (2004) Cellular targets and mechanisms of nitros(yl)ation: an insight into their nature and kinetics in vivo. Proc Natl Acad Sci U S A 101:4308–4313

Burns KA, Vanden Heuvel JP (2007) Modulation of PPAR activity via phosphorylation. Biochim Biophys Acta 1771:952–960

Bürzle M, Hediger MA (2012) Functional and physiological role of vitamin C transporters. Curr Top Membr 70:357–375

Butler AR, Megson IL, Wright PG (1998) Diffusion of nitric oxide and scavenging by blood in the vasculature. Biochim Biophys Acta 1425:168–176

Butterfield DA, Poon HF, St Clair D, Keller JN, Pierce WM, Klein JB, Markesbery WR (2006) Redox proteomics identification of oxidatively modified hippocampal proteins in mild cognitive impairment: insights into the development of Alzheimer's disease. Neurobiol Dis 22:223–232

Caesar I, Jonson M, Nilsson KPR, Thor S, Hammarström P (2012) Curcumin promotes A-beta fibrillation and reduces neurotoxicity in transgenic Drosophila. PLoS One 7, e31424

Cai Z, Yan LJ (2013) Protein oxidative modifications: beneficial roles in disease and health. J Biochem Pharmacol Res 1:15–26

Calabrese V, Butterfield DA, Stella AM (2003) Nutritional antioxidants and the heme oxygenase pathway of stress tolerance: novel targets for neuroprotection in Alzheimer's disease. Ital J Biochem 52:177–181

Carvalho C, Correia SC, Santos RX, Cardoso S, Moreira PI, Clark TA, Zhu X, Smith MA, Perry G (2009) Role of mitochondrial-mediated signaling pathways in Alzheimer disease and hypoxia. J Bioenerg Biomembr 41:433–440

Ceni E, Crabb DW, Foschi M, Mello T, Tarocchi M, Patussi V, Moraldi L, Moretti R, Milani S, Surrenti C, Galli A (2006) Acetaldehyde inhibits PPARγ via H_2O_2-mediated c-Abl activation in human hepatic stellate cells. Gastroenterology 131:1235–1252

Chami L, Buggia-Prévot V, Duplan E, Delprete D, Chami M, Peyron J-F, Checter F (2012) Nuclear factor-κB regulates βAPP and β- and γ-secretases differently at physiological and supraphysiological Aβ concentrations. J Biol Chem 287:24573–24584

Chen SS, Michael A, Butler-Mannuel A (2012) Advances in the treatment of ovarian cancer: a potential role of anti-inflammatory phytochemicals. Discov Med 13:7–17

Choi JS, Kim SY, Cha JH, Choi YS, Sung KW, Oh ST, Kim ON, Chung JW, Chun MH, Lee SB, Lee MY (2003) Upregulation of gp130 and STAT3 activation in the rat hippocampus following transient forebrain ischemia. Glia 41:237–246

Chung HS, Wang SB, Venkatraman V, Murray CI, Van Eyk JE (2013) Cysteine oxidative posttranslational modifications: emerging regulation in the cardiovascular system. Circ Res 112:382–392

Cooke MS, Evans MD, Dizdaroglu M, Lunec J (2003) Oxidative DNA damage: mechanisms, mutation, and disease. FASEB J 17:1195–1214

Cooke MS, Olinski R, Evans MD (2006) Does measurement of oxidative damage to DNA have clinical significance? Clin Chim Acta 365:30–49

Cooke MS, Olinski R, Loft S (2008) Measurement and meaning of oxidatively modified DNA lesions in urine. Cancer Epidemiol Biomarkers Prev 17:3–14

Curtis JM, Hahn WS, Stone MD, Inda JJ, Droullard DJ, Kuzmicic JP, Donoghue MA, Long EK, Armien AG, Lavandero S, Arriaga E, Griffin TJ, Bernlohr DA (2012) Protein carbonylation and adipocyte mitochondrial function. J Biol Chem 287:32967–329680

Damsma GE, Cramer P (2009) Molecular basis of transcriptional mutagenesis at 8-oxoguanine. J Biol Chem 284:31658–31663

Darnell JE Jr, Kerr IM, Stark GR (1994) Jak-STAT pathways and transcriptional activation in response to IFNs and other extracellular signaling proteins. Science 264:1415–1421

Deane R, Du Yan S, Submamaryan RK, LaRue B, Jovanovic S, Hogg E, Welch D, Manness L, Lin C, Yu J, Zhu H, Ghiso J, Frangione B, Stern A, Schmidt AM, Armstrong DL, Arnold B, Liliensiek B, Nawroth P, Hofman F, Kindy M, Stern D, Zlokovic B (2003) RAGE mediates amyloid beta peptide transport across the blood-brain barrier and accumulation in brain. Nat Med 29:907–913

Denner LA, Rodriguez-Rivera J, Haidacher SJ, Jahrling JB, Carmical JR, Hernandez CM, Zhao Y, Sadygov RG, Starkey JM, Spratt H, Luxon BA, Wood TG, Dineley KT (2012) Cognitive enhancement with rosiglitazone links the hippocampal PPARγ and ERK MAPK signaling pathways. J Neurosci 32:16725–16735

Desvergne B, Wahli W (1999) Peroxisome proliferator-activated receptors: nuclear control of metabolism. Endocr Rev 20:649–688

Dhalla NS, Temsah RM, Netticadan T (2000) Role of oxidative stress in cardiovascular diseases. J Hypertens 18:655–673

Dickey C, Kraft C, Jinwal U, Koren J, Johnson A, Anderson L, Lebson L, Lee D, Dickson D, de Silva R, Binder LI, Morgan D, Lewis J (2009) Aging analysis reveals slowed tau turnover and enhanced stress response in a mouse model of tauopathy. Am J Pathol 174:228–238

Ding Q, Markesbery WR, Chen Q, Li F, Keller JN (2005) Ribosome dysfunction is an early event in Alzheimer's disease. J Neurosci 25:9171–9175

Dringen R (2000) Metabolism and functions of glutathione in brain. Prog Neurobiol 62:649–671

Droge W (2002) Free radicals in the physiological control of cell function. Physiol Rev 82:47–95

Dröge W, Schipper HM (2007) Oxidative stress and aberrant signaling in aging and cognitive decline. Aging Cell 6:361–370

Duan W, Chang Y, Li R, Xu Q, Lei J, Yin C, Li T, Wu Y, Ma Q, Li X (2014) Curcumin inhibits hypoxia inducible factor-1α-induced epithelial-mesenchymal transition in HepG2 hepatocellular carcinoma cells. Mol Med Rep 10:2505–2510

Escribano L, Simón AM, Pérez-Mediavilla A, Salazar-Colocho P, Del Río J, Frechilla D (2009) Rosiglitazone reverses memory decline and hippocampal glucocorticoid receptor downregulation in an Alzheimer's disease mouse model. Biochem Biophys Res Commun 379:406–410

Fang F, Lue LF, Yan S, Xu H, Luddy JS, Chen D, Walker DG, Stern DM, Yan S, Schmidt AM, Chen JX, Yan SS (2010) RAGE-dependent signaling in microglia contributes to neuroinflammation, A-beta accumulation, and impaired learning/memory in a mouse model of Alzheimer's disease. FASEB J 24:1043–1055

Farooqui AA (2010) Neurochemical aspects of neurotraumatic and neurodegenerative diseases. Springer, New York

Farooqui AA (2011) Lipid mediators and their metabolism in the brain. Springer, New York

Farooqui AA (2012a) Generation of reactive oxygen species in the brain: signaling for neural cell survival or suicide. In: Farooqui T, Farooqui AA (eds) Oxidative stress in vertebrates and invertebrates: molecular aspects of cell signaling. Wiley, Hoboken, NJ, pp 3–15

Farooqui AA (2012b) Lipid mediators and their metabolism in the nucleus: implications for Alzheimer's disease. J Alzheimers Dis 30(Suppl 2):S163–S178

Farooqui AA (2013) Metabolic syndrome: an important risk factor for stroke, Alzheimer disease and depression. Springer, New York, Heidelberg, Dordrecht, London

Farooqui AA (2014) Inflammation and oxidative stress in neurological disorders. Springer, Switzerland

Farooqui AA, Horrocks LA (2007) Glycerophospholipids in brain. Springer, New York

Farooqui AA, Horrocks LA, Farooqui T (2000) Deacylation and reacylation of neural membrane glycerophospholipids. J Mol Neurosci 14:123–135

Farout L, Friguet B (2006) Proteasome function in aging and oxidative stress: implications in protein maintenance failure. Antioxid Redox Signal 8:205–216

Feeney MB, Schoneich C (2012) Tyrosine modifications in aging. Antioxid Redox Signal 17:1571–1579

Finkel T (2011) Signal transduction by reactive oxygen species. J Cell Biol 194:7–15

Fleming AM, Alshykhly OR, Zhu J, Muller JG, Burrows CJ (2015) Rates of chemical cleavage of DNA and RNA oligomers containing guanine oxidation products. Chem Res Toxicol 28:1292–1300

Forman HJ, Zhang H, Rinna A (2009) Glutathione: overview of its protective roles, measurement, and biosynthesis. Mol Aspects Med 30:1–12

Foster MW, Hess DT, Stamler JS (2009) Protein S-nitrosylation in health and disease: a current perspective. Trends Mol Med 15:391–404

Gackowski D, Rozalski R, Siomek A, Dziaman T, Nicpon K, Klimarczyk M, Araszkiewicz A, Olinski R (2008) Oxidative stress and oxidative DNA damage is characteristic for mixed Alzheimer disease/vascular dementia. J Neurol Sci 266:57–62

Galasko D, Bell J, Mancuso JY, Kupiec JW, Sabbagh MN, van Dyck C, Thomas RG, Aisen PS; Alzheimer's Disease Cooperative Study (2014) Clinical trial of an inhibitor of RAGE-Aβ interactions in Alzheimer disease. Neurology 82:1536–1542

Galli A, Stewart M, Dorris R, Crabb D (1998) High-level expression of RXRα and the presence of endogenous ligands contribute to expression of a peroxisome proliferator-activated receptor-responsive gene in hepatoma cells. Arch Biochem Biophys 354:288–294

Garcia-Alloza M, Borrelli LA, Rozkalne A, Hyman BT, Bacskai BJ (2007) Curcumin labels amyloid pathology in vivo, disrupts existing plaques, and partially restores distorted neurites in an Alzheimer mouse model. J Neurochem 102:1095–1104

Garcia-Ruiz C, Fernández-Checa JC (2006) Mitochondrial glutathione: hepatocellular survival-death switch. J Gastroenterol Hepatol 21:S3–S6

Ghosh S, Banerjee S, Sil PC (2015) The beneficial role of curcumin on inflammation, diabetes and neurodegenerative disease: a recent update. Food Chem Toxicol 83:111–124

Gilgun-Sherki Y, Melamed E, Offen D (2001) Oxidative stress induced-neurodegenerative diseases: the need for antioxidants that penetrate the blood brain barrier. Neuropharmacology 40:959–975

Gohil K, Vasu VT, Cross CE (2010) Dietary alpha-tocopherol and neuromuscular health: search for optimal dose and molecular mechanisms continues! Mol Nutr Food Res 54:693–709

Goldin A, Beckman JA, Schmidt AM, Creager M (2006) Advanced glycation end products:sparking the development of diabetic vascular injury. Circulation 114:597–605

González C, Farías G, Maccioni RB (1998) Modification of tau to an Alzheimer's type protein interferes with its interaction with microtubules. Cell Mol Biol 44:1117–1127

Granic I, Dolga AM, Nijholt IM, van Dijk G, Eisel UL (2009) Inflammation and NF-kappaB in Alzheimer's disease and diabetes. J Alzheimers Dis 16:809–821

Grune T, Jung T, Merker K, Davies KJ (2004) Decreased proteolysis caused by protein aggregates, inclusion bodies, plaques, lipofuscin, ceroid, and 'aggresomes' during oxidative stress, aging, and disease. Int J Biochem Cell Biol 36:2519–2530

Halliwell B (2001) Role of free radicals in the neurodegenerative diseases: therapeutic implications for antioxidant treatment. Drugs Aging 18:685–716

Halliwell B (2007) Biochemistry of oxidative stress. Biochem Soc Trans 35:1147–1150

Hamaguchi T, Ono K, Yamada M (2010) REVIEW: Curcumin and Alzheimer's disease. CNS Neurosci Ther 16:285–297

Han D, Antunes F, Canali R, Rettori D, Cadenas E (2003) Voltage-dependent anion channels control the release of the superoxide anion from mitochondria to cytosol. J Biol Chem 278:5557–5563

Harrington CR, Colaco CALS (1994) A glycation connection. Nature 370:247–248

He HJ, Wang XS, Pan R, Wang DL, Liu MN, He RQ (2009) The proline-rich domain of Tau plays a role in interactions with actin. BMC Cell Biol 10:81

Hess DT, Matsumoto A, Kim SO, Marshall HE, Stamler JS (2005) Protein S-nitrosylation: purview and parameters. Nat Rev Mol Cell Biol 6:150–166

Hishikawa N, Takahashi Y, Amakusa Y, Tanno Y, Tuji Y, Niwa H, Murakami N, Krishna UK (2012) Effects of turmeric on Alzheimer's disease with behavioral and psychological symptoms of dementia. Ayu 33:499–504

Hoefer M, Allison SC, Schauer GF, Neuhaus JM, Hall J, Dang JN, Weiner MW, Miller BL, Rosen HJ (2008) Fear conditioning in frontotemporal lobar degeneration and Alzheimer's disease. Brain 131:1646–1657

Holmgren A, Johansson C, Berndt C, Lonn ME, Hudemann C, Lillig CH (2005) Thiol redox control via thioredoxin and glutaredoxin systems. Biochem Soc Trans 33:1375–1377

Honda K, Smithi MA, Zhu X, Baus D, Merrick WC, Tartakoff AM, Hattier T, Harris PL, Siedlak SL, Fujioka H, Liu Q, Moreira PI, Miller FP, Nunomura A, Shimohama S, Perry G (2005) Ribosomal RNA in Alzheimer disease is oxidized by bound redox-active iron. J Biol Chem 280:20978–20986

Hooper C, Killick R, Lovestone S (2008) The GSK3 hypothesis of Alzheimer's disease. J Neurochem 104:1433–1439

Hoover BR, Reed MN, Su J, Penrod RD, Kotilinek LA, Grant MK, Pitstick R, Carlson GA, Lanier LM, Yuan LL, Ashe KH, Liao D (2010) Tau mislocalization to dendritic spines mediates synaptic dysfunction independently of neurodegeneration. Neuron 68:1067–1081

Hort J, Laczó J, Vyhnálek M, Bojar M, Bures J, Vlcek K (2007) Spatial navigation deficit in amnestic mild cognitive impairment. Proc Natl Acad Sci U S A 104:4042–4047

Horvath EM, Magenheim R, Kugler E, Vacz G, Szigethy A, Levardi F, Kollai M, Szabo C, Lacza Z (2009) Nitrative stress and poly(ADP-ribose) polymerase activation in healthy and gestational diabetic pregnancies. Diabetologia 52:1935–1943

Hotulainen P, Hoogenraad CC (2010) Actin in dendritic spines. Connecting dynamics to function. J Cell Biol 189:619–629

Hsieh HL, Yang CM (2013) Role of redox signaling in neuroinflammation and neurodegenerative diseases. Biomed Res Int 2013:484613

Hwang C, Sinskey AJ, Lodish HF (1992) Oxidized redox state of glutathione in the endoplasmic reticulum. Science 257:1496–1502

Jahrling JB, Hernandez CM, Denner L, Dineley KT (2014) PPARγ recruitment to active ERK during memory consolidation is required for Alzheimer's disease-related cognitive enhancement. J Neurosci 34:4054–4063

Jang JH, Surh YJ (2003) Potentiation of cellular antioxidant capacity by Bcl-2: implications for its antiapoptotic function. Biochem Pharmacol 13:1371–1379

Jayaprakasha GK, Rao LJ, Sakariah KK (2006) Antioxidant activities of curcumin, demethoxycurcumin and bisdemethoxycurcumin. Food Chem 98:720–724

Jellinger KA (2009) Recent advances in our understanding of neurodegeneration. J Neural Transm 116:1111–1162

Jeong GS, Pal HO, Jeong SO, Kim YL, Shin MK, Seo BY, Han SY, Lee HS, Jeong JG, Koh JS, Chung HT (2006) Comparative effects of curcuminoids on endothelial heme oxygenase-1 expression: Ortho-methoxy groups are essential to enhance heme oxygenase activity and protection. Exp Mol Med 38:393–400

Joe B, Lokesh BR (1994) Role of capsaicin, curcumin and dietary n-3 fatty acids in lowering the generation of reactive oxygen species in rat peritoneal macrophages. Biochim Biophys Acta 1224:255–263

Joe B, Vijaykumar M, Lokesh BR (2004) Biological properties of curcumin--cellular and molecular mechanisms of action. Crit Rev Food Sci Nutr 44:97–111

Jones DP, Park Y, Gletsu-Miller N, Liang Y, Yu T, Accardi CJ, Ziegler TR (2011) Dietary sulfur amino acid effects on fasting plasma cysteine/cystine redox potential in humans. Nutrition 27:199–205

Jordan PA, Gibbins JM (2006) Extracellular disulfide exchange and the regulation of cellular function. Antioxid Redox Signal 8:312–324

Joshi G, Johnson JA (2012) The Nrf2-ARE pathway: a valuable therapeutic target for the treatment of neurodegenerative diseases. Recent Pat CNS Drug Discov 7:218–229

Joshi G, Gan KA, Johnson DA, Johnson JA (2015) Increased Alzheimer's disease-like pathology in the APP/PS1ΔE9 mouse model lacking Nrf2 through modulation of autophagy. Neurobiol Aging 36:664–679

Jovanovic SV, Boone CW, Steenken S, Trinoga M, Kaskey RB (2001) How curcumin works preferentially with water soluble antioxidants. J Am Chem Soc 123:3064–3068

Jung KK, Lee HS, Cho JY, Shin WC, Rhee MH, Kim TG, Kang JH, Kim SH, Hong S, Kang SY (2006) Inhibitory effect of curcumin on nitric oxide production from lipopolysaccharide-activated primary microglia. Life Sci 79:2022–2031

Jung T, Catalgol B, Grune T (2009) The proteasomal system. Mol Aspects Med 30:191–296

Justicia C, Gabriel C, Planas AM (2000) Activation of the JAK/STAT pathway following transient focal cerebral ischemia: signaling through Jak1 and Stat3 in astrocytes. Glia 30:253–270

Kamata H, Hirata H (1999) Redox regulation of cellular signalling. Cell Signal 11:1–14

Kang Y, Viswanath V, Jha N, Qiao X, Mo JQ, Andersen JK (1999) Brain gamma-glutamyl cysteine synthetase (GCS) mRNA expression patterns correlate with regional-specific enzyme activities and glutathione levels. J Neurosci Res 58:436–441

Kang ES, Woo IS, Kim HJ, Eun SY, Paek KS, Kim HJ, Chang KC, Lee JH, Lee HT, Kim JH, Nishinaka T, Yabe-Nishimura C, Seo HG (2007) Up-regulation of aldose reductase expression mediated by phosphatidylinositol 3-kinase/Akt and Nrf2 is involved in the protective effect of curcumin against oxidative damage. Free Radic Biol Med 43:535–545

Kawanishi S, Hiraku Y, Oikawa S (2001) Mechanism of guanine-specific DNA damage by oxidative stress and its role in carcinogenesis and aging. Mutat Res 488:65–76

Kim DSHL, Park SY, Kim JY (2001) Curcuminoids from Curcuma longa L. (Zingiberaceae) that protect PC12 rat pheochromocytoma and normal human umbilical vein endothelial cells from Aβ(1- 42) insult. Neurosci Lett 303:57–61

Kim E, Kim JH, Shin HY, Lee H, Yang JM, Kim J, Sohn JH, Kim H, Yun CO (2003) Ad-mTERT-delta19, a conditional replication-competent adenovirus driven by the human telomerase promoter, selectively replicates in and elicits cytopathic effect in a cancer cell-specific manner. Hum Gene Ther 14:1415–1428

Kim GY, Kim KH, Lee SH, Yoon MS, Lee HJ, Moon DO (2005) Curcumin inhibits immunostimulatory function of dendritic cells: MAPKs and translocation of NF-B as potential targets. J Immunol 174:8116–8124

Kyriakis JM, Avruch J (2012) Mammalian MAPK signal transduction pathways activated by stress and inflammation: a 10-year update. Physiol Rev 92:689–737

La Fata G, Weber P, Mohajeri MH (2014) Effects of Vitamin E on cognitive performance during ageing and in Alzheimer's disease. Nutrients 6:5453–5472

Lauderback CM, Hackett JM, Huang FF, Keller JN, Szweda LI, Markesbery WR, Butterfield DA (2001) The glial glutamate transporter, GLT-1, is oxidatively modified by 4-hydroxy-2-nonenal in the Alzheimer's disease brain: the role of Aβ(1–42). J Neurochem 78:413–416

Le Belle JE, Orozco NM, Paucar AA, Saxe JP, Mottahedeh J, Pyle AD, Wu H, Kornblum HI (2011) Proliferative neural stem cells have high endogenous ROS levels that regulate self-renewal and neurogenesis in a PI3K/Akt-dependant manner. Cell Stem Cell 8:59–71

Lee SH, Blair IA (2001) Oxidative DNA damage and cardiovascular disease. Trends Cardiovasc Med 11:148–155

Lee JS, Surh YJ (2005) Nrf2 as a novel molecular target for chemoprevention. Cancer Lett 224:171–184

Lim GP, Chu T, Yang F, Beech W, Frautschy SA, Cole GM (2001) The curry spice curcumin reduces oxidative damage and amyloid pathology in an Alzheimer transgenic mouse. J Neurosci 21:8370–8377

Limoli CL, Rola R, Giedzinski E, Mantha S, Huang TT, Fike JR (2004) Cell-density-dependent regulation of neural precursor cell function. Proc Natl Acad Sci U S A 101:16052–16057

Liu D, Croteau DL, Souza-Pinto N, Pitta M, Tian J, Wu C, Jiang H, Mustafa K, Keijzers G, Bohr VA, Mattson MP (2010) Evidence that OGG1 glycosylase protects neurons against oxidative DNA damage and cell death under ischemic conditions. J Cereb Blood Flow Metab 31:680–692

Lovell MA, Markesbery WR (2007) Oxidative DNA damage in mild cognitive impairment and late-stage Alzheimer's disease. Nucleic Acids Res 35:7497–7504

Lu SC (2009) Regulation of glutathione synthesis. Mol Aspects Med 30:42–59

Lu J, Holmgren A (2014) The thioredoxin antioxidant system. Free Radic Biol Med 66:75–87

Lu J, Liu Y (2010) Deletion of Ogg1 DNA glycosylase results in telomere base damage and length alteration in yeast. EMBO J 29:398–409

Ma QL, Yang FS, Rosario ER, Ubeda OJ, Beech W, Gant DJ, Chen PP, Hudspeth B, Chen CR, Zhao YL, Vinters HV, Frautschy SA, Cole GM (2009) beta-amyloid oligomers induce phosphorylation of tau and inactivation of insulin receptor substrate via c-jun N-terminal kinase signaling suppression by omega-3 fatty acids and curcumin. J Neurosci 29:9078–9089

Ma Q-L, Zuo X, Yang F, Ubeda OJ, Gant DJ, Alaverdyan M, Teng E, Hu S, Chen P-P, Maiti P, Teter B, Cole GM, Frautschy SA (2013) Curcumin suppresses soluble tau dimers and corrects molecular chaperone, synaptic, and behavioral deficits in aged human tau transgenic mice. J Biol Chem 288:4056–4065

Maccarrone M, Brüne B (2009) Redox regulation in acute and chronic inflammation. Cell Death Differ 16:1184–1186

Mahoney CW, Azzi A (1988) Vitamin E inhibits protein kinase C activity. Biochem Biophys Res Commun 154:694–697

Marinho HS, Real C, Cyrne L, Soares H, Antunes F (2014) Hydrogen peroxide sensing, signaling and regulation of transcription factors. Redox Biol 2:535–562

Martin L, Latypova X, Wilson CM, Magnaudeix A, Perrin ML, Terro F (2013) Tau protein phosphatases in Alzheimer's disease: the leading role of PP2A. Ageing Res Rev 12:39–49

Martínez MC, Andriantsitohaina R (2009) Reactive nitrogen species: molecular mechanisms and potential significance in health and disease. Antioxid Redox Signal 11:669–702

Masuda T, Maekawa T, Hidaka K, Bando H, Takeda Y, Yamaguchi H (2001) Chemical studies on antioxidant mechanisms of curcumin: analysis of oxidative coupling products from curcumin and linoleate. J Agric Food Chem 49:2539–2547

Mattson MP, Meffert MK (2006) Roles for NF-kappaB in nerve cell survival, plasticity, and disease. Cell Death Differ 13:852–860

Mejía S, Giraldo M, Pineda D, Ardila A, Lopera F (2003) Nongenetic factors as modifiers of the age of onset of familial Alzheimer's disease. Int Psychogeriatr 15:337–349

Memet S (2006) NF-kappaB functions in the nervous system: from development to disease. Biochem Pharmacol 72:1180–1195

Meyer M, Schreck R, Baeuerle PA (1993) H2O2 and antioxidants have opposite effects on activation of NF-kappa B and AP-1 in intact cells: AP-1 as secondary antioxidant-responsive factor. EMBO J 12:2005–2015

Mishra S, Palanivelu K (2008) The effect of curcumin (turmeric) on Alzheimer's disease: an overview. Ann Indian Acad Neurol 11:13–19

Misiaszek R, Crean C, Joffe A, Geacintov NE, Shafirovich V (2004) Oxidative DNA damage associated with combination of guanine and superoxide radicals and repair mechanisms via radical trapping. J Biol Chem 279:32106–32115

Morales-Ruiz T, Birincioglu M, Jaruga P, Rodriguez H, Roldan-Arjona T, Dizdaroglu M (2003) Arabidopsis thaliana Ogg1 protein excises 8-hydroxyguanine and 2,6-diamino-4-hydroxy-5-formamidopyrimidine from oxidatively damaged DNA containing multiple lesions. Biochemistry 42:3089–3095

Morgan PE, Dean RT, Davies MJ (2002) Inactivation of cellular enzymes by carbonyls and protein-bound glycation/glycoxidation products. Arch Biochem Biophys 403:259–269

Mosley CA, Liotta DC, Snyder JP (2007) Highly active anticancer curcumin analogues. Adv Exp Med Biol 595:77–103

Mukhopadhyay A, Banerjee S, Stafford LJ, Xia C, Liu M, Aggarwal BB (2002) Curcumin induced suppression of cell proliferation correlates with downregulation of cyclin D1 expression and CDK4-mediated retinoblastoma protein phosphorylation. Oncogene 21:8852–8861

Muller DP (2010) Vitamin E and neurological function. Mol Nutr Food Res 54:710–718

Münch G, Westcott B, Menini T, Gugliucci A (2012) Advanced glycation endproducts and their pathogenic roles in neurological disorders. Amino Acids 42:1221–1236

Murakami K, Murata N, Noda Y (2011) SOD1 (copper/zinc superoxide dismutase) deficiency drives amyloid β protein oligomerization and memory loss in mouse model of Alzheimer disease. J Biol Chem 286:44557–44568

Naito Y, Shimozawab M, Kurodab M, Nakabeb N, Manabeb H, Katadab H, Kokurac S, Ichikawad H, Yoshidaa N, Noguchie N, Yoshikawa T (2005) Tocotrienols reduce 25-hydroxycholesterol-induced monocyte-endothelial cell interaction by inhibiting the surface expression of adhesion molecules. Atherosclerosis 180:19–25

Nakabeppu Y, Tsuchimoto D, Yamaguchi H, Sakumi K (2007) Oxidative damage in nucleic acids and Parkinson's disease. J Neurosci Res 85:919–934

Naksuriya O, Okonogi S, Schiffelers RM, Hennink WE (2014) Curcumin nanoformulations: a review of pharmaceutical properties and preclinical studies and clinical data related to cancer treatment. Biomaterials 35:3365–3383

Narlawar R, Pickhardt M, Leuchtenberger S, Baumann K, Krause S, Dyrks T, Weggen S, Mandelkow E, Schmidt B (2008) Curcumin-derived pyrazoles and isoxazoles: Swiss army knives or blunt tools for Alzheimer's disease? ChemMedChem 3:165–172

Neeley WL, Essigmann JM (2006) Mechanisms of formation, genotoxicity, and mutation of guanine oxidation products. Chem Res Toxicol 19:491–505

Nicolakakis N, Aboulkassim T, Ongali B, Lecrux C, Fernandes P, Rosa-Neto P, Tong XK, Hamel E (2008) Complete rescue of cerebrovascular function in aged Alzheimer's disease transgenic mice by antioxidants and pioglitazone, a peroxisome proliferator-activated receptor γ agonist. J Neurosci 28:9287–9296

Nicolas CS, Peineau S, Amici M, Csaba Z, Fafouri A, Javalet C, Collett VJ, Hildebrandt L, Seaton G, Choi SL, Sim SE, Bradley C, Lee K, Zhuo M, Kaang BK, Gressens P, Dournaud P, Fitzjohn SM, Bortolotto ZA, Cho K, Collingridge GL (2012) The Jak/STAT pathway is involved in synaptic plasticity. Neuron 73:374–390

Nicolas CS, Amici M, Bortolotto ZA, Doherty A, Csaba Z, Fafouri A, Dournaud P, Gressens P, Collingridge GL, Peineau S (2013) The role of JAK-STAT signaling within the CNS. JAKSTAT 2, e22925

Nishida Y, Yokota T, Takahashi T, Uchihara T, Jishage K-I, Mizusawa H (2006) Deletion of vitamin E enhances phenotype of Alzheimer disease model mouse. Biochem Biophys Res Commun 350:530–536

Nishida Y, Ito S, Ohtsuki S, Yamamoto N, Takahashi T, Iwata N, Jishage K, Yamada H, Sasaguri H, Yokota S, Piao W, Tomimitsu H, Saido TC, Yanagisawa K, Terasaki T, Mizusawa H, Yokota T (2009) Depletion of vitamin E increases amyloid beta accumulation by decreasing its clearances from brain and blood in a mouse model of Alzheimer disease. J Biol Chem 284:33400–33408

Niu J, Shi Y, Iwai K, Wu ZH (2011) LUBAC regulates NF-κB activation upon genotoxic stress by promoting linear ubiquitination of NEMO. EMBO J 30:3741–3753

Nunomura A, Hofer T, Moreira PI, Castellani RJ, Smith MA, Perry G (2009) RNA oxidation in Alzheimer disease and related neurodegenerative disorders. Acta Neuropath 118:151–166

Ohshima H, Tatemichi M, Sawa T (2003) Chemical basis of inflammation-induced carcinogenesis. Arch Biochem Biophys 417:3–11

Olinski R, Rozalski R, Gackowski D, Foksinski M, Siomek A, Cooke MS (2006) Urinary measurement of 8-oxodg, 8-oxogua, and 5HMUra: a noninvasive assessment of oxidative damage to DNA. Antioxid Redox Signal 8:1011–1019

Ono K, Hasegawa K, Naiki H, Yamada M (2004) Curcumin has potent anti-amyloidogenic effects for Alzheimer's β-amyloid fibrils in vitro. J Neurosci Res 75:742–750

Orlando RA, Gonzales AM, Royer RE, Deck LM, Vander Jagt DL (2012) A chemical analog of curcumin as an improved inhibitor of amyloid abeta oligomerization. PLoS One 7, e31869

Ortega AL, Mena S, Estrela JM (2011) Glutathione in cancer cell death. Cancers (Basel) 3:1285–1310

Oyinbo CA (2011) Secondary injury mechanisms in traumatic spinal cord injury: a nugget of this multiply cascade. Acta Neurobiol Exp (Wars) 71:281–299

Park SY, Kim DS (2002) Discovery of natural products from *Curcuma longa* that protect cells from β-amyloid insult: a drug discovery effort against Alzheimer's disease. J Nat Prod 65:1227–1231

Pérez-de-Puig I, Martín A, Gorina R, de la Rosa X, Martinez E, Planas AM (2013) Induction of hemeoxygenase-1 expression after inhibition of hemeoxygenase activity promotes inflammation and worsens ischemic brain damage in mice. Neuroscience 243:22–32

Phillis JW, Horrocks LA, Farooqui AA (2006) Cyclooxygenases, lipoxygenases, and epoxygenases in CNS: their role and involvement in neurological disorders. Brain Res Rev 52:201–243

Picklo MJ, Montine TJ, Amarnath V, Neely MD (2002) Carbonyl toxicology and Alzheimer's disease. Toxicol Appl Pharmacol 184:187–197

Pinti M, Cevenini E, Nasi M, De Biasi S, Salvioli S, Monti D, Benatti S, Gibellini L, Cotichini R, Stazi MA, Trenti T, Franceschi C, Cossarizza A (2014) Circulating mitochondrial DNA increases with age and is a familiar trait: implications for "inflammaging". Eur J Immunol 44:1552–1562

Pocernich CB, Butterfield DA (2012) Elevation of glutathione as a therapeutic strategy in Alzheimer disease. Biochim Biophys Acta 1822:625–630

Poli G, Leonarduzzi G, Biasi F, Chiarpotto E (2004) Oxidative stress and cell signalling. Curr Med Chem 11:1163–1182

Popa-Wagner A, Mitran S, Sivanesan S, Chang E, Buga A-M (2013) ROS and brain diseases: the good, the bad, and the ugly. Oxid Med Cell Longev 2013:963520

Powers ET, Morimoto RI, Dillin A, Kelly JW, Balch WE (2009) Biological and chemical approaches to diseases of proteostasis deficiency. Annu Rev Biochem 78:959–991

Prabhakar NR, Semenza GL (2012) Adaptive and maladaptive cardiorespiratory responses to continuous and intermittent hypoxia mediated by hypoxia-inducible factors 1 and 2. Physiol Rev 92:967–1003

Price JL, McKeel DW Jr, Buckles VD, Roe CM, Xiong C, Grundman M, Hansen LA, Petersen RC, Parisi JE, Dickson DW, Smith CD, Davis DG, Schmitt FA, Markesbery WR, Kaye J, Kurlan R, Hulette C, Kurland BF, Higdon R, Kukull W, Morris JC (2009) Neuropathology of nondemented aging: presumptive evidence for preclinical Alzheimer disease. Neurobiol Aging 30:1026–1036

Priyadarsini KI, Maity DK, Naik GH, Kumar MS, Unnikrishnan MK, Satav JG, Mohan H (2003) Role of phenolic O-H and methylene hydrogen on the free radical reactions and antioxidant activity of curcumin. Free Radic Biol Med 35:475e484

Prokai L, Yan LJ, Vera-Serrano JL, Stevens SM Jr, Fordter MJ (2007) Mass spectrometry-based survey of age-associated protein carbonylation in rat brain mitochondria. J Mass Spectrom 42:1583–1589

Pryor WA (1986) Oxy-radicals and related species: their formation, lifetimes, and reactions. Annu Rev Physiol 48:657–667

Pugazhenthi S, Akhov L, Selvaraj G, Wang M, Alam J (2007) Regulation of heme oxygenase-1 expression by demethoxy curcuminoids through Nrf2 by a PI3-kinase/Akt-mediated pathway in mouse beta-cells. Am J Physiol Endocrinol Metab 293:E645–E655

Purkayastha S, Cai D (2013) Neuroinflammatory basis of metabolic syndrome. Mol Metab 2:356–363

Radak Z, Boldogh I (2010) 8-Oxo-7,8-dihydroguanine: links to gene expression, aging, and defense against oxidative stress. Free Radic Biol Med 49:587–596

Radi R (2013) Protein tyrosine nitration: biochemical mechanisms and structural basis of functional effects. Acc Chem Res 46:550–559

Rao RS, Moller IM (2011) Pattern of occurrence and occupancy of carbonylation sites in proteins. Proteomics 11:4166–4173

Rathore P, Dohare P, Varma S, Ray A, Sharma U, Jagannathan NR, Jaganathanan NR, Ray M (2008) Curcuma oil: reduces early accumulation of oxidative product and is anti-apoptogenic in transient focal ischemia in rat brain. Neurochem Res 33:1672–1682

Reddie KG, Carroll KS (2008) Expanding the functional diversity of proteins through cysteine oxidation. Curr Opin Chem Biol 12:746–754

Reddy PH, Tripathi R, Troung Q, Tirumala K, Reddy TP, Anekonda V, Shirendeb UP, Calkins MJ, Reddy AP, Mao P, Manczak M (2012) Abnormal mitochondrial dynamics and synaptic degeneration as early events in Alzheimer's disease: implications to mitochondria-targeted antioxidant therapeutics. Biochim Biophys Acta 1822:639–649

Reiter E, Jiang Q, Christen S (2007) Anti-inflammatory properties of alpha- and gamma-tocopherol. Mol Aspects Med 28:668–691

Rhee Y, Valentine MR, Termini J (1995) Oxidative base damage in RNA detected by reverse transcriptase. Nucleic Acids Res 23(16):3275–3282

Ricciarelli R, Tasinato A, Clément S, Ozer NK, Boscoboinik D, Azzi A (1998) Alpha-Tocopherol specifically inactivates cellular protein kinase C alpha by changing its phosphorylation state. Biochem J 334:243–249

Ringman JM, Frautschy SA, Teng E, Begum AN, Bardens J, Beigi M, Gylys KH, Badmaev V, Heath DD, Apostolova LG, Porter V, Vanek Z, Marshall GA, Hellemann G, Sugar C, Masterman DL, Montine TJ, Cummings JL, Cole GM (2012) Oral curcumin for Alzheimer's disease: tolerability and efficacy in a 24-week randomized, double blind, placebo-controlled study. Alzheimers Res Ther 4:43

Ristow M, Zarse K (2010) How increased oxidative stress promotes longevity and metabolic health: the concept of mitochondrial hormesis (mitohormesis). Exp Gerontol 45:410–418

Rodriguez-Rivera J, Denner L, Dineley KT (2011) Rosiglitazone reversal of Tg2576 cognitive deficits is independent of peripheral gluco-regulatory status. Behav Brain Res 216:255–261

Rota C, Rimbach G, Minihane A-M, Stoecklin E, Barella L (2005) Dietary vitamin E modulates differential gene expression in the rat hippocampus: potential implications for its neuroprotective properties. Nutr Neurosci 8:21–29

Rubbo H, Trostchansky A, Botti H, Batthyány C (2002) Interactions of nitric oxide and peroxynitrite with low-density lipoprotein. Biol Chem 383:547–552

Rubbo H, Trostchansky A, O'Donnell VB (2009) Peroxynitrite-mediated lipid oxidation and nitration: mechanisms and consequences. Arch Biochem Biophys 484:167–172

Sabbagh MN, Agro A, Bell J, Aisen PS, Schweizer E, Galasko D (2011) PF-04494700, an oral inhibitor of receptor for advanced glycation end products (RAGE), in Alzheimer disease. Alzheimer Dis Assoc Disord 25:206–212

Saitoh M, Nishitoh H, Fujii M, Takeda K, Tobiume K, Sawada Y, Kawabata M, Miyazono K, Ichijo H (1998) Mammalian thioredoxin is a direct inhibitor of apoptosis signal-regulating kinase (ASK) 1. EMBO J 17:2596–2606

Salvemini D, Doyle TM, Cuzzocrea S (2006) Superoxide, peroxynitrite and oxidative/nitrative stress in inflammation. Biochem Soc Trans 34:965–970

Sayre LM, Smith MA, Perry G (2001) Chemistry and biochemistry of oxidative stress in neurodegenerative disease. Curr Med Chem 8:721–738

Schiavone S, Jacquet V, Trabace L, Krause KH (2012) Severe life stress and oxidative stress in the brain: from animal models to human pathology. Antioxid Redox Signal 18:1475–1490

Schieke SM, von Montfort C, Buchczyk DP, Timmer A, Grether-Beck S, Krutmann J, Holbrook NJ, Klotz LO (2004) Singlet oxygen-induced attenuation of growth factor signaling: possible role of ceramides. Free Radic Res 38:729–737

Schmidt AM, Sahagan B, Nelson RB, Selmer J, Rothlein R, Bell JM (2009) The role of RAGE in amyloid-beta peptide-mediated pathology in Alzheimer's disease. Curr Opin Investig Drugs 10:672–680

Schubert D, Soucek T, Blouw B (2009) The induction of HIF-1 reduces astrocyte activation by amyloid beta peptide. Eur J Neurosci 29:1323–1334

Selcher JC, Nekrasova T, Paylor R, Landreth GE, Sweatt JD (2001) Mice lacking the ERK1 isoform of MAP kinase are unimpaired in emotional learning. Learn Mem 8:11–19

Shan X, Lin C-LG (2006) Quantification of oxidized RNAs in Alzheimer's disease. Neurobiol Aging 27:657–662

Sharma RA, Gescher AJ, Steward WP (2005) Curcumin: the story so far. Eur J Cancer 41:1955–1968

Shen HM, Liu ZG (2006) JNK signaling pathway is a key modulator in cell death mediated by reactive oxygen and nitrogen species. Free Radic Biol Med 40:928–939

Shimizu S, Eguchi Y, Kosaka H, Kamiike W, Matsuda H, Tsujimoto Y (1995) Prevention of hypoxia-induced cell death by Bcl-2 and Bcl-xL. Nature 13:811–813

Siddiq A, Aminova LR, Ratan RR (2007) Hypoxia inducible factor prolyl 4-hydroxylase enzymes: center stage in the battle against hypoxia, metabolic compromise and oxidative stress. Neurochem Res 32:931–946

Singh SP, Wishnok JS, Keshive M, Deen WM, Tannenbaum SR (1996) The chemistry of the S-nitrosoglutathione/glutathione system. Proc Natl Acad Sci U S A 93:14428–14433

Singh S, Vrishni S, Singh BK, Rahman I, Kakkar P (2010) Nrf2-ARE stress response mechanism: a control point in oxidative stress-mediated dysfunctions and chronic inflammatory diseases. Free Radic Res 44:1267–1288

Solecki GM, Groh IA, Kajzar J, Haushofer C, Scherhag A, Schrenk D, Esselen M (2013) Genotoxic properties of cyclopentenone prostaglandins and the onset of glutathione depletion. Chem Res Toxicol 26:252–261

Stadtman ER (2001) Protein oxidation in aging and age-related diseases. Ann N Y Acad Sci 928:22–38

Stadtman ER (2006) Protein oxidation and aging. Free Radic Res 40:1250–1258

Stadtman ER, Berlett BS (1991) Metal-catalyzed oxidation of proteins. Physiological consequences. J Biol Chem 266:17201–17211

Stahl N, Boulton TG, Farruggella T, Ip NY, Davis S, Witthuhn BA, Quelle FW, Silvennoinen O, Barbieri G, Pellegrini S (1994) Association and activation of Jak-Tyk kinases by CNTF-LIF-OSM-IL-6 beta receptor components. Science 263:92–95

Steenken S, Jovanovic SV (1997) How easily oxidizable is DNA? One-electron reduction potentials of adenosine and guanosine radicals in aqueous solution. J Am Chem Soc 119:617–618

Sugiura H, Ichinose M (2011) Nitrative stress in inflammatory lung diseases. Nitric Oxide 25:138–144

Szabó C (2003) Multiple pathways of peroxynitrite cytotoxicity. Toxicol Lett 140–141:105–112

Szweda LI, Uchida K, Tsai L, Stadtman ER (1993) Inactivation of glucose-6-phosphate dehydrogenase by 4-hydroxy-2-nonenal. Selective modification of an active-site lysine. J Biol Chem 268:3342

Tanaka T, Williams RL, Rabbitts TH (2007) Tumour prevention by a single antibody domain targeting the interaction of signal transduction proteins with RAS. EMBO J 26:3250–3259

Thannickal VJ, Fanburg B (2003) The concept of compartmentalization in signaling by reactive oxygen species. In: Forman HJ, Fukuto J, Torres M (eds) Signal transduction by reactive oxygen and nitrogen species: pathways and chemical principles. Kluwer Academic Publishers, Dordrecht, pp 291–310

Thies E, Mandelkow EM (2007) Missorting of Tau in neurons causes degeneration of synapses that can be rescued by the kinase MARK2/Par-1. J Neurosci 27:2896–2907

Thippeswamy T, McKay JS, Quinn JP, Morris R (2006) Nitric oxide, a biological double-faced janus—is this good or bad? Histol Histopathol 21:445–458

Thom SR, Bhopale VM, Milovanova TN, Yang M, Bogush M (2012) Thioredoxin reductase linked to cytoskeleton by focal adhesion kinase reverses actin S-nitrosylation and restores neutrophil beta(2) integrin function. J Biol Chem 287:30346–30357

Thomas SL, Zhong D, Zhou W, Malik S, Liotta D, Snyder JP, Hamel E, Giannakakou P (2008) EF24, a novel curcumin analog, disrupts the microtubule cytoskeleton and inhibits HIF-1. Cell Cycle 7:2409–2417

Townsend M, Mehta T, Selkoe DJ (2007) Soluble Abeta inhibits specific signal transduction cascades common to the insulin receptor pathway. J Biol Chem 282:33305–33312

Tsai YM, Chien CF, Lin LC, Tsai TH (2011) Curcumin and its nanoformulation: the kinetics of tissue distribution and blood–brain barrier penetration. Int J Pharm 416:331–338

Turpaev KT (2002) Reactive oxygen species and regulation of gene expression. Biochemistry (Mosc) 67:281–292

Turrens JF (2003) Mitochondrial formation of reactive oxygen species. J Physiol 552:335–344

Uchida K, Stadtman ER (1992) Selective cleavage of thioether linkage in proteins modified with 4-hydroxynonenal. Proc Natl Acad Sci U S A 89:5611

Valko M, Leibfritz D, Moncol J, Cronin MT, Mazur M, Telser J (2007) Free radicals and antioxidants in normal physiological functions and human disease. Int J Biochem Cell Biol 39:44–84

van Himbergen TM, Beiser AS, Ai M, Seshadri S, Otokozawa S, Au R, Thongtang N, Wolf PA, Schaefer EJ (2012) Biomarkers for insulin resistance and inflammation and the risk for all-cause dementia and Alzheimer disease: results from the Framingham Heart Study. Arch Neurol 69:594–600

Varga T, Czimmerer Z, Nagy L (2011) PPARs are a unique set of fatty acid regulated transcription factors controlling both lipid metabolism and inflammation. Biochim Biophys Acta 1812:1007–1022

Venugopal R, Jaiswal AK (1998) Nrf2 and Nrf1 in association with Jun proteins regulate antioxidant response element-mediated expression and coordinated induction of genes encoding detoxifying enzymes. Oncogene 17:3145–3156

Walsh DM, Hartley DM, Condron MM, Selkoe DJ, Teplow DB (2001) In vitro studies of amyloid β-protein fibril assembly and toxicity provide clues to the aetiology of Flemish variant (Ala692->Gly) Alzheimer's disease. Biochem J 355:869–877

Wang X, Su B, Lee H-G, Perry G, Smith MA, Zhu X (2009) Impaired balance of mitochondrial fission and fusion in Alzheimer's disease. J Neurosci 29:9090–9103

Wang Y, Yin H, Wang L, Shuboy A, Lou J, Han B, Zhang X, Li J (2013) Curcumin as a potential treatment for Alzheimer's disease: a study of the effect of curcumin on hippocampal expression of glial fibrillary acidic protein. Am J Chin Med 41:1–12

Wang P, Su C, Li R, Wang H, Ren Y, Sun H, Yang J, Sun J, Shi J, Tian J, Jiang S (2014) Mechanisms and effects of curcumin on spatial learning and memory improvement in APPswe/PS1dE9 mice. J Neurosci Res 92:218–231

Watson BD (1993) Evaluation of the concomitance of lipid peroxidation in experimental models of cerebral ischemia and stroke. Prog Brain Res 96:69–95

Wei P, Li R, Wang H, Ren Y, Sun H, Yang J, Wang P (2012) Effect of curcumin on synapse-related protein expression of APP/PS1 double transgenic mice. Zhongguo Zhong Yao Za Zhi 37:1818–1821

Winterbourn CC, Hampton MB (2008) Thiol chemistry and specificity in redox signaling. Free Radic Biol Med 45:549–561

Wu JX, Zhang LY, Chen YL, Yu SS, Zhao Y, Zhao J (2015) Curcumin pretreatment and post-treatment both improve the antioxidative ability of neurons with oxygen-glucose deprivation. Neural Regen Res 10:481–489

Wurtmann EJ, Wolin SL (2009) RNA under attack: cellular handling of RNA damage. Crit Rev Biochem Mol Biol 44:34–49

Xiao Z, Zhang A, Lin J, Zheng Z, Shi X, Di W, Zhu Y, Zhou G, Fang Y (2014) Telomerase: a target for therapeutic effects of curcumin and a curcumin derivative in Aβ1-42 insult in vitro. PLos One 9, e101251

Xiong Z, Hongmei Z, Lu S, Yu L (2011) Curcumin mediates presenilin-1 activity to reduce β-amyloid production in a model of Alzheimer's disease. Pharmacol Rep 63:1101–1108

Yamagishi S, Amano S, Inagaki Y, Okamoto T, Koga K, Sasaki N, Yamamoto H, Takeuchi M, Makita Z (2002) Advanced glycation end products-induced apoptosis and overexpression of vascular endothelial growth factor in bovine retinal pericytes. Biochem Biophys Res Commun 290:973–978

Yan LJ, Sohal RS (1998) Mitochondrial adenine nucleotide translocase is modified oxidatively during aging. Proc Natl Acad Sci U S A 95:12896–12901

Yan SD, Chen X, Schmidt AM, Brett J, Godman G, Zou YS, Scott CW, Caputo C, Frappier T, Smith MA (1994) Glycated tau protein in Alzheimer disease: a mechanism for induction of oxidant stress. Proc Natl Acad Sci U S A 91:7787–7791

Yang F, Lim GP, Begum AN, Ubeda OJ, Simmons MR, Ambegaokar SS, Chen PP, Kayed R, Glabe CG, Frautschy SA, Cole GM (2005) Curcumin inhibits formation of amyloid β oligomers and fibrils, binds plaques, and reduces amyloid in vivo. J Biol Chem 280:5892–5901

Yang J, Chen H, Vlahov IR, Cheng JX, Low PS (2006) Evaluation of disulfide reduction during receptor-mediated endocytosis by using FRET imaging. Proc Natl Acad Sci U S A 103:13872–13877

Yang C, Zhang X, Fan H, Liu Y (2009) Curcumin upregulates transcription factor Nrf2, HO-1 expression and protects rat brains against focal ischemia. Brain Res 1282:133–141

Yao M, Nguyen TV, Pike CJ (2005) Beta-amyloid-induced neuronal apoptosis involves c-Jun N-terminal kinase-dependent downregulation of Bcl-w. J Neurosci 25:1149–1158

Yu BP (1994) Cellular defenses against damage from reactive oxygen species. Physiol Rev 74:139–162

Zhang L, Fiala M, Cashman J, Sayre J, Espinosa A, Mahanian M, Zaghi J, Badmaev V, Graves MC, Bernard G, Rosenthal M (2006) Curcuminoids enhance amyloid-β uptake by macrophages of Alzheimer's disease patients. J Alzheimers Dis 10:1–7

Zhang X, Zhou K, Wang R, Cui J, Lipton SA, Liao FF, Xu H, Zhang YW (2007) Hypoxia-inducible factor 1alpha (HIF-1alpha)-mediated hypoxia increases BACE1 expression and beta-amyloid generation. J Biol Chem 282:10873–10880

Zhang C, Browne A, Child D, Tanzi RE (2010) Curcumin decreases amyloid-β peptide levels by attenuating the maturation of amyloid-β precursor protein. J Biol Chem 285:28472–28480

Zhao WQ, De Felice FG, Fernandez S, Chen H, Lambert MP, Quon MJ, Krafft GA, Klein WL (2008) Amyloid beta oligomers induce impairment of neuronal insulin receptors. FASEB J 22:246–260

Zimmermann R, Panzenböck U, Wintersperger A, Levak-Frank S, Graier W, Glatter O, Fritz G, Kostner GM, Zechner R (2001) Lipoprotein lipase mediates the uptake of glycated LDL in fibroblasts, endothelial cells, and macrophages. Diabetes 50:1643–1653

Zingg J-M (2007) Modulation of signal transduction by vitamin E. Mol Aspects Med 28:481–506

Chapter 7
Effects of Curcumin on Neuroinflammation in Animal Models and in Patients with Alzheimer Disease

7.1 Introduction

Curcumin is a natural polyphenol in the rhizomes of turmeric (*Curcuma longa* L.). It has been traditionally used as food flavoring in Asia and Far East. In recent decades lot of work has been published on pharmacological properties of curcumin. Commercially available preparations of curcumin contain at least three curcumin compounds including curcumin, demethoxycurcumin and bisdemethoxycurcumin. Curcumin is the major component and demethoxycurcumin and bisdemethoxycurcumin are minor constituents. The ratio of curcumin:demethoxy-curcumin:bisemethoxycurcumin in commercially available preparation is about 66:23:11. Curcumin has multifunctional pharmacological properties (Gupta et al. 2013), such as antioxidant, antiinflammatory, anti-depressant like, free radical scavenging, antidiabetic, anti-hyperalgesic, anticarcinogenic, antimicrobial, hepatoprotective, thrombosuppressive, and antinociceptive properties (Aggarwal 2010; Ghorbani et al. 2014; Zhu et al. 2014). Converging evidence suggests that curcumin is not only a potent anti-aging agent, but a potent ancient drug, which delays the onset of age-related diseases, such as cardio- and cerebrovascular, neurodegenerative diseases as well as various types of cancers. Aging as well as almost all age-related diseases are accompanied by increase in chronic low grade inflammation as well as oxidative stress (Farooqui 2010, 2014). At the molecular level curcumin not only decreases the expression of NADPH oxidase subunits gp91^{phox} and gp47^{phox}, down-regulates Rac1 activity, and suppresses generation and accumulation of reactive oxygen species (ROS) and advanced glycation end products (AGEs) along with decrease in the expression of receptor for AGE (RAGE) (Tang and Chen 2014), but also lowers chronic inflammation leading to delay in the onset of age-related diseases, such as stroke, Alzheimer disease (AD), and various types of cancers (Farooqui 2012). Curcumin also acts by inhibiting the activity of NFκB and reducing the levels of proinflammatory cytokines (tumor necrosis factor-α, TNF-α; interleukin-1β, IL-1β; interleukin-6, IL-6; and interleukin-8, IL-8). Curcumin not only downregulates activities of inducible nitric oxide synthase (iNOS), matrix metalloproteinase-9

© Springer International Publishing Switzerland 2016
A.A. Farooqui, *Therapeutic Potentials of Curcumin for Alzheimer Disease*,
DOI 10.1007/978-3-319-15889-1_7

(MMP-9), cyclooxygenase-2 (COX-2), and 5-lipoxygenase (5-LOX), and adhesion molecules, but also by simultaneously increasing the level and activity of proteins involved in antioxidative defense (heme oxygenase-1, HO-1) and heat shock proteins, a class of molecular chaperones that bind with nonnative proteins and assist them to acquire native structure and thus prevent misfolding and the aggregation process during the conditions of stress (Surh et al. 2001; Chang 2001; Paul and Mahanta 2014). Various constituents of curcumin with acetylcholinesterase activity and modulate spatial memory with different degree of effectiveness (Ahmed et al. 2010; Ahmed and Gilani 2009). Curcumin produces some inhibitory effect on acetylcholinesterase activity, but produces remarkable effect in scopolamine-induced amnesia. It effectively rescues from Aβ-mediated impairment in LTP (Ahmed and Gilani 2011; Ahmed et al. 2011), supporting the view that each constituent has its own effect on various parameters of memory formation, storage, and anti-inflammatory gene expression.

Curcumin also produces beneficial effects in the vascular system through the elevation of the level of NOS, and sirtuins, enzymes leading to the modulation of cellular homeostasis in the vascular system (Yang et al. 2013). Curcumin also induces pro-survival signaling pathway (Akt/GSK3β) in brains of patients with diabetes (Liu et al. 2010a, b; Wang et al. 2009a), a condition, which is a major risk factor for stroke, AD, and depression (Farooqui 2013). Finally, curcumin can also bind with the vanilloid receptor TRPV1 (transient receptor potential cation channel, subfamily V, member 1) and the aryl hydrocarbon receptor AHR (Yeon et al. 2010; Ciolino et al. 1998). It is reported that curcumin acts by inhibiting the activity of cytochrome P450 family member A1 (CYP1A1) (Rinaldi et al. 2002). Curcumin also increases apoptosis and differentiation of vitamin D-treated tumor cells (Mosieniak et al. 2006). Direct binding of curcumin to the vitamin D receptor (VDR) results in heterodimerization with the retinoic X receptor. This results in translocation of heteromer to the nucleus. In the nucleus, curcumin activates gene transcription of vitamin D target genes (Bartik et al. 2010).

7.2 Neurochemical Aspects of Neuroinflammation in the Brain

Neuroinflammation is an innate immune response of the brain to diverse stimuli, including stress, injury and infection. Neuroinflammation is a neuroprotective process. However, when prolonged neuroinflammation overrides the bounds of physiological control, it becomes destructive. It involves immune cells (lymphocytes, monocytes), and neuroglial cells (microglia, and astrocytes) (Stoll and Jander 1999). These cells interact with each other to clear out the causes of the injury, limiting ongoing infection and/or compartmentalize damaged tissue, and to initiate tissue repair aimed at restoring normal physiological function (Nathan 2002). Neuroinflammation involves processes are responsible for sensing, transducing, amplifying, and turning off the mechanisms that involve the participation of various lipid mediators. Neuroinflammation not only isolates the injured brain tissue from uninjured areas, destroys injured cells and rebuilds the extracellular matrix (Minghetti 2005), but also disrupts the blood-brain barrier (BBB), allowing cells from the hematopoietic system to leave the blood stream and come in contact to the injury site (Lossinsky and Shivers 2004).

Among neural cells, microglial cells provide brain with constant support to surrounding neurons. Microglial cells have a high level of plasticity allowing them to change their shape and function in response to environmental cues (Saijo and Glass 2011). After injury or over time with the aging, the morphology of glial cells is progressively altered. Activation of microglia in the brain is accompanied by the release of proinflammatory mediators where as activation of astrocytes is believed to be necessary for containing the immune response, repairing the BBB, and attenuating further neuronal death (Lossinsky and Shivers 2004; Bush et al. 1999; Franceschi et al. 2000; Farooqui et al. 2007). A major objective of neuroinflammation is to neutralize the causative factors (stress, injury and infection) and ultimately bring about healing of the lesion by recruiting lymphocytes, monocytes and macrophages of the hematopoietic system and activating glial cells (microglia and astrocytes) (Minghetti 2005). The activated microglia and astrocytes not only produce inflammatory molecules (cytokines and chemokines), but also secret neurotransmitters, ROS, and nitric oxide (NO) (Farooqui 2014). The released cytokines (TNF-α, IL-1β, and IL-6) and chemokines (CXCL3, CX3CL1 CD47, CXCL12, CXCR4, and CD200; and MCP1) are the major effectors of the neuroinflammatory signals (Allan and Rothwell 2003; Batlle et al. 2015). These cytokines and chemokines have been reported to modulate neurochemical and neurophysiologic mechanisms associated with cognitive function and memory formation (Gemma and Bickford 2007). Cytokines and chemokines establish a feedback loop, to activate more astrocytes and microglia leading to further generation of arachidonic acid-derived enzymic and non-enzymic lipid mediators, such as prostaglandins, leukotrienes, and thromboxanes; isoprostanes, 4-hydroxynonenals, acrolein, and malondialdehyde (Farooqui and Horrocks 2006). Microglial cells express a repertoire of various receptors such as TREM2, FcγRs, MHC-II, CD200R, RAGE, CX3CR1 (fractalkaline), CXCR3 and 4, purinergic receptors, Toll-like receptors 2 and 4, galectins 1 and 3, scavenger receptors (e.g., CD36), CD47, integrins and SIRPα (Hu et al. 2015). Through these receptors, microglial cells provide both pro-inflammatory and anti-inflammatory response, in a varying range depending on the signals dictated by their environment (Hu et al. 2014). As stated above, the secreted inflammatory molecules also recruit other cells such as monocytes and lymphocytes to cross the blood brain barrier (BBB) to enhance neuroinflammation in the CNS (Das and Basu 2008). Among others, Toll-like receptors (TLRs) on macrophages are involved in pathogenic pattern recognition to generate immunologically relevant responses (Takeuchi and Akira 2010). The generation of inflammatory mediators ROS, NO, and prostaglandin E_2 along with the release of TNF-α is the prototypical response underlying acute phase of inflammation (Farooqui and Horrocks 2006). Healthy neurons and glial cells contain millimolar concentrations of ATP within presynaptic vesicles and granules (Abbracchio et al. 2009). Neuronal ATP serves as a neurotransmitter while astrocytic ATP allows distant astrocytes to communicate with each other and modulate neuronal response. However, the release of ATP from neurons or astrocytes is very low (nanomolar range). This steady state balance is disrupted in neurological disorders when damaged neurons and chronically activated glial cells release dramatic levels of ATP, uridine triphosphate (UTP) and other intracellular nucleotides leading to the activation of P_2 type of purinergic receptors on microglial cells (Abbracchio et al. 2009). According to the two-hypothesis of neuronal injury "an initial localized neuroinflammation is induced and supported by the constitutively active COX-1 and generation of PGE_2 (first hit) but the changes in the growth factor-, cytokine levels during injury lead not only to the activation of inducible COX-2 resulting in the massive release of PGE_2, but also in the simultaneous release of ATP from injured cells (second hit), which significantly enhances the inflammatory response through increased synthesis of PGE_2" (Fiebich et al. 2014; Lourbopoulos et al. 2015). Collectively, these studies suggest that neuroinflammation is a multiple factorial process, which is mediated by the activation of inflammatory and immune cells.

Neuroinflammation is different from the peripheral inflammation not only in the initiation, but also in tissue sensitivity. The brain is immune privileged organ because of the presence of blood-brain barrier (BBB), which only allows certain cells and molecules to enter and exit the brain (Griffiths et al. 2007; Hickey 2001). The inflammatory response in the brain is not only supported by an increase in ROS generation and the release of pro-inflammatory cytokines and chemokines. The release of above factors not only disrupts the BBB, but also promote sustain recruitment of activated leukocytes and microglia leading to the creation of a positive feedback loop, which prolongs inflammation and contributes to ongoing neuronal damage (Iadecola and Anrather 2011; Ziebell and Morganti-Kossmann 2010). Dependent upon the time and severity of the inflammation, it is apparent that the inflammatory response in the acute phase is widely shown to be detrimental while inflammation at the chronic phase may be essential for repair and regeneration (Bowen et al. 2006).

Inflammation in the brain and visceral tissue can be induced not only by infection or trauma, but also by the chronic over-nutrition (Fig. 7.1) (Cai 2013). Over-nutrition results in persistent low grade inflammation in the circulation and peripheral metabolic tissues disrupting the metabolic homeostasis of the body. Thus, under certain conditions, when inflammatory response is unable to repair the tissue damage. For example chronic increase in glucocorticoid receptors activation can turn into a chronic condition, with continuous high levels of triglycerides and cholesterol. These biochemical changes contribute to obesity, insulin resistance, glucose intolerance, hyperlipidemia, atherosclerosis and hypertension. Collectively, this condition is known as metabolic syndrome (Cai 2009; Lumeng and Saltiel 2011; Farooqui 2013). Hypothalamus plays an important regulatory role in maintaining metabolic homeostasis and over-nutrition-mediated chronic inflammatory changes in the brain. These changes can disrupt neurohormone- and neurotransmitter-mediated central regulatory functions to propagate insulin resistance, obesity, and chronic peripheral and neurological disorders (Thaler and Schwartz 2010; Cai 2009, 2013; Cai and Liu 2012; Farooqui 2013). At the molecular level, overnutrition-mediated

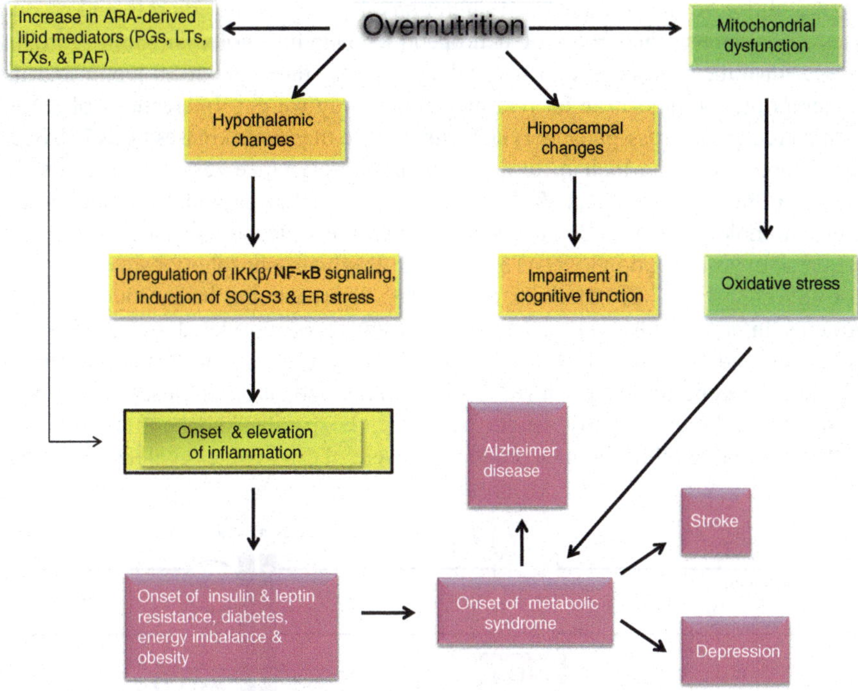

Fig. 7.1 Overnutrition-induced changes in hypothalamus and hippocampus. Arachidonic acid (*ARA*); prostaglandins (*PGs*); leukotrienes (*LTs*); thromboxanes (*TXs*); platelet activating factor (*PAF*); endoplasmic reticulum (*ER*); inhibitor of nuclear factor kappa-B kinase subunit beta (*IKKβ*); nuclear factor kappa-light-chain-enhancer of activated B cell (*NF-κB*); suppressor of cytokine signaling 3 (*SOCS3*)

proinflammatory axis involves upregulation of IκB kinase-β (IKKβ) and its downstream nuclear transcription factor NF-κB (IKKβ/NF-κB signaling) in the hypothalamus leading to chronic energy imbalance and changes in fat mass and body weight (Fig. 7.1) (Li et al. 2012; Oh et al. 2010; Farooqui 2013). This supports the view that NF-κB is a centrally important regulator of transcription that mediates immune cell communication and inflammatory responses (Gabuzda and Yankner 2013; Zhang et al. 2013). The activation of NF-κB in microglial cells promotes the secretion of TNF-α, which not only stimulates NF-κB signaling in neurons of the medial basal hypothalamus, but also feeds the forward loop sets up of a new homeostatic state in the aging hypothalamus, leading to epigenetic changes in neuroendocrine genes. NF-κB-mediated epigenetic repression of the gonadotropin-releasing hormone (GnRH) gene may contribute to multiple systemic alterations, such as decline in muscle strength, skin atrophy and reduction in neurogenesis along with memory impairment (Fig. 7.2). These changes in homeostatic state may be closely associated with the release of proinflammatory cytokines (Zhang et al. 2013; Farooqui 2014, 2015).

In addition to above mentioned alterations, hypothalamus controls and modulates the synthesis and roles of neuropeptides (leptin, cholecystokinin, ghrelin, orexin, insulin, neuropeptide Y) signaling. Among these peptides, leptin and its receptors play a major role in food intake and body weight. Interactions of leptin with its receptors cross-phosphorylate and activate the Janus kinases (JAK1, JAK2, JAK3, and TYK2), which in turn phosphorylates tyrosine residues in cytosolic domain (Ghilardi and Skoda 1997). This phosphorylation provides binding motifs for src homology 2 (SH2)-domain containing proteins such as signal transducer and activator of transcription3 (STAT3) and SH-2-domain-phosphotyrosine phosphatase (SHP-2) (Fig. 7.2) (Bjorbaek et al. 1999; Bjorbaek and Kahn 2004). STAT3 proteins interact with Y1138, become tyrosine phosphorylated by JAK2, then dissociate and form dimers in the cytoplasm, finally migrating to the nucleus to regulate the gene transcription (Li 2008). STAT3 activation is closely associated with the regulation of body weight by leptin, as specific knockout (KO) of the Y1138 residue of ObRb in mice producing severe obesity (Bates et al. 2003, 2005).

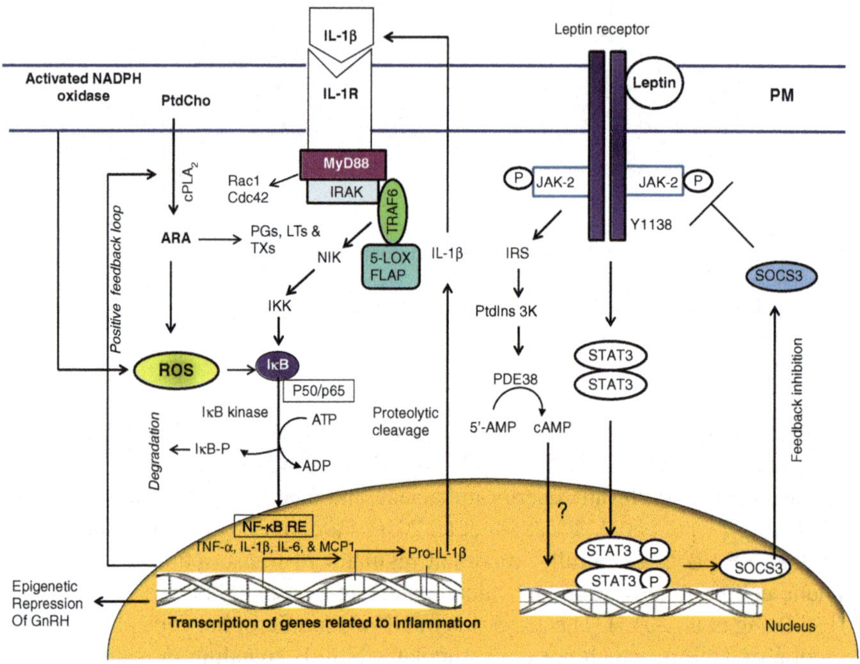

Fig. 7.2 Interactions between IL-1β and IL-1β receptors along with modulation of leptin signaling in the hypothalamus. Plasma membrane (*PM*); phosphatidylcholine (*PtdCho*); cytosolic phospholipase A₂ (*cPLA₂*); arachidonic acid (*ARA*); adaptor protein (*MyD88*); 5-lipoxygenase (*5-LOX*); 5-LOX activating protein (*FLAP*); reactive oxygen species (*ROS*); tumor necrosis factor receptor-associated factor adaptor protein 6 (*TRAF6*); NF-κB-inducing kinase (*NIK*); IκB kinase (*IKK*); NF-kappaB (*NF-κB*); NF-kappaB response element (*NF-κB-RE*); inhibitory subunit of NF-κB (*IκB*); tumor necrosis factor-alpha (*TNF-α*); interleukin-1beta (*IL-1β*); interleukin-6 (*IL-6*); Janus kinase 2 (*JAK2*); and signal transducer and activator of transcription 3 (*STAT3*). phosphatidylinositol-3-kinases (*PtdIns 3K*); and phosphodiesterase 3β (*PDE3β*)

Leptin and JAK signaling is closely associated with regulation of food intake as it relates to caloric and nutrition requirements (Dietrich and Horvath 2009; Blouet and Schwartz 2010; Williams et al. 2011). Collectively, these studies suggest that food intake is regulated by a complex signaling network involving hypothalamic as well as extra-hypothalamic structures in the brain (Williams et al. 2011). Peripheral signals from adipose stores communicate with neurons within the arcuate nucleus of the hypothalamus and modulate food intake. When fat stores are reduced and energy levels are low, hunger signals are induced through elevation in the ghrelin, and decrease in levels of insulin, glucose, leptin and cholecystokinin. In contrast, high levels of glucose, insulin, cholecystokinin and reduction in levels of ghrelin result in an increase in pro-opiomelanocortin (POMC). POMC increases α-melanocyte-stimulating hormone (α-MSH) leading to termination of feeding signals (Coll et al. 2007; Valassi et al. 2008; Abizaid and Horvath 2008). These observations suggest that interactions between hypothalamic and gut neuropeptides control food intake.

Overnutrition-mediated inflammation in hypothalamus is also supported by the activation of Toll-like receptor4 (TLR4) (see below), induction of endoplasmic reticulum stress, and induction of SOCS3 along with other intracellular inflammatory signals associated with high levels of circulating n-6 fatty acid (arachidonic acid) and saturated fatty acids (palmitic acid) (Fessler et al. 2009; Zhang et al. 2008; Thaler et al. 2012). These fatty acids exacerbate the inflammatory response through the production of inflammatory cytokines (Farooqui 2013). Proinflammatory cytokines also signal through the JAK/STAT pathway, which leads to DNA binding and transcriptional modulation by the activated STAT proteins (Murray 2007). STAT proteins have ability to recruit DNA binding and transcriptional regulatory molecules to specific DNA sequences. This may be another signaling pathway through which cytokine exposure may modulate epigenetic modifications across the genome (Vahedi et al. 2012; Hedrich et al. 2014; Li et al. 2014). It is interesting to note that onset and development of inflammation in visceral tissues as a result of overnutrition take weeks to months. In contrast, the onset of hypothalamic inflammation in rats and mice occurs within 1–3 days after overnutrition prior to substantial weight gain. Hypothalamic inflammation is also accompanied by reactive gliosis involving both microglial and astroglial cell populations along with increase in markers of neuron injury (TNF-α, IL1-β, and IL-6) within a week (Horvath et al. 2010; Thaler et al. 2012; Farooqui 2013). The presence of high levels of n-6 fatty acids in phospholipids may contribute to the generation of high levels of proinflammatory eicosanoids (prostaglandins, leukotrienes, and thromboxanes) and platelet activating factor, which along with increased expression and secretion of proinflammatory cytokines. These mediators may support and intensify inflammation in the hypothalamus. A recent study has indicated that hypothalamic inflammation accelerates aging and shortens lifespan in mice (Zhang et al. 2013). Long term consumption of high calorie diet (western diet) produces persistance inflammation and gliosis in the mediobasal hypothalamic region (Thaler and Schwartz 2010; Thaler et al. 2012; Farooqui 2015). This information in rodents is supported by MRI studies in humans, which indicate that increase in inflammation and gliosis also occurs in the mediobasal hypothalamic region of obese humans (Thaler et al. 2012). Converging evidence suggests that interplay among over-nutrition-mediated chronic neuroinflammation, mitochondrial dysfunction, increase in ROS production, and activation of the NLRP3 "inflammasome" along with decline in GnRH level may lead to the release of more proinflammatory cytokines (TNF-α, IL-1β) contributing to the pathogenesis of age-related neurological disorders (Thaler and Schwartz 2010; Thaler et al. 2012; Newsholme et al. 2010; Zhang et al. 2013; Farooqui 2014, 2015).

Overnutrition also produces changes in hippocampal morphology/plasticity and impairment of cognitive function in normal rats (Granholm et al. 2008; Stranahan et al. 2008). Overnutrition-mediated changes in hippocampal morphology/plasticity are of considerable interest because this region is involved in learning and memory formation. At the molecular level these impairments may involve changes in glutamate-receptor subtypes, second-messenger systems, and protein kinases (Farooqui 2013). Furthermore, overnutrition-induced neuroinflammation inhibits neurogenesis in the adult hippocampus (Monje et al. 2003). The molecular mechanism, function and significance of the modulation of neurogenesis during inflammatory processes are not fully understood. However, it is suggested that immune cells release interleukins and NO, which negatively regulate adult neurogenesis and may contribute and support the molecular mechanisms of inflammatory reactions on adult neurogenesis (Packer et al. 2003). Without inflammatory responses, brain tissue will be damaged by the generation of above mentioned lipid mediators. Two types of neuroinflammation (acute inflammation and chronic inflammation) have been reported to occur in the brain.

7.2.1 Acute Neuroinflammation and Brain Damage

The acute neuroinflammation develops rapidly with the onset of pain. It involves rapid activation of microglia and astrocytes, damage to the BBB and acute upregulation of proinflammatory cytokines along with recruitment of neutrophils, and monocytes (Table 7.1). At the molecular level, acute inflammation is accompanied by rapid increase in glutamate, overactivation of glutamate receptors, rapid increase in intracellular calcium, stimulation of calcium-dependent and calcium independent phospholipases A_2 (PLA_2), rapid increase in levels of prostaglandins (PGs), leukotrienes (LTs), and thromboxanes (TXs) due to increase in activities of cyclooxygenase-2 (COX-2) and 5-lipoxygenase (5-LOX), the enzymes, which oxidize arachidonic acid into inflammatory lipid mediators. During inflammatory response, activated microglia and astrocytes rapidly assemble around the injury site in an attempt to protect and repair the injured brain through the process of resolution (Serhan 2009; Farooqui 2011).

The resolution of neuroinflammation is complex process, which involves several distinct cellular mechanisms (Serhan 2009; Farooqui 2011). Clearance of damaged cells is critical for resolution. It is mediated in part via the non-phlogistic recruitment of monocytes that, as macrophages, participate in the phagocytosis of apoptotic cells

Table 7.1 Differences between acute and chronic inflammation

Features	Acute inflammation	Chronic inflammation
Onset	Rapid onset	Delayed onset, persistent and long-lasting
Duration	Short duration (days) with heat, swelling, loss of function in the affected area, host-protective, normally self-limited	Long duration (months, years), below the limit of detection, sustained detrimental
Causative agents	Pathogens invasion, ischemia, trauma, immune response	Persistent infection, long term consumption of high calorie, diet, autoimmunity
Vascular changes	Vasodilation	–
Cellular events	Activation of neutrophils and polymorphonuclear leukocytes, vascular leakage, edema	T cells, B cells, macrophages, fibroblast
Primary mediators	Vasoactive amines (serotonin, histamine) PGs, LTs, TXs, and their receptors	TNF-α, IL-1β, and their receptors, ROS, RNS

Table 7.2 Factors and mediators associated with the resolution of neuroinflammation

Factors	Reference
Lipoxins	Serhan et al. (2008) and Recchiuti and Serhan (2012)
Resolvins	Serhan et al. (2008) and Recchiuti and Serhan (2012)
Protectins	Serhan et al. (2008) and Recchiuti and Serhan (2012)
Maresins	Serhan et al. (2008) and Recchiuti and Serhan (2012)
Annexin A1	El Kebir and Fileps (2013) and Headland and Norling (2015)
Galactins	Norling et al. (2009) and Headland and Norling (2015)
Hydrogen sulfide	Wallace et al. (2012) and Headland and Norling (2015)
Adenosine	Desai and Leitinger (2014) and Headland and Norling (2015)

(Schwab et al. 2007). At the molecular level, the resolution of inflammation is orchestrated not only by the generation of lipid mediators derived from docosahexaenoic acid, Ca^{2+} and phospholipid-binding proteins (PLA_2 inhibitory protein), but also by gaseous mediators (H_2S), a purine, and other neuromodulators, which are under the control of the vagus nerve (Table 7.2) (Serhan 2009; Farooqui 2011; Headland and Norling 2015). Neuroinflammation is also counteracted by naturally occurring lipid signaling molecules such as the N-acylethanolamines, N-arachidonoylethanolamine (an endocannabinoid), and its congener N-palmitoylethanolamine (palmitoylethanolamide or PEA). PEA contributes to the maintenance of cellular homeostasis when faced with external stressors provoked by neuroinflammation (Skaper et al. 2015). Furthermore, neuroinflammation is also modulated by the synthesis of resolvins, neuroprotectins, and maresins, a group of lipid mediators derived from 15-LOX-mediated oxidation of docosahexaenoic acid (Lawrence and Gilroy 2007; Serhan et al. 2008). These mediators also reduce oxidative stress-induced apoptosis by modulating pro-inflammatory gene expression and the Bcl-2 family of proteins (Farooqui 2011; Bazan et al. 2012). Neuroprotectin D1 also downregulates cyclooxgenase-2 expression and limit neuroinflammation-mediated damage during epileptogenesis (Farooqui 2011; Bazan et al. 2012). Lipoxins, a group of arachidonic acid-derived lipid mediators also inhibit LPS-mediated production of NO, IL-1β and TNF-α in a concentration-dependent manner via NF-κB, ERK, p38 MAPK and AP-1 signaling pathways in LPS-activated microglia (Wang et al. 2011). Lipoxins also block IL-1β and TNF-α-mediated upregulation of intercellular cell adhesion molecule-1 (ICAM-1) and promote resolution of inflammation (Chinthamani et al. 2012).

Acute inflammation occurs in neurotraumatic diseases, such as stroke, traumatic brain injury, and spinal cord injury (Farooqui 2014). As stated above, both microglial cells and astrocytes participate in acute neuroinflammation. Microglial cells account for 10 % of total glial cell population in the brain. They are referred as resident macrophages and representative of the brains innate immune system. Their expression of MHC antigens, T- and B-lymphocyte markers and other immune cell antigens in the relatively immune privileged brain couples microglia to the adaptive immunity mediated by lymphocytes. Under physiological conditions, they play essential roles in brain circuit maturation during development participating in the precise refinement of synaptic connections. In the developing brain and in areas of remodeling, microglial cells are responsible for the phagocytosis of cellular debris, resulting from apoptosis and normal cell death (Aloisi 2001). Microglia contribute to phagocytosis of debris present in the extracellular space, including damaged cells, plaques, and foreign matter. For microglia surrounding neurons, subtypes of microglia can provide trophic support to neurons through the release of nerve growth factors, BDNF, and other neurotrophic factors (Elkabes et al. 1996; Parkhurst et al. 2013) supporting the view that microglia are capable of assisting in the synaptic plasticity. In addition, microglia also promote the refining the neuronal circuit by pruning synapses and axonal terminals (Tremblay and Majewska 2011; Parkhurst et al. 2013; Salter and Beggs 2014). These studies support the view that in addition to immune surveillance and response, microglial cells perform a number of additional distinct functions compared to immune cells in the blood. Under pathological conditions, microglial cells are dysregulated and their overactivation can be detrimental for neuronal cells due to the release excitatory amino acids and cytokines (Block and Hong 2005). Uncontrolled microglia responses may be harmful to survival of injured neurons if their activation supersedes threshold of tolerability, resulting in damage rather than a defensive sentinel role afflicted by excessive neuroinflammation (Farooqui 2011, 2014). These events briefly involve uncontrolled phagocytosis, induction of T-cell response, secretion of pro-inflammatory neurotoxic molecules and short-lived potentially cytotoxic species, NO and ROS which inevitably contribute to oxidative stress and mitochondrial dysfunction. Microglial cells also secrete a variety of inflammatory mediators including cytokines (TNF-α, and interleukins IL-1β and IL-6) and chemokines (CXCL3, CX3CL1 CD47, CXCL12, CXCR4, and CD200; and MCP1), which initiate, promote and maintain excessive neuroinflammation (Block and Hong 2005; Tansey and Wyss-Coray 2008). Overactivation of microglial cells and excessive neuroinflammation not only result into uncontrolled phagocytosis, induction of T-cell response, but also secretion of pro-inflammatory neurotoxic molecules, which initiate and contribute to oxidative stress and mitochondrial dysfunction.

Astrocytes are specialized neural cells that outnumber neurons by over fivefold. They perform many essential complex functions in the healthy brain. Astrocytes respond to all forms of brain insults through a process called as reactive astrogliosis, which is a pathological hallmark of brain structural lesions. In response to acute injury, astrocytes undergo cellular alterations including swelling, hypertrophy (astrogliosis) and proliferation (astrocytosis), characterized by increased expression of glial fibrillary acidic protein (GFAP). Increased expression of GFAP leads to the formation of a glial scar (Bush et al. 1999; Schiff et al. 2012; Reali et al. 2005; Szmydynger-Chodobska et al. 2012). Microarray analysis of the astrocyte transcriptome in the aging brain has not only indicated the dysregulation of genes involved in proliferation and apoptosis, but also alteration in regulation of intracellular signaling pathways including insulin growth factor signaling, phosphatidylinositol 3-kinase (PtdIns 3 K)/Akt, and mitogen-activated protein kinase (MAPK) pathways (Zhou et al. 2007; Simpson et al. 2011). Astrocytes also release another factor called glia calcium-binding protein S100B. S100B controls microglial cell activity (van Eldik and Wainwright 2003). Under normal conditions at low levels S100B acts as a neurotrophic factor not only by interacting with a wide array of proteins, such as enzymes, scaffold/adaptor proteins, transcription factors, ubiquitin E3 ligases, but also by stimulating the glutamate uptake (Tramontina et al. 2006). At high concentrations S100B binds with RAGE and induces microglial activation that leads to brain damage (Blais and Rivest 2004), supporting the view that secreted S100B may be involved in astrocyte-microglia interactions and cross talk, with an important role in the initial phase of brain injury. Many studies have indicated that reactive astrocytes exert both pro- and anti-inflammatory regulatory functions in vivo through the involvement of specific molecular signaling pathways (Sofroniew 2009). For example, reactive astrocytes may exert pro-inflammatory roles at early times after insults and in the center or immediate vicinity of lesions, but produce anti-inflammatory effects at later times and at the borders between lesions and healthy tissue. In cases of severe injury, reactive astrocytes form scars, which act as cell migration barriers around the borders of areas where intense inflammation is needed and thereby restrict the spread of inflammatory cells and infectious agents into adjacent healthy tissue (Sofroniew 2005, 2009).

During acute neuroinflammation, both activated microglial and astroglial cells also release excessive amount of NO, which causes neuronal cell death by damaging the mitochondrial electron transport chain function in neurons leading to disruption in neuronal ATP synthesis and increase in generation of ROS (Moncada and Bolaños 2006). The activation of NADPH oxidase in microglia results in the generation of both superoxide ($O_2^{\cdot-}$) and releases proinflammatory TNF-α. NO produced in microglia or astrocytes may react with $O_2^{\cdot-}$ to generate the neurotoxic peroxynitrite radical ($ONOO^-$) (Bal-Price et al. 2002). Thiol oxidation and nitration of tyrosine residues are the major mechanism by which $ONOO^-$ induces conformational change in proteins (Bal-Price et al. 2002; Abramov et al. 2005; Brown and Bal-Price 2003; Jang and Han 2006). Peroxynitrite also causes oxidative damage to mitochondrial structural proteins and enzymes and peroxidative damage to lipids within membranes leading to profound changes in function and membrane integrity

(Bal-Price et al. 2002; Abramov et al. 2005; Brown and Bal-Price 2003; Jang and Han 2006). Peroxynitrite inhibits mitochondrial respiration by inactivation of ETC I and III (Boczkowski et al. 2001). ONOO⁻ also disrupts the ferrous-sulfur active site of the tricarboxylic acid cycle enzyme aconitase, leading to its inhibition and impairing ATP production (Han et al. 2005). The enzyme nicotinamide nucleotide transhydrogenase catalyzes the reduction of NAD, which is another crucial mitochondrial enzyme readily inactivated by peroxynitrite-mediated nitration and oxidation (Forsmark-Andrée et al. 1996). Inactivation of mitochondrial electron transport enzymes increases the mitochondrial production of superoxide and hydrogen peroxide creating adaptive and synergistic damage. ONOO⁻ also inhibits induces caspase-dependent neuronal apoptosis, and promotes glutamate release resulting in excitotoxicity and neuronal death (Bal-Price et al. 2002; Abramov et al. 2005; Brown and Bal-Price 2003).

$O_2^{\cdot-}$ and/or ONOO⁻ also modulate COX enzymes (COX-1 and COX-2) in the development of inflammatory pain sensitivity (Mollace et al. 2005). Although the molecular mechanisms by which NO activates COX enzymes are not fully understood. It is suggested that ONOO⁻ is involved in this activation through the oxidative inactivation and/or modification of key amino acids residues in the COX polypeptide (Landino et al. 1996). In addition, ONOO- also increases the production of PGs from macrophages by acting post-transcriptionally or translationally to increase COX-2 protein levels or to increase its mRNA stability, at least in part through $O_2^{\cdot-}$ and the p38 MAPK pathway (Habib et al. 1997). Furthermore, iNOS binds COX-2, and iNOS-derived NO increases the catalytic activity of COX-2 through S-nitrosylation in a macrophage cell line (Kim et al. 2005a). The induction and onset of neuroinflammation are not only associated with interactions among microglia, astrocytes, neurons, PMN, and endothelial cells but also cross talk among various proinflammatory mediators derived from enzymic and nonenzymic degradation of phospholipids, sphingolipid, and cholesterol (Farooqui et al. 2007). Collective evidence suggests that neuroinflammation is a double edge sword. On one side, it produces beneficial effects by promoting neuronal survival, tissue homeostasis and resolution through the generation of lipoxins, neuroprotectins, resolvins, and marsins and on the other side the onset of intense neuroinflammation produces brain damage through the over-production of inflammatory mediators (PGs, LTs, TXs, PAF, TNF-α, IL-1β, and IFN-γ). These inflammatory mediators bind to their own receptors on glial and neuronal cells and with the support from inflammatory redox sensitive transcription factor (NF-κB) intensify and maintain neuroinflammation (Farooqui 2014).

7.2.2 Chronic Neuroinflammation and Brain Damage

Chronic neuroinflammation lingers for years causing damage to brain tissues because immune system is unable to detect it. Chronic neuroinflammation is also supported by activities of microglia and astrocytes. In mammals, during aging microglial cells undergo replicative senescences characterized by telomere shortening and changes in homogenous distribution of microglial cells throughout the brain (Franceschi et al. 2000; Hefendehl et al. 2014). These processes not only lead to dysregulated response of microglial cells to injuries, changes in neuroprotective functions, and an increase in neurotoxic inflammatory responses, but also in the formation of glial lipofuscin granules (Damani et al. 2011; Hefendehl et al. 2014). In the mature brain, microglia exist in a resting state characterized by ramified morphology and are responsible for host defense and tissue repair (Nimmerjahn et al. 2005). However, activation of microglia by brain injury or immunological stimuli is accompanied by morphological changes from ramified into an amoeboid type. Overactivated microglia express proinflammatory cytokines (TNF-α, IL-1β, IL-6), ROS and RNS. Multiple signaling pathways such as protein kinases (mitogen-activated protein kinases (MAPKs), protein kinase C (PKC) and phosphoinositide 3 kinase (PtdIns 3 K)/Akt) and activation of NF-κB contribute to the expression of proinflammatory cytokines in microglial activation. As stated above, activation of NFκB promotes transcription of proinflammatory genes that include cell adhesion molecules, such as intercellular adhesion molecule-1 and vascular cell adhesion molecule 1 (VCAM-1), enzymes, such as iNOS, COX-2, and proinflammatory cytokines. High levels of proinflammatory cytokines contribute to neuronal damage and neurodegenerative processes (Block et al. 2007). Activated microglial cells are the major source of inflammation in the brain of Alzheimer disease (AD) patients (Van Eldik et al. 2007). In animal models of AD, accumulation of Aβ oligomer not only triggers an increase in proinflammatory cytokine TNF-α levels, but also orchestrate neuronal stress mechanisms, which impair brain insulin signaling (Fig. 7.3) (Bomfim et al. 2012; Farooqui 2013), initiate synaptic alterations, and induce cognitive dysfunction (Lourenco et al. 2013). This cascade involves stress kinases (PKR, JNK, IKKα) and initiation factor, eukaryotic Initiation Factor 2 (eIF2) in the brains animal models of AD (Lourenco et al. 2013). The activation of neuronal stress-related protein kinases results in excessive phosphorylation of eIF2α, which plays a key role in control of protein translation, causing synapse dysfunction and memory loss. The precise mechanisms linking stress kinases-mediated metabolic stress at the synapses is not fully understood. However, increase in phosphorylation of eIF2α-P, impairments LTP (Ma et al. 2013) and synapse loss (Lourenco et al. 2013) in mice models of AD have been reported to occur. Restoration of normal brain level of eIF2α-P results in abrogation of synaptic proteins and cognition (Lourenco et al. 2013; Ma et al. 2013), supporting the view that there is a tight connection between phosphorylation of eIF2α-and synapse/memory integrity. Reduction in neuroinflammation has been reported to counteract memory deficits in AD mouse models (Kiyota et al. 2010; Bachstetter et al. 2012).

In AD, activated microglia diffusely scattered throughout the cerebral cortex, and focally concentrated around Aβ plaques (Streit 2004). Microglial cells express a large number of receptors allowing them to respond to several cytokines and chemokines signals from other cells circulating in blood and tissue. These cytokines and chemokines play an important role in maintaining local brain homeostasis and preventing Aβ_{42}-mediated synaptic and inflammatory injury. Notably, clearance of

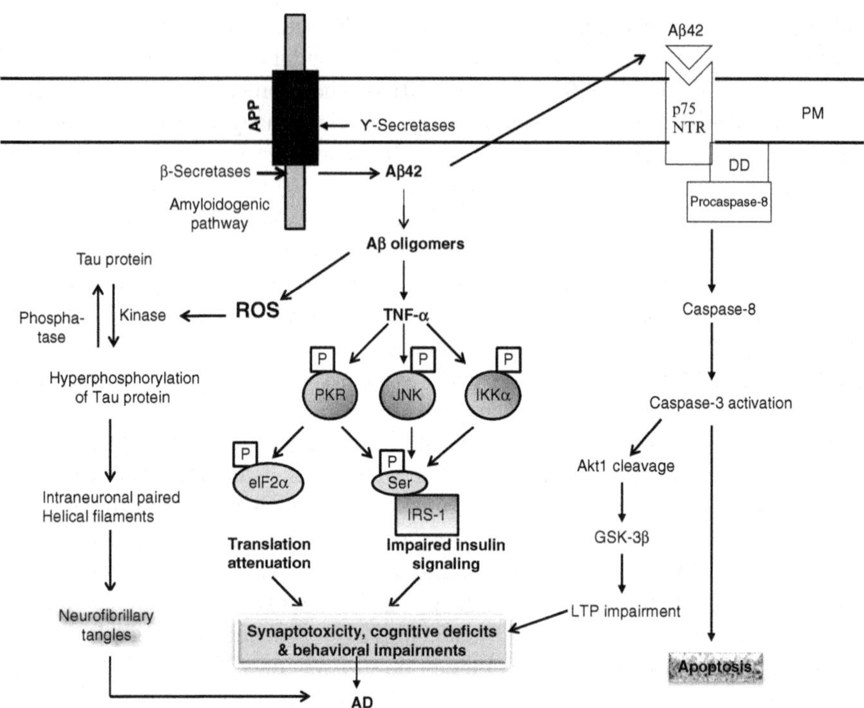

Fig. 7.3 Aβ oligomer, TNF-α, and stress kinase-mediated metabolic changes in brain of animal model of AD. Plasma membrane (*PM*); Amyloid precursor protein (*APP*); monomeric β-amyloid peptide (*Aβ*); tumor necrosis factor-α (*TNF-α*); double-stranded RNA-dependent protein kinase (*PKR*); c-jun N-terminal kinase (*JNK*); IkappaB kinaseα (*IKKα*); eukaryotic initiation factor 2 (*elf2*); insulin receptor substrate-1 (*IRS-1*); death domain (*DD*); and p75 neurotrophin receptor ($p75^{NTR}$)

accumulating Aβ is dependent on effective sensing by microglia (mediated by chemokines), followed by the removal and degradation of Aβ. The prolonged exposure of microglia to proinflammatory cytokines along with accumulation of Aβ peptide causes microglia to lose their normal abilities to clear toxic proteins and control inflammation (Hickman et al. 2008; Krabbe et al. 2013). Thus, microglia cells have emerged as critical regulators of innate immune responses in AD (Nimmerjahn et al. 2005). Recent studies have indicated more sophisticated functions of microglial cells going beyond immune surveillance. Of particular interest is the involvement of microglia in plasticity and maintenance of the adult brain by secreting BDNF (Parkhurst et al. 2013) and refining the neuronal circuit by pruning synapses and axonal terminals (Tremblay and Majewska 2011; Parkhurst et al. 2013; Salter and Beggs 2014). Similar to neuronal BDNF, microglial BDNF also acts on neuronal TrkB and modulate glutamatergic synaptic transmission and plasticity (Rex et al. 2007). Microglial BDNF may also affect inhibitory synaptic transmission via TrkB signaling in the hippocampus (Zheng et al. 2011). An important function of BDNF release from microglia is the modulation of ATP binding to the purinergic receptor (P2X$_4$R). It is suggested that ATP, which is released at sites of active synaptic transmission (Khakh and North 2012), acts as a robust chemoattractant for microglial processes (see below) (Davalos et al. 2005). Microglial cells are also involved in the assembly and activation of the inflammasomes, which are intracellular signaling platforms that detect a set of substances emerging during tissue damage, metabolic imbalances and infections and proteolytically activate the highly proinflammatory cytokines IL-1β and IL-18 (Nunes and de Souza 2013). These multimeric protein complexes usually consist of three partners: an inflammasome sensor protein, which can be a PAMP- or DAMP-detecting module in the form of a NLR, such as NLRP3 and NLRC4, or an endogenous DNA (released from mitochondria) detecting module such as AIM2 (absent in melanoma 2), the adaptor protein caspase-activating recruitment domain protein (ASC) and caspase-1 that enzymatically processes pro-IL-1β and pro-IL-18 for their activation. Activation of inflammasomes also causes pyroptosis, which corresponds to a rapid and proinflammatory form of cell death (Chakraborty et al. 2010; Nunes and de Souza 2013; Latz et al. 2013). Several types of inflammasomes are found in the brain tissue (Kummer et al. 2007). Among NLR proteins, NRLP1 is expressed in neurons, while NLRP3 and ice protease-activating factor (IPAF) are expressed in microglial cells (Jamilloux et al. 2013; Masters et al. 2005; Abulafia et al. 2009; Minkiewicz et al. 2013; Halle et al. 2008; Hanamsagar et al. 2011). NLRP2 and IPAF inflammasome are expressed in primary astrocytes. The physiological significance of differential distribution of inflammasomes in various types of neural cell is not fully understood. However, it is suggested that differential distribution of inflammasomes may be related to the microenvironment necessary for the survival of neurons and maintenance of BBB. Among the NLR sensors, NLRP3 is associated with TLR-mediated activation of Aβ production in AD (Halle et al. 2008; Tan et al. 2013). While all inflammasomes recognize certain pathogens, activation of NLRP3 inflammasome is the most versatile and important clinically. NLRP3 orchestrates the activation of precursors of pro-inflammatory caspases, which, in turn, cleave precursor forms of interleukin (IL)-1β, IL-18 and IL-33 into their active forms (Fig. 7.4) (Schroder and Tschopp 2010). In addition to the production and secretion of IL-1β and IL-18, the inflammasome/caspase-1 complexes may also target different effector molecules to regulate processes, such as pyroptosis and neural repair (Miao et al. 2011). During inflammasome activation, NLRP3 undergoes oligomerization through the central nucleotide-binding domain leading recruitment of an adaptor protein, ASC with the pyrin domain (PYD) and an amino-terminal caspase-recruitment-and-activation domain (CARD domain). NLRP3 binds with the PYD domain of ASC through its own PYD domains, whereas the CARD domain of ASC recruits procaspase-1. Assembly of the inflammasome initiates self-cleavage of caspase-1 and the formation of the active heterotetrameric caspase-1. Active caspase-1 further proteolytically processes pro-IL-1β and pro-IL-18 to their mature forms (Latz et al. 2013).

As stated above, during activation microglial cells undergo morphological transformation from the ramified state to the amoeboid state (Schilling et al. 2004). The exact molecular mechanism remains to be determined. However, it is suggested that lyso-PtdCho promotes activation of microglia through P2X7R signaling regulation

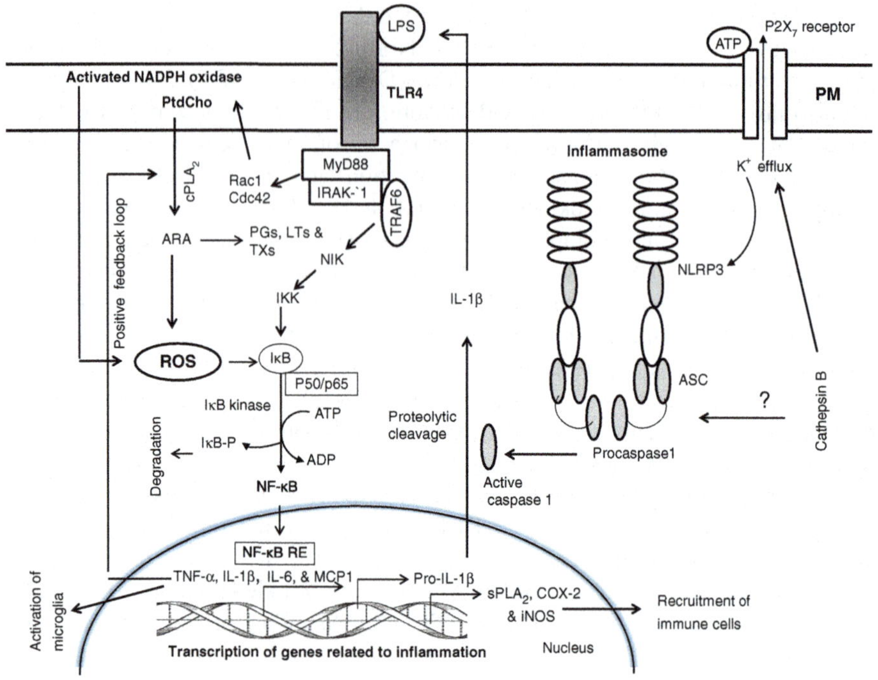

Fig. 7.4 Contribution of neuroinflammasomes. Plasma membrane (*PM*); phosphatidylcholine (*PtdCho*); cytosolic phospholipase A$_2$ (*cPLA$_2$*); arachidonic acid (*ARA*); Toll-like receptors (*TLRs*); proinflammatory endotoxin lipopolysaccharide (*LPS*); adaptor protein (*MyD88*); reactive oxygen species (*ROS*); tumor necrosis factor receptor-associated factor adaptor protein 6 (*TRAF6*); NF-κB-inducing kinase (*NIK*); IκB kinase (*IKK*); NF-kappaB (*NF-κB*); NF-kappaB response element (*NF-κB-RE*); inhibitory subunit of NF-κB (*I-κB*); tumor necrosis factor-alpha (*TNF-α*); interleukin-1beta (*IL-1β*); interleukin-6 (*IL-6*); pro-interleukin-1β (*pro-IL-1β*); NLR family, pyrin domain containing 3 (*NLRP3*) and apoptosis-associated speck-like protein containing a CARD (*ASC*), collectively known as the NLRP3 inflammasome

(Fig. 7.4) (Takenouchi et al. 2007). As stated above, microglial cells also release BNDF. An important function of microglial BDNF is the modulation of ATP binding to the purinergic receptor P2X$_4$R. It is suggested that ATP, which is released at sites of active synaptic transmission (Khakh and North 2012), acts as a robust chemoattractant for microglial processes (see below) (Davalos et al. 2005). Microglial BDNF is also involved in inhibitory synaptic transmission via TrkB signaling in the hippocampus (Zheng et al. 2011). Microglial and astrocytes express TLRs, which are crucial as a first line of defense against exogenous and endogenous molecules (Lehnardt 2010). Microglia express 9 types of TLRs, which contribute to neurotoxicity in both mice and human beings (Hanke and Kielian 2011; Olson and Miller 2004). The majority of TLRs are linked with the central adaptor molecule called Myeloid differentiation primary response gene 88 (MyD88), with the exception of TLR3, to bridge the receptor to downstream signaling intermediates (Takeuchi and Akira 2010; Coll and O'Neill 2010). The activation of TLR results in the recruitment MyD88, which is associated with the activity of serine/threonine kinase interleukin-1 receptor-associated kinase (IRAK) (Fig. 7.4) IRAK subsequently interacts with TNF receptor-associated factor (TRAF) adaptor protein TRAF6, and provides a link to NF-κB-inducing kinase (NIK). NIK phosphorylates I-κB kinase (IKK), leading to I-κB phosphorylation. I-κB phosphorylation targets the protein for ubiquitination and proteasome-mediated degradation, resulting in the release and migration of NF-κB to the nucleus, where it binds with NF-κB-RE and modulates the expression of numerous immune response genes. These MyD88-independent adaptors include TIR domain-containing adaptor inducing interferon-β (TRIF) and TRIF-related adaptor molecule (TRAM), which are pivotal for the expression of IFN-inducible genes following TLR4 activation (Takeuchi and Akira 2010; Yamamoto et al. 2003, 2004). Astrocytes, and microglia, display an array of receptors involved in innate immunity, including Toll-like receptors (TLRs), nucleotide-binding oligomerization domains, double-stranded RNA dependent protein kinase, mannose receptor, and components of the complement system (Farfara et al. 2008). Another common feature of neurodegenerative diseases is the presence of a large number of activated astrocytes and microglia, which contribute to gliosis, especially astrogliosis, when stimulated with various factors including lipopolysaccharide (LPS), interleukin-1β (IL-1β), and tumor necrosis factor-(TNF-α) (Ridet et al. 1997).

Overnutrition also results in the release of proinflammatory cytokines (IL-1β, IL-18 and IL-33). These cytokines stimulates PLA$_2$, COX-2, and 5-LOX leading to the generation of lyso-phosphatidylcholine (lyso-PtdCho), platelet-activating factor (PAF), eicosanoids (PGs, LTs, and TXs) along with the production of ROS, protein-ases, and complement proteins. The generation of these factors contributes to a potent response. Lyso-PtdCho promotes the transformation of ramified resting microglia (characterized by small cell bodies and long processes with secondary branching) into activated, deramified amoeboid microglia, which plays a major role in the maintenance of brain immune function (Farooqui 2011) and PGs, LTs, TXs, and PAF modulate the intensity of neuroinflammation. Converging evidence suggests that alterations in the expression of inflammasome-derived mediators may promote a variety of innate immune processes, which contribute neurodegeneration and cognitive impairment in neurotraumatic, neurodegenerative, and neuropsychiatric diseases (Chakraborty et al. 2010; Farooqui 2010).

Chronic neuroinflammation differs from acute inflammation in that it is below the threshold of pain perception. As a result, the immune system continues to attack at the cellular level. Chronic inflammation lingers for years causing continued insult to the brain tissue reaching the threshold of detection (Wood 1998; Tansey et al. 2007) and initiating the pathogenesis of chronic visceral and brain diseases. Onset of prolonged chronic inflammatory state produces detrimental health effects and predisposes to a wide variety of chronic diseases, especially those that are more prevalent with advanced age, such as cardiovascular diseases, diabetes, and neurodegenerative diseases (Farooqui et al. 2007). Chronic inflammation is also a strong predictor of both disability and mortality in the elderly—even in the absence of clinical disease (Farooqui 2013). Significant information is available on the generation and release of proinflammatory mediators. However, few studies have been performed on internal and external factors, which modulate the dynamic and kinetic aspects of chronic neuroinflammation. Depending on its timing and magnitude in brain tissue, neuroinflammation serves multiple purposes. As stated above, neuroinflammation not only protects uninjured neurons and promotes the removal of degenerating neuronal debris, but also in assisting in repair and recovery processes of uninjured neurons in the area surrounding the injury site (Farooqui 2010, 2011). It is not yet known whether neuroinflammatory events precede neurodegenerative diseases, or neuroinflammation is a direct consequence of the damage that occurs in the pathology of neurodegenerative diseases. For example, Aβ plaques have been shown to induce proinflammatory effects in animal models of AD (Halliday et al. 2000; Tuppo and Arias 2005) supporting the view that neuroinflammatory events initiate or even aid in the progression of AD (Heneka and O'Banion 2007; Bales et al. 2000).

7.3 Effect of Neuroinflammation on Telomere Length and Cognition

Chronic inflammation and oxidative stress are known to accelerate the shorting of telomeres, which are DNA-protein structures at the end of linear chromosomes that contribute to chromosomal stability (Kananen et al. 2010; Thomas et al. 2008; Lukens et al. 2009).

Telomeres are considered as a marker of cellular aging (Blackburn 2010). They are made up of several thousand repetitive DNA sequences (TTAGGG) coated by capping proteins. Telomeres are essential for maintaining chromosomal integrity during replication (Lin et al. 2012; Zanni and Wick 2011). Telomerase is the enzyme responsible for the maintenance of the length of telomeres by addition of guanine-rich repetitive sequences. Shorter telomere length and lower telomerase activity may reduce cell survival and induce cognitive impairment in several age-related diseases (Lin et al. 2012; Zanni and Wick 2011; Kananen et al. 2010), including AD (Kananen et al. 2010; Thomas et al. 2008; Lukens et al. 2009). Many factors such as neurotransmitter, oxidative stress, neuroendocrine factors (Mosley 1996), hormones, and abnormal protein aggregation modulate microglial cell functions (Flanary and Streit 2005). For example, Aβ has been shown to accelerate microglia senescence (Flanary and Streit 2005). It is proposed that chronic inflammation and oxidative stress may accelerate telomere degradation (Kaszubowska 2008). Inflammation and oxidative stress-mediated shorting of telomeres may be an important mechanism underlying cognitive decline and neurodegeneration in AD (Gorelick 2010). Systemic chronic inflammatory markers, such as IL-6, TNF-α, and hsCRP are closely linked with the loss of cerebral volume, and large population-based studies have consistently shown high blood levels of these inflammatory indices, to predict cognitive decline (Gorelick 2010).

7.4 Role of Redox Signaling in Neuroinflammation in the Brain

Reduction/oxidation ("redox") reactions are associated with the transfer of electrons between molecules, in which the reduced form of a molecule is oxidized after electron(s) are transferred to another molecule. The reduced and oxidized forms of the same molecule are named redox couples and their inter-conversion usually requires a second redox couple, which also provides and accepts electrons. These processes of electron transfer are catalyzed by enzymes, and are associated with the generation of ROS. Examples of redox reactions are reactions involving NADPH/ NAD, thioredoxin-1, thioredoxin-2, GSH/GSSG and cysteine/cystine (Cys/CySS).

Under physiological conditions 'redox homeostasis' is maintained by well coordinated balance between the ubiquitous generation and efficient removal of ROS/ RNS by the cellular antioxidant defense systems which is supported by enzymes such as superoxide dismutase, catalase, glutathione peroxidase and glutathione S-transferase and proteins that efficiently sequester and neutralize them such as reduced glutathione, vitamins E and C. When an imbalance in redox status due to the generation of high levels of ROS/RNS occurs, oxidative stress develops.

Excessive generation of ROS causes oxidative damage to cellular lipids, proteins, and nucleic acids. However, low levels of ROS are essential for cell signaling through different mechanisms. Thus, it is not surprising that regulation of redox and generation of ROS are multifaceted, which differ among tissues and cellular compartments (Go and Jones 2008).

As stated in Chap. 6, $O_2^{\bullet-}$, the less reactive ROS, which are generated through mitochondrial dysfunction, uncontrolled ARA cascade, and activation of NADPH oxidases. $O_2^{\bullet-}$ are readily converted by oxidoreduction reactions with transition metals or other redox cycling compounds into more aggressive radical species ($OH^{\bullet}$ and H_2O_2) that cause extensive cellular damage to lipids and DNA (Fig. 7.5) (Hancock et al. 2001; Beal 2005). $NO^{\bullet}$, which is generated by the action of nitric oxide synthase on arginine reacts with $O_2^{\bullet-}$ to generate the neurotoxic $ONOO^-$ (Bal-Price et al. 2002). ROS/RNS play an important role in cell signaling through redox signaling. To maintain proper cellular homeostasis and normal neural cell function, a balance must occur between ROS/RNS production and oxygen consumption and $NO^{\bullet}$. Excessive ROS/RNS have to be either quenched by converting them into metabolically nondestructive molecules or be scavenged/neutralized right after their formation. This protective mechanism is called the antioxidant defense system preventing ROS/RNS mediated damage of cells leading to various diseases and aging (Yu 1994; Winterbourn and Hampton 2008). Another mechanism of redox signaling involves H_2O_2-mediated oxidation of cysteine residues within proteins (Rhee 2006). H_2O_2 is generated from superoxide produced by mitochondria and NADPH oxidases (Fig. 7.5) (Sun et al. 2007; Brand 2010). The glutathione thiol/disulfide redox couple (GSH/GSSG) is another predominant mechanism for maintaining the intracellular microenvironment in a highly reduced state that is essential for antioxidant/detoxification capacity, redox enzyme regulation, cell

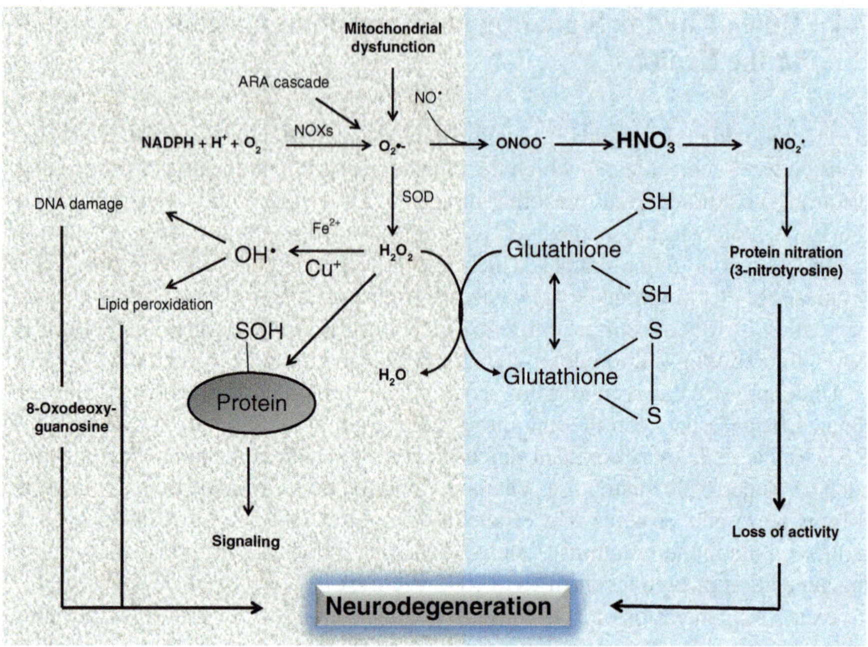

Fig. 7.5 Generation of ROS/RNS and coupling and neutralization of oxidative stress by glutathione. Superoxide ($O_2^{\cdot-}$); hydroxyl radical ($OH^{\cdot}$); nitric oxide ($NO^{\cdot}$); peroxynitrite ($ONOO^-$); hydrogen peroxide (H_2O_2); and superoxide dismutase (SOD)

cycle progression, and transcription of antioxidant response elements (ARE) (Biswas et al. 2006; Fratelli et al. 2005).

$$2GSH + O_2 \rightarrow GSSG + 2H_2O_2$$

$$2GSH + 2H_2O_2 \rightarrow GSSG + 2H_2O \ (GSH \ peroxidase)$$

$$2GSSG + NADH \rightarrow 2GSH + NADP \ (GSSG \ reductase)$$

Due to their susceptibility towards redox modification, protein thiols represent primary targets for the modulation of protein activity, conformation and oligomerization. This type of thiol redox modification occurs only in specific proteins of endoplasmic reticulum (ER) (Poole 2015), a subcellular organelle, which has oxidizing environment. ER contains the entire machinery for introducing disulfides for proper folding and functioning of secreted proteins. Similarly, the extent of steady-state thiol modifications (e.g. glutathionylation and cysteinylation) is modulated by the ambient redox potential of the cellular compartment. A shift in the steady-state redox potential is not only driven by metabolic changes, but also by the onset of oxidative stress. These processes can impact the distribution of reduced versus oxidized protein thiols (Banerjee 2012). Thus, thiol systems are important for protecting against oxidative damage and serving in redox signaling mechanisms necessary for repairing the cellular damage.

Neuroinflammation, which is caused by increased expression of cytokines (TNF-α, IL1-β, and IL-10) and ROS-mediated activation and translocation of NF-κB to the nucleus are closely intertwined processes. Neuroinflammation is connected to oxidative stress by at least two different mechanisms: production of high levels of ROS by activated glia such as microglia and astrocytes and arachidonic-acid signaling through the activation of cyclooxygenase and lipoxygenase pathways (Dringen et al. 2005). The complex interplay between inflammatory mediators and markers for oxidative stress caused by the overnutrition may regulate the progression of Alzheimer disease (Farooqui 2010, 2013). Thus, the most common hypotheses to explain the pathogenesis of AD include interactions among neuroinflammation, oxidative stress, mitochondrial dysfunction, alterations in calcium homeostasis, neuronal cell cycle induction, proteasomal dysfunction, protein aggregation, decrease in blood flow, and alterations in BBB (Golde 2009; Farooqui 2010). However, placing these pathways in the proper relationship to the onset, time course, and progress of neurodegeneration and its relationship to cytoskeletal pathology of AD are challenging issues that are not fully understood (Golde 2009). Despite some support for this hypothesis, prospective clinical trials targeting oxidative stress and inflammation in AD have failed. With regard to oxidation, the largest prospective clinical trials have tested combinations of the monoamine oxidase inhibitor, selegiline and alpha-tocopherol or the cholinesterase inhibitor, donepezil and α-tocopherol, either in patients with severe AD or in subjects with mild cognitive impairment (Sano et al. 1997; Petersen et al. 2005). NO beneficial effects have been observed in between antioxidant-treated and controlled groups and in terms of disease progression or on cognitive assessment at the end of intervention. Recently, the clinical trial of Vitamin E and Memantine in AD (TEAM-AD) has indicated a reduction in functional cognitive decline in those receiving α-tocopherol compared to placebo (Dysken et al. 2014; Craft et al., 2014). However, these results have been criticized due to the relatively high dose of α-tocopherol used, and the fact memantine alone or in combination with vitamin E did not produce similar neuroprotective effects (Corbett and Ballard 2014). Similarly, placebo-controlled trials for AD using anti-inflammatory agents produce very little benefit leading to patient dropout from the study, although majority of these trials used a relatively short treatment window before trial termination or cessation (McGeer et al. 1996). Thus, more human clinical trials are needed on drugs with antioxidant and anti-inflammatory activities.

7.5   Effect of Age on Neuroinflammation

Aging is accompanied by gradual deterioration of molecular and cellular activities leading to decline in cognitive function along with memory impairment and an increased susceptibility to neurodegenerative disorders (Wong 2013). During aging activated glial cells produce high levels of ROS (H_2O_2, $^•OH$, $NO^•$ and $ONOO^-$) (Lynch 1998; Lynch et al. 2010; Streit et al. 2004). These cytotoxic molecules can diffuse to other parts of the neural cell damaging proteins, nucleic acids and lipids, producing oxidative stress, and mitochondrial dysfunction leading to apoptotic cell death. In addition, oxidative metabolism in glial cells also produces advanced glycation end-products (AGEs), which are proteins that form non-enzymatically in a reaction between reducing sugars, amino groups of proteins and other compounds. Aging increases levels of AGEs in humans and rodents (Kimura et al. 1998) and binding of AGE with AGE receptor (RAGE) on microglial cells can activate the NF-κB signaling pathways that is involved in pro-inflammatory responses (Kimura et al. 1998). Aging also decreases the production of glutathione in astrocytes, which supply this antioxidant to neurons (Butterfield and Sultana 2007). Such age-related abnormalities in anti-oxidant protection may partly contribute to neuroinflammation through production of pro-inflammatory and cytotoxic factors (Fuller et al. 2009).

Numerous studies link elevated levels of cytokines with hippocampal memory deficits. Infact, cytokine-mediated memory impairments can be reversed by pharmacological inhibition of cytokines (Gibertini et al. 1995; Pugh et al. 1999). In adult and aged rodents, proinflammatory cytokines have been reported to alter long-term potentiation (LTP), a cellular model of synaptic plasticity, which contribute to learning and memory (Murray and Lynch 1998). The inhibition of microglial activation and increase in cytokines expression by minocycline restores hippocampal LTP in aged rats (Griffin et al. 2006) and improves memory in an animal model of AD (Choi et al. 2007). Accumulating evidence suggests that neuroinflammation is an important component of cognitive dysfunction caused by aging process.

7.6 Effects of Curcumin on Inflammation in Animal Models and Patients with Alzheimer Disease

Neurochemical mechanisms associated with the pathogenesis of AD remain unknown. However, it is becoming increasingly evident that abnormalities in multiple signal transduction pathways, mitochondrial dysfunction, accumulation of Aβ oligomers and hyperphosphorylated Tau, loss of synapses along with onset of neuroinflammation, and oxidative stress may contribute to cognitive deficits and memory loss in AD (Lesné et al. 2006; Farooqui 2010). Mounting evidence suggests that increase in levels of inflammatory molecules (PGs, LTs, TXs, PAF, IL-1β, IL-6 and TNF-α) and activation of microglial cells and astrocytes, which surround the senile plaques in brains of AD patients and AD transgenic animal models is closely associated with the pathogenesis of AD (Farooqui 2010). Concomitantly, a down-regulation in anti-inflammatory molecules (CD200, CD200 receptor, vitamin D receptor, peroxisome proliferator-activated receptors) has been reported to occur in post mortem brains from AD patients (Walker et al. 2009). Under normal conditions, anti-inflammatory systems regulate inflammatory responses and protect from uncontrolled inflammatory damage (Griffiths et al. 2007). However, their activities become deficient with aging and further defective under sustained inflammatory stimulation (Frank et al. 2006; Lue et al. 2010). These observations implicate neuroinflammation in the pathogenesis of AD. High levels of IL-1β, IL-6 and TNF-α may

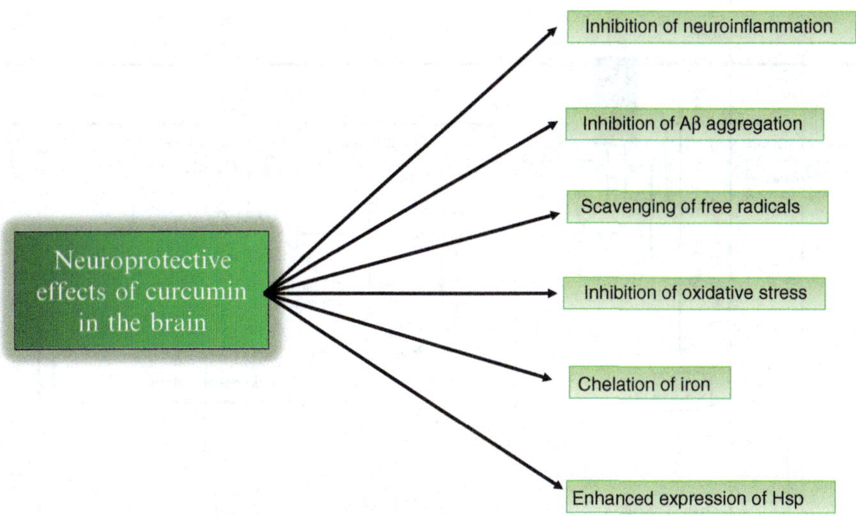

Fig. 7.6 Processes associated with neuroprotective effects of curcumin in the brain

also contribute to abnormalities in neurophysiologic mechanisms related to cognition and memory (Gemma and Bickford 2007). Cytokines establish a feedback loop, to activate more astrocytes and microglia leading to further generation of inflammatory molecules. In addition, the secreted inflammatory molecules also recruit other cells such as monocytes and lymphocytes to cross the BBB to enhance neuroinflammation in the CNS (Das and Basu 2008).

Studies in animal models of AD have indicated that curcumin produces beneficial effects not only by decreasing oxidative and inflammatory damage, but also by preventing deposition of Aβ (Fig. 7.6) (Yang et al. 2005). It is suggested that at the molecular level, curcumin acts not only by targeting transcription factor (NF-κB), but also by interacting with enzymes (COX2, 5-LOX, iNOS, HO-1, and JNK), and inhibiting the expression of proinflammatory cytokines (TNF-α, IL-1β, IL-6) (Shishodia et al. 2005) (Fig. 7.7). Modulation of above mentioned transcription factors, enzymes, and cytokines by curcumin results in neuroprotection through antiinflammatory, antioxidant, and anti-protein aggregate and neurogenic effects of curcumin in AD models (Yang et al. 2005; Cole and Frautschy 2007; Ma et al. 2009). Curcumin produces its beneficial effects in several ways (Quitschke et al. 2013). The enol form of curcumin binds with Aβ fibrils resulting in staining of amyloid plaques and neurofibrillary tangles in brain sections in vitro (Yanagisawa et al. 2010; Mohorko et al. 2010; Mutsuga et al. 2012), but also in vivo (Garcia-Alloza et al. 2007; Yang et al. 2005). Curcumin also retards the formation of Aβ fibril by promoting disaggregation of existing fibrils in vitro with IC_{50} values of 0.19 to 1 μM (Yang et al. 2005; Ono et al. 2004; Kim et al. 2005b). However, other studies have indicated IC_{50} values in the 10 to 12 μM. Intravenous injections curcumin have been reported to prevent plaque formation by disrupting existing plaques in Alzheimer transgenic mouse models both by oral uptake and through injections

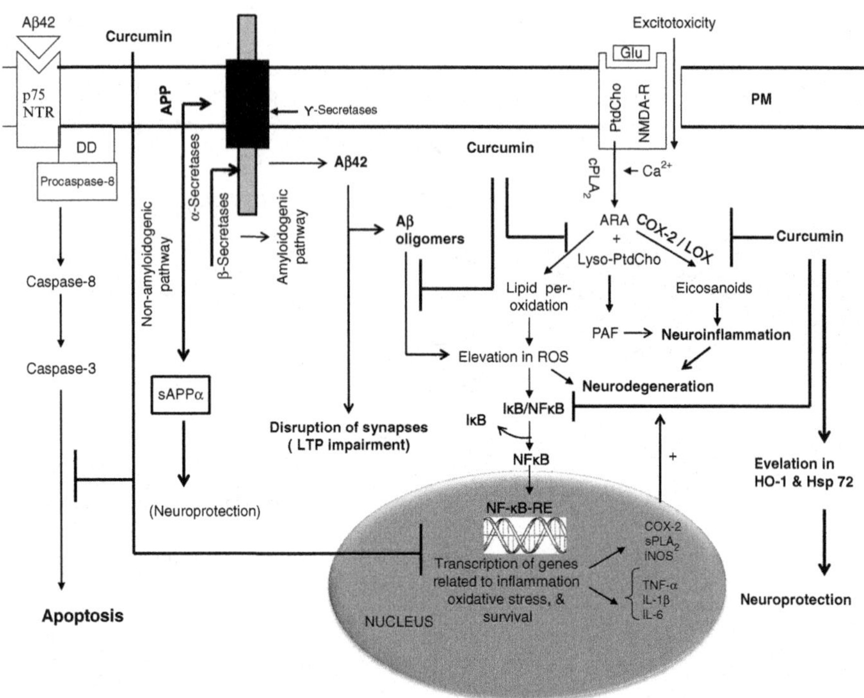

Fig. 7.7 A hypothetical signal transduction diagram showing the effect of curcumin on signal transduction processes in AD. Amyloid precursor protein (*APP*); monomeric β-amyloid peptide42 (*Aβ42*); *N*-methyl-D-aspartate receptor (*NMDA-R*); Glutamate (*Glu*); Phosphatidylcholine (*PtdCho*); cytosolic phospholipase A$_2$ (*cPLA$_2$*); lysophosphatidylcholine (*lyso-PtdCho*); cyclooxygenase (*COX*); lipoxygenase (*LOX*); arachidonic acid (*ARA*); platelet activating factor (*PAF*); reactive oxygen species (*ROS*); nuclear factor-κB (*NF-κB*); nuclear factor-kappaB-response element (*NF-κB-RE*); inhibitory subunit of NF-κB (*I-κB*); tumor necrosis factor-α (*TNF-α*); interleukin-1β (*IL-1β*); interleukin-6 (*IL-6*); inducible nitric oxide synthase (*iNOS*); secretory phospholipase A$_2$ (*sPLA$_2$*); death domain (*DD*); heme oxygenase (*HO-1*); heat shock protein 72 (*Hsp72*); long term potentiation (*LTP*); and soluble amyloid precursor proteinα (*sAPPα*). *Positive sign* indicates stimulation

(Lim et al. 2001; Yang et al. 2005; Begum et al. 2008; Wang et al. 2009b). It is also reported that curcumin produces its beneficial effect not only by modulating APP processing and modulating gene expression of proteins associated by apoptosis and neuroinflammation, but also by increasing the Aβ uptake by macrophages (Zhang et al. 2006; Liu et al. 2010b; Xiong et al. 2011; Ahmed and Gilani 2011). Curcumin has been reported to induce heat shock proteins and reduce protein misfolding and aggregation (Kato et al. 1998). Target genes of curcumin also include JUN, STAT3, APP and GSK3B, suggesting a regulatory effect of curcumin in the formation of amyloid. CEBPB, a C/EBP family member, has been shown to be involved in astrocytes and microglial activation (Ejarque-Ortiz et al. 2007). Translocation of STAT1 from cytosol to nucleus may also be involved in inflammatory activation in AD brains (Kitamura et al. 1997). Therefore, the interaction of curcumin with CEBPB and STAT1 may have anti-inflammatory effect.

Curcumin inhibits Aβ-induced expression of Egr-1 protein and Egr-1 DNA-binding activity in THP-1 monocytic cells. Egr-1 plays an important role in amyloid peptide-induced cytochemokine gene expression in monocytes. By inhibiting Egr-1 DNA-binding activity, curcumin mediates a reduction in the inflammation. The chemotaxis of monocytes, which can occur in response to chemokines from activated microglia and astrocytes in the brain, can be decreased by curcumin (Pendurthi and Rao 2000; Giri et al. 2003, 2004). In addition, curcumin also contributes to reduction in senile plaque deposition and suppression of JNK/IRS-1/tau signaling pathway in 3xTg-AD mice (Ma et al. 2009). Thus, treatment of the 3xTg-AD mice on high-fat diet with curcumin for 4 months not only reduces levels of phosphorylated JNK, IRS-1, and tau, but also prevents the degradation of total IRS-1. These neurochemical changes are accompanied by improvement in Y-maze performance (Ma et al. 2009). Mice consuming curcumin for 1 month show more significant effects on Y-maze performance and the combination of curcumin and fish oil show more significant inhibition of JNK, IRS-1, and tau phosphorylation. Collective evidence suggests that curcumin not only reduces Aβ-mediated (a) increase in the level of ROS, (b) decrease in mitochondrial membrane potential, and (c) caspase activation. In addition, curcumin protects human neurons from oligomeric amyloid-β-induced toxicity as well (Zhang et al. 2010; Mishra et al. 2011). Curcumin confers neuroprotection against Aβ-induced toxicity not only by inhibiting the formation of Aβ oligomers and fibrils, but also binding to plaques and restoring homeostasis of the inflammatory system, boosting the heat shock system to enhance clearance of toxic aggregates, scavenging free radicals, chelating iron and inducing anti-oxidant response elements (Zhang et al. 2010; Mishra et al. 2011; Hu et al. 2015).

As stated in Chap. 6, six small clinical studies on treatment of AD patients or mild cognitive impairment with curcumin have been carried out. They have been listed on the clinical trial.gov website. Amongst these, two are still recruiting, one of them has unknown status, other has been terminated due to unstated reasons, and two have been completed. The first one was a 6-month randomized, double blind, placebo-controlled study supplementing 1 or 4 g/d of oral curcumin in 34 patients with AD (Baum et al. 2008). In this trial patients received simultaneous supplementation with 120 mg/d ginkgo leaf extract. Six month supplementation of curcumin produces no beneficial effects in these patients (Baum and Ng 2004). The second study was 24 weeks randomized, double blind, placebo-controlled study with 2 or 4 g/d oral supplementation of curcumin in 36 patients with mild-to-moderate AD. Like the first study, this study did not produce was any clinical or biochemical evidence of the neuroprotective efficacy of curcumin (Ringman et al. 2012). One small study reported from Japan has indicated that three patients with AD whose behavioral symptoms improved remarkably as a result of curcumin treatment (Hishikawa et al. 2012). After 12 weeks of curcumin supplementation one patient gave better scores on the minimental state examination (MMSE). In the other two cases, no significant change is observed in the MMSE; however, patient came to recognize their family within 1 year of treatment. However, patients were able to tolerate high doses of curcumin without any harmful effects. At present, four clinical trials have been conducted to test the usefulness of curcumin for the treatment of

AD. Two trials have been performed in China and USA, but no significant differences are observed in cognitive function between placebo and curcumin groups. No results have been reported on two other remaining clinical trials. Additional trials are necessary to determine the clinical usefulness and efficacy of curcumin in the prevention and treatment of AD.

7.7   Conclusion

Neuroinflammation (the inflammatory response in the brain) is a well coordinated and protective mechanism, which is characterized by redness, pain, swelling and temperature. At the cellular level, neuroinflammation is accompanied by activation of microglia and astrocytes. Microglia have very little ability to remove $A\beta$ and are unable to effectively phagocytate high concentrations (or insoluble conformations) of it, resulting in aberrantly activated microglia that associate with both $A\beta$ plaques and neurofibrillary tangles. $A\beta$ stimulates microglia and astrocytes leading to the secretion of proinflammatory cytokines and chemokines. At the molecular level, neuroinflammation is supported by the generation of high levels of ARA-derived lipid mediators (PGs, LTs, and TXs) and increased expression of proinflammatory cytokines (tumor necrosis factor-α (TNF-α), interleukin-1β (IL-1β), IL-6 and chemokines [IL-8; CXC chemokine receptor 4 (CXCR4)], which act by further recruiting non-neural inflammatory cells (macrophages, neutrophils, monocytes, dendritic and mast cells) to the site of brain damage and producing more reactive species. Inflammatory mediators, including inflammatory cytokines, ROS, and nitric oxide (NO), impair mitochondrial function by inducing the accumulation of mtDNA mutations and inhibiting mitochondrial respiratory chain and energy production. Above mentioned inflammatory mediators along with activation of transcription factors, NF-κB, STAT3, HIF1-α, AP-1, NFAT, and Nrf2 modulate signal transduction cascades, which either promote or inhibits cellular stress responses. Induction of COX-2, iNOS, and aberrant expression of inflammatory cytokines and chemokines has also been reported to play a critical role in oxidative stress-mediated inflammation. This sustained inflammatory/oxidative environment leads to a vicious circle, which can damage healthy neighboring neural cells over a long period of time may leading to neurodegeneration. Neuroinflammation is an early event in the amyloid pathology and precedes plaque deposition in experimental models of AD. Amyloid deposition results in microglial cell activation and the production of proinflammatory mediators that contribute to disease pathogenesis. However, microglia cells may also play a beneficial role in restricting senile plaque formation by clearing $A\beta$ deposits and secreting neuroprotective factors.

Curcumin is an excellent antioxidant antiinflammatory agent. It inhibits cyclooxygenase and lipoxygenase activities leading to decrease in generation of ARA-derived PGs, LTs, and TXs. Curcumin also inhibits the production of NO, TNF-α, and interleukin-1β. Curcumin mediates its beneficial effect not only by modulating APP processing and modulating gene expression of proteins associated with apoptosis and neuroinflammation, but also by increasing Aβ uptake by macrophages. Curcumin also induces heat shock proteins and reduces protein misfolding and aggregation. Nevertheless, in studies, daily administration of curcumin 1 g for 6 months has no effect in AD patients. This drawback of curcumin may be due to several factors including its insolubility in water, poor absorption and rapid metabolism that need to be overcome for a more meaningful therapeutic intervention with curcumin.

References

Abbracchio MP, Burnstock G, Verkhratsky A, Zimmermann H (2009) Purinergic signalling in the nervous system: an overview. Trends Neurosci 32:19–29

Abizaid A, Horvath TL (2008) Brain circuits regulating energy homeostasis. Regul Pept 149:3–10

Abramov AY, Jacobson J, Wientjes F, Hothersall J, Canevari L, Duchen MR (2005) Expression and modulation of an NADPH oxidase in mammalian astrocytes. J Neurosci 25:9176–9184

Abulafia DP, de Rivero Vaccari JP, Lozano JD, Lotocki G, Keane RW, Dietrich WD (2009) Inhibition of the inflammasome complex reduces the inflammatory response after thromboembolic stroke in mice. J Cereb Blood Flow Metab 29:534–544

Aggarwal BB (2010) Targeting inflammation-induced obesity and metabolic diseases by curcumin and other nutraceuticals. Annu Rev Nutr 30:173–199

Ahmed T, Gilani AH (2009) Inhibitory effect of curcuminoids on acetylcholinesterase activity and attenuation of scopolamine-induced amnesia may explain medicinal use of turmeric in Alzheimer's disease. Pharmacol Biochem Behav 91:554–559

Ahmed T, Gilani AH (2011) A comparative study of curcuminoids to measure their effect on inflammatory and apoptotic gene expression in an Aβ plus ibotenic acid-infused rat model of Alzheimer's disease. Brain Res 1400:1–18

Ahmed T, Enam SA, Gilani AH (2010) Curcuminoids enhance memory in an amyloid-infused rat model of Alzheimer's disease. Neuroscience 169:1296–1306

Ahmed T, Gilani AH, Hosseinmardi N, Semnanian S, Enam SA, Fathollahi Y (2011) Curcuminoids rescue LTP impaired by amyloid peptide in rat hippocampal slices. Synapse 65:572–582

Allan SM, Rothwell NJ (2003) Inflammation in central nervous system injury. Philos Trans R Soc Lond B Biol Sci 358:1669–1677

Aloisi F (2001) Immune function of microglia. Glia 36:165–179

Bachstetter AD, Norris CM, Sompol P, Wilcock DM, Goulding D, Neltner JH, St Clair D, Watterson DM, Van Eldik LJ (2012) Early stage drug treatment that normalizes proinflammatory cytokine production attenuates synaptic dysfunction in a mouse model that exhibits age-dependent progression of Alzheimer's disease-related pathology. J Neurosci 32:10201–10210

Bales KR, Du Y, Holtzman D, Cordell B, Paul SM (2000) Neuroinflammation and Alzheimer's disease: critical roles for cytokine/Abetainduced glial activation, NF-kappaB, and apolipoprotein E. Neurobiol Aging 21:427–432, discussion 451–423

Bal-Price A, Matthias A, Brown GC (2002) Stimulation of the NADPH oxidase in activated rat microglia removes nitric oxide but induces peroxynitrite production. J Neurochem 80:73–80

Banerjee R (2012) Redox outside the box: linking extracellular redox remodeling with intracellular redox metabolism. J Biol Chem 287:4397–4402

Bartik L, Whitfield GK, Kaczmarska M, Lowmiller CL, Moffet EW, Furmick JK, Hernandez Z, Haussler CA, Haussler MR, Jurutka PW (2010) Curcumin: a novel nutritionally derived ligand of the vitamin D receptor with implications for colon cancer chemoprevention. J Nutr Biochem 21:1153–1161

Bates SH, Stearns WH, Dundon TA, Schubert M, Tso AW, Wang Y, Banks AS, Lavery HJ, Haq AK, Maratos-Flier E, Neel BG, Schwartz MW, Myers MG Jr (2003) STAT3 signalling is required for leptin regulation of energy balance but not reproduction. Nature 421:856–859

Bates SH, Kulkarni RN, Seifert M, Myers MG Jr (2005) Roles for leptin receptor/STAT3-dependent and -independent signals in the regulation of glucose homeostasis. Cell Metab 1:169–178

Batlle M, Ferri L, Andrade C, Ortega FJ, Vidal-Taboada JM, Pugliese M, Mahy N, Rodríguez MJ (2015) Astroglia-microglia cross talk during neurodegeneration in the rat hippocampus. Biomed Res Int 2015:102419

Baum L, Ng A (2004) Curcumin interaction with copper and iron suggests one possible mechanism of action in Alzheimer's disease animal models. J Alzheimers Dis 6:367–377

Baum L, Lam CW, Cheung SK, Kwok T, Lui V, Tsoh J, Lam L, Leung V, Hui E, Ng C, Woo J, Chiu HF, Goggins WB, Zee BC, Cheng KF, Fong CY, Wong A, Mok H, Chow MS, Ho PC, Ip SP, Ho CS, Yu XW, Lai CY, Chan MH, Szeto S, Chan IH, Mok V (2008) Six-month randomized, placebo-controlled, double-blind, pilot clinical trial of curcumin in patients with Alzheimer disease. J Clin Psychopharmacol 28:110–113

Bazan NG, Stark DT, Petasis NA (2012) Lipid mediators: eicosanoids, docosanoids and platelet-activating factor. In: Siegel G, Albers RW, Brady ST, Price DL (eds) Basic neurochemistry: molecular, cellular and medical aspects, 8th edn. pp 643–662

Beal MF (2005) Mitochondria take center stage in aging and neurodegeneration. Ann Neurol 58:495–505

Begum AN, Jones MR, Lim GP, Morihara T, Kim P, Heath DD, Rock CL, Pruitt MA, Yang F, Hudspeth B, Hu S, Faull KF, Teter B, Cole GM, Frautschy SA (2008) Curcumin structure-function, bioavailability, and efficacy in models of neuroinflammation and Alzheimer's disease. J Pharmacol Exp Ther 326:196–208

Biswas S, Chida AS, Rahman I (2006) Redox modifications of protein-thiols: emerging roles in cell signaling. Biochem Pharmacol 71:551–564

Bjorbaek C, Kahn BB (2004) Leptin signaling in the central nervous system and the periphery. Recent Prog Horm Res 59:305–331

Bjorbaek C, El-Haschimi K, Frantz JD, Flier JS (1999) The role of SOCS-3 in leptin signaling and leptin resistance. J Biol Chem 274:30059–30065

Blackburn EH (2010) Telomeres and telomerase: the means to the end (Nobel Lecture). Angew Chem Int Ed 49:7405–7421

Blais V, Rivest S (2004) Effects of TNF-α and IFN-γ on nitric oxide-induced neurotoxicity in the mouse brain. J Immunol 172:7043–7052

Block ML, Hong JS (2005) Microglia and inflammation-mediated neurodegeneration: multiple triggers with a common mechanism. Prog Neurobiol 76:77–98

Block ML, Zecca L, Hong JS (2007) Microglia-mediated neurotoxicity: uncovering the molecular mechanisms. Nat Rev Neurosci 8:57–69

Blouet C, Schwartz GJ (2010) Hypothalamic nutrient sensing in the control of energy homeostasis. Behav Brain Res 209:1–12

Boczkowski J, Lisdero CL, Lanone S, Carreras MC, Aubier M, Poderoso JJ (2001) Peroxynitrite-mediated mitochondrial dysfunction. Biol Signals Recept 10:66–80

Bomfim TR, Forny-Germano L, Sathler LB, Brito-Moreira J, Houzel JC, Decker H, Silverman MA, Kazi H, Melo HM, McClean PL, Holscher C, Arnold SE, Talbot K, Klein WL, Munoz DP, Ferreira ST, De Felice FG (2012) An anti-diabetes agent protects the mouse brain from defective insulin signaling caused by Alzheimer's disease-associated Aβ oligomers. J Clin Invest 122:1339–1353

Bowen KK, Naylor M, Vemuganti R (2006) Prevention of inflammation is a mechanism of preconditioning-induced neuroprotection against focal cerebral ischemia. Neurochem Int 49:127–135

Brand M (2010) The sites and topology of mitochondrial superoxide production. Exp Gerontol 45:466–472

Brown GC, Bal-Price A (2003) Inflammatory neurodegeneration mediated by nitric oxide, glutamate, and mitochondria. Mol Neurobiol 27:325–355

Bush TG, Puvanachandra N, Horner CH, Polito A, Ostenfeld T, Svendsen CN, Mucke L, Johnson MH, Sofroniew MV (1999) Leukocyte infiltration, neuronal degeneration, and neurite outgrowth after ablation of scar-forming, reactive astrocytes in adult transgenic mice. Neuron 23:297–308

Butterfield DA, Sultana R (2007) Redox proteomics identification of oxidatively modified brain proteins in Alzheimer's disease and mild cognitive impairments: insights into the progression of this dementing disorder. J Alzheimers Dis 12:61–72

Cai D (2009) NFkappaB-mediated metabolic inflammation in peripheral tissues versus central nervous system. Cell Cycle 8:2542–2548

Cai D (2013) Neuroinflammation and neurodegeneration in overnutrition-induced diseases. Trends Endocrinol Metab 24:40–47

Cai D, Liu T (2012) Inflammatory cause of metabolic syndrome via brain stress and NF-kappaB. Aging (Albany, NY) 4:98–115

Chakraborty S, Kaushik DK, Gupta M, Basu A (2010) Inflammasome signaling at the heart of central nervous system pathology. J Neurosci Res 88:1615–1631

Chang DM (2001) Curcumin: a heat shock response inducer and potential cytoprotector. Crit Care Med 29:2231–2232

Chinthamani S, Odusanwo O, Mondal N, Nelson J, Neelamegham S, Baker OJ (2012) Lipoxin A4 inhibits immune cell binding to salivary epithelium and vascular endothelium. Am J Physiol Cell Physiol 302:C968–C978

Choi Y, Kim HS, Shin KY, Kim EM, Kim M (2007) Minocycline attenuates neuronal cell death and improves cognitive impairment in Alzheimer's disease models. Neuropsychopharmacology 32:2393–2404

Ciolino HP, Daschner PJ, Wang TT, Yeh GC (1998) Effect of curcumin on the aryl hydrocarbon receptor and cytochrome P450 1A1 in MCF-7 human breast carcinoma cells. Biochem Pharmacol 56:197–206

Cole GM, Frautschy SA (2007) The role of insulin and neurotrophic factor signaling in brain aging and Alzheimer's disease. Exp Gerontol 42:10–21

Coll RC, O'Neill LA (2010) New insights into the regulation of signalling by toll-like receptors and nod-like receptors. J Innate Immun 2:406–421

Coll AP, Farooqi IS, O'Rahilly S (2007) The hormonal control of food intake. Cell 129:251–262

Corbett A, Ballard C (2014) The value of vitamin E as a treatment for Alzheimer's disease remains unproven despite functional improvement, due to a lack of established effect on cognition or other outcomes from RCTs. Evid Based Med 19:140

Craft S, Thielke S, Turvey CL, Woodman C, Monnell KA, Gordon K, Tomaska J, Segal Y, Peduzzi PN, Guarino PD (2014) Effect of vitamin E and memantine on functional decline in Alzheimer disease: the TEAM-AD VA cooperative randomized trial. JAMA 311:33–44

Damani MR, Zhao L, Fontainhas AM, Amaral J, Fariss RN, Wong WT (2011) Age-related alterations in the dynamic behavior of microglia. Aging Cell 10:263–276

Das S, Basu A (2008) Inflammation: a new candidate in modulating adult neurogenesis. J Neurosci Res 86:1199–1208

Davalos D, Grutzendler J, Yang G, Kim JV, Zuo Y, Jung S, Littman DR, Dustin ML, Gan WB (2005) ATP mediates rapid microglial response to local brain injury in vivo. Nat Neurosci 8:752–758

Desai BN, Leitinger N (2014) Purinergic and calcium signaling in macrophage function and plasticity. Front Immunol 5:580

Dietrich M, Horvath T (2009) Feeding signals and brain circuitry. Eur J Neurosci 30:1688–1696

Dringen R, Pawlowski PG, Hirrlinger J (2005) Peroxide detoxification by brain cells. J Neurosci Res 79:157–165

Dysken MW, Sano M, Asthana S, Vertrees JE, Pallaki M, Llorente M, Love S, Schellenberg GD, McCarten JR, Malphurs J, Prieto S, Chen P, Loreck DJ, Trapp G, Bakshi RS, Mintzer JE, Heidebrink JL, Vidal-Cardona A, Arroyo LM, Cruz AR, Zachariah S, Kowall NW, Chopra MP, Craft S, Thielke S, Turvey CL, Woodman C, Monnell KA, Gordon K, Tomaska J, Segal Y, Peduzzi PN, Guarino PD (2014) Effect of vitamin E and memantine on functional decline in Alzheimer disease: the TEAM-AD VA cooperative randomized trial. JAMA 311:33–44

Ejarque-Ortiz A, Medina MG, Tusell JM, Perez-Gonzalez AP, Serratosa J, Saura J (2007) Upregulation of CCAAT/enhancer binding protein beta in activated astrocytes and microglia. Glia 55:178–188

El Kebir D, Filep JG (2013) Targeting neutrophil apoptosis for enhancing the resolution of inflammation. Cells 2:330–348

Elkabes S, DiCicco-Bloom EM, Black IB (1996) Brain microglia/macrophages express neurotrophins that selectively regulate microglial proliferation and function. J Neurosci 16:2508–2521

Farfara D, Lifshitz V, Frenkel D (2008) Neuroprotective and neurotoxic properties of glial cells in the pathogenesis of Alzheimer's disease: Alzheimer's review series. J Cell Mol Med 12:762–780

Farooqui AA (2010) Neurochemical aspects of neurotraumatic and neurodegenerative diseases. Springer, New York

Farooqui AA (2011) Lipid mediators and their metabolism in the brain. Springer, New York

Farooqui AA (2012) Phytochemicals, signal transduction, and neurological disorders. Springer, New York

Farooqui AA (2013) Metabolic syndrome: an important risk factor for stroke, Alzheimer disease, and depression. Springer, New York

Farooqui AA (2014) Inflammation and oxidative stress in neurological disorders. Springer, Switzerland

Farooqui AA (2015) High calorie diet and the human brain. Springer, Switzerland

Farooqui AA, Horrocks LA (2006) Phospholipase A_2-generated lipid mediators in the brain: the good, the bad, and the ugly. Neuroscientist 12:245–260

Farooqui AA, Horrocks LA, Farooqui T (2007) Modulation of inflammation in brain: a matter of fat. J Neurochem 101:577–599

Fessler MB, Rudel LL, Brown JM (2009) Toll-like receptor signaling links dietary fatty acids to the metabolic syndrome. Curr Opin Lipidol 20:379–385

Fiebich BL, Akter S, Akundi RS (2014) The two-hit hypothesis for neuroinflammation: role of exogenous ATP in modulating inflammation in the brain. Front Cell Neurosci 8:260

Flanary BE, Streit WJ (2005) Effects of axotomy on telomere length, telomerase activity, and protein in activated microglia. J Neurosci Res 82:160–171

Forsmark-Andrée P, Persson B, Radi R, Dallner G, Ernster L (1996) Oxidative modification of nicotinamide nucleotide transhydrogenase in submitochondrial particles: effect of endogenous ubiquinol. Arch Biochem Biophys 336:113–120

Franceschi C, Bonafe M, Valensin S, Olivieri F, De Luca M, Ottaviani E, De Benedictis G (2000) Inflamm-aging. An evolutionary perspective on immunosenescence. Ann N Y Acad Sci 908:244–254

Frank MG, Barrientos RM, Biedenkapp JC, Rudy JW, Watkins LR, Maier SF (2006) mRNA upregulation of MHC II and pivotal pro-inflammatory genes in normal brain aging. Neurobiol Aging 27:717–722

Fratelli M, Goodwin LO, Ørom UA et al (2005) Gene expression profiling reveals a signaling role of glutathione in redox regulation. Proc Natl Acad Sci U S A 102:13998–14003

Fuller S, Münch G, Steele M (2009) Activated astrocytes: a therapeutic target in Alzheimer's disease? Expert Rev Neurother 9:1585–1595

Gabuzda D, Yankner BA (2013) Inflammation links ageing to the brain. Nature 497:197–198

Garcia-Alloza M, Borrelli LA, Rozkalne A, Hyman BT, Bacskai BJ (2007) Curcumin labels amyloid pathology *in vivo*, disrupts existing plaques, and partially restores distorted neurites in an Alzheimer mouse model. J Neurochem 102:1095–1104

Gemma C, Bickford PC (2007) Interleukin-1beta and caspase-1: players in the regulation of age-related cognitive dysfunction. Rev Neurosci 18:137–148

Ghilardi N, Skoda RC (1997) The leptin receptor activates Janus kinase 2 and signals for proliferation in a factor-dependent cell line. Mol Endocrinol 11:393–399

Ghorbani Z, Hekmatdoost A, Mirmiran P (2014) Anti-hyperglycemic and insulin sensitizer effects of turmeric and its principle constituent curcumin. Int J Endocrinol Metab 12, e18081

Gibertini M, Newton C, Friedman H, Klein TW (1995) IL-1 beta and TNF alpha modulate delta 9-tetrahydrocannabinol-induced catalepsy in mice. Pharmacol Biochem Behav 50:141–146

Giri RK, Selvaraj SK, Kalra VK (2003) Amyloid peptide-induced cytokine and chemokine expression in THP-1 monocytes is blocked by small inhibitory RNA duplexes for early growth response-1 messenger RNA. J Immunol 170:5281–5294

Giri RK, Rajagopal V, Kalra VK (2004) Curcumin, the active constituent of turmeric, inhibits amyloid peptide-induced cytochemokine gene expression and CCR5-mediated chemotaxis of THP-1 monocytes by modulating early growth response-1 transcription factor. J Neurochem 91:1199–1210

Go Y-M, Jones DP (2008) Redox compartmentalization in eukaryotic cells. Biochem Biophys Acta 1780:1273–1290

Golde TE (2009) The therapeutic importance of understanding mechanisms of neuronal cell death in neurodegenerative disease. Mol Neurodegener 4:8

Gorelick PB (2010) Role of inflammation in cognitive impairment: results of observational epidemiological studies and clinical trials. Ann N Y Acad Sci 1207:155–162

Granholm AC, Bimonte-Nelson HA, Moore AB, Nelson ME, Freeman LR, Sambamurti K (2008) Effects of a saturated fat and high cholesterol diet on memory and hippocampal morphology in the middle-aged rat. J Alzheimers Dis 14:133–145

Griffin R, Nally R, Nolan Y, McCartney Y, Linden J, Lynch MA (2006) The age-related attenuation in long-term potentiation is associated with microglial activation. J Neurochem 99:1263–1272

Griffiths M, Neal JW, Gasque P (2007) Innate immunity and protective neuroinflammation: new emphasis on the role of neuroimmune regulatory proteins. Int Rev Neurobiol 82:29–55

Gupta SC, Kismali G, Aggarwal BB (2013) Curcumin, a component of turmeric: from farm to pharmacy. Biofactors 39:2–13

Habib A, Bernard C, Lebret M, Creminon C, Esposito B, Tedgui A, Maclouf J (1997) Regulation of the expression of cyclooxygenase-2 by nitric oxide in rat peritoneal macrophages. J Immunol 158:3845–3851

Halle A, Hornung V, Petzold GC, Stewart CR, Monks BG, Reinheckel T, Fitzgerald KA, Latz E, Moore KJ, Golenbock DT (2008) The NALP3 inflammasome is involved in the innate immune response to amyloid-beta. Nat Immunol 9:857–865

Halliday G, Robinson SR, Shepherd C, Kril J (2000) Alzheimer's disease and inflammation: a review of cellular and therapeutic mechanisms. Clin Exp Pharmacol Physiol 27:1–8

Han D, Canali R, Garcia J, Aguilera R, Gallaher TK, Cadenas E (2005) Sites and mechanisms of aconitase inactivation by peroxynitrite: modulation by citrate and glutathione. Biochemistry 44:11986–11996

Hanamsagar R, Torres V, Kielian T (2011) Inflammasome activation and IL-1beta/IL-18 processing are influenced by distinct pathways in microglia. J Neurochem 119:736–748

Hancock JT, Desikan R, Neill SJ (2001) Role of reactive oxygen species in cell signalling pathways. Biochem Soc Trans 29:345–350

Hanke ML, Kielian T (2011) Toll-like receptors in health and disease in the brain: mechanisms and therapeutic potential. Clin Sci (Lond) 121:367–387

Headland SE, Norling LV (2015) The resolution of inflammation: principles and challenges. Semin Immunol. pii: S1044-5323(15)00022-6

Hedrich CM, Rauen T, Apostolidis SA, Grammatikos AP, Rodriguez Rodriguez N, Ioannidis C, Kyttaris VC, Crispin JC, Tsokos GC (2014) Stat3 promotes IL-10 expression in lupus T cells through trans-activation and chromatin remodeling. Proc Natl Acad Sci U S A 111:13457–13462

Hefendehl JK, Neher JJ, Suhs RB, Kohsaka S, Skodras A, Jucker M (2014) Homeostatic and injury-induced microglia behavior in the aging brain. Aging Cell 13:60–69

Heneka MT, O'Banion MK (2007) Inflammatory processes in Alzheimer's disease. J Neuroimmunol 184:69–91

Hickey WF (2001) Basic principles of immunological surveillance of the normal central nervous system. Glia 36:118–124

Hickman SE, Allison EK, El Khoury J (2008) Microglial dysfunction and defective beta-amyloid clearance pathways in aging Alzheimer's disease mice. J Neurosci 28:8354–8360

Hishikawa N, Takahashi Y, Amakusa Y, Tanno Y, Tuji Y, Niwa H, Murakami N, Krishna UK (2012) Effects of turmeric on Alzheimer's disease with behavioral and psychological symptoms of dementia. Ayu 33:499–504

Horvath TL, Sarman B, Garcia-Caceres C, Enriori PJ, Sotonyi P, Shanabrough M, Borok E, Argente J, Chowen JA, Perez-Tilve D (2010) Synaptic input organization of the melanocortin system predicts diet-induced hypothalamic reactive gliosis and obesity. Proc Natl Acad Sci U S A 107:14875–14880

Hu X, Liou AK, Leak RK, Xu M, An C, Suenaga J, Shi Y, Gao Y, Zheng P, Chen J (2014) Neurobiology of microglial action in CNS injuries: receptor-mediated signaling mechanisms and functional roles. Prog Neurobiol 119–120:60–84

Hu S, Maiti P, Ma Q, Zuo X, Jones MR, Cole GM, Frautschy SA (2015) Clinical development of curcumin in neurodegenerative disease. Expert Rev Neurother 15:629–637

Iadecola C, Anrather J (2011) The immunology of stroke: from mechanisms to translation. Nat Med 17:796–808

Jamilloux Y, Pierini R, Querenet M, Juruj C, Fauchais AL, Jauberteau MO, Jarraud S, Lina G, Etienne J, Roy CR, Henry T, Davoust N, Ader F (2013) Inflammasome activation restricts Legionella pneumophila replication in primary microglial cells through flagellin detection. Glia 61:539–549

Jang B, Han S (2006) Biochemical properties of cytochrome c nitrated by peroxynitrite. Biochimie 88:53–58

Kananen L, Surakka I, Pirkola S, Suvisaari J, Lonnqvist J, Peltonen L et al (2010) Childhood adversities are associated with shorter telomere length at adult age both in individuals with an anxiety disorder and controls. PLoS One 5, e10826

Kaszubowska L (2008) Telomere shortening and ageing of the immune system. J Physiol Pharmacol 59:169–186

Kato K, Ito H, Kamei K, Iwamoto I (1998) Stimulation of the stress-induced expression of stress proteins by curcumin in cultured cells and in rat tissues in vivo. Cell Stress Chaperones 3:152–160

Khakh BS, North RA (2012) Neuromodulation by extracellular ATP and P2X receptors in the CNS. Neuron 76:51–69

Kim SF, Huri DA, Snyder SH (2005a) Inducible nitric oxide synthase binds, S-nitrosylates, and activates cyclooxygenase-2. Science 310:1966–1970

Kim H, Park BS, Lee KG, Choi CY, Jang SS, Kim YH, Lee SE (2005b) Effects of naturally occurring compounds on fibril formation and oxidative stress of β-amyloid. J Agric Food Chem 53:8537–8541

Kimura T, Takamatsu J, Miyata T, Miyakawa T, Horiuchi S (1998) Localization of identified advanced glycation end-product structures, Nε(carboxymethyl)lysine and pentosidine, in age-related inclusions in human brains. Pathol Int 48:575–579

Kitamura Y, Shimohama S, Ota T, Matsuoka Y, Nomura Y, Taniguchi T (1997) Alteration of transcription factors NF-kappaB and STAT1 in Alzheimer's disease brains. Neurosci Lett 237:17–20

Kiyota T, Okuyama S, Swan RJ, Jacobsen MT, Gendelman HE, Ikezu T (2010) CNS expression of anti-inflammatory cytokine interleukin-4 attenuates Alzheimer's disease-like pathogenesis in APP+PS1 bigenic mice. FASEB J 24:3093–3102

Krabbe G, Halle A, Matyash V, Rinnenthal JL, Eom GD, Bernhardt U, Miller KR, Prokop S, Kettenmann H, Heppner FL (2013) Functional impairment of microglia coincides with beta-amyloid deposition in mice with Alzheimer-like pathology. PLoS One 8, e60921

Kummer JA, Broekhuizen R, Everett H, Agostini L, Kuijk L, Martinon F, van Bruggen R, Tschopp J (2007) Inflammasome components NALP 1 and 3 show distinct but separate expression profiles in human tissues suggesting a site-specific role in the inflammatory response. J Histochem Cytochem 55:443–452

Landino LM, Crews BC, Timmons MD, Morrow JD, Marnett LJ (1996) Peroxynitrite, the coupling product of nitric oxide and superoxide, activates prostaglandin biosynthesis. Proc Natl Acad Sci U S A 93:15069–15074

Latz E, Xiao TS, Stutz A (2013) Activation and regulation of the inflammasomes. Nat Rev Immunol 13:397–411

Lawrence T, Gilroy DW (2007) Chronic inflammation: a failure of resolution? Int J Exp Pathol 88:85–94

Lehnardt S (2010) Innate immunity and neuroinflammation in the CNS: the role of microglia in Toll-like receptor-mediated neuronal injury. Glia 58:253–263

Lesné S, Koh MT, Kotilinek L, Kayed R, Glabe CG, Yang A, Gallagher M, Ashe KH (2006) A specific amyloid-beta protein assembly in the brain impairs memory. Nature 440:352–357

Li WX (2008) Canonical and non-canonical JAK-STAT signaling. Trends Cell Biol 18:545–551

Li J, Tang Y, Cai D (2012) IKKbeta/NF-kappaB disrupts adult hypothalamic neural stem cells to mediate a neurodegenerative mechanism of dietary obesity and pre-diabetes. Nat Cell Biol 14:999–1012

Li P, Spolski R, Liao W, Leonard WJ (2014) Complex interactions of transcription factors in mediating cytokine biology in T cells. Immunol Rev 261:141–156

Lim GP, Chu T, Yang F, Beech W, Frautschy SA, Cole GM (2001) The curry spice curcumin reduces oxidative damage and amyloid pathology in an Alzheimer transgenic mouse. J Neurosci 21:8370–8377

Lin J, Epel E, Blackburn E (2012) Telomeres and lifestyle factors: roles in cellular aging. Mutat Res 730:85–89

Liu Y, Tanabe K, Baronnier D, Patel S, Woodgett J, Cras-Méneur C, Permutt MA (2010a) Conditional ablation of Gsk-3beta in islet beta cells results in expanded mass and resistance to fat feeding-induced diabetes in mice. Diabetologia 53:2600–2610

Liu H, Li Z, Qiu D, Gu Q, Lei Q, Mao L (2010b) The inhibitory effects of different curcuminoids on β-amyloid protein, β-amyloid precursor protein and β-site amyloid precursor protein cleaving enzyme 1 in swAPP HEK293 cells. Neurosci Lett 485:83–88

Lossinsky AS, Shivers RR (2004) Structural pathways for macromolecular and cellular transport across the blood-brain barrier during inflammatory conditions. Histol Histopathol 19:535–564

Lourbopoulos A, Ertürk A, Hellal F (2015) Microglia in action: how aging and injury can change the brain's guardians. Front Cell Neurosci 9:54

Lourenco MV, Clarke JR, Frozza RL, Bomfim TR, Forny-Germano L, Batista AF, Sathler LB, Brito-Moreira J, Amaral OB, Silva CA, Freitas-Correa L, Espírito-Santo S, Campello-Costa P, Houzel JC, Klein WL, Holscher C, Carvalheira JB, Silva AM, Velloso LA, Munoz DP, Ferreira ST, De Felice FG (2013) TNF-α mediates PKR-dependent memory impairment and brain IRS-1 inhibition induced by Alzheimer's β-amyloid oligomers in mice and monkeys. Cell Metab 18:831–843

Lue LF, Kuo YM, Beach T, Walker DG (2010) Microglia activation and anti-inflammatory regulation in Alzheimer's disease. Mol Neurobiol 41:115–128

Lukens JN, Van Deerlin V, Clark CM, Xie SX, Johnson FB (2009) Comparisons of telomere lengths in peripheral blood and cerebellum in Alzheimer's disease. Alzheimers Dement 5:463–469

Lumeng CN, Saltiel AR (2011) Inflammatory links between obesity and metabolic disease. J Clin Invest 121:2111–2117

Lynch MA (1998) Analysis of the mechanisms underlying the age-related impairment in long-term potentiation in the rat. Rev Neurosci 9:169–201

Lynch AM, Murphy KJ, Deighan BF, O'Reilly JA, Gun'ko YK, Cowley TR, Gonzalez-Reyes RE, Lynch MA (2010) The impact of glial activation in the aging brain. Aging Dis 1:262–278

Ma QL, Yang F, Rosario ER, Ubeda OJ, Beech W, Gant DJ, Chen PP, Hudspeth B, Chen C, Zhao Y, Vinters HV, Frautschy SA, Cole GM (2009) Beta-amyloid oligomers induce phosphorylation of tau and inactivation of insulin receptor substrate via c-Jun N-terminal kinase signaling: suppression by omega-3 fatty acids and curcumin. J Neurosci 29:9078–9089

Ma T, Trinh MA, Wexler AJ, Bourbon C, Gatti E, Pierre P, Cavener DR, Klann E (2013) Suppression of eIF2α kinases alleviates Alzheimer's disease-related plasticity and memory deficits. Nat Neurosci 16:1299–1305

Masters SL, Dunne A, Subramanian SL, Hull RL, Tannahill GM, Sharp FA, Becker C, Franchi L, Yoshihara E, Chen Z, Mullooly N, Mielke LA, Harris J, Coll RC, Mills KH, Mok KH, Minghetti L (2005) Role of inflammation in neurodegenerative diseases. Curr Opin Neurol 18:315–321

McGeer PL, Schulzer M, McGeer EG (1996) Arthritis and anti-inflammatory agents as possible protective factors for Alzheimer's disease: a review of 17 epidemiologic studies. Neurology 47:425–432

Miao EA, Rajan JV, Aderem A (2011) Caspase-1-induced pyroptotic cell death. Immunol Rev 243:206–214

Minghetti L (2005) Role of inflammation in neurodegenerative diseases. Curr Opin Neurol 18:315–321

Minkiewicz J, de Rivero Vaccari JP, Keane RW (2013) Human astrocytes express a novel NLRP2 inflammasome. Glia 61:1113–1121

Mishra S, Mishra M, Seth P, Kumar S (2011) Tetrahydrocurcumin confers protection against amyloid beta-induced toxicity. Neuroreport 22:23–27

Mohorko N, Repovs G, Popovic M, Kovacs GG, Bresjanac M (2010) Curcumin labeling of neuronal fibrillar tau inclusions in human brain samples. J Neuropathol Exp Neurol 69:405–414

Mollace V, Muscoli C, Masini E, Cuzzocrea S, Salvemini D (2005) Modulation of prostaglandin biosynthesis by nitric oxide and nitric oxide donors. Pharmacol Rev 57:217–252

Moncada S, Bolaños JP (2006) Nitric oxide, cell bioenergetics and neurodegeneration. J Neurochem 97:1676–1689

Monje ML, Toda H, Palmer TD (2003) Inflammatory blockade restores adult hippocampal neurogenesis. Science 302:1760–1765

Mosieniak G, Sliwinska M, Piwocka K, Sikora E (2006) Curcumin abolishes apoptosis resistance of calcitriol-differentiated HL-60 cells. FEBS Lett 580:4653–4660

Mosley RL (1996) Aging, immunity and neuroendocrine hormones. Adv Neuroimmunol 6:419–432

Murray PJ (2007) The JAK-STAT signaling pathway: input and output integration. J Immunol 178:2623–2629

Murray CA, Lynch MA (1998) Evidence that increased hippocampal expression of the cytokine interleukin-1 beta is a common trigger for age-and stress-induced impairments in long-term potentiation. J Neurosci 18:2974–2981

Mutsuga M, Chambers JK, Uchida K, Tei M, Makibuchi T, Mizorogi T, Takashima A, Nakayama H (2012) Binding of curcumin to senile plaques and cerebral amyloid angiopathy in the aged brain of various animals and to neurofibrillary tangles in Alzheimer's brain. J Vet Med Sci 74:51–57

Nathan C (2002) Points of control in inflammation. Nature 420:846–852

Newsholme P, Nunez G, Yodoi J, Kahn SE, Lavelle EC, O'Neill LA (2010) Activation of the NLRP3 inflammasome by islet amyloid polypeptide provides a mechanism for enhanced IL-1beta in type 2 diabetes. Nat Immunol 11:897–904

Nimmerjahn A, Kirchhoff F, Helmchen F (2005) Resting microglial cells are highly dynamic surveillants of brain parenchyma in vivo. Science 308:1314–1318

Norling LV, Perretti M, Cooper D (2009) Endogenous galectins and the control of the host inflammatory response. J Endocrinol 201:169–184

Nunes T, de Souza HS (2013) Inflammasome in intestinal inflammation and cancer. Mediators Inflamm 2013:8

Oh I, Thaler JP, Ogimoto K, Wisse BE, Morton GJ, Schwartz MW (2010) Central administration of interleukin-4 exacerbates hypothalamic inflammation and weight gain during high-fat feeding. Am J Physiol Endocrinol Metab 299:E47–E53

Olson JK, Miller SD (2004) Microglia initiate central nervous system innate and adaptive immune responses through multiple TLRs. J Immunol 173:3916–3924

Ono K, Hasegawa K, Naiki H, Yamada M (2004) Curcumin has potent anti-amyloidogenic effects for Alzheimer's β-amyloid fibrils *in vitro*. J Neurosci Res 75:742–750

Packer MA, Stasiv Y, Benraiss A, Chmielnicki E, Grinberg A, Westphal H, Goldman SA, Enikolopov G (2003) Nitric oxide negatively regulates mammalian adult neurogenesis. Proc Natl Acad Sci U S A 100:9566–9571

Parkhurst CN, Yang G, Ninan I, Savas JN, Yates JR 3rd, Lafaille JJ, Hempstead BL, Littman DR, Gan WB (2013) Microglia promote learning-dependent synapse formation through brain-derived neurotrophic factor. Cell 155:1596–1609

Paul S, Mahanta S (2014) Association of heat-shock proteins in various neurodegenerative disorders: is it a master key to open the therapeutic door? Mol Cell Biochem 386:45–61

Pendurthi UR, Rao LV (2000) Suppression of transcription factor Egr-1 by curcumin. Thromb Res 97:179–189

Petersen RC, Thomas RG, Grundman D, Bennett R, Doody S, Ferris D, Galasko D, Jin S, Kaye J, Levey A, Pfeiffer E, Sano M, van Dyck CH, Thal LJ; Alzheimer's Disease Cooperative Study Group (2005) Vitamin E and donepezil for the treatment of mild cognitive impairment. N Engl J Med 352:2379–2388

Poole LB (2015) The basics of thiols and cysteines in redox biology and chemistry. Free Radic Biol Med 80:148–157

Pugh CR, Nguyen KT, Gonyea JL, Fleshner M, Wakins LR, Maier SF, Rudy JW (1999) Role of interleukin-1 beta in impairment of contextual fear conditioning caused by social isolation. Behav Brain Res 106:109–118

Quitschke WW, Steinhauff N, Rooney J (2013) The effect of cyclodextrin-solubilized curcuminoids on amyloid plaques in Alzheimer transgenic mice: brain uptake and metabolism after intravenous and subcutaneous injection. Alzheimers Res Ther 5:16

Reali C, Scintu F, Pillai R, Donato R, Michetti F, Sogos V (2005) S100β counteracts effects of the neurotoxicant trimethyltin on astrocytes and microglia. J Neurosci Res 8:677–686

Recchiuti A, Serhan CN (2012) Pro-resolving lipid mediators (SPMs) and their actions in regulating miRNA in novel resolution circuits in inflammation. Front Immunol 3:298

Rex CS, Lin CY, Kramár EA, Chen LY, Gall CM, Lynch G (2007) Brain-derived neurotrophic factor promotes long-term potentiation-related cytoskeletal changes in adult hippocampus. J Neurosci 27:3017–3029

Rhee S (2006) Cell signaling. H2O2, a necessary evil for cell signaling. Science 312:1882–1883

Ridet JL, Malhotra SK, Privat A, Gage FH (1997) Reactive astrocytes: cellular and molecular cues to biological function. Trends Neurosci 20:570–577

Rinaldi AL, Morse MA, Fields HW, Rothas DA, Pei P, Rodrigo KA, Renner RJ, Mallery SR (2002) Curcumin activates the aryl hydrocarbon receptor yet significantly inhibits (-)-benzo(a) pyrene-7R-trans- 7,8-dihydrodiol bioactivation in oral squamous cell carcinoma cells and oral mucosa. Cancer Res 62:5451–5456

Ringman JM, Frautschy SA, Teng E, Begum AN, Bardens J, Beigi M, Gylys KH, Badmaev V, Heath DD, Apostolova LG, Porter V, Vanek Z, Marshall GA, Hellemann G, Sugar C, Masterman DL, Montine TJ, Cummings JL, Cole GM (2012) Oral curcumin for Alzheimer's disease: tolerability and efficacy in a 24-week randomized, double blind, placebo-controlled study. Alzheimers Res Ther 4:43

Saijo K, Glass CK (2011) Microglial cell origin and phenotypes in health and disease. Nat Rev Immunol 11:775–787

Salter MW, Beggs S (2014) Sublime microglia: expanding roles for the guardians of the CNS. Cell 158:15–24

Sano M, Ernesto RG, Thomas MR, Klauber K, Schafer M, Grunderson P, Woodbury P, Growdon J, Cotman CW, Pfeiffer E, Schneider LS, Thal LJ (1997) A controlled trial of selegiline, alpha-tocopherol, or both as treatment for Alzheimer's disease. The Alzheimer's Disease Cooperative Study. N Engl J Med 336:1216–1222

Schiff L, Hadker N, Weiser S, Rausch C (2012) A literature review of the feasibility of glial fibrillary acidic protein as a biomarker for stroke and traumatic brain injury. Mol Diagn Ther 16:79–92

Schilling T, Lehmann F, Ruckert B, Eder C (2004) Physiological mechanisms of lysophosphatidylcholine-induced de-ramification of murine microglia. J Physiol 557:105–120

Schroder K, Tschopp J (2010) The inflammasomes. Cell 140:821–832

Schwab JM, Chiang N, Arita M, Serhan CN (2007) Resolvin E1 and protectin D1 activate inflammation-resolution programmes. Nature 447:869–874

Serhan CN (2009) Systems approach to inflammation resolution: identification of novel anti-inflammatory and pro-resolving mediators. J Thromb Haemost 7(Suppl 1):44–48

Serhan CN, Chiang N, Van Dyke TE (2008) Resolving inflammation: dual anti-inflammatory and pro-resolution lipid mediators. Nat Rev Immunol 8:349–361

Shishodia S, Sethi G, Aggarwal BB (2005) Curcumin: getting back to the roots. Ann N Y Acad Sci 1056:206–217

Simpson JE, Ince PG, Shaw PJ, Heath PR, Raman R, Garwood CJ, Gelsthorpe C, Baxter L, Forster G, Matthews FE, Brayne C, Wharton SB; MRC Cognitive Function and Ageing Neuropathology Study Group (2011) microarray analysis of the astrocyte transcriptome in the aging brain: relationship to Alzheimer's pathology and APOE genotype. Neurobiol Aging 32:1795–1807

Skaper SD, Facci L, Barbierato M, Zusso M, Bruschetta G, Impellizzeri D, Cuzzocrea S, Giusti P (2015) N-Palmitoylethanolamine and neuroinflammation: a novel therapeutic strategy of resolution. Mol Neurobiol 52:1034–1042

Sofroniew MV (2005) Reactive astrocytes in neural repair and protection. Neuroscientist 5:400–407

Sofroniew MV (2009) Molecular dissection of reactive astrogliosis and glial scar formation. Trends Neurosci 32:638–647

Stoll G, Jander S (1999) The role of microglia and macrophages in the pathophysiology of the CNS. Prog Neurobiol 58:233–247

Stranahan AM, Norman ED, Lee K, Cutler RG, Telljohann RS, Egan JM, Mattson MP (2008) Diet-induced insulin resistance impairs hippocampal synaptic plasticity and cognition in middle-aged rats. Hippocampus 18:1085–1088

Streit WJ (2004) Microglia and Alzheimer's disease pathogenesis. J Neurosci Res 77:1–8

Streit WJ, Mrak RE, Griffin WS (2004) Microglia and neuroinflammation: a pathological perspective. J Neuroinflammation 1:14

Sun GY, Horrocks LA, Farooqui AA (2007) The roles of NADPH oxidase and phospholipases A_2 in oxidative and inflammatory responses in neurodegenerative diseases. J Neurochem 103:1–16

Surh YJ, Chun KS, Cha HH, Han SS, Keum YS et al (2001) Molecular mechanisms underlying chemopreventive activities of anti-inflammatory phytochemicals: down-regulation of COX-2 and iNOS through suppression of NF-kappa B activation. Mutat Res 480–481:243–268

Szmydynger-Chodobska J, Strazielle N, Gandy JR, Keefe TH, Zink BJ, Ghersi-Egea JF, Chodobski A (2012) Posttraumatic invasion of monocytes across the blood-cerebrospinal fluid barrier. J Cereb Blood Flow Metab 32:93–104

Takenouchi T, Sato M, Kitani H (2007) Lysophosphatidylcholine potentiates Ca2+ influx, pore formation and p44/42 MAP kinase phosphorylation mediated by P2X7 receptor activation in mouse microglial cells. J Neurochem 102:1518–1532

Takeuchi O, Akira S (2010) Pattern recognition receptors and inflammation. Cell 140:805–820

Tan MS, Yu JT, Jiang T, Zhu XC, Tan L (2013) The NLRP3 inflammasome in Alzheimer's disease. Mol Neurobiol 48:875–882

Tang Y, Chen A (2014) Curcumin eliminates the effect of advanced glycation end-products (AGEs) on the divergent regulation of gene expression of receptors of AGEs by interrupting leptin signaling. Lab Invest 94:503–516

Tansey MG, Wyss-Coray T (2008) Cytokines in CNS inflammation and disease. In: Lane TE, Carson M, Bergmann C, Wyss-Coray T (eds) Central nervous system diseases and inflammation, 1st edn. Springer, New York, pp 59–106

Tansey MG, McCoy MK, Frank-Cannon TC (2007) Neuroinflammatory mechanisms in Parkinson's disease: potential environmental triggers, pathways, and targets for early therapeutic intervention. Exp Neurol 208:1–25

Thaler JP, Schwartz MW (2010) Minireview: inflammation and obesity pathogenesis: the hypothalamus heats up. Endocrinology 151:4109–4115

Thaler JP, Yi CX, Schur EA, Guyenet SJ, Hwang BH, Dietrich MO, Zhao X, Sarruf DA, Izgur V, Maravilla KR (2012) Obesity is associated with hypothalamic injury in rodents and humans. J Clin Invest 122:153–162

Thomas P, Callaghan NJO, Fenech M (2008) Telomere length in white blood cells, buccal cells and brain tissue and its variation with ageing and Alzheimer's disease. Mech Ageing Dev 129:183–190

Tramontina F, Tramontina AC, Souza DF, Leite MC, Gottfried C, Souza DO, Wofchuk ST, Gonçalves CA (2006) Glutamate uptake is stimulated by extracellular S100B in hippocampal astrocytes. Cell Mol Neurobiol 26:81–86

Tremblay ME, Majewska AK (2011) A role for microglia in synaptic plasticity? Commun Integr Biol 4:220–222

Tuppo EE, Arias HR (2005) The role of inflammation in Alzheimer's disease. Int J Biochem Cell Biol 37:289–305

Vahedi G, Takahashi H, Nakayamada S, Sun HW, Sartorelli V, Kanno Y, O'Shea JJ (2012) STATs shape the active enhancer landscape of T cell populations. Cell 151:981–993

Valassi E, Scacchi M, Cavagnini F (2008) Neuroendocrine control of food intake. Nutr Metab Cardiovasc Dis 18:158–168

van Eldik LJ, Wainwright MS (2003) The Janus face of glial-derived S100B: beneficial and detrimental functions in the brain. Restor Neurol Neurosci 21:97–108

Van Eldik LJ, Thompson WL, Ralay Ranaivo H, Behanna HA, Martin Watterson D (2007) Glia proinflammatory cytokine upregulation as a therapeutic target for neurodegenerative diseases: function-based and target-based discovery approaches. Int Rev Neurobiol 82:277–296

Walker DG, Dalsing-Hernandez JE, Campbell NA, Lue LF (2009) Decreased expression of CD200 and CD200 receptor in Alzheimer's disease: a potential mechanism leading to chronic inflammation. Exp Neurol 215:5–19

Wallace JL, Ferraz JG, Muscara MN (2012) Hydrogen sulfide: an endogenous mediator of resolution of inflammation and injury. Antioxid Redox Signal 17:58–67

Wang Y, Feng W, Xue W, Tan Y, Hein DW, Li XK, Cai L (2009a) Inactivation of GSK-3beta by metallothionein prevents diabetes-related changes in cardiac energy metabolism, inflammation, nitrosative damage, and remodeling. Diabetes 58:1391–1402

Wang YJ, Thomas P, Zhong JH, Bi FF, Kosaraju S, Pollard A, Fenech M, Zhou XF (2009b) Consumption of grape seed extract prevents amyloid-β deposition and attenuates inflammation in brain of an Alzheimer's disease mouse. Neurotox Res 15:3–14

Wang YP, Wu Y, Li LY, Zheng J, Liu RG, Zhou JP, Yuan SY, Shang Y, Yao SL (2011) Aspirin-triggered Lipoxin A4 attenuates LPS-induced pro-inflammatory responses by inhibiting activation of NF-κB and MAPKs in BV-2 microglial cells. J Neuroinflammation 8:95

Williams KW, Scott MM, Elmquist JK (2011) Modulation of the central melanocortin system by leptin, insulin, and serotonin: co-ordinated actions in a dispersed neuronal network. Eur J Pharmacol 660:2–12

Winterbourn CC, Hampton MB (2008) Thiol chemistry and specificity in redox signaling. Free Radic Biol Med 45:549–561

Wong WT (2013) Microglial aging in the healthy CNS: phenotypes, drivers, and rejuvenation. Front Cell Neurosci 7:22

Wood PL (1998) Neuroinflammation: mechanisms and management. Humana Press, Totowa, NJ

Xiong Z, Hongmei Z, Lu S, Yu L (2011) Curcumin mediates presenilin-1 activity to reduce β-amyloid production in a model of Alzheimer's disease. Pharmacol Rep 63:1101–1108

Yamamoto M, Sato S, Hemmi H, Hoshino K, Kaisho T, Sanjo H, Takeuchi O, Sugiyama M, Okabe M, Takeda K, Akira S (2003) Role of adaptor TRIF in the MyD88-independent toll-like receptor signaling pathway. Science 301:640–643

Yamamoto M, Takeda K, Akira S (2004) TIR domain-containing adaptors define the specificity of TLR signaling. Mol Immunol 40:861–868

Yanagisawa D, Shirai N, Amatsubo T, Taguchi H, Hirao K, Urushitani M, Morikawa S, Inubushi T, Kato M, Kato F, Morino K, Kimura H, Nakano I, Yoshida C, Okada T, Sano M, Wada Y, Wada KN, Yamamoto A, Tooyama I (2010) Relationship between the tautomeric structures of curcumin derivatives and their Aβ-binding activities in the context of therapies for Alzheimer's disease. Biomaterials 31:4179–4185

Yang F, Lim GP, Begum AN, Ubeda OJ, Simmons MR, Ambegaokar SS, Chen PP, Kayed R, Glabe CG, Frautschy SA, Cole GM (2005) Curcumin inhibits formation of amyloid β oligomers and fibrils, binds plaques, and reduces amyloid *in vivo*. J Biol Chem 280:5892–5901

Yang Y, Duan W, Lin Y, Yi W, Liang Z, Yan J, Wang N, Deng C, Zhang S, Li Y, Chen W, Yu S, Yi D, Jin Z (2013) SIRT1 activation by curcumin pretreatment attenuates mitochondrial oxidative damage induced by myocardial ischemia reperfusion injury. Free Radic Biol Med 65C:667–679

Yeon KY, Kim SA, Kim YH, Lee MK, Ahn DK, Kim HJ, Kim JS, Jung SJ, Oh SB (2010) Curcumin produces an antihyperalgesic effect via antagonism of TRPV1. J Dent Res 89:170–174

Yu BP (1994) Cellular defenses against damage from reactive oxygen species. Physiol Rev 74:139–162

Zanni GR, Wick JY (2011) Telomeres: unlocking the mystery of cell division and aging. Consult Pharm 26:78–90

Zhang L, Fiala M, Cashman J, Sayre J, Espinosa A, Mahanian M, Zaghi J, Badmaev V, Graves MC, Bernard G, Rosenthal M (2006) Curcuminoids enhance amyloid-β uptake by macrophages of Alzheimer's disease patients. J Alzheimers Dis 10:1–7

Zhang X, Zhang G, Zhang H, Karin M, Bai H, Cai D (2008) Hypothalamic IKKbeta/NF-kappaB and ER stress link overnutrition to energy imbalance and obesity. Cell 135:61–73

Zhang C, Browne A, Child D, Tanzi RE (2010) Curcumin decreases amyloid-beta peptide levels by attenuating the maturation of amyloid-beta precursor protein. J Biol Chem 285:28472–28480

Zhang GLJ, Purkayashtha S, Tang Y, Zhang H, Yin Y, Li B, Liu G, Cai D (2013) Hypothalamic control of systemic aging through IKKB/NF-KB activation and GnRH decline. Nature 497:211–216

Zheng K, An JJ, Yang F, Xu W, Xu ZQ, Wu J, Hökfelt TG, Fisahn A, Xu B, Lu B (2011) TrkB signaling in parvalbumin-positive interneurons is critical for gamma-band network synchronization in hippocampus. Proc Natl Acad Sci U S A 108:17201–17206

Zhou Y, Ling EA, Dheen ST (2007) Dexamethasone suppresses monocyte chemoattractant protein-1 production via mitogen activated protein kinase phosphatase-1 dependent inhibition of Jun N-terminal kinase and p38 mitogen-activated protein kinase in activated rat microglia. J Neurochem 102:667–678

Zhu HT, Bian C, Yuan JC, Chu WH, Xiang X, Chen F, Wang CS, Feng H, Lin JK (2014) Curcumin attenuates acute inflammatory injury by inhibiting the TLR4/MyD88/NF-κB signaling pathway in experimental traumatic brain injury. J Neuroinflammation 11:59

Ziebell JM, Morganti-Kossmann MC (2010) Involvement of pro- and anti-inflammatory cytokines and chemokines in the pathophysiology of traumatic brain injury. Neurotherapeutics 7:22–30

Chapter 8
Therapeutic Importance of Curcumin in Neurological Disorders Other Than Alzheimer Disease

8.1 Introduction

Curcumin (diferuloylmethane) is one of the active components of dietary spice turmeric (Curcuma longa Linn). It has antioxidant, anti-inflammatory and cancer chemo-preventive properties (McNally et al. 2007; Goel et al. 2008). Curcumin reduces oxidative damage and improves cognitive functions related to aging process. Both in vitro and in vivo studies have indicated that curcumin binds with Aβ and inhibits its aggregation (Yang et al. 2005; Hong et al. 2009), as well as fibril and oligomer formation (Yang et al. 2005). In vivo studies have shown that dietary curcumin not only crosses the blood-brain barrier (BBB) and significantly decreases Aβ deposition and plaque burden in AD transgenic mice (Yang et al. 2005; Lim et al. 2001; Wang et al. 2009; Garcia-Alloza et al. 2007), but markedly inhibits Tau phosphorylation (Ma et al. 2009). Furthermore, curcumin promotes stem cell-mediated recovery from ischemic and spinal cord injury (Ormond et al. 2014). In this regard, curcumin may enhance proliferation of stem cells for swift regeneration (Ormond et al. 2014; Kim et al. 2008). Curcumin mediates biphasic effects on the proliferation of some stem cells including spinal cord neural progenitor cells (Son et al. 2014), embryonic neural progenitor cells (Kim et al. 2008) and 3T3-L1 preadipocytes (Kim et al. 2011). These effects of curcumin are mediated by activation of extracellular signal-regulated kinases (ERKs) and p38 kinases, which are associated the regulation of neuronal plasticity and stress responses (Kim et al. 2008).

Curcumin attenuates neuroinflammation by inhibiting cyclooxygenase (COX-2), phospholipases (PLA$_2$s), transcription factor and enzymes associated with metabolism of neural membrane phospholipids into prostaglandins. The reduction of the release of ROS by stimulated neutrophils, and inhibition of AP-1 and NF-κB blocks the activation of the pro-inflammatory cytokines TNF-α and IL-1β (Kim et al. 2005). Antioxidant activity of curcumin is mediated through the modulation of Nrf2-keap1 pathway, and reduction in genomic instability events (Yang et al. 2005; Wang et al. 2009; Thomas et al. 2009). As stated in Chap. 3, Nrf2 is present primarily in the

© Springer International Publishing Switzerland 2016

A.A. Farooqui, *Therapeutic Potentials of Curcumin for Alzheimer Disease*,

DOI 10.1007/978-3-319-15889-1_8

cytoplasm, where it remains bound to the BTB-Kelch-like ECH-associated protein 1 (Keap1), which acts as a receptor of electrophilic compounds and promotes Nrf2 ubiquitination for subsequent degradation by 26S proteasome complex (Anderica-Romero et al. 2013). Modification of Keap 1 by oxidation or binding with curcumin releases Nrf2, which then migrates into the nucleus where it binds as a heterodimer to the antioxidant responsive element in DNA to initiate target gene expression. Nrf2-regulated genes include antioxidants enzymes, molecular chaperones, DNA repair enzymes, and anti-inflammatory response proteins (Bryan et al. 2013). These proteins promote the reduction in ROS generation whilst increasing the ability of the cell to repair any subsequent damage (Bryan et al. 2013; Gupta et al. 2011). Curcumin also suppress pro-inflammatory pathways by blocking both the production of TNF-α, IL-1β, and other proinflammatory cytokines including IL-8, MIP-1β, and MCP-1 in astrocytes and microglial cells. Curcumin promotes reduction in GFAP expression and improves spatial memory in the Aβ-induced rat model of AD. Curcumin retards COX-2 and glial fibrillary acidic protein expression in Aβ treated astrocytes (Wang et al. 2013). The anti-neuroinflammatory effects of curcumin can be blocked by a PPARγ antagonist GW9662 supporting the view that curcumin may act as a PPARγ agonist to inhibit Aβ induced neuroinflammation (Wang et al. 2010a). The treatment of athymic mice with NanoCurc™ (a polymeric nanoparticle encapsulated curcumin) not only decreases in the levels of H_2O_2, but also results in caspase-3 and caspase-7 activities in the brain. These neurochemical changes are accompanied by increase in levels of reduced glutathione (GSH) (Ray et al. 2011), suggesting that curcumin retards oxidative stress. In addition, curcumin mediates the induction of heat shock proteins (Kato et al. 1998). It also modulates several target genes including JUN, STAT3, APP and GSK3B, once again supporting the view that curcumin may modulate the deposition of Aβ and inflammatory responses in the brain. Curcumin also regulates apoptotic cell death through its interaction with TNFRSF1A, CASP7 and CASP8. In addition, curcumin has been reported to improve learning and memory (Pan et al. 2008) by regulating many genes associated with memory formation and cognitive function such as CREBBP, EP300, HDAC1 and NR3C1. Curcumin interacts with steroid hormone receptors (AR, ESR1, and NR3C1), which not only protect neurons from Aβ toxicity, may also increase their survival in a variety of coincidental insults including AD-associated neurotoxicity (Liang et al. 2012). Collective evidence suggests that curcumin produces its beneficial effects in animal models of AD through above mentioned mechanisms. Curcumin also mediates its beneficial effects in neurological disorders other than AD through its antioxidant, anti-inflammatory, and antidiabetic effects.

8.2   Neurological Disorders

Diseases of brain, spinal cord, and nerves are called as neurological disorders. More than 600 neurological disorders have been described in the literature. Structural, neurochemical, and electrophysiological abnormalities in the brain, spinal cord or

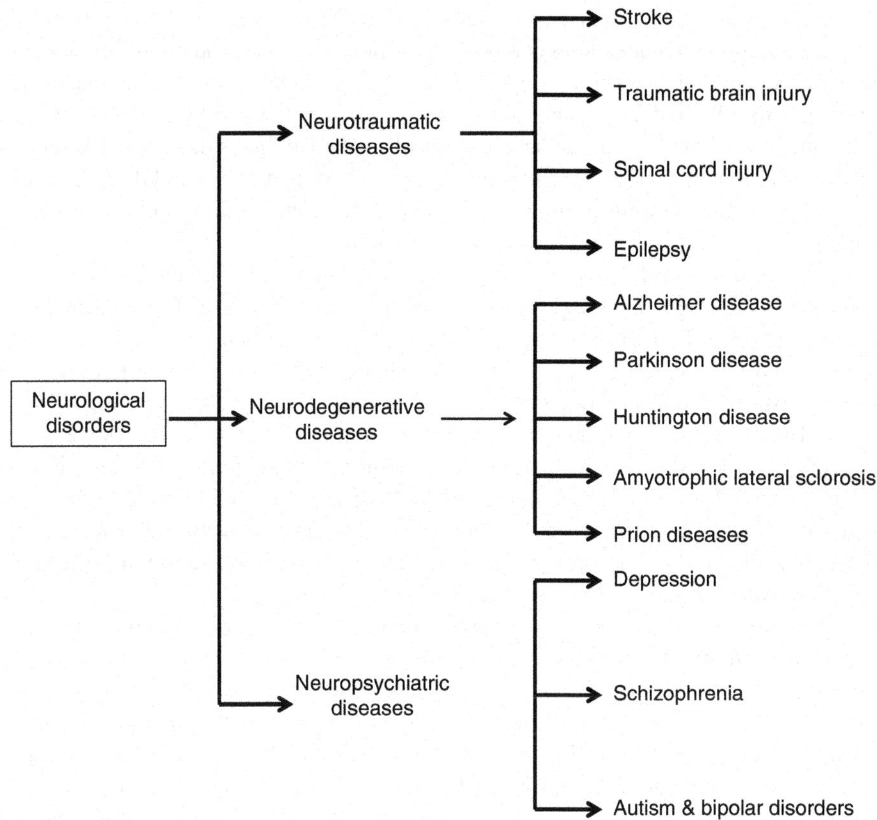

Fig. 8.1 Classification of neurological disorders

other nerves may cause a variety of symptoms such as paralysis, muscle weakness, poor coordination, seizures, confusion, pain and altered levels of consciousness (Farooqui 2010; Deleidi et al. 2015). For the sake of simplicity, I will classify neurological disorders into three classes: neurotraumatic, neurodegenerative, and neuropsychiatric diseases. Common neurotraumatic diseases are strokes, traumatic brain injury (TBI), spinal cord injury (SCI), and epilepsy (Farooqui 2010). Common neurodegenerative diseases include Alzheimer disease (AD), Parkinson disease (PD), Huntington disease (HD), amyotrophic lateral sclerosis (ALS), multiple sclerosis (MS), and prion diseases (Fig. 8.1) (Farooqui 2010). Neuropsychiatric diseases include both neurodevelopmental disorders and behavioral or psychological difficulties associated with some neurological disorders. Examples of neuropsychiatric disorders are depression, schizophrenia, some forms of bipolar affective disorders, autism, mood disorders, attention-deficit disorder, dementia, tardive dyskinesia, and chronic fatigue syndrome. Neuropsychiatric diseases involve the abnormalities in cerebral cortex and limbic system (thalamus, hypothalamus, hippocampus, and amygdale) (Farooqui 2010). In addition, various types of brain tumors also fall under neurological disorders.

Among neurotraumatic diseases, stroke is a metabolic insult, which is caused by severe reduction or blockade in cerebral blood flow. This blockade not only results in the deficiency of oxygen and reduction in glucose metabolism, but also leads to reduction in ATP and accumulation of toxic products. TBI and SCI are caused by falls and motor cycle and car accidents, and shaken infant syndrome (Farooqui 2010; Deleidi et al. 2015). Neurotraumatic and neurodegenerative and diseases also involve muscle dystrophy leading to a decline in neuronal and muscular functions which often limit quality of life as well as life span.

Neurological disorders share oxidative stress and neuroinflammation as common mechanisms of brain injury and neural cell death. In addition to oxidative stress, and neuroinflammation, neurotraumatic diseases involve ischemia, marked increase in intracellular Ca^{2+} and marked reduction in ATP, and rapid loss of ion homeostasis. In contrast, neurodegenerative diseases are accompanied by the accumulation of misfolded proteins, mitochondrial and proteasomal dysfunction, loss of synapses, and premature and slow death of certain neuronal populations in brain tissue (Graeber and Moran 2002; Farooqui 2010; Deleidi et al. 2015). For example in AD, neuronal degeneration occurs in the nucleus basalis, whereas in PD, neurons die in the substantia nigra. The most severely affected neurons in Huntington disease (HD) are striatal medium spiny neurons (Farooqui 2010).

The most important risk factors for stroke and neurodegenerative diseases are old age, race/ethnicity, a positive family history, unhealthy lifestyle, and endogenous factors (Farooqui and Farooqui 2009). The onset of stroke and neurodegenerative diseases is often subtle and usually occurs in mid to late life. Their progression depends not only on genetic, but also on environmental factors (Graeber and Moran 2002). The onset of neurological diseases may occur when neurons fail to respond adaptively to age-related increase in oxidative and nitrosative stress and neuroinflammation. The persistence presence of oxidative and nitrosative stress and neuroinflammation induces the accumulation of damaged proteins, DNA, and membrane fragments.

Like neurotraumatic and neurodegenerative diseases, neurochemical changes in neuropsychiatric diseases involve mild oxidative and neuroinflammation (Morris and Berk 2015). At the nuclear level, abnormalities in neuropsychiatric diseases may be regulated by overexpression or underexpression of genes and alterations in neurotransmitters that modulate behavioral symptoms, such as thoughts or actions, delusions, delirium, and hallucinations. These behavioral abnormalities are the hallmarks of many neuropsychiatric diseases. In addition to signal transduction processes associated with dopamine, glutamate and γ-aminobutyric acid (GABA) receptors-mediated behavioral abnormalities (Luscher et al. 2011), neuropsychiatric disorders also involve gray matter atrophy caused by reduction in neuronal and glial size, increase in cellular packing density, disruption in neuronal connectivity, particularly in the dorsolateral prefrontal cortex, and distortions in neuronal orientation (Blitzer et al. 2005). These neurochemical and morphological changes may simultaneously mediate alterations within a single microcircuit in more than one region. Changes in microcircuits and neurotransmitters (synthesis and transport) may not only vary on a region-by-region basis but also from one neuropsychiatric disease to another. Both macro- and microcircuitry within the specific brain system (such as limbic system) may serve as "triggers" for the onset of neuropsychiatric condition (Benes 2000; Harrison 1999; Morris and Berk 2015). Neurochemical and neuroimaging studies have also indicated alterations in cerebral blood flow and glucose utilization in the limbic system and prefrontal cortex of patients with major depression and other neuropsychiatric diseases (Ito et al. 1996; Kimbrell et al. 2002; Morris and Berk 2015). Converging evidence suggests that neuropsychiatric diseases are mediated by genetic factors, and physiological conditions alterations in blood flow, disruption of cellular connectivity, decrease in neurogenesis, alterations in microcircuitry, decrease in neuroplasticity along with mild oxidative stress, and mild neuroinflammation.

8.3 Therapeutic Importance of Curcumin in Neurotraumatic Diseases

Stroke, TBI, and SCI are accompanied by the induction oxidative stress, neuroinflammation, and alterations in neuroplasticity. Significant information has been provided on neurochemical aspects of oxidative stress and neuroinflammation in Chaps. 6 and 7 respectively. Neuroplasticity is defined as a process associated with reorganization or adaptation in the brain tissue that leads to the formation of new neural connections. Neuroplasticity allows the neuron not only to compensate for neuronal injury, but also promotes neurons to respond and adjust normal homeostasis in their environment (Thickbroom and Mastaglia 2009). Neuroplasticity is supported by neurons, glia, and vascular cells. Neuroplasticity not only restores memory formation, but also promotes cognitive function, which are progressively lost in neurotraumatic diseases (Farooqui 2010).

Curcumin not only reduces oxidative stress and neuroinflammation caused by neurotraumatic injury, but also reverses the behavioral deficits mediated by metabolic and traumatic injuries (Fig. 8.2). The mechanism associated with the beneficial effects of curcumin is not fully understood. However, recent studies have indicated that curcumin acts not only by modulating hypothalamus-pituitary-adrenal disturbances, lowering inflammation and protecting against oxidative stress, mitochondrial damage, neuroprogression and intestinal hyperpermeability, but also by promoting and restoring memory deficits in a dose dependent manner (Bergman et al. 2013; Lopresti et al. 2012). In addition, curcumin also modulates levels of norepinephrine, dopamine, and serotonin and enhances neurogenesis, notably in the frontal cortex and hippocampal regions of the brain (KulKarni et al. 2008; Kulkarni et al. 2009; Xu et al. 2007).

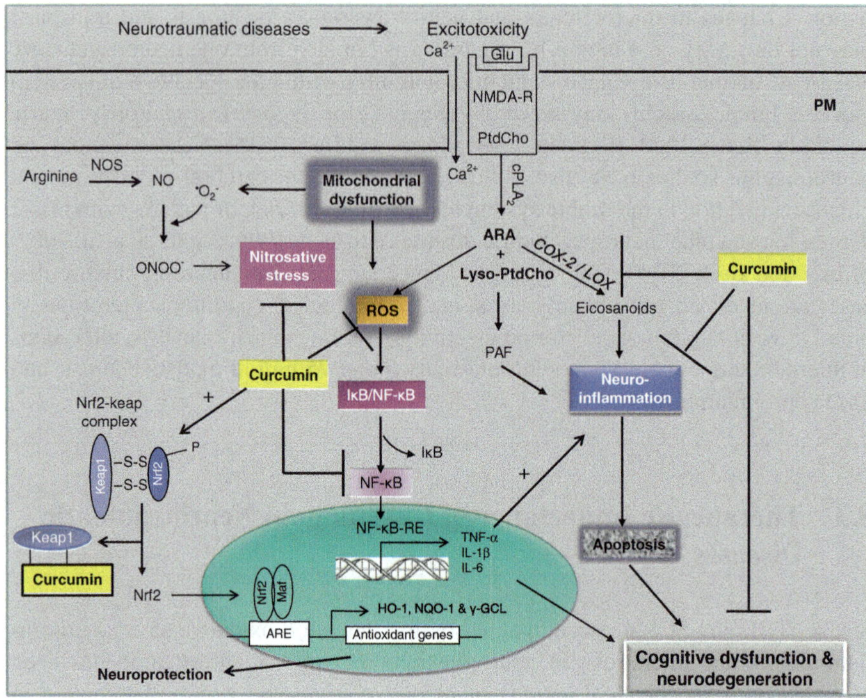

Fig. 8.2 A hypothetical signal transduction diagram showing the sites of actions of curcumin on signal transduction processes in neurotraumatic diseases. Plasma membrane (*PM*); *N*-methyl-D-aspartate receptor (*NMDA-R*); Glutamate (*Glu*); Phosphatidylcholine (*PtdCho*); cytosolic phospholipase A₂ (*cPLA₂*); lysophosphatidylcholine (*lyso-PtdCho*); cyclooxygenase (*COX*); lipoxygenase (*LOX*); arachidonic acid (*ARA*); platelet activating factor (*PAF*); reactive oxygen species (*ROS*); nuclear factor-κB (*NF-κB*); nuclear factor-κB-response element (*NF-κB-RE*); inhibitory subunit of NF-κB (*I-κB*); tumor necrosis factor-α (*TNF-α*); interleukin-1β (*IL-1β*); interleukin-6 (*IL-6*); nitric oxide (*NO*); Nuclear factor (erythroid-derived 2)-like 2 (*Nrf2*); and Kelch like-ECH-associated protein 1 (*Keap1*). *Blocked arrow* indicates sites of curcumin action. *Positive sign* indicates stimulation

8.3.1   Effects of Curcumin on Stroke-Mediated Neuronal Injury

Stroke is one of the most important causes of mortality and morbidity in Western countries. New or recurrent strokes affect about 780,000 Americans every year. On average, someone in the United States has a stroke every 40 s. While age is the major risk factor, patients who have a stroke are likely to have more than one risk factor (high blood pressure, atherosclerosis, and metabolic syndrome) (Farooqui 2010; Manea et al. 2015). Approximately 85 % of strokes are ischemic, the remainder 15 % being haemorrhagic. More than half of the stroke patients are left with a major disability (Adamson et al. 2004). As stated above, stroke is caused by reduction or block in blood flow due to formation of a clot leading to reversible and irreversible alterations in normal cellular function (Farooqui 2010). Different brain regions are more sensitive to stroke-mediated neuronal injury than others. The white matter is more resilient to ischemic injury than gray matter (Mattson et al. 2001). In addition, certain populations of neurons are selectively more vulnerable to ischemic injury than other neurons. For example, in the hippocampus, CA1 pyramidal neurons are highly susceptible to ischemic injury, whereas dentate granule neurons are more resistant (Mattson et al. 2001). Stroke impairs the neurologic functions through the breakdown of cellular and subcellular integrity mediated by excitotoxic glutamatergic signaling, Ca^{2+}-influx, alterations in ionic balance and redox, and free-radical generation (Farooqui et al. 2008; Farooqui 2010). These processes involve stimulation of phospholipases A_2, C, and D (PLA_2, PLC and PLD), calcium/calmodulin-dependent kinases (CaMKs), mitogen-activated protein kinases (MAPKs) such as extracellular signal-regulated kinase (ERK), p38, and c-Jun N-terminal kinase (JNK), nitric oxide synthases (NOS), calpains, calcinurin, and endonucleases. Stimulation of these enzymes bring them in contact with appropriate substrates and modulates cell survival/degeneration mechanisms (Hou and MacManus 2002; Farooqui and Horrocks 2007). Stroke-mediated brain injury also exerts a potent suppressive effect on lymphoid organs, which promotes intercurrent infections, a major determinant of stroke morbidity and mortality (Meisel et al. 2005; Urra et al. 2009). Therefore, the immune system is closely related to critical events determining the fate of the ischemic brain and the survival of stroke patients. Thus, the molecular events caused by cerebral ischemia not only activate components of innate immunity, but also promote inflammatory signaling that contributes to brain damage. Inflammatory signaling contributes to early molecular events triggered by the arterial occlusion and culminating in the invasion of the brain by blood-borne leukocytes. The ultimate goal of neuroinflammation is to reestablish homeostasis. Acute neuroinflammation inflicts considerable damage to the metastable penumbral tissue. Adaptive immunity is deeply involved in the central and peripheral events triggered by stroke, but evidence that a classical autoimmune response against brain antigens unveiled by tissue damage contributes to the acute phase of the damage is lacking (Meisel et al. 2005; Urra et al. 2009).

A single injection of curcumin 30 min after focal cerebral ischemic/reperfusion injury in rats produces beneficial effects not only by decreasing infarct volume, improving neurological deficit, and reducing mortality, but also by reducing the water content of the brain in a dose-dependent manner (Jiang et al. 2007). In cultured astrocytes, curcumin significantly inhibits iNOS expression. It is proposed that curcumin also acts not only by preventing peroxynitrite mediated BBB alterations, but also by decreasing lipid peroxidation-mediated damage, preventing glial cell activation, and retarding apoptotic cell death (Jiang et al. 2007). Curcumin also induces expression of genes encoding phase II drug-metabolizing enzymes such as NAD(P)H:quinone oxidoreductase1 (NQO1), glutathione S-transferase (GST), aldoketo-reductase (AR), and hemeoxygenase-1 (HO-1) (Scapagnini et al. 2004, 2006; Yang et al. 2009; Wu et al. 2013). Collectively, these studies suggest that neuroprotective activity of curcumin in cerebral ischemia is mediated through its antioxidant, anti-inflammatory, and apoptotic activities (Zhao et al. 2010).

8.3.2   Effects of Curcumin on Traumatic Brain Injury (TBI)-Mediated Injury

Traumatic brain injury (TBI), which is caused by fall or motor cycle or car accidents is accompanied by mechanical trauma to head (primary injury), which involves rapid deformation of brain tissue and rupture of neural cell membranes leading to the release of intracellular contents, disruption of blood flow, breakdown of the blood brain barrier, intracranial hemorrhage, brain edema, and axonal shearing, in which the axons of neurons are stretched and torn. The primary injury is followed secondary injury, which at the cellular level involves activation of microglial cells and astrocytes, and demyelination involving oligodendroglia. At the molecular level TBI involves a complex cascade of neurochemical processes, such as onset of oxidative stress, excitotoxicity, and neuroinflammation along with increase in inflammatory mediators, free radical damage, thrombosis, and macromolecule extravasation (Raghupathi 2004; Maas et al. 2008; Donkin and Vink 2010; Chauhan 2014). Importantly, all of these mechanisms contribute to the development of cerebral edema and compromised blood–brain barrier (BBB) function. Mitochondrial dysfunction at the neuronal/astrocytic level has been reported to be a key characteristic of TBI pathophysiology (Motori et al. 2013; Signoretti et al. 2004). TBI can lead to cognitive, behavioral, and motor deficits. Adult brain responds to TBI by inducing reactive gliosis and decreasing levels of brain-derived neurotrophic factor (BDNF), leading to cognitive impairment. TBI-mediated decrease in BDNF can be counteracted by exercise-mediated increase in BDNF level (Griesbach et al. 2009). Processes that mediate induction of BDNF and activation of its intracellular receptors can produce neural regeneration, reconnection, and dendritic sprouting, and can improve synaptic efficacy (Rostami et al. 2014). It is important to note that multiple studies support the hypothesis that single moderate-severe TBI is an important risk factor for AD (Fleminger et al. 2003), PD (Bower et al. 2003), and amyotrophic lateral sclerosis (ALS) (Chen et al. 2007; Schmidt et al. 2010). TBI not only increases incidence of seizures, sleep disorders, and neuroendocrine dysregulation, but also contributes to non-neurological disorders such as sexual dysfunction, bladder and bowel incontinence, and systemic metabolic dysregulation that may arise and/or persist for many months post-injury (Xiong et al. 2013).

Curcumin treatment in TBI not only dramatically reduces oxidative damage and neuroinflammation, but also normalizes levels of BDNF, synapsin I, decreases the levels of AMP-activated protein kinase (AMPK), ubiquitous mitochondrial creatine kinase (uMtCK) and cytochrome c oxidase II and induce changes in CREB (Wu et al. 2006; Sharma et al. 2009). Additionally, curcumin and pyrazole curcumin derivative reduce increase calcium-independent phospholipase A_2 (iPLA$_2$) activity and decrease levels of 4-hydroxynonenal. Furthermore, curcumin supplementation counteracts the cognitive impairment induced by TBI and effectively restores parameters of membrane homeostasis. Converging evidence suggests that curcumin increases expression of BDNF, which modulates synaptic plasticity, membrane homeostasis, and cognition in the injured brain (Wu et al. 2006; Sharma et al. 2009, 2010).

8.3.3   Effects of Curcumin on Spinal Cord Injury (SCI)

Trauma to spinal cord due to fall or motor cycle and car accidents results in autode-structive neurochemical changes that lead to varying degrees of apoptotic and necrotic cell death along with paralysis. Like TBI, SCI consists of two broadly defined events: a primary event, which is caused by the mechanical insult, and a secondary event, which involves a series of systemic and local neurochemical changes that occur in spinal cord after the initial traumatic insult (Klussmann and Martin-Villalba 2005). Primary event is instantaneous and beyond therapeutic man-agement, but the secondary event develops over the hours and days after SCI, caus-ing neurochemical alterations resulting in behavioral and functional impairments. Neurochemical alterations in SCI involve a complex cascade of neurochemical events, such as the release of glutamate, induction excitotoxicity, influx of calcium ions, activation of calcium-dependent enzymes (phospholipase A_2, nitric oxide syn-thases, proteases, endonucleases, and matrix metalloproteinase), release of proin-flammatory cytokines and chemokines, generation of proinflammatory lipid mediators (eicosanoids) and onset of neuroinflammation, and oxidative stress. These neurochemical processes are accompanied by the activation of microglial cells, recruitment of neutrophils, and activation of macrophages and vascular endo-thelial cells and T cells leading to the onset of acute neuroinflammation, and oxida-tive stress (Bruce et al. 2000). Production of ROS downregulates proteins of tight junctions and activates matrix metalloproteinases (MMPs). These processes pro-mote the opening of BBB (Fehlings and Tighe 2008). Moreover, loosening of the vasculature and perivascular unit by oxidative stress not only results in further acti-vation of MMPs, but also facilitate opening of aquaporin fluid channel promoting vascular or cellular fluid edema, and enhancing the leakiness of the BBB. These processes contribute to a failure in normal neural function and spinal shock, and represent a generalized failure of circuitry in the spinal neural network. Hemorrhaging, localized edema, thrombosis, and vasospasm further exacerbate the neural injury. Inhibitory elements (neurite out growth inhibitor, myelin-associated glycoprotein, oligodendrocyte-myelin glycoprotein, and chondroitin sulfate proteo-glycan) in the spinal cord tissue inhibit damaged nerve fibers to exhibit regenerative sprouting (Filbin 2003).

SCI is also accompanied by the up-regulation of cell cycle proteins, which may contribute not only to apoptotic cell death of neurons and oligodendrocytes, but also to post-traumatic gliosis and microglial activation (Wu et al. 2011). Significant neuronal loss occurs in the hippocampus at 12 weeks but not 8 days after the SCI (Wu et al. 2014). Detailed investigations have indicated that SCI increases in expression of cell-cycle-related gene (cyclins A1, A2, D1, E2F1, and PCNA) and protein (cyclin D1 and CDK4) in hippocampus and cerebral cortex. Systemic administration of the selective cyclin-dependent kinase inhibitor CR8 after SCI sig-nificantly reduces cell cycle gene and protein expression along with microglial activation and neurodegeneration in the brain, which may contribute to cognitive decline, and depression. These studies support the view that SCI can initiate a chronic brain neuro-degenerative response, likely related to delayed, sustained induction of M1-type microglia and related cell cycle activation, which result in cognitive deficits and physiological depression (Wu et al. 2014). SCI also leads to the formation of glial scar, which is a strong physical and chemical barrier that inhibits axonal regeneration. However, axonal sprouting occurs at a site distal to the glial scar, which lays a foundation for reconstruction of the neural circuit based on injured neural compensatory connections (Fouad and Tse 2008). It should be noted that understanding of extent of SCI, and the time since its occurrence is critical when trying to assess long-term reorganization and the potential for recovery of function from SCI.

Curcumin produces beneficial effects in SCI not only by inhibiting oxidative stress, neuroinflammation, and apoptotic cell death, but also quenching astrocyte activation leading to significant improvements in neurologic deficit, and restoration of cellular homeostasis and normalization of redox equilibrium around the injury site (Lin et al. 2011a). Curcumin not only reduces the expression of proinflammatory cytokines, chemokines, and the glial fibrillary acidic protein (GFAP), but also suppresses the reactive gliosis by inhibiting the generation of TGF-β1, TGF-β2, SOX-9, and by decreasing the deposition of chondroitin sulfate proteoglycan, and improving the microenvironment for nerve growth.

Curcumin also inhibits the activation of signal transducer and activator of transcription-3 (STAT-3) and NF-κB in the injured spinal cord (Wang et al. 2014a). As mentioned above, curcumin treatment greatly reduces the astrogliosis in SCI mice and significantly decreases the expression of IL-1β and NO, as well as the number of Iba1$^+$ inflammatory cells at the lesion site (Wang et al. 2014a). Converging evidence suggests that curcumin increases neuronal survival not only by inhibiting neuroinflammation and oxidative stress, but also by attenuating astrocyte reactivation along with suppression of glial scar formation (Lin et al. 2011a, b; Wang et al. 2014a; Yuan et al. 2015). Curcumin also increases tissue levels of glutathione and glutathione peroxidase and catalase, which may be beneficial for neuronal survival (Cemil et al. 2010). Curcumin also activates Nrf2 target genes, and particularly HO-1, in astrocytes and neurons leading to its neuroprotective effects (Scapagnini et al. 2011).

8.3.4 Effects of Curcumin on Epilepsy

Epilepsy is a common neurological condition, which is generally associated with certain psychiatric comorbidities. In epilepsy, a variety of structures of the brain such as the hippocampus, the amygdala and the piriform cortex are susceptible to trigger electrical discharges, which initiate brain damage through the epileptogenic mechanisms (Blümcke et al. 1999). These discharges not only produce cellular death in the CA1, but also cause mossy fiber sprouting and dispersion in the granule cell layer altering the formation of recurrent excitatory circuits associated with seizure susceptibility (Heck et al. 2004). Epilepsy is characterized by a dominant symptom (seizures), the disruption of network homeostasis, and changes in glial cell physiology (David et al. 2009; Boison et al. 2013; Devinsky et al. 2013). Most information about mechanisms of epilepsy has been obtained from animal models. More recently, these studies are being validated in human tissue using cellular and molecular approaches applied to either surgical biopsy material or autopsy samples from epileptic patients. Currently available models for adult-onset and development of epilepsies include rat models of status epilepticus-induced, post-traumatic, and post-stroke epilepsies (Pitkanen et al. 2007; Dichter 2006). Disruption of signaling network homeostasis is probably caused by the dysregulation of several signal transduction pathways including alterations in water/K^+, glutamate, and adenosine homeostasis simultaneously (Boison et al. 2013). Converging evidence suggests that epilepsy is a complex temporal and spatial abnormalities in neural network structure and activity mediated by post-translational modifications of proteins, activation of immediate early genes (IEGs), and other alterations in profiles of gene expression and function that eventually lead to deregulation of neural circuits with a predisposition for synchronous electrical activity (Rakhade and Jensen 2009). At the molecular level, epileptogenesis involves depolarization of GABA receptors, hyperexcitability of AMPA, and NMDA receptors, and disruption of adenosine homeostasis along with changes in adenosine receptor-mediated pathways including mitochondrial bioenergetics, and adenosine receptor-independent changes to the epigenome (Silverstein and Jensen 2007; Qureshi and Mehler 2010a, b; Boison et al. 2013). At the cellular level, activation of microglia, production of proinflammatory cytokines, and onset of oxidative stress in the brain play an active role in the pathogenesis of epilepsy. Damage in the hippocampus is associated with temporal lobe epilepsy, a common form of epilepsy in humans, damaging hippocampus (Martinc et al. 2012). Neurochemical changes in pentylenetetrazole model of epilepsy include marked cognitive deficits, significant decrease in NADH:cytochrome-c reductase (complex I), and cytochrome-c oxidase (complex IV) activities along with an increase in ROS, lipid peroxidation and protein carbonyls.

Curcumin possesses anticonvulsant properties. It produces anticonvulsant effects in several animal models of epilepsy. Thus, acute administration of curcumin in pentylenetetrazole (PTZ)-induced kindling in mice produces anticonvulsant effects in kindling mice (Agrawal et al. 2011; Sharma et al. 2010; Kaur et al. 2014). Administration of curcumin (100 mg/kg, p.o.) results in amelioration of cognitive deficits, restoration of mitochondrial complex activity, and significant reduction in ROS generation, lipid peroxidation and protein carbonyls. Moreover, glutathione levels are also restored in rats supplemented with curcumin. Curcumin protected mitochondria from seizure induced structural alterations. Collective evidence suggests that curcumin supplementation not only prevents mitochondrial dysfunctions and oxidative stress, but also improve cognitive functions in a chronic model of epilepsy (Kaur et al. 2014). It is also reported that curcumin may act by inhibiting nitric oxide synthase activity, lipid peroxidation, and elevating levels of reduced glutathione. Furthermore, curcumin may also act through the modulation of central monoaminergic pathway and acetylcholinesterase activity (Choudhary et al. 2013).

8.4 Therapeutic Importance of Curcumin in Neurodegenerative Diseases Other Than Alzheimer Disease

Neurodegenerative diseases are a debilitating group of diseases associated with site specific premature and slow death of specific neuronal populations and synapses in brain and spinal cord that modulate thinking, skilled movements, decision making, cognition, and memory (Soto and Estrada 2008). The molecular mechanisms associated with pathogenesis of neurodegenerative diseases remain elusive. However, it is becoming increasingly evident that the accumulation of misfolded proteins, induction of chronic oxidative stress, onset of chronic neuroinflammation may contribute to synaptic dysfunction, neuronal apoptosis, and brain damage in neurodegenerative diseases. Converging evidence suggests that neurodegenerative diseases are very complex and multi-faceted pathological conditions, which are also promoted the interplay between peripheral and resident CNS immunity (Amor et al. 2014). Neurodegenerative diseases include AD, PD, HD, ALS, and prion diseases. Information on neurochemistry of AD and effect of curcumin on AD has been described in Chaps. 1, 6, and 7. Other neurodegenerative diseases are PD, HD, ALS, and prion diseases. Characteristic feature of many neurodegenerative diseases is the accumulation of disease-specific proteins, such as accumulation of Aβ and its aggregates in the cerebral cortex and hippocampal region in AD, α-synuclein and its aggregates in the brain stem in PD, huntingtin and its aggregates in striatal medium spiny neurons in HD, abnormalities in Cu/Zn-superoxide dismutase in ALS, and misfolded PrP^{sc} polymerized amyloid fibril is involved in neurodegeneration in prion diseases (Fig. 8.3) (Selkoe 2003; Bates 2003; Beckman et al. 2001; Farooqui 2010) along with induction of oxidative stress and neuroinflammation. A causative link between protein aggregate formation and neurodegenerative diseases has not yet been established. However, it is becoming increasingly evident that the toxic action of soluble oligomers and protofibrillar derivatives of misfolded proteins may contribute to neurodegeneration in neurodegenerative diseases (Jellinger 2009). This suggestion is supported by the observation that a single-domain antibody can recognize a common conformational epitope, which is displayed by several disease associated proteins, including Aβ, α-synuclein, τ-protein, prions, and polyglutamine (polyQ)-containing peptides (Jellinger 2009). The burden of these biologic mechanisms (protein aggregation, oxidative stress and neuroinflammation) on degenerative pathophysiology is mutable, influenced by environmental factors and behavioral determinants, such as diet and exercise (Gomez-Pinilla 2011; Farooqui 2013, 2015). In fact, it is becoming increasingly evident that certain dietary compounds have potential therapeutic applications for numerous neurodegenerative diseases (Farooqui 2012). As stated above, most neurodegenerative diseases are accompanied by the progressive cognitive dysfunction and motor disabilities with devastating consequences to patients. In older individuals and animals, age-related changes may result in loss of motor and decrease in cognitive performance. These processes may be caused by enhanced rate (upregulation) of interplay among

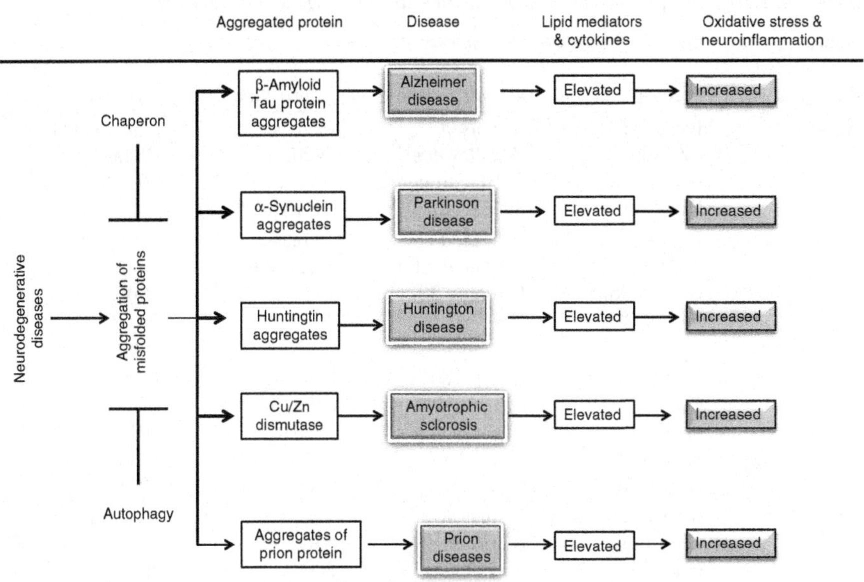

Fig. 8.3 Chart showing abnormal aggregated proteins that accumulate in neurodegenerative diseases and consequences of their accumulation

excitotoxicity, oxidative stress, and neuroinflammation and may be a common mechanism of brain damage in neurotraumatic and neurodegenerative diseases (Farooqui and Horrocks 2007; Farooqui et al. 2007; Farooqui 2010, 2011; Deleidi et al. 2015). Most neurodegenerative diseases are accompanied by elevations in levels of lipid mediators (Table 8.1) (Farooqui and Farooqui 2011). Converging evidence suggests that diet, genetic, exercise, and environmental factors may also contribute to the increase in the vulnerability of neurons to neurotraumatic and neurodegenerative diseases (Kidd 2005; Farooqui 2010, 2013, 2015).

8.4.1 Effects of Curcumin in Parkinson Disease

Parkinson Disease (PD) is a chronic and progressive neurological disorder characterized by the selective loss of dopaminergic neurons of the substantia nigra pars compacta. Clinical diagnosis has indicated that cardinal behavioral features of PD include rigidity, dyskinesia, gait imbalance, and tremor at rest (Jankovic 2008). The pathogenesis of PD remains unknown. However, experimental studies have indicated the involvement of oxidative stress, mitochondria dysfunction, apoptosis, and inflammation either separately or cooperatively to induce neurodegeneration in PD (Dauer and Przedborski 2003). The degeneration of dopaminergic neurons results in the depletion of dopamine leading to abnormal dopaminergic neurotransmission in

Table 8.1 Levels of lipid mediators and cytokines in neurodegenerative diseases

Neuro-degenerative disease	Eicosanoids	4-Hydroxynonel	Isoprostane	Cytokines
AD	Increased (Montine et al. 1998; Farooqui 2011)	Increased (Bradley et al. 2010; Farooqui 2011)	Increased (Miller et al. 2014; Farooqui 2011)	Upregulated (Wang et al. 2014b; Farooqui 2011)
PD	Increased (Phillis et al. 2006; Farooqui 2011)	Increased (Qin et al. 2007; Farooqui 2011)	Increased (Seet et al. 2010)	Upregulated (Nagatsu et al. 2000; Milyukhina et al. 2015)
HD	Increased (Kumar et al. 2011; Kalonia and Kumar 2011)	Increased (Lee et al. 2011)	Increased (Montine et al. 1999)	Upregulated (Kumar et al. 2011; Kalonia and Kumar 2011)
ALS	Increased (Yasojima et al. 2001; Farooqui 2011)	Increased (Periluigi et al. 2005)	Increased (Greco et al. 2000)	Upregulated (Furukawa et al. 2014; Italiani et al. 2014)
Prion disease	Increased (Minghetti and Pocchiari 2007: Farooqui 2011)	Increased (Andreoletti et al. 2002)	Increased (Minghetti et al. 2000)	Upregulated (Van Everbroeck et al. 2002)

the basal ganglia motor circuit causing resting tremor, rigidity, bradykinesia, posture and ambulating difficulty. PD does not manifest clinically until 80 % of striatal dopamine is depleted, thus most neuronal damage occurs before the patient presents clinical symptoms. The degeneration of dopaminergic neurons in the substantia nigra pars compacta is due to monoamine oxidase (MAO)-mediated abnormal dopamine metabolism and hydrogen peroxide generation leading to oxidative stress. Excessive oxidative stress causes the oxidative modification of macromolecules (lipids, proteins, and DNA) leading to cell damage and even cell death. The pathological effects of ROS also contribute to reduction in ATP (adenosine triphosphate) production, in an increase of iron levels, and in an increase of intracellular calcium levels along with alterations in mitochondrial respiratory chain complexes function. Apart from DA depletion, a profound reduction in specific neurochemical markers such as tyrosine hydrolase (TH), DA transporter (DAT), and vesicular monoamine transporter 2 (VMAT2) has been reported in PD (Heikkila and Sonsalla 1992; Miller et al. 1999). The risk of PD is increased by exposure to neurotoxins and pesticides and through oxidative damage to neurons by a decrease in the mitochondrial complex I activity and glutathione levels (Srinivas et al. 2008).

PD is also characterized by the presence of α-synuclein aggregates as Lewy body inclusions in specific regions of the brain such as substantia nigra, thalamus and

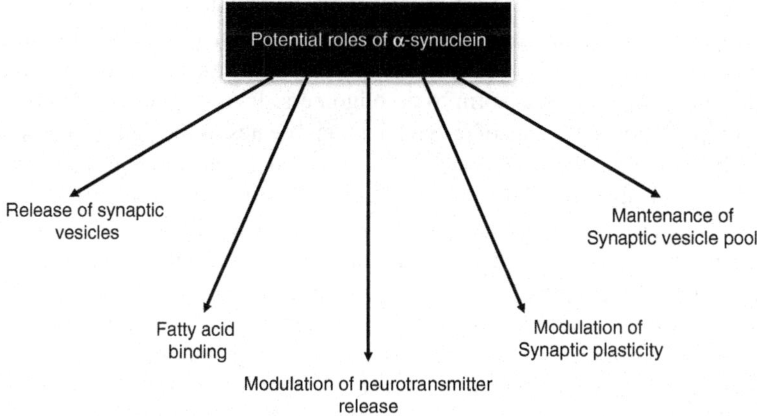

Fig. 8.4   Potential roles of α-synuclein in the brain

neocortex (Uversky and Eliezer 2009). α-Synuclein is a140-amino acid protein (mol mass 14 kDa), which is encoded by a single gene consisting of seven exons located in chromosome 4 (Chen et al. 1995). α-Synuclein is primarily localized at the presynaptic terminals of neurons (Iwai et al. 1995). This protein lacks both cysteine and tryptophan residues. α-Synuclein is present in high concentration at presynaptic terminals and is found in both soluble and membrane-associated fractions of the brain (Lee et al. 2002). Very little is known about the role of α-synuclein in the brain. However, it is suggested that this protein plays an important role in the regulation of synaptic vesicle release and trafficking, maintenance of synaptic vesicle pools, fatty acid binding, neurotransmitter release, synaptic plasticity, and neuronal survival (Fig. 8.4) (Uversky and Eliezer 2009). Similar to β-amyloid in AD, α-synuclein undergoes rapid self-aggregation in vivo with an accelerated rate in the presence of transition metal ions, DA, proteins and lipids. Other factors, which promote α-synuclein aggregation in vitro include subtle changes in the environment (i.e. increase in temperature, decrease in pH), addition of amphipathic molecules such as herbicides, presence of external metal ions (industrial pollutants) and the interactions with membranes and other proteins (Bisaglia et al. 2009; Uversky et al. 2001a, b). Oxidative stress up-regulates the expression of α-synuclein, and promotes its fibrillization and aggregation (Vila et al. 2000). Conversely, a high degree of fibrillization and aggregation of α-synuclein results in an increase of reactive oxygen species (ROS) and neurotoxicity (Hsu et al. 2000). This vicious cycle between α-synuclein and oxidative stress may contribute to the progression of loss of substantia nigra pars compactal dopaminergic neurons in PD. It is also reported that oxidative stress not only cause nuclear membrane modifications, but also promotes the translocation of α-synuclein to the nucleus where it can form complexes with histones leading to its oligomerization into insoluble fibrils (Zhou et al. 2013). Aggregation and high levels of α-synuclein have been shown to induce oxidant production or increase the level of oxidative stress. Within cells, α-synuclein normally adopts an α-helical conformation. However, under high levels of oxidative stress α-synuclein undergoes a profound conformational transition to a β-sheet-rich structure that polymerizes to form toxic oligomers. Involvement of soluble oligomeric and protofibrillar forms of α-synuclein aggregates in the pathogenesis of PD is not only supported by the consistent detection of α-synuclein deposits in affected brain areas, but also by pathogenic mutations affecting the α-synuclein gene in familial PD and association of the α-synuclein locus with idiopathic PD in genome-wide association studies. Furthermore, in vitro studies on cell culture systems and animal models also support association of α-synuclein with PD (Irvine et al. 2008; Simon-Sanchez et al. 2009). Recent studies on neurodegenerative potency of α-synuclein fibrils have indicated that toxicity of α-synuclein fibrils may be due to its ability to penetrate neural cell membranes (Volles et al. 2001; Pieri et al. 2012). Thus, compounds that inhibit α-synuclein aggregation and fibrillization and stabilize it in a non-toxic state can therefore serve as therapeutic molecules for both prevention of accumulation of aggregated α-synuclein and maintenance of normal physiological concentrations of α-synuclein (Li et al. 2004).

Curcumin provides protection against α-synuclein-mediated cell death in tissue culture system inhibits mitochondrial toxicity and blocks the formation of ROS (Liu et al. 2011; Wang et al. 2010b). Curcumin not only interacts with α-synuclein, but also inhibits and reverses the formation of toxic α-synuclein aggregate species (Fig. 8.5) (Pandey et al. 2008). Detailed investigations indicate that curcumin binds to the non-amyloid-β component (NAC) domain of α-synuclein to shield hydrophobic residues that drive self-aggregation, thus promoting and stabilizing non-aggregate forms of the protein in vitro (Ahmad and Lapidus 2012). Thus, in animal models of PD, curcumin produces its beneficial effects by promoting the disaggregation of higher-order oligomeric and fibrillar forms of the α-synuclein, which are responsible for the toxicity of α-synuclein. Curcumin also inhibits neuroinflammation in animal models of PD (Liu et al. 2011). Studies on the effect of chronic and acute curcumin treatment in the Syn-GFP mouse line showing the overexpression of wild-type human α-synuclein protein, indicating that curcumin containing diet significantly improves gait impairments and results in an increase in phosphorylated forms of α-synuclein at cortical presynaptic terminals (Spinelli et al. 2015). Acute curcumin treatment also results in an increase in phosphorylated α-synuclein in terminals, but has no direct effect on α-synuclein aggregation, as measured by in vivo multiphoton imaging and Proteinase-K digestion (Spinelli et al. 2015). These studies support the view that even at low concentration dietary curcumin intervention correlates with significant behavioral and molecular changes in a genetic synucleinopathy mouse model that mimics human disease (Spinelli et al. 2015). Curcumin also provides protection against α-synuclein-induced cytotoxicity in SH-SY5Y neuroblastoma cells by decreasing cytotoxicity of aggregated α-synuclein, reducing intracellular ROS, inhibiting caspase-3 activation and ameliorating signs of apoptosis (Wang et al. 2010b). Recently, curcumin derivatives (CNB-001, pyrazole curcumin derivative, and curcumin-bis-α D-glucoside) have

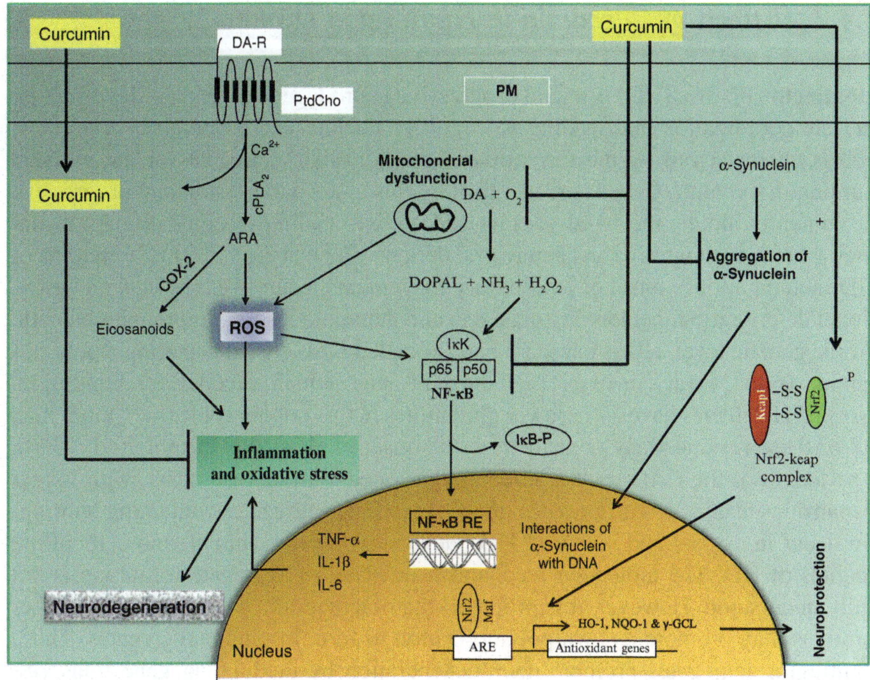

Fig. 8.5 A hypothetical signal transduction diagram showing the sites of actions of curcumin on signal transduction processes in Parkinson disease. Plasma membrane (*PM*); dopamine (*DA*); dopamine receptor (*DA-R*); Phosphatidylcholine (*PtdCho*); cytosolic phospholipase A_2 (*cPLA$_2$*); lysophosphatidylcholine (*lyso-PtdCho*); cyclooxygenase (*COX*); lipoxygenase (*LOX*); arachidonic acid (*ARA*); platelet activating factor (*PAF*); reactive oxygen species (*ROS*); hydrogen peroxide (*H_2O_2*); nuclear factor-κB (*NF-κB*) with its subunits (p65 and p50); nuclear factor-κB-response element (*NF-κB-RE*); inhibitory subunit of NF-κB (*I-κB*); tumor necrosis factor-α (*TNF-α*); interleukin-1β (*IL-1β*); interleukin-6 (*IL-6*); Nuclear factor (erythroid-derived 2)-like 2 (*Nrf2*); and Kelch like-ECH-associated protein 1 (*Keap1*). *Blocked arrow* indicates sites of curcumin action. *Positive sign* indicates stimulation

been synthesized. These derivatives not only protect DA-ergic neurons against MPTP toxicity by regulating various molecular and cellular events, but also protect against excitotoxicity, glucose-starvation assay (Liu et al. 2008), and exhibit strong antioxidant and antiapoptotic properties (Jayaraj et al. 2013). The therapeutic potential of these derivatives is supported by their ability to cross blood BBB, restoration of behavioral impairments, reduction in oxidative stress, restoration of mitochondrial deficits, and enhancement in expressions of TH, DAT, and VMAT2 in animal model of PD (Gadad et al. 2012; Jayaraj et al. 2014; Ahsan et al. 2015). In 1-methyl-4-phenylpyridinium ion-induced apoptosis in PC12 cells, curcumin also mediates its effect through the modulation of Bcl-2-mitochondria-ROS-iNOS pathway (Chen et al. 2006). This information can be used for paving road for human trials.

8.4.2 Effects of Curcumin in Huntington Disease

Huntington disease (HD) is a fatal progressive neurodegenerative disorder affecting muscle coordination and leading to cognitive decline and psychiatric symptoms. HD is characterized by the formation of amyloid-like aggregates of the mutated huntingtin protein (Cepeda et al. 2001). This autosomal dominant progressive prominently affects the basal ganglia and cortex, leading to clinically significant motor function, cognitive and behavioral deficits (Cepeda et al. 2001). Symptoms of HD include midlife onset of involuntary movements, cognitive, physical and emotional deterioration, personality changes, and dementia leading to premature death. At the genetic level, HD is caused by an expanded CAG repeat encoding a polyglutamine (polyQ) tract in exon 1 of the HD gene, which encodes for huntingtin. Normal HD alleles have 37 or fewer glutamines in this polymorphic tract, more than 37 of these residues may contribute to the onset of HD (Rubinsztein et al. 1996). The length of the CAG tract is directly correlated with disease onset, with longer expansions leading to earlier onset of HD. Insoluble aggregates containing huntingtin occur in cytosol and nuclei of HD patients, transgenic animal, and cell culture models of HD. The molecular mechanism involved in aggregate formation is not fully understood. However, it is proposed that interactions of huntingtin with other proteins may promote its own polymerization to form insoluble aggregates. These huntingtin aggregates promote neurodegeneration by modulating gene transcription, protein interactions, protein transport inside the nucleus and cytoplasm as well as vesicular transport along with proteosomal dysfunction, axonal transport deficit, and excitotoxicity (Scherzinger et al. 1997; Bonilla 2000; Cowan and Raymond 2006; Gil and Rego 2008). It is also reported that the accumulation of mutated huntingtin inclusions is not a consequence of direct proteasomal inhibition but rather results from the gross failure of protein quality control systems in association with the sequestration of molecular chaperones (Hipp et al. 2012). Studies on vertebrate and invertebrate models of HD have indicated that expression of the polyQ-expanded form of huntingtin may not only result in mitochondrial dysfunction, but also in mutant huntingtin-mediated alterations in activity of the NMDA type glutamate receptor especially in the striatum. Proteolysis and migration of huntingtin to the nucleus also occur relatively early in HD, while formation of ubiquitinated aggregates of huntingtin and transcriptional dysregulation occur as late effects of the gene mutation (Bonilla 2000; Cowan and Raymond 2006). Other neurochemical changes in HD include decrease in levels of GABA, dynorphin, substance P, reduction in expression of both glutamate transporters, GLAST, GLT-1, glutamate uptake, and increase in somatostatin and neuropeptide Y (Faideau et al. 2010). In addition, hippocalcin (a neuronal calcium sensor protein) is highly expressed in the medium spiny striatal output neurons that degenerate selectively in HD. Decrease in hippocalcin expression occurs in parallel with the onset of disease in mouse models of HD. In situ hybridization histochemistry studies have indicated that hippocalcin RNA is diminished by 63 % in human HD brain (Rudinskiy et al. 2009).

CAG140 knock-in (KI) mice are a slowly progressing mouse model of HD that exhibit pathological, molecular and behavioral deficits as early as 2 years before developing spontaneous motor deficits which is itself reminiscent of the clinically manifest phase of HD (Hickey et al. 2008). Like HD patients, these mice display abnormal aggregates of mutant huntingtin and striatal transcriptional deficits, as well as early motor, cognitive and affective abnormalities, many months prior to exhibiting spontaneous gait deficits, decreased striatal volume, and neuronal loss. Supplementation of curcumin in diet for several months ameliorates three aspects of HD in CAG140 KI mice, with the most notable effect on the huntingtin aggregates. Curcumin supplementation not only leads to partially behavioral changes, but also improves transcriptional deficits and observed in HD. It is proposed that behavioral improvements may be related to improvements in the level of the striatal transcripts for the D1 dopaminergic receptor and DARRP-32 (Hickey et al. 2012; Svenningsson et al. 2004).

8.4.3   Effects of Curcumin on Prion Diseases

Prion diseases are characterized by the accumulation of abnormal isoforms of cellular prion protein (PrP^C) and onset of motor dysfunctions, dementia and neuropathological changes such as spongiosis, astroglyosis and neuronal loss (Prusiner 2001; Grossman et al. 2003). PrP^C is a copper binding glycoprotein, which is abundantly expressed in neurons and glial cells within the brain tissue. The expression of PrP^c is particularly high in the hippocampus, striatum, and frontal cortex, with apparently wide subcellular distribution, including synaptic sites (Fournier et al. 2000). Involvement of PrP^C in synaptic function is supported by studies on PrP-null mice. It is shown that PrP^c deficits is associated with spatial learning (Criado et al. 2005), alterations in long-term potentiation (Collinge et al. 1994; Maglio et al. 2006), and increased excitability of hippocampal neurons (Collinge et al. 1994; Mallucci et al. 2002). In the prion diseases the α-helical form of PrP^C is transformed into β-stranded form (PrP^{Sc}), which causes scrapie in goats and sheep, bovine spongiform encephalopathy (mad cow disease) in cattle, and Creutzfeldt-Jakob disease (CJD), kuru, and Gerstmann-Sträussler-Scheinker syndrome in humans (Prusiner 2001; Grossman et al. 2003). PrP^{Sc} aggregates are stable and proteinase K-resistant, and have ability to form fibrils. The molecular mechanisms underlying the transformation of PrP^c to pathological isoforms of PrP^{sc} are poorly understood. Conversion of PrP^c is initiated by interaction with PrP^{sc} (or "agent") resulting in refolding of PrP^c into new pathological PrP^{sc} ("agent replication"). PrP^c interacts/binds with many different molecules including Cu^{2+} ions, nucleic acids, several (receptor) proteins, laminin γ1-chain, Aβ oligomers, and the prion protein itself (Beraldo et al. 2011; Kessels et al. 2010; Linden et al. 2008; Laurén et al. 2009). The processes underlying agent replication (normal to abnormal PrP conversion) are initiated by selective interaction between PrP^c molecules and potentially modulated by chaperone molecules. Accumulation of such prion fibrils in the brain is believed to cause neuronal cell death and onset of disease.

Earlier studies have indicated that prion diseases differ from other amyloid-associated protein misfolding diseases, such as AD, in that prion diseases are naturally transmitted between individuals and involve spread of protein aggregation between tissues (Speare et al. 2010). Recent studies indicate that oligomeric forms of Aβ complexes with the glycosylphosphatidyl-inositol-anchored membrane protein and PrP^c. These complexes play an important role in the progression of AD pathogenesis. Anti-PrP^c antibodies inhibit Aβ oligomer binding to PrP^c and rescue synaptic plasticity in hippocampal slices from oligomeric Aβ toxicity. Based on these results, it is proposed that PrP^c mediates Aβ-oligomer-induced synaptic dysfunction. These results also support the view that there are symptomatic pathologies and mechanistic overlap between AD and CJD (Laurén et al. 2009; Nygaard and Strittmatter 2009; Gunther and Strittmatter 2010). Both conditions are dementing disorders, which are associated with the deposition of extracellular protein aggregates leading to neurodegeneration. AD is accompanied by the buildup of Aβ-containing plaques in the neuropil along with deposition of neurofibrillary tangles composed of hyperphosphorylated tau protein. On the other hand, CJD is accompanied by deposition of PrP^{Sc}. Treatment of rat cortical neurons with either synthetic Aβ or PrP 106–126 peptides results in increased tau protein phosphorylation at Ser202/Thr205 and cell death in a Cdk5-dependent manner (Lopes et al. 2007). It is interesting to note that PrP^C is a negative regulator of β-secretase (Parkin et al. 2007). It is localized in specialized membrane regions containing contain APOE, amyloid beta precursor protein (APP), disintegrin, and metalloprotease domain 23 (Schmitt-Ulms et al. 2004). Recent preclinical studies have demonstrated that the endogenous PrP^C protects against excitotoxicity by downregulating a subpopulation of NMDARs, suggesting that progressive misfolding of PrP^C into the disease-associated form of the protein (PrP^{Sc}) may result in the loss of this neuroprotective function and subsequent neurodegeneration in CJD (Khosravani et al. 2008). Thus, high levels of glutamate chronically activate both NMDARs and metabotropic glutamate receptor 5 (mGluR5), resulting in the elevation of intracellular Ca^{2+} associated with AD that promotes neuronal injury and cell death (Farooqui et al. 2008; Ong et al. 2013). Elevated expression of mGluR5 has been reported to occur in astrocytes, which are found clustered around Aβ plaques (Casley et al. 2009). mGluR5 contributes to the AD pathogenesis by acting as scaffolds for the PrP^c/Aβ oligomer complex, enabling the propagation of neurotoxic signaling in AD. In addition, PrP^c and Aβ oligomer signaling via NMDARs may also contribute to AD pathology (Fig. 8.6) (Um et al. 2013; Hamilton et al. 2015).

Many compounds have been identified as inhibitors of PrP^{Sc} formation, but none of these compounds are known to be safe or effective for use in humans and animals (Caughey et al. 2003). Recent efforts have identified curcumin as an inhibitor

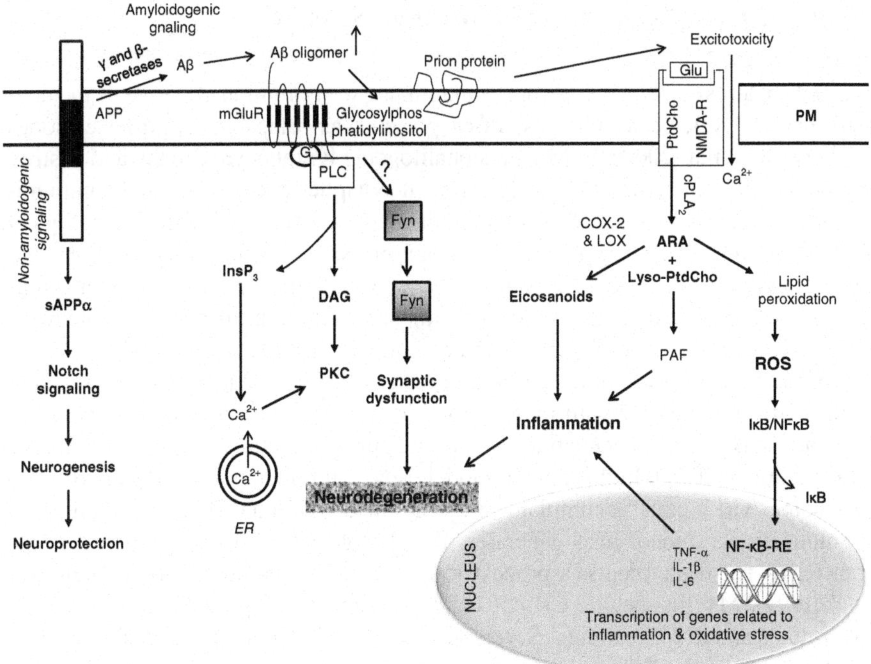

Fig. 8.6 A hypothetical signal transduction diagram showing generation of Aβ and interactions among Aβ, glycosylphosphatidylinositol, and prion protein. Amyloid precursor protein (*APP*); β-amyloid (*Aβ*); cellular prion protein (*PrP^C*); glutamate (*Glu*); metabotropic glutamate receptor5 (*mGluR5*); phospholipase C (*PLC*); diacylglycerol (*DAG*); protein kinase C (*PKC*); inositol 1,4,5-triphosphate (*InsP₃*); Src kinase (*Fyn*); phosphatidylcholine (*PtdCho*); Lyso-phosphatidylcholine (*Lyso-PtdCho*); cytosolic phospholipase A₂ (*cPLA₂*); lyso-phosphatidylcholine (*lyso-PtdCho*); cyclooxygenase (*COX*); lipoxygenase (*LOX*); arachidonic acid (*ARA*); platelet-activating factor (*PAF*); reactive oxygen species (*ROS*); nuclear factor-κB (*NF-κB*); nuclear factor-κB-response element (*NF-κB-RE*); inhibitory subunit of NF-κB (*I-κB*); tumor necrosis factor-α (*TNF-α*); interleukin-1β (*IL-1β*); interleukin-6 (*IL-6*)

of prion fibril formation (Caughey et al. 2003; Hafner-Bratkovic et al. 2008). Curcumin inhibits in vitro conversion of PrP^c protease resistant PrP^sc in neuroblastoma cell lines (Caughey et al. 2003). It recognizes the converted beta-form of the PrP^sc both as oligomers and fibrils but not the native form. Curcumin interacts with the prion fibrils in the left-handed chiral arrangement as determined by circular dichroism (Hafner-Bratkovic et al. 2008). It can label the plaques of the brain sections of variant Creutzfeld-Jakob disease cases and stains the same structures as antibodies against the PrP^sc. However, in vivo studies on the effect of dietary or injected curcumin show no significant effect on the onset of scrapie in hamsters (Caughey et al. 2003).

8.4.4 Effect of Curcumin in Multiple Sclerosis

Multiple sclerosis (MS) is a chronic inflammatory autoimmune disease of the central nervous system, which most often presents as relapsing-remitting episodes (Noseworthy et al. 2000). In MS, encephalitogenic lymphocytes attack and destroy myelinated neurons, thereby interfering with synaptic transmission and communication between neurons. MS is accompanied by a variety of pathophysiological features such as breakdown of the BBB, autoimmune attack, injury of axons and break down of myelin sheaths (Noseworthy et al. 2000). The most common symptoms of MS are weakness in one or more limbs, sensory disturbances, optic neuritis, ataxia, bladder dysfunction, fatigue and cognitive deficits (O'Connor 2002).

Experimental autoimmune encephalomyelitis (EAE), induced in various rodent models by immunization with spinal cord or brain homogenate in complete Freund's adjuvant, has provided valuable insights into immunopathogenic mechanisms of MS (Steinman 2009). In this animal model, IFN-γ–producing Th1 cells and IL-17–producing Th17 cells accumulate in the brain and initiate demyelination. This immunization protocol also generates T cells that cause disease upon adoptive transfer to naive recipients, a process termed "passive immunization" (Glabinski et al. 1997). Myelin-specific CD4$^+$ T cells are found in peripheral blood of healthy individuals and in MS patients (Severson and Hafler 2010). Potentially autoreactive CD4$^+$ and CD8$^+$ T cells have been detected at autopsy in brain tissues from individuals with MS but not in relevant controls (Dornmair et al. 2003). It is proposed that, in MS as in EAE, disease-causing T cells are initially activated in peripheral lymphoid organs, where they undergo differentiation and expansion.

Very little is known about the effect of curcumin on various facets of the immune response, including its effect on lymphoid cell populations, antigen presentation, humoral and cell-mediated immunity, and cytokine production (Gautam et al. 2007; Bright 2007). C57BL/6 mice induced to develop EAE express elevated levels of interferon (IFN) γ, interleukin (IL)-12, IL-17, and IL-23 in the brain and lymphoid organs. Ex vivo and in vitro treatment with curcumin also results in a dose-dependent decrease in the secretion of IFNγ, IL-17, IL-12 and IL-23 in culture. The inhibition of EAE by curcumin is also associated with an up-regulation of IL-10, peroxisome proliferator activated receptor γ and CD4$^+$CD25$^+$ Treg cells in the CNS and lymphoid organs. These findings highlight that curcumin differentially regulates CD4$^+$ T helper cell responses in EAE (Kanakasabai et al. 2012). Similarly, studies on the effect of curcumin in Lewis rats have indicated that curcumin treatment results in a dramatic reduction in the number of inflammatory cells infiltration in the spinal cord. The proliferation of the MBP-reaction lymphocyte also is reduced in a curcumin dose-dependent manner. Furthermore, quantitative reverse-transcription polymerase chain reaction studies have indicated that treatment with curcumin results in inhibition of mRNA expression for IL-17, TGF-β, IL-6, IL-21, and STAT3 (Xie et al. 2009). These observations support the view that curcumin ameliorate EAE not only by inhibiting differentiation and development of Th17 cells, which down-regulates the expression of IL-6, IL-21, RORγt signaling, but also by inhibiting STAT3-phosphorylation (Xie et al. 2009).

8.5   Therapeutic Importance of Curcumin in Neuropsychiatric Diseases

Neuropsychiatric disorders begin early in life and are characterized by abnormalities in cerebral cortex and limbic system (thalamus, hypothalamus, hippocampus, and amygdale) (Farooqui 2010). Examples of neuropsychiatric diseases are schizophrenia, major depression, bipolar disorder, and autism. The symptoms of neuropsychiatric disorders include hallucinations, delusions, sadness, and guilt. The pathogenesis of depression includes disturbance in neurotransmitters (dopamine, serotonin), increase in inflammatory processes, defect in neurogenesis and abnormalities in synaptic plasticity (decrease in BDNF), mitochondrial dysfunction, and redox imbalance. At the molecular level, curcumin mediates its effects not only by inhibiting monoamine oxidase enzymes, elevating brain serotonin, and BDNF levels, but also by modulating the hypothalamus-pituitary-adrenal axis (Kulkarni et al. 2008; Bergman et al. 2013). As stated above, BDNF plays a central role in brain development and plasticity by opposing neuronal damage and promoting neurogenesis and cell survival (Takahashi et al. 1999; Banasr et al. 2004). BDNF-tyrosine kinase B signaling leads to phosphorylation and activation of transcription factors such as CREB, which induces gene expression and long-lasting synaptic changes (Yoshii and Constantine-Paton 2010; Bramham and Messaoudi 2005). In addition, curcumin also acts as strong antioxidants in vitro through the numerous mechanisms, such as radical scavenging, metal ions (Fe^{3+}, Cu^{2+}, and others) chelation, and the modulation of antioxidant enzyme activities. In the scavenging ability the position and the number of phenolic –OH groups play a role through donation of a hydrogen atom from their hydroxyl groups to radicals, resulting in radical moiety elimination. During this reaction, phenoxyl radical is formed that can form stable compound and terminates radical reaction via reaction with another radical (Hegde et al. 2011). Curcumin can easily cross BBB and directly induce neuroprotection probably through its antioxidant ability to inhibit lipid peroxidation and neutralize ROS and RNS (Sreejayan and Rao 1997). In addition, curcumin can affect number of cellular pathways on molecular level and via anti-inflammatory properties. Curcumin inhibits cyclooxygenase 1 and cyclooxygenase 2 and influences many other signaling pathways leading to cell protection and enhancement of cell survival (Scapagnini et al. 2011).

8.5.1   Curcumin and Depression

Depression is a multifactorial devastating neuropsychiatric disorder caused by genetic, environmental, psychological, and biological factors and characterized by anorexia, weight loss, fatigue, lethargy, sleep disorders (insomnia or hypersomnia), hyperalgesia, reduction of locomotor activity, failure to concentrate, feelings of worthlessness or guilt, diminished cognitive functioning, and recurrent thoughts of death (Davidson et al. 2002). Genetic, epigenetic, environmental, psychological, and biological factors, which contribute to the etiology of depression are mutually connected through signal transduction processes that modulate the severity of depression. Several molecular mechanisms have been reported to contribute to the pathogenesis of depression Petersen and Posner (2012). Among these mechanisms oxidative stress and neuroinflammation have been reported to play a major role. Reactive oxygen and nitrogen species have been shown to modulate levels and activities of noradrenaline (norepinephrine), serotonin, dopamine and glutamate, the principal neurotransmitters involved in the neurobiology of depression (Scapagnini et al. 2012). Depression is associated with abnormalities in the metabolism of neurotransmitters (serotonin and dopamine) caused by alterations in activities of monoamine oxidase along with depletion in levels of tryptophan (Leonard and Maes 2012; Lopresti et al. 2013). The pathogenesis of major depression is also supported by neuroinflammatory processes involving elevated levels of proinflammatory cytokines (TNF-α, IL-1β, IFN-γ), decrease in ratio of n-3 to n-6 long-chain polyunsaturated fatty acid in diet, reduction in neurogenesis, and dysfunction of hypothalamic-pituitary-adrenal axis (Fig. 8.7) (Parker et al. 2006; Leonard and Maes 2012; Kupfer et al. 2012; Gałecki et al. 2015). It is also demonstrated that in rodents, chronic stress and the stress hormone cortisol also produces oxidative damage to mitochondrial function. Mitochondria play a key role in synaptic

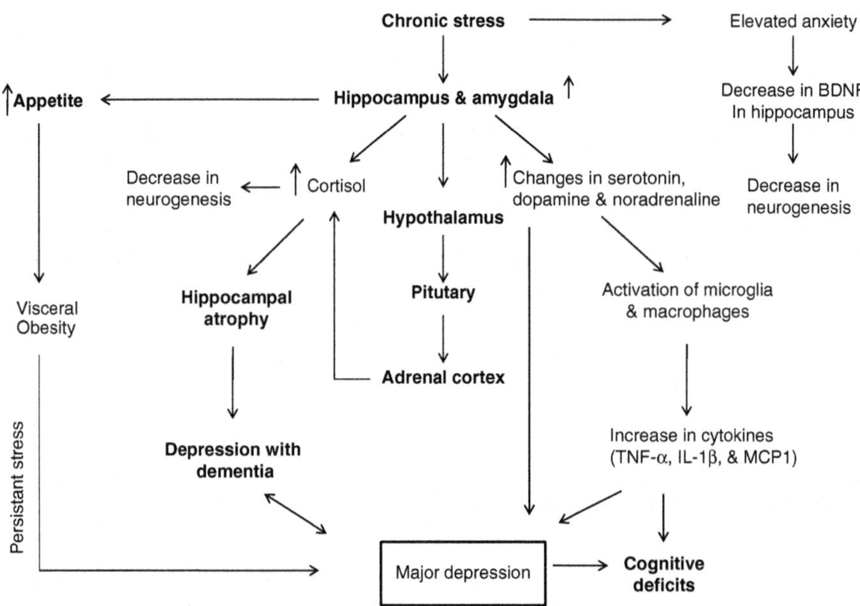

Fig. 8.7 Effects of chronic stress on the depression. Tumor necrosis factor-α (*TNF-α*); interleukin-1beta (*IL-1β*); monocyte chemoattractant protein-1 (*MCP-1*); brain-derived neurotrophic factor (*BDNF*). *Upward arrow* indicates increase

neurotransmitter signaling by providing ATP, mediating lipid and protein synthesis, buffering intracellular calcium, and regulating apoptotic and resilience pathways. It should be noted that low antioxidant capacity and abnormal early morning cortisol levels are closely associated with depression. Depression is frequently associated with cardiovascular diseases, stroke; chronic age-related neurological disorders (AD, PD, ALS, and dementia), inflammatory visceral diseases (chronic arthritis), and metabolic syndrome (Farooqui 2013, 2015). High calorie diet or diet containing refined sugar promotes chronic neuroinflammation through the induction of insulin resistance. Refined sugar also suppresses activity of BDNF, which promotes neuro-plasticity. BDNF levels are critically low in both depression and schizophrenia, which animal models suggest might actually be causative (Pistell et al. 2010; Farooqui 2015).

Curcumin produces mild antidepressant effects in rodents. The antidepressant effects and profiles of curcumin as tested by tail suspension test (TST) and the forced swimming test (FST) are comparable with classical antidepressants, such as fluoxetine and imipramine (Xu et al. 2005; Kulkarni et al. 2008; Wang et al. 2008; Mei et al. 2011; Sanmukhani et al. 2011). In animal models of depression caused by environmental or pharmacological manipulations, curcumin acts by reversing the increase in immobility and in MAO activity as well as the depletion of brain mono-amines (Bhutani et al. 2009). In the same way, curcumin ameliorates the low sucrose consumption (an anedonia-like behavior) and the increases in serum corticosterone levels caused by this same paradigm (Li et al. 2009). The molecular mechanisms associated with antidepressant effects of curcumin are not fully understood. However, it is becoming increasingly evident that curcumin acts by modulating hippocampal and cortical neurogenesis through the involvement of BDNF and pCREB/CREB ratio (Xu et al. 2006a, 2007; Gomez-Pinilla and Nguyen 2012). Chronic curcumin administration reverses stress-mediated alterations by inducing neurotrophin levels in a manner comparable to the tricyclic antidepressant imipramine (Xu et al. 2006b). Curcumin not only boosts neuronal function, but it can also protect against neurodegenerative effects. Beneficial effects of curcumin are also observed in a pain depression dyad model (Arora et al. 2011). In this model, rats treated with reserpine show a decrease in nociceptive threshold and an increase in immobility time accompanied by a decrease in dopamine, serotonin and norepinephrine along with increase in substance P concentration and inflammatory cytokines in cortex and hippocampus. Curcumin ameliorates these behavioral and biochemical alterations induced by reserpine (Arora et al. 2011). Administration of curcumin (1000 mg/day) in 60 patients with major depression in randomized and placebo controlled for 6 weeks results in an effective and safe modality for treatment of depressive patients without concurrent suicidal ideation or other psychotic disorders (Sanmukhani et al. 2014). Contrary to these results no significant differences are observed between the groups of patients with major depression following administration of curcumin (500 mg) and placebo for 5 weeks in randomized, double-blind, and placebo controlled clinical study (Bergman et al. 2013) probably due to low daily doses used. However, the patients in the curcumin group demonstrated a trend to a more rapid relief of depressive symptoms in comparison to those in the placebo group (Bergman et al. 2013).

8.5.2   Curcumin and Bipolar Disorders

Bipolar disorders are chronic and debilitating illnesses characterized by recurrent manic and depressive episodes (Ketter 2010). The pathophysiology of bipolar disorders remains elusive. However, it is becoming increasingly evident that bipolar disorders are accompanied by neurostructural changes related to neuroplasticity, neuronal interconnectivity, and cognitive deterioration (Martinez-Aran et al. 2004; Berk et al. 2010). In addition to brain abnormalities, bipolar disorders have also been linked to several metabolic alterations, including obesity, arterial hypertension, and changes in glucose metabolism (McIntyre et al. 2010; Fagiolini et al. 2008).

Curcumin has been reported to produce neuroprotective effects in an animal model of mania induced by ketamine administration in rats (Gazal et al. 2014). It is reported that ketamine treatment (25 mg/kg, for 8 days) results in hyperlocomotion in the open-field test and oxidative damage in prefrontal cortex (PFC) and hippocampus (HP) along with increase in lipid peroxidation and decrease in total thiol content. Pretreatment of rats with curcumin (20 and 50 mg/kg, for 14 days) or with lithium chloride (45 mg/kg, positive control) prevents behavioral and pro-oxidant effects induced by ketamine. These observations support the view that curcumin may mediate intervention in bipolar disorders not only by reducing the episode relapse, but also by retarding the oxidative damage associated with the manic phase of this illness (Gazal et al. 2014).

Studies on the effect of curcumin in animal model of traumatic memory formation in posttraumatic stress disorder (PTSD) have indicated that diet enriched with 1.5 % curcumin not only prevents the training-related elevation in the expression of the immediate early genes (IEGs) Arc/Arg3.1 and Egr-1 in the lateral amygdala, but also impairs the 'consolidation' of an auditory Pavlovian fear memory. Short-term memory (STM) remains intact. However, long-term memory is significantly impaired (Monsey et al. 2015). Additional studies revealed that dietary curcumin is effective at the reconsolidation of older memories (Monsey et al. 2015). Collectively, these observations indicate that a diet enriched with curcumin is capable of impairing fear memory consolidation and reconsolidation processes, findings that may have important clinical implications for the treatment of disorders such as PTSD that are characterized by unusually strong and persistently reactivated fear memories (Monsey et al. 2015).

8.6   Conclusion

In addition to Alzheimer disease, curcumin produces neuroprotective effects in neurotraumatic disorders including stroke, spinal cord injury, traumatic brain injury, and epilepsy. Curcumin not only protects neurons in PD, HD, and prion diseases, but also promotes beneficial effects in neuropsychological disorders, such as depression, bipolar disorders, and tardive dyskinesia. Similar to Alzheimer disease, the molecular mechanism associated with neuroprotective action of curcumin is not fully understood. However, it is becoming increasingly evident that anti-inflammatory and antioxidant properties of curcumin may be responsible for neuroprotective effects. Anti-inflammatory and antioxidant properties of curcumin can be attributed to its chemical structure, which appears to influence multiple signaling pathways activated in neurotraumatic, neurodegenerative, and neuropsychological disorders. Some of these signaling pathways involve activities of inhibition of phospholipases A_2, lipooxygenases, cyclooxygenase-2, reduction in the generation of prostaglandins, leukotrienes, thromboxane, nitric oxide, and increased expression of MCP-1, interferon-inducible protein, TNF-α, interleukin-12 (IL-12), and prevention of NF-κB stimulation. In addition, curcumin is a potent inhibitor of reactive astrocyte expression and thus may prevent cell death indirectly. Curcumin also modulates various neurotransmitter levels in the brain.

References

Adamson J, Beswick A, Ebrahim S (2004) Stroke and disability. J Stroke Cerebrovasc Dis 13:171–177

Agrawal R, Tyagi E, Shukla R, Nath C (2011) Insulin receptor signaling in rat hippocampus: a study in STZ (ICV) induced memory deficit model. Eur Neuropsychopharmacol 21:261–273

Ahmad B, Lapidus LJ (2012) Curcumin prevents aggregation in α-synuclein by increasing reconfiguration rate. J Biol Chem 287:9193–9199

Ahsan N, Mishra S, Jain MK, Surolia A, Gupta S (2015) Curcumin pyrazole and its derivative (N-(3-Nitrophenylpyrazole) curcumin inhibit aggregation, disrupt fibrils and modulate toxicity of wild type and mutant α-synuclein. Sci Rep 5:9862

Amor S, Peferoen LA, Vogel DY, Breur M, van der Valk P, Baker D, van Noort JM (2014) Inflammation in neurodegenerative diseases—an update. Immunology 142:151–166

Anderica-Romero AC, Gonzalez-Herrera IG, Santamaria A, Pedraza-Chaverri J (2013) Cullin 3 as a novel target in diverse pathologies. Redox Biol 1:366–372

Andreoletti O, Levavasseur E, Uro-Coste E, Tabouret G, Sarradin P, Delisle MB, Berthon P, Salvayre R, Schelcher F, Negre-Salvayre A (2002) Astrocytes accumulate 4-hydroxynonenal adducts in murine scrapie and human Creutzfeldt-Jakob disease. Neurobiol Dis 11:386–393

Arora V, Kuhad A, Tiwari V, Chopra K (2011) Curcumin ameliorates reserpine induced pain-depression dyad: behavioural, biochemical, neurochemical and molecular evidences. Psychoneuroendocrinology 36:1570–1581

Banasr M, Hery M, Printemps R, Daszuta A (2004) Serotonin-induced increases in adult cell proliferation and neurogenesis are mediated through different and common 5-HT receptor subtypes in the dentate gyrus and the subventricular zone. Neuropsychopharmacology 29:450–460

Bates G (2003) Huntingtin aggregation and toxicity in Huntington's disease. Lancet 361: 1642–1644

Beckman JS, Estévez AG, Crow JP, Barbeito L (2001) Superoxide dismutase and the death of motoneurons in ALS. Trends Neurosci 24(11 Suppl):S15–S20

Benes FM (2000) Emerging principles of altered neural circuitry in schizophrenia. Brain Res Brain Res Rev 31:251–269

Beraldo FH, Arantes CP, Santos TG, Machado CF, Roffe M, Hajj GN, Lee KS, Magalhães AC, Caetano FA, Mancini GL, Lopes MH, Américo TA, Magdesian MH, Ferguson SS, Linden R, Prado MA, Martins VR (2011) Metabotropic glutamate receptors transduce signals for neurite outgrowth after binding of the prion protein to laminin γ1 chain. FASEB J 25:265–279

Bergman J, Miodownik C, Bersudsky Y, Sokolik S, Lerner PP, Kreinin A, Polakiewicz J, Lerner V (2013) Curcumin as an add-on to antidepressive treatment: a randomized, double-blind, placebo-controlled, pilot clinical study. Clin Neuropharmacol 36:73–77

Berk M, Conus P, Kapczinski F, Andreazza AC, Yücel M, Wood SJ, Pantelis C, Malhi GS, Dodd S, Bechdolf A, Amminger GP, Hickie IB, McGorry PD (2010) From neuroprogression to neuroprotection: implications for clinical care. Med J Aust 193:S36–S40

Bhutani MK, Bishnoi M, Kulkarni SK (2009) Anti-depressant like effect of curcumin and its combination with piperine in unpredictable chronic stress-induced behavioral, biochemical and neurochemical changes. Pharmacol Biochem Behav 92:39–43

Bisaglia M, Tessari I, Mammi S, Bubacco L (2009) Interaction between a-synuclein and metal ions still looking for a role in the pathogenesis of Parkinson's disease. Neuromolecular Med 11:239–251

Blitzer RD, Iyengar R, Landau EM (2005) Postsynaptic signaling networks: cellular cogwheels underlying long-term plasticity. Biol Psychiatry 57:113–119

Blümcke I, Beck H, Lie AA, Wiestler OD (1999) Molecular neuropathology of human mesial temporal lobe epilepsy. Epilepsy Res 36:205–223

Boison D, Sandau US, Ruskin DN, Kawamura M Jr, Masino SA (2013) Homeostatic control of brain function – new approaches to understand epileptogenesis. Front Cell Neurosci 7:109

Bonilla E (2000) Huntington disease. A review. Invest Clin 41:117–141

Bower JH, Maraganore DM, Peterson BJ, McDonnell SK, Ahlskog JE, Rocca WA (2003) Head trauma preceding PD: a case-control study. Neurology 60:1610–1615

Bradley MA, Markesbery WR, Lovell MA (2010) Increased levels of 4-hydroxynonenal and acrolein in the brain in preclinical Alzheimer disease. Free Radic Biol Med 48:1570–1576

Bramham CR, Messaoudi E (2005) BDNF function in adult synaptic plasticity: the synaptic consolidation hypothesis. Prog Neurobiol 76:99–125

Bright JJ (2007) Curcumin and autoimmune disease. Adv Exp Med Biol 595:425–451

Bruce JH, Norenberg MD, Kraydieh S, Puckett W, Marcillo A, Dietrich D (2000) Schwannosis: role of gliosis and proteoglycan in human spinal cord injury. J Neurotrauma 17:781–788

Bryan HK, Olayanju A, Goldring CE, Park BK (2013) The Nrf2 cell defence pathway: Keap1-dependent and-independent mechanisms of regulation. Biochem Pharmacol 85:705–717

Casley CS, Lakics V, Lee H-G, Broad LM, Day TA, Cluett T, Smith MA, O'Neill MJ, Kingston AE (2009) Up-regulation of astrocyte metabotropic glutamate receptor 5 by amyloid-beta peptide. Brain Res 1260:65–75

Caughey B, Raymond LD, Raymond GJ, Maxson L, Silveira J, Baron GS (2003) Inhibition of protease-resistant prion protein accumulation in vitro by curcumin. J Virol 77:5499–5502

Cemil B, Topuz K, Demircan MN, Kurt G, Tun K, Kutlay M, Ipcioglu O, Kucukodaci Z (2010) Curcumin improves early functional results after experimental spinal cord injury. Acta Neurochir (Wien) 152:1583–1590

Cepeda C, Ariano MA, Calvert CR, Flores-Hernandez J, Chandler SH, Leavitt BR, Hayden MR, Levine MS (2001) NMDA receptor function in mouse models of Huntington disease. J Neurosci Res 66:525–539

Chauhan NB (2014) Chronic neurodegenerative consequences of traumatic brain injury. Restor Neurol Neurosci 32:337–365

Chen X, de Silva HA, Pettenati MJ, Rao PN, St George-Hyslop P, Roses AD Xia Y, Horsburgh K, Ueda K, Saitoh T (1995) The human NACP/alpha-synuclein gene: chromosome assignment to 4q21.3–q22 and TaqI RFLP analysis. Genomics 26: 425–427

Chen J, Tang XQ, Zhi JL, Cui Y, Yu HM, Tang EH, Sun SN, Feng JQ, Chen PX (2006) Curcumin protects PC12 cells against 1-methyl-4-phenylpyridinium ion-induced apoptosis by Bcl-2-mitochondria-ROS-iNOS pathway. Apoptosis 11:943–953

Chen H, Richard M, Sandler DP, Umbach DM, Kamel F (2007) Head injury and amyotrophic lateral sclerosis. Am J Epidemiol 166:810–816

Choudhary KM, Mishra A, Poroikov VV, Goel RK (2013) Ameliorative effect of curcumin on seizure severity, depression like behavior, learning and memory deficit in post-pentylenetetrazole-kindled mice. Eur J Pharmacol 704:33–40

Collinge J, Whittington MA, Sidle KC, Smith CJ, Palmer MS, Clarke AR, Jefferys JG (1994) Prion protein is necessary for normal synaptic function. Nature 370:295–297

Cowan CM, Raymond LA (2006) Selective neuronal degeneration in Huntington's disease. Curr Top Dev Biol 75:25–71

Criado JR, Sanchez-Alavez M, Conti B, Giacchino JL, Wills DN, Henriksen SJ, Race R, Manson JC, Chesebro B, Oldstone MB (2005) Mice devoid of prion protein have cognitive deficits that are rescued by reconstitution of PrP in neurons. Neurobiol Dis 19:255–265

Dauer W, Przedborski S (2003) Parkinson's disease: mechanisms and models. Neuron 39:889–909

David Y, Cacheaux LP, Ivens S, Lapilover E, Heinemann U, Kaufer D, Friedman A (2009) Astrocytic dysfunction in epileptogenesis: consequence of altered potassium and glutamate homeostasis? J Neurosci 29:10588–10599

Davidson RJ, Pizzagalli D, Nitschike JB, Putnam K (2002) Depression: perspectives from affective neuroscience. Annu Rev Psychol 53:545–574

Deleidi M, Jäggle M, Rubino G (2015) Immune aging, dysmetabolism, and inflammation in neurological diseases. Front Neurosci 9:172

Devinsky O, Vezzani A, Najjar S, De Lanerolle NC, Rogawski MA (2013) Glia and epilepsy: excitability and inflammation. Trends Neurosci 36:174–184

Dichter MA (2006) Models of epileptogenesis in adult animals available for antiepileptogenesis drug screening. Epilepsy Res 68:31–35

Donkin JJ, Vink R (2010) Mechanisms of cerebral edema in traumatic brain injury: therapeutic developments. Curr Opin Neurol 23:293–299

Dornmair K, Goebels N, Weltzien HU, Wekerle H, Hohlfeld R (2003) T-cell-mediated autoimmunity: novel techniques to characterize autoreactive T-cell receptors. Am J Pathol 163:1215–1226

Fagiolini A, Chengappa KN, Soreca I, Chang J (2008) Bipolar disorder and the metabolic syndrome: causal factors, psychiatric outcomes and economic burden. CNS Drugs 22:655–669

Faideau M, Kim J, Cormier K, Gilmore R, Welch M, Auregan G, Dufour N, Guillermier M, Brouillet E, Hantraye P, Déglon N, Ferrante RJ, Bonvento G (2010) In vivo expression of polyglutamine-expanded huntingtin by mouse striatal astrocytes impairs glutamate transport: a correlation with Huntington's disease subjects. Hum Mol Genet 19:3053–3067

Farooqui AA (2010) Neurochemical aspects of neurotraumatic and neurodegenerative diseases. Springer, New York

Farooqui AA (2011) Lipid mediators and their metabolism in the brain. Springer, New York

Farooqui AA (2012) Phytochemicals, signal transduction, and neurological disorders. Springer, New York

Farooqui AA (2013) Metabolic syndrome: an important risk factor for stroke, Alzheimer disease, and depression. Springer, New York

Farooqui AA (2015) High calorie diet and the human brain: metabolic consequences of long-term consumption. Springer, Switzerland

Farooqui T, Farooqui AA (2009) Aging: an important factor for the pathogenesis of neurodegenerative diseases. Mech Ageing Dev 130:203–215

Farooqui T, Farooqui AA (2011) Lipid-mediated oxidative stress and inflammation in the pathogenesis of Parkinson's disease. Parkinsons Dis 2011:247467

Farooqui AA, Horrocks LA (2007) Glycerophospholipids in the brain: phospholipases A_2 in neurological disorders. Springer, New York

Farooqui AA, Horrocks LA, Farooqui T (2007) Modulation of inflammation in brain: a matter of fat. J Neurochem 101:577–599

Farooqui AA, Ong WY, Horrocks LA (2008) Neurochemical aspects of excitotoxicity. Springer, New York

Fehlings MG, Tighe A (2008) Spinal cord injury: the promise of translational research. Neurosurg Focus 25:E1

Filbin MT (2003) Myelin-associated inhibitors of axonal regeneration in the adult mammalian CNS. Nat Rev Neurosci 4:703–713

Fleminger S, Oliver DL, Lovestone S, Rabe-Hesketh S, Giora A (2003) Head injury as a risk factor for Alzheimer's disease: the evidence 10 years on; a partial replication. J Neurol Neurosurg Psychiatry 74:857–862

Fouad K, Tse A (2008) Adaptive changes in the injured spinal cord and their role in promoting functional recovery. Neurol Res 30:17–27

Fournier JG, Escaig-Haye F, Grigoriev V (2000) Ultrastructural localization of prion proteins: physiological and pathological implications. Microsc Res Tech 50:76–88

Furukawa T, Matsui N, Fujita K, Miyashiro A, Nodera H, Izumi Y, Shimizu F, Miyamoto K, Takahashi Y, Kanda T, Kusunoki S, Kaji R (2014) Increased proinflammatory cytokines in sera of patients with multifocal motor neuropathy. J Neurol Sci 346:75–79

Gadad BS, Subramanya PK, Pullabhatla S, Shantharam IS, Rao KS (2012) Curcumin-glucoside, a novel synthetic derivative of curcumin, inhibits α-synuclein oligomer formation: relevance to Parkinson's disease. Curr Pharm Des 18:76–84

Gałecki P, Talarowska M, Anderson G, Berk M, Maes M (2015) Mechanisms underlying neuro-cognitive dysfunctions in recurrent major depression. Med Sci Monit 21:1535–1547

Garcia-Alloza M, Borrelli LA, Rozkalne A, Hyman BT, Bacskai BJ (2007) Curcumin labels amyloid pathology in vivo, disrupts existing plaques, and partially restores distorted neurites in an Alzheimer mouse model. J Neurochem 102:1095–1104

Gautam SC, Gao X, Dutchavsky S (2007) Immunomodulation by curcumin. Adv Exp Med Biol 595:321–341

Gazal M, Valente MR, Acosta BA, Kaufmann FN, Braganhol E, Lencina CL, Stefanello FM, Ghisleni G, Kaster MP (2014) Neuroprotective and antioxidant effects of curcumin in a ketamine-induced model of mania in rats. Eur J Pharmacol 724:132–139

Gil JM, Rego AC (2008) Mechanisms of neurodegeneration in Huntington's disease. Eur J Neurosci 27:2803–2820

Glabinski AR, Tani M, Tuohy VK, Ransohoff RM (1997) Murine experimental autoimmune encephalomyelitis: a model of immune-mediated inflammation and multiple sclerosis. Methods Enzymol 288:182–190

Goel A, Kunnumakkara AB, Aggarwal BB (2008) Curcumin as "Curecumin": from kitchen to clinic. Biochem Pharmacol 75:787–809

Gomez-Pinilla F (2011) The combined effects of exercise and foods in preventing neurological and cognitive disorders. Prev Med 52(Suppl 1):S75–S80

Gomez-Pinilla F, Nguyen TT (2012) Natural mood foods: the actions of polyphenols against psychiatric and cognitive disorders. Nutr Neurosci 15:127–133

Graeber MB, Moran LB (2002) Mechanisms of cell death in neurodegenerative diseases: fashion, fiction, and facts. Brain Pathol 12:385–390

Greco A, Minghetti L, Levi G (2000) Isoprostanes, novel markers of oxidative injury, help understanding the pathogenesis of neurodegenerative diseases. Neurochem Res 25:1357–1364

Griesbach GS, Hovda DA, Gomez-Pinilla F (2009) Exercise-induced improvement in cognitive performance after traumatic brain injury in rats is dependent on BDNF activation. Brain Res 1288:105–115

Grossman A, Zeiler B, Sapirstein V (2003) Prion protein interactions with nucleic acid: possible models for prion disease and prion function. Neurochem Res 28:955–963

Gunther EC, Strittmatter SM (2010) Beta-amyloid oligomers and cellular prion protein in Alzheimer's disease. J Mol Med (Berl) 88:331–338

Gupta SC, Prasad S, Kim JH, Patchva S, Webb LJ, Priyadarsini IK, Aggarwal BB (2011) Multitargeting by curcumin as revealed by molecular interaction studies. Nat Prod Rep 28:1937–1955

Hafner-Bratkovic I, Gaspersic J, Smid LM, Bresjanac M, Jerala R (2008) Curcumin binds to the alpha-helical intermediate and to the amyloid form of prion protein – a new mechanism for the inhibition of PrP(Sc) accumulation. J Neurochem 104:1553–1564

Hamilton A, Zamponi GW, Ferguson SS (2015) Glutamate receptors function as scaffolds for the regulation of β-amyloid and cellular prion protein signaling complexes. Mol Brain 8:18

Harrison PJ (1999) Neurochemical alterations in schizophrenia affecting the putative receptor targets of atypical antipsychotics. Focus on dopamine (D1, D3, D4) and 5-HT2a receptors. Brain 122:593–624

Heck N, Garwood J, Loeffler JP, Larmet Y, Faissner A (2004) Differential upregulation of extracellular matrix molecules associated with the appearance of granule cell dispersion and mossy fiber sprouting during epileptogenesis in a murine model of temporal lobe epilepsy. Neuroscience 129:309–324

Hegde ML, Hegde PM, Rao KS, Mitra S (2011) Oxidative genome damage and its repair in neurodegenerative diseases: function of transition metals as a double-edged sword. J Alzheimers Dis 24(Suppl 2):183–198

Heikkila RE, Sonsalla PK (1992) The MTPT-treated mouse as a model of Parkinsonism: how good is it? Neurochem Int 20(Suppl 1):299S–303S

Hickey MA, Kosmalska A, Enayati J, Cohen R, Zeitlin S, Levine MS, Chesselet MF (2008) Extensive early motor and non-motor behavioral deficits are followed by striatal neuronal loss in knock-in Huntington's disease mice. Neuroscience 157:280–295

Hickey MA, Zhu C, Medvedeva V, Lerner RP, Patassini S, Franich NR, Maiti P, Frautschy SA, Zeitlin S, Levine MS, Chesselet MF (2012) Improvement of neuropathology and transcriptional deficits in CAG 140 knock-in mice supports a beneficial effect of dietary curcumin in Huntington's disease. Mol Neurodegener 7:12

Hipp MS, Patel CN, Bersuker K, Riley BE, Kaiser SE, Shaler TA, Brandeis M, Kopito RR (2012) Indirect inhibition of 26S proteasome activity in a cellular model of Huntington's disease. J Cell Biol 196:573–587

Hong HS, Rana S, Barrigan L, Shi A, Zhang Y, Zhou F, Jin LW, Hua DH (2009) Inhibition of Alzheimer's amyloid toxicity with a tricyclic pyrone molecule in vitro and in vivo. J Neurochem 108:1097–1108

Hou ST, MacManus JP (2002) Molecular mechanisms of cerebral ischemia-induced neuronal death. Int Rev Cytol 221:93–148

Hsu LJ, Sagara Y, Arroyo A, Rockenstein E, Sisk A, Mallory M, Wong J, Takenouchi T, Hashimoto M, Masliah E (2000) Alpha-synuclein promotes mitochondrial deficit and oxidative stress. Am J Pathol 157:401–410

Irvine GB, El Agnaf OM, Shankar GM, Walsh DM (2008) Protein aggregation in the brain: the molecular basis for Alzheimer's and Parkinson's diseases. Mol Med 14:451–464

Italiani P, Carlesi C, Giungato P, Puxeddu I, Borroni B, Bossù P, Migliorini P, Siciliano G, Boraschi D (2014) Evaluating the levels of interleukin-1 family cytokines in sporadic amyotrophic lateral sclerosis. J Neuroinflammation 11:94

Ito H, Kawashima R, Awata S, Ono S, Sato K, Goto R, Koyama M, Sato M, Fukuda H (1996) Hypoperfusion in the limbic system and prefrontal cortex in depression: SPECT with anatomic standardization technique. J Nucl Med 37:410–414

Iwai A, Masliah E, Yoshimoto M, Ge N, Flanagan L, de Silva HA, Kittel A, Saitoh T (1995) The precursor protein of non-A beta component of Alzheimer's disease amyloid is a presynaptic protein of the central nervous system. Neuron 14:467–475

Jankovic J (2008) Parkinson's disease: clinical features and diagnosis. J Neurol Neurosurg Psychitry 79:368–376

Jayaraj RL, Tamilselvam K, Manivasagam T, Elangovan N (2013) Neuroprotective effect of CNB-001, a novel pyrazole derivative of curcumin on biochemical and apoptotic markers against rotenone-induced SK-N-SH cellular model of Parkinson's disease. J Mol Neurosci 51:863–870

Jayaraj RL, Elangovan N, Manigandan K, Singh S, Shukla S (2014) CNB-001 a novel curcumin derivative, guards dopamine neurons in MPTP model of Parkinson's disease. Biomed Res Int 2014:236182

Jellinger KA (2009) Recent advances in our understanding of neurodegeneration. J Neural Transm 116:1111–1162

Jiang J, Wang W, Sun Y, Hu M, Li F, Zhu DY (2007) Neuroprotective effect of curcumin on focal cerebral ischemic rats by preventing blood-brain barrier damage. Eur J Pharmacol 561:54–62

Kalonia H, Kumar A (2011) Suppressing inflammatory cascade by cyclo-oxygenase inhibitors attenuates quinolinic acid induced Huntington's disease-like alterations in rats. Life Sci 88:784–791

Kanakasabai S, Casalini E, Walline CC, Mo C, Chearwae W, Bright JJ (2012) Differential regulation of CD4(+) T helper cell responses by curcumin in experimental autoimmune encephalomyelitis. J Nutr Biochem 23:1498–1507

Kato K, Ito H, Kamei K, Iwamoto I (1998) Stimulation of the stress-induced expression of stress proteins by curcumin in cultured cells and in rat tissues in vivo. Cell Stress Chaperones 3:152–160

Kaur H, Bal A, Sandhir R (2014) Curcumin supplementation improves mitochondrial and behavioral deficits in experimental model of chronic epilepsy. Pharmacol Biochem Behav 125:55–64

Kessels HW, Nguyen LN, Nabavi S, Malinow R (2010) The prion protein as a receptor for amyloid-beta. Nature 466:E3–E4

Ketter TA (2010) Diagnostic features, prevalence, and impact of bipolar disorder. J Clin Psychiatry 71, e14

Khosravani H, Zhang Y, Tsutsui S, Hameed S, Altier C, Hamid J, Chen L, Villemaire M, Ali Z, Jirik FR, Zamponi GW (2008) Prion protein attenuates excitotoxicity by inhibiting NMDA receptors. J Cell Biol 181(3):551–565

Kidd PM (2005) Neurodegeneration from mitochondrial insufficiency: nutrients, stem cells, growth factors, and prospects for brain rebuilding using integrative management. Altern Med Rev 10:268–293

Kim GY, Kim KH, Lee SH, Yoon MS, Lee HJ, Moon DO (2005) Curcumin inhibits immunostimulatory function of dendritic cells: MAPKs and translocation of NF-B as potential targets. J Immunol 174:8116–8124

Kim SJ, Son TG, Park HR, Park M, Kim M-S, Kim HS, Chung HY, Mattson MP, Lee J (2008) Curcumin stimulates proliferation of embryonic neural progenitor cells and neurogenesis in the adult hippocampus. J Biol Chem 283:14497–14505

Kim J-H, Park S-H, Nam S-W, Kwon H-J, Kim B-W, Kim W-J et al (2011) Curcumin stimulates proliferation, stemness acting signals and migration of 3T3-L1 preadipocytes. Int J Mol Med 28:429–435

Kimbrell TA, Ketter TA, George MS, Little JT, Benson BE, Willis MW, Herscovitch P, Post RM (2002) Regional cerebral glucose utilization in patients with a range of severities of unipolar depression. Biol Psychiatry 51:237–252

Klussmann S, Martin-Villalba A (2005) Molecular targets in spinal cord injury. J Mol Med 83:657–671

Kulkarni SK, Bhutani MK, Bishnoi M (2008) Antidepressant activity of curcumin: involvement of serotonin and dopamine system. Psychopharmacol (Berl) 201:435–442

Kulkarni S, Dhir A, Akula KK (2009) Potentials of curcumin as an antidepressant. ScientificWorldJournal 9:1233–1241

Kumar P, Kalonia H, Kumar A (2011) Role of LOX/COX pathways in 3-nitropropionic acid-induced Huntington's disease-like symptoms in rats: protective effect of licofelone. Br J Pharmacol 164:644–654

Kupfer DJ, Frank E, Phillips ML (2012) Major depressive disorder: new clinical, neurobiological, and treatment perspectives. Lancet 379:1045–1055

Laurén J, Gimbel DA, Nygaard HB, Gilbert JW, Stephen M (2009) Cellular prion protein mediates impairment of synaptic plasticity by amyloid-B oligomers. Nature 457:1128–1132

Lee HJ, Choi C, Lee SJ (2002) Membrane-bound α-synuclein has a high aggregation propensity and the ability to seed the aggregation of the cytosolic form. J Biol Chem 277:671–678

Lee J, Kosaras B, Del Signore SJ, Cormier K, McKee A, Ratan RR, Kowall NW, Ryu H (2011) Modulation of lipid peroxidation and mitochondrial function improves neuropathology in Huntington's disease mice. Acta Neuropathol 121:487–498

Leonard B, Maes M (2012) Mechanistic explanations how cell-mediated immune activation, inflammation and oxidative and nitrosative stress pathways and their sequels and concomitants play a role in the pathophysiology of unipolar depression. Neurosci Biobehav Rev 36: 764–785

Li J, Zhu M, Rajamani S, Uversky VN, Fink AL (2004) Rifampicin inhibits alpha-synuclein fibrillation and disaggregates fibrils. Chem Biol 11:1513–1521

Li YC, Wang FM, Pan Y, Qiang LQ, Cheng G, Zhang WY, Kong LD (2009) Antidepressant-like effects of curcumin on serotonergic receptor-coupled AC-cAMP pathway in chronic unpredictable mild stress of rats. Prog Neuropsychopharmacol Biol Psychiatry 33:435–449

Liang D, Han G, Feng X, Sun J, Duan Y, Lei H (2012) Concerted perturbation observed in a hub network in Alzheimer's disease. PLoS One 7, e40498

Lim GP, Chu T, Yang F, Beech W, Frautschy SA, Cole GM (2001) The curry spice curcumin reduces oxidative damage and amyloid pathology in an Alzheimer transgenic mouse. J Neurosci 21:8370–8377

Lin MS, Lee YH, Chiu WT, Hung KS (2011a) Curcumin provides neuroprotection after spinal cord injury. J Surg Res 166:280–289

Lin MS, Sun YY, Chiu WT, Chang CY, Hung CC, Shie FS, Tsai SH, Lin JW, Hung KS, Lee YH (2011b) Curcumin attenuates the expression and secretion of RANTES following spinal cord injury in vivo and lipopolysaccharide-induced astrocyte reactivation in vitro. J Neurotrauma 28:1259–1269

Linden R, Martins VR, Prado MA, Cammarota M, Izquierdo I, Brentani RR (2008) Physiology of the prion protein. Physiol Rev 88:673–728

Liu Y, Dargusch R, Maher P, Schubert D (2008) A broadly neuroprotective derivative of curcumin. J Neurochem 105:1336–1345

Liu Z, Yu Y, Li X, Ross CA, Smith WW (2011) Curcumin protects against A53T alpha-synuclein-induced toxicity in a PC12 inducible cell model for Parkinsonism. Pharmacol Res 63: 439–444

Lopes JP, Oliveira CR, Agostinho P (2007) Role of cyclin-dependent kinase 5 in the neurodegenerative process triggered by amyloid-beta and Prion peptides: implications for Alzheimer's disease and Prion-related encephalopathies. Cell Mol Neurobiol 27:943–957

Lopresti AL, Hood SD, Drummond PD (2012) Multiple antidepressant potential modes of action of curcumin: a review of its anti-inflammatory, monoaminergic, antioxidant, immune-modulating and neuroprotective effects. J Psychopharmacol 26:1512–1524

Lopresti AL, Hood SD, Drummond PD (2013) A review of lifestyle factors that contribute to important pathways associated with major depression: diet, sleep and exercise. J Affect Disord 148:12–27

Luscher B, Shen Q, Sahir N (2011) The GABAergic deficit hypothesis of major depressive disorder. Mol Psychiatry 16:383–406

Ma QL, Yang F, Rosario ER, Ubeda OJ, Beech W, Gant DJ, Chen PP, Hudspeth B, Chen C, Zhao Y, Vinters HV, Frautschy SA, Cole GM (2009) β-Amyloid oligomers induce phosphorylation of tau and inactivation of insulin receptor substrate via c-Jun N-terminal kinase signaling: suppression by omega-3 fatty acids and curcumin. J Neurosci 29:9078–9089

Maas AI, Stocchetti N, Bullock R (2008) Moderate and severe traumatic brain injury in adults. Lancet Neurol 7:728–741

Maglio LE, Martins VR, Izquierdo I, Ramirez OA (2006) Role of cellular prion protein on LTP expression in aged mice. Brain Res 1097:11–18

Mallucci GR, Ratté S, Asante EA, Linehan J, Gowland I, Jefferys JG, Collinge J (2002) Post-natal knockout of prion protein alters hippocampal CA1 properties, but does not result in neurodegeneration. EMBO J 21:202–210

Manea MM, Comsa M, Minca A, Dragos D, Popa C (2015) Brain-heart axis – review article. J Med Life 8:266–271

Martinc B, Grabnar I, Vovk T (2012) The role of reactive species in epileptogenesis and influence of antiepileptic drug therapy on oxidative stress. Curr Neuropharmacol 10:328–343

Martinez-Aran A, Vieta E, Colom F, Goikolea JM, Colom F, Martínez-Arán A, Benabarre A (2004) Cognitive impairment in euthymic bipolar patients: implications for clinical and functional outcome. Bipolar Disord 6:224–232

Mattson MP, Duan W, Pedersen WA, Culmsee C (2001) Neurodegenerative disorders and ischemic brain diseases. Apoptosis 6:69–81

McIntyre RS, Danilewitz M, Liauw SS, Kemp DE, Nguyen HT, Kahn LS, Kucyi A, Soczynska JK, Woldeyohannes HO, Lachowski A, Kim B, Nathanson J, Alsuwaidan M, Taylor VH (2010) Bipolar disorder and metabolic syndrome: an international perspective. J Affect Disord 126:366–387

McNally SJ, Harrison EM, Ross JA, Garden OJ, Wigmore SJ (2007) Curcumin induces heme oxygenase 1 through generation of reactive oxygen species, p38 activation and phosphatase inhibition. Int J Mol Med 19:165–172

Mei X, Xu D, Xu S, Zheng Y (2011) Gastroprotective and antidepressant effects of a new zinc(II)-curcumin complex in rodent models of gastric ulcer and depression induced by stresses. Pharmacol Biochem Behav 99:66–74

Meisel C, Schwab J, Prass K, Meisel A, Dirnagl U (2005) Central nervous system injury-induced immune deficiency syndrome. Nat Rev Neurosci 6:775–786

Miller GW, Erickson JD, Perez JT, Penland SN, Mash DC, Rye DB, Levey AI (1999) Immunochemical analysis of vesicular monoamine transporter (VMAT2) protein in Parkinson's disease. Exp Neurol 156:138–148

Miller E, Morel A, Saso L, Saluk J (2014) Isoprostanes and neuroprostanes as biomarkers of oxidative stress in neurodegenerative diseases. Oxid Med Cell Longev 2014:572491

Milyukhina IV, Karpenko MN, Klimenko VM (2015) Clinical parameters and the level of certain cytokines in blood and cerebrospinal fluid of patients with Parkinson's disease. Klin Med (Mosk) 93:51–55

Minghetti L, Pocchiari M (2007) Cyclooxygenase-2, prostaglandin E2, and microglial activation in prion diseases. Int Rev Neurobiol 82:265–275

Minghetti L, Greco A, Cardone F, Puopolo M, Ladogana A, Almonti S, Cunningham C, Perry VH, Pocchiari M, Levi G (2000) Increased brain synthesis of prostaglandin E2 and F2-isoprostane in human and experimental transmissible spongiform encephalopathies. J Neuropathol Exp Neurol 59:866–871

Monsey MS, Gerhard DM, Boyle LM, Briones MA, Seligsohn M, Schafe GE (2015) A diet enriched with curcumin impairs newly acquired and reactivated fear memories. Neuropsychopharmacology 40:1278–1288

Montine TJ, Sidell KR, Crews BC, Markesbery WR, Marnett LJ, Roberts LJ 2nd, Morrow JD (1998) Elevated cerebrospinal fluid prostaglandin E2 levels in patients with probable Alzheimer's disease. Neurology 53:1495–1498

Montine TJ, Beal MF, Robertson D, Cudkowicz ME, Biaggioni I, O'Donnell H, Zackert WE, Roberts LJ, Morrow JD (1999) Cerebrospinal fluid F2-isoprostanes are elevated in Huntington's disease. Neurology 52:1104–1105

Morris G, Berk M (2015) The many roads to mitochondrial dysfunction in neuroimmune and neuropsychiatric disorders. BMC Med 13:68

Motori E, Puyal J, Toni N, Ghanem A, Angeloni C, Malaguti M, Cantelli-Forti G, Berninger B, Conzelmann KK, Götz M, Winklhofer KF, Hrelia S, Bergami M (2013) Inflammation-induced alteration of astrocyte mitochondrial dynamics requires autophagy for mitochondrial network maintenance. Cell Metab 18:844–859

Nagatsu T, Mogi M, Ichinose H, Togari A (2000) Cytokines in Parkinson's disease. J Neural Transm Suppl 58:143–151

Noseworthy JH, Lucchinetti C, Rodriguez M, Weinshenker BG (2000) Multiple sclerosis. N Engl J Med 343:938–952

Nygaard HB, Strittmatter SM (2009) Cellular prion protein mediates the toxicity of beta-amyloid oligomers: implications for Alzheimer disease. Arch Neurol 66:1325–1328

O'Connor P (2002) Key issues in the diagnosis and treatment of multiple sclerosis. An overview. Neurology 59:S1–S33

Ong WY, Tanaka K, Dawe GS, Ittner LM, Farooqui AA (2013) Slow excitotoxicity in Alzheimer's disease. J Alzheimers Dis 35:643–668

Ormond DR, Shannon C, Oppenheim J, Zeman R, Das K, Murali R, Jhanwar-Uniyal M (2014) Stem cell therapy and curcumin synergistically enhance recovery from spinal cord injury. PLoS One 9, e88916

Pan R, Qiu S, Lu DX, Dong J (2008) Curcumin improves learning and memory ability and its neuroprotective mechanism in mice. Chin Med J (Engl) 121:832–839

Pandey N, Strider J, Nolan WC, Yan SX, Galvin JE (2008) Curcumin inhibits aggregation of α-synuclein. Acta Neuropathol 115:479–489

Parker G, Gibson NA, Brotchie H, Heruc G, Rees AM, Hadzi-Pavlovic D (2006) Omega-3 fatty acids and mood disorders. Am J Psychiatry 163:969–978

Parkin ET, Watt NT, Hussain I, Eckman EA, Eckman CB, Manson JC, Baybutt HN, Turner AJ, Hooper NM (2007) Cellular prion protein regulates beta-secretase cleavage of the Alzheimer's amyloid precursor protein. Proc Natl Acad Sci U S A 104:11062–11067

Periluigi M, Fai Poon H, Hensley K, Pieree WM, Klein JB, Calabrese V, De Marco C, Butterfield DA (2005) Proteomic analysis of 4-hydroxy-2-nonenal-modified proteins in G93A-SOD1 transgenic mice—a model of familial amyotrophic lateral sclerosis. Free Radic Biol Med 38:960–968

Petersen SE, Posner MI (2012) The attention system of the human brain: 20 years after. Annu Rev Neurosci 35:73–89

Phillis JW, Horrocks LA, Farooqui AA (2006) Cyclooxygenases, lipoxygenases, and epoxygenases in CNS: their role and involvement in neurological disorders. Brain Res Rev 52:201–243

Pieri L, Madiona K, Bousset L, Melki R (2012) Fibrillar alpha-synuclein and huntingtin exon 1 assemblies are toxic to the cells. Biophys J 102:2894–2905

Pistell PJ, Morrison CD, Gupta S, Knight AG, Keller JN, Ingram DK, Bruce-Keller AJ (2010) Cognitive impairment following high fat diet consumption is associated with brain inflammation. J Neuroimmunol 219:25–32

Pitkanen A, Kharatishvili I, Karhunen H, Lukasiuk K, Immonen R, Nairismägi J, Gröhn O, Nissinen J (2007) Epileptogenesis in experimental models. Epilepsia 48(Suppl 2):13–20

Prusiner SB (2001) Shattuck lecture—neurodegenerative diseases and prions. N Engl J Med 344:1516–1526

Qin Z, Hu D, Han S, Reaney SH, Di Monte DA, Fink AL (2007) Effect of 4-hydroxy-2-nonenal modification on alpha-synuclein aggregation. J Biol Chem 282:5862–5870

Qureshi IA, Mehler MF (2010a) Epigenetic mechanisms underlying human epileptic disorders and the process of epileptogenesis. Neurobiol Dis 39:53–60

Qureshi IA, Mehler MF (2010b) The emerging role of epigenetics in stroke II. RNA regulatory circuitry. Arch Neurol 67:1435–1441

Raghupathi R (2004) Cell death mechanisms following traumatic brain injury. Brain Pathol 14:215–222

Rakhade SN, Jensen FE (2009) Epileptogenesis in the immature brain: emerging mechanisms. Nat Rev Neurol 5:380–391

Ray B, Bisht S, Maitra A, Naitra A, Lahiri DK (2011) Neuroprotective and neurorescue effects of a novel polymeric nanoparticle formulation of curcumin (NanoCurc®) in the neuronal cell culture and animal model: implications for Alzheimer's disease. J Alzheimers Dis 23:61–77

Rostami E, Krueger F, Plantman S, Davidsson J, Agoston D, Grafman J, Risling M (2014) Alteration in BDNF and its receptors, full-length and truncated TrkB and p75(NTR) following penetrating traumatic brain injury. Brain Res 1542:195–205

Rubinsztein DC, Leggo J, Coles R, Almqvist E, Biancalana V, Cassiman JJ, Chotai K, Connarty M, Crauford D, Curtis A, Curtis D, Davidson MJ, Differ AM, Dode C, Dodge A, Frontali M, Ranen NG, Stine OC, Sherr M, Abbott MH, Franz ML, Graham CA, Harper PS, Hedreen JC,

Hayden MR (1996) Phenotypic characterization of individuals with 30-40 CAG repeats in the Huntington disease (HD) gene reveals HD cases with 36 repeats and apparently normal elderly individuals with 36-39 repeats. Am J Hum Genet 59:16–22

Rudinskiy N, Kaneko YA, Beesen AA, Gokce O, Regulier E, Deglon N, Luthi-Carter R (2009) Diminished hippocalcin expression in Huntington's disease brain does not account for increased striatal neuron vulnerability as assessed in primary neurons. J Neurochem 111:460–472

Sanmukhani J, Anovadiya A, Tripathi CB (2011) Evaluation of antidepressant like activity of curcumin and its combination with fluoxetine and imipramine: an acute and chronic study. Acta Pol Pharm 68:769–775

Sanmukhani J, Satodia V, Trivedi J, Patel T, Tiwari D, Panchal B, Goel A, Tripathi CB (2014) Efficacy and safety of curcumin in major depressive disorder: a randomized controlled trial. Phytother Res 28:579–585

Scapagnini G, Butterfield DA, Colombrita C, Sultana R, Pascale A, Calabrese V (2004) Ethyl ferulate, a lipophilic polyphenol, induces HO-1 and protects rat neurons against oxidative stress. Antioxid Redox Signal 6:811–818

Scapagnini G, Colombrita C, Amadio M, D'Agata V, Arcelli E, Sapienza M, Quattrone A, Calabrese V (2006) Curcumin activates defensive genes and protects neurons against oxidative stress. Antioxid Redox Signal 8:395–403

Scapagnini G, Vasto S, Abraham NG, Caruso C, Zella D, Fabio G (2011) Modulation of Nrf2/ARE pathway by food polyphenols: a nutritional neuroprotective strategy for cognitive and neurodegenerative disorders. Mol Neurobiol 44:192–201

Scapagnini G, Davinelli S, Drago F, De Lorenzo A, Oriani G (2012) Antioxidants as antidepressants: fact or fiction? CNS Drugs 26:477–490

Scherzinger E, Lurz R, Turmaine M, Mangiarini L, Hollenbach B, Hasenbank R, Bates GP, Davies SW, Lehrach H, Wanker EE (1997) Huntingtin-encoded polyglutamine expansions form amyloid-like protein aggregates in vitro and in vivo. Cell 90:549–558

Schmidt S, Kwee LC, Allen KD, Oddone EZ (2010) Association of ALS with head injury, cigarette smoking and APOE genotypes. J Neurol Sci 291:22–29

Schmitt-Ulms G, Hansen K, Liu J, Cowdrey C, Yang J, DeArmond SJ, Cohen FE, Prusiner SB, Baldwin MA (2004) Time-controlled transcardiac perfusion cross-linking for the study of protein interactions in complex tissues. Nat Biotechnol 22:724–731

Seet RC, Lee CY, Lim EC, Tan JJ, Quek AM, Chong WL, Looi WF, Huang SH, Wang H, Chan YH, Halliwell B (2010) Oxidative damage in Parkinson disease: measurement using accurate biomarkers. Free Radic Biol Med 48:560–566

Selkoe DJ (2003) Folding proteins in fatal ways. Nature 426:900–904

Severson C, Hafler DA (2010) T-cells in multiple sclerosis. Results Probl Cell Differ 51:75–98

Sharma S, Zhuang Y, Ying Z, Wu A, Gomez-Pinilla F (2009) Dietary curcumin supplementation counteracts reduction in levels of molecules involved in energy homeostasis after brain trauma. Neuroscience 161:1037–1044

Sharma S, Ying Z, Gomez-Pinilla F (2010) A pyrazole curcumin derivative restores membrane homeostasis disrupted after brain trauma. Exp Neurol 226:191–199

Signoretti S, Marmarou A, Tavazzi B, Dunbar J, Amorini AM, Lazzarino G, Vagnozzi R (2004) The protective effect of cyclosporin A upon N-acetylaspartate and mitochondrial dysfunction following experimental diffuse traumatic brain injury. J Neurotrauma 21:1154–1167

Silverstein FS, Jensen FE (2007) Neonatal seizures. Ann Neurol 62:112–120

Simon-Sanchez J, Schulte C, Bras JM, Sharma M, Gibbs JR, Berg D, Paisan-Ruiz C, Lichtner P, Scholz SW, Hernandez DG, Kruger R, Federoff M, Klein C, Goate A, Perlmutter J, Bonin M, Nalls MA, Illig T, Gieger C, Houlden H, Steffens M, Okun MS, Racette BA, Cookson MR, Foote KD, Fernandez HH, Traynor BJ, Schreiber S, Arepalli S, Zonozi R, Gwinn K, van der Brug M, Lopez G, Chanock SJ, Schatzkin A, Park Y, Hollenbeck A, Gao J, Huang X, Wood NW, Lorenz D, Deuschl G, Chen H, Riess O, Hardy JA, Singleton AB, Gasser T (2009) Genome-wide association study reveals genetic risk underlying Parkinson's disease. Nat Genet 41:1308–1312

Son S, Kim K-T, Cho D-C, Kim H-J, Sung J-K, Bae J-S (2014) Curcumin stimulates proliferation of spinal cord neural progenitor cells via a mitogen-activated protein kinase signaling pathway. J Korean Neurosurg Soc 56:1–4

Soto C, Estrada LD (2008) Protein misfolding and neurodegeneration. Arch Neurol 65:184–189

Speare JO, Offerdahl DK, Hasenkrug A, Carmody AB, Baron GS (2010) GPI anchoring facilitates propagation and spread of misfolded Sup35 aggregates in mammalian cells. EMBO J 29:782–794

Spinelli KJ, Osterberg VR, Meshul CK, Soumyanath A, Unni VK (2015) Curcumin treatment improves motor behavior in α-synuclein transgenic mice. PLoS One 10, e0128510

Sreejayan, Rao MN (1997) Nitric oxide scavenging by curcuminoids. J Pharm Pharmacol 49:105–107

Srinivas BMM, Mythri R, Jagatha B, Vali S (2008) Neuroprotective effect of curcumin against inhibition of mitochondrial complex I in vitro and in vivo. Implications for Parkinson's disease explained via in silico studies. J Neurochem 106:9–28

Steinman L (2009) A molecular trio in relapse and remission in multiple sclerosis. Nat Rev Immunol 9:440–447

Svenningsson P, Nishi A, Fisone G, Girault JA, Nairn AC, Greengard P (2004) DARPP-32: an integrator of neurotransmission. Annu Rev Pharmacol Toxicol 44:269–296

Takahashi J, Palmer TD, Gage FH (1999) Retinoic acid and neurotrophins collaborate to regulate neurogenesis in adult-derived neural stem cell cultures. J Neurobiol 38:65–81

Thickbroom GW, Mastaglia FL (2009) Plasticity in neurological disorders and challenges for non-invasive brain stimulation (NBS). J Neuroeng Rehabil 6:4–10

Thomas P, Wang YJ, Zhong JH, Kosaraju S, O'Callaghan NJ, Zhou XF, Fenech M (2009) Grape seed polyphenols and curcumin reduce genomic instability events in a transgenic mouse model for Alzheimer's disease. Mutat Res 661:25–34

Um JW, Kaufman AC, Kostylev M, Heiss JK, Stagi M, Takahashi H, Kerrisk ME, Vortmeyer A, Wisniewski T, Koleske AJ, Gunther EC, Nygaard HB, Strittmatter SM (2013) Metabotropic glutamate receptor 5 is a coreceptor for Alzheimer aβ oligomer bound to cellular prion protein. Neuron 79:887–902

Urra X, Cervera A, Villamor N, Planas AM, Chamorro A (2009) Harms and benefits of lymphocyte subpopulations in patients with acute stroke. Neuroscience 158:1174–1183

Uversky VN, Eliezer D (2009) Biophysics of Parkinson's disease: structure and aggregation of alpha-synuclein. Curr Protein Pept Sci 10:483–499

Uversky VN, Li J, Fink AL (2001a) Metal-triggered structural transformations, aggregation, and fibrillation of human α-synuclein. A possible molecular NK between Parkinson's disease and heavy metal exposure. J Biol Chem 276:44284–44296

Uversky VN, Li J, Fink AL (2001b) Pesticides directly accelerate the rate of alpha-synuclein fibril formation a possible factor in Parkinson's disease. FEBS Lett 500:105–108

Van Everbroeck B, Dewulf E, Pals P, Lübke U, Martin JJ, Cras P (2002) The role of cytokines, astrocytes, microglia and apoptosis in Creutzfeldt-Jakob disease. Neurobiol Aging 23:59–64

Vila M, Vukosavic S, Jackson-Lewis V, Neystat M, Jakowec M, Przedborski S (2000) α-Synuclein up-regulation in substantia nigra dopaminergic neurons following administration of the Parkinsonian toxin MPTP. J Neurochem 74:721–729

Volles MJ, Lee SJ, Rochet JC, Shtilerman MD, Ding TT, Kessler JC, Lansbury PT Jr (2001) Vesicle permeabilization by protofibrillar alpha-synuclein: implications for the pathogenesis and treatment of Parkinson's disease. Biochemistry 40:7812–7819

Wang R, Xu Y, Wu HL, Li YB, Li YH, Guo JB, Li XJ (2008) The antidepressant effects of curcumin in the forced swimming test involve 5-HT1 and 5-HT2 receptors. Eur J Pharmacol 578:43–50

Wang YJ, Thomas P, Zhong JH, Bi FF, Kosaraju S, Pollard A, Fenech M, Zhou XF (2009) Consumption of grape seed extract prevents amyloid-β deposition and attenuates inflammation in brain of an Alzheimer's disease mouse. Neurotox Res 15:3–14

Wang HM, Zhao YX, Zhang S, Liu GD, Kang WY, Tang HD, Ding JQ, Chen SD (2010a) PPARgamma agonist curcumin reduces the amyloid-beta-stimulated inflammatory responses in primary astrocytes. J Alzheimers Dis 20:1189–1199

Wang MS, Boddapati S, Emadi S, Sierks MR (2010b) Curcumin reduces α-synuclein induced cytotoxicity in Parkinson's disease cell model. BMC Neurosci 11:57

Wang Y, Yin H, Wang L, Shuboy A, Lou J, Han B, Zhang X, Li J (2013) Curcumin as a potential treatment for Alzheimer's disease: a study of the effects of curcumin on hippocampal expression of glial fibrillary acidic protein. Am J Chin Med 41:59–70

Wang YF, Zu JN, Li J, Chen C, Xi CY, Yan JL (2014a) Curcumin promotes the spinal cord repair via inhibition of glial scar formation and inflammation. Neurosci Lett 560:51–56

Wang P, Guan PP, Wang T, Yu X, Guo JJ, Wang ZY (2014b) Aggravation of Alzheimer's disease due to the COX-2-mediated reciprocal regulation of IL-1β and Aβ between glial and neuron cells. Aging Cell 13:605–615

Wu A, Ying Z, Gomez-Pinilla F (2006) Dietary curcumin counteracts the outcome of traumatic brain injury on oxidative stress, synaptic plasticity, and cognition. Exp Neurol 197:309–317

Wu J, Stoica BA, Faden AI (2011) Cell cycle activation and spinal cord injury. Neurotherapeutics 8:221–228

Wu J, Li Q, Wang X, Yu S, Li L, Wu X, Chen Y, Zhao J, Zhao Y (2013) Neuroprotection by curcumin in ischemic brain injury involves the Akt/Nrf2 pathway. PLoS One 8, e59843

Wu J, Zhao Z, Sabirzhanov B, Stoica BA, Kumar A, Luo T, Skovira J, Faden AI (2014) Spinal cord injury causes brain inflammation associated with cognitive and affective changes: role of cell cycle pathways. J Neurosci 34:10989–11006

Xie L, Li XK, Funeshima-Fuji N, Kimura H, Matsumoto Y, Isaka Y, Takahara S (2009) Amelioration of experimental autoimmune encephalomyelitis by curcumin treatment through inhibition of IL-17 production. Int Immunopharmacol 9:575–581

Xiong Y, Mahmood A, Chopp M (2013) Animal models of traumatic brain injury. Nat Rev Neurosci 14:128–142

Xu Y, Ku BS, Yao HY, Lin YH, Ma X, Zhang YH, Li XJ (2005) The effects of curcumin on depressive-like behaviors in mice. Eur J Pharmacol 518:40–46

Xu Y, Ku B, Tie L, Yao H, Jiang W, Ma X, Li X (2006a) Curcumin reverses the effects of chronic stress on behavior, the HPA axis, BDNF expression and phosphorylation of CREB. Brain Res 1122:56–64

Xu F, Plummer MR, Len GW, Nakazawa T, Yamamoto T, Black IB, Wu K (2006b) Brain-derived neurotrophic factor rapidly increases NMDA receptor channel activity through Fyn-mediated phosphorylation. Brain Res 1121:22–34

Xu Y, Ku B, Cui L, Li X, Barish PA, Foster TC, Ogle WO (2007) Curcumin reverses impaired hippocampal neurogenesis and increases serotonin receptor 1A mRNA and brain-derived neurotrophic factor expression in chronically stressed rats. Brain Res 1162:9–18

Yang F, Lim GP, Begum AN, Ubeda OJ, Simmons MR, Ambegaokar SS, Chen PP, Kayed R, Glabe CG, Frautschy SA, Cole GM (2005) Curcumin inhibits formation of amyloid β oligomers and fibrils, binds plaques, and reduces amyloid in vivo. J Biol Chem 280:5892–5901

Yang C, Zhang X, Fan H, Liu Y (2009) Curcumin upregulates transcription factor Nrf2, HO-1 expression and protects rat brains against focal ischemia. Brain Res 1282:133–141

Yasojima K, Tourtellotte WW, McGeer EG, McGeer PL (2001) Marked increase in cyclooxygenase-2 in ALS spinal cord: implications for therapy. Neurology 57:952–956

Yoshii A, Constantine-Paton M (2010) Postsynaptic BDNF-TrkB signaling in synapse maturation, plasticity, and disease. Dev Neurobiol 70:304–322

Yuan J, Zou M, Xiang X, Zhu H, Chu W, Liu W, Chen F, Lin J (2015) Curcumin improves neural function after spinal cord injury by the joint inhibition of the intracellular and extracellular components of glial scar. J Surg Res 195:235–245

Zhao J, Yu S, Zheng W, Feng G, Luo G, Wang L, Zhao Y (2010) Curcumin improves outcomes and attenuates focal cerebral ischemic injury via antiapoptotic mechanisms in rats. Neurochem Res 35:374–379

Zhou M, Xu S, Mi J, Uéda K, Chan P (2013) Nuclear translocation of alpha-synuclein increases susceptibility of MES23.5 cells to oxidative stress. Brain Res 1500:19–27

Chapter 9
Treatment of Alzheimer Disease with Phytochemicals Other Than Curcumin

9.1 Introduction

Phytochemicals are heterogeneous group of bioactive compounds, which are produced by plants and are consumed by humans to produce their health-promoting effects in normal individuals and patients with acute and chronic diseases. Unlike vitamins and minerals, phytochemicals are not needed for the maintenance of cell viability, but play a vital role in protecting neural cells from inflammation and oxidative stress associated with normal age-related visceral and brain diseases (Kannappan et al. 2011; Farooqui 2012). In addition to antioxidant and anti-inflammatory effects, phytochemicals have ability to stimulate detoxification enzymes, immune and hormonal responses, and epigenetic effects in humans (Kannappan et al. 2011; Farooqui 2012). More than 7000 phytochemicals have been identified, which possess antiproliferative, anti-inflammatory, antioxidant, antiviral, and hypo-cholesterolemic activities (Fig. 9.1).

Roots, stems, leaves, fruits and seeds of plants contain phytochemicals. The consumption of plants in the form of vegetables, nuts, and fruits results in beneficial effects on human body. Examples of phytochemicals are flavonoids in various fruits, catechins in green tea, resveratrol in red wine, ginkgo biloba, isoflavones in soy, and sulfur compounds in garlic. Plants produce these phytochemicals and store them in vulnerable regions (the skin, seeds and leaves) in order to discourage insects and other organisms from eating and killing the plant. In addition, phytochemicals also function in chemical defense against environmental stress and contribute to repair wound healing process in the plant.

Epidemiological studies have indicated that incidences of neurological disorders among people living in Asia are lower than the western world (Chandra et al. 2001; Chen et al. 2011). This may be due to the regular consumption of phytochemicals in the form of spices. Extensive research over the last 10 years has indicated that phytochemicals derived from various spices (turmeric, red pepper, black pepper, licorice, clove, ginger, garlic, coriander, and cinnamon); fruits, and nuts target

© Springer International Publishing Switzerland 2016

A.A. Farooqui, *Therapeutic Potentials of Curcumin for Alzheimer Disease*,
DOI 10.1007/978-3-319-15889-1_9

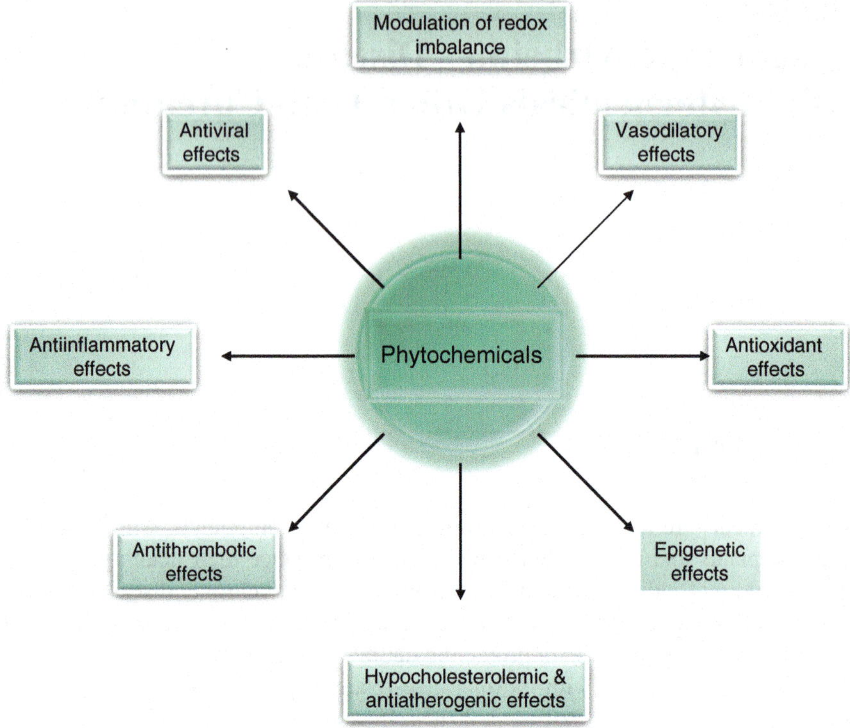

Fig. 9.1 Biological effects of phytochemicals

inflammatory and oxidative stress pathways and retard or delay the onset of neurological diseases (Kannappan et al. 2011; Farooqui 2012). In addition, phytochemicals also stimulate the synthesis of adaptive enzymes involved in synaptic plasticity and neuronal repair (Eggler et al. 2008; Spencer 2009; Spencer 2010; Williams and Spencer 2012). Phytochemicals also modulate angiogenesis, ionic homeostasis, cerebral blood flow, and redox imbalance through cross-talk signaling networks involved in controlling energy metabolism (Farooqui 2012). Some phytochemicals (curcumin and grape derived polyphenols) inhibit neuropathological processes (deposition of Aβ aggregates) in brain regions typically involved in the pathogenesis of AD (Wang et al. 2008; Farooqui 2012).

Phytochemicals present in plant foods have poor bioavailability not only because of their poor absorption, but also due to rapid excretion. The bioavailability of most phytochemicals in the peripheral tissues is higher than the brain due to the blood-brain barrier (BBB) (Farooqui 2012). However, small amounts of phytochemicals still enter and exert anti-inflammatory, antioxidant and anticarcinogenic effects at low doses in the brain (Spencer 2009; Farooqui 2012). The lipophilicity and charge of the phytochemicals is a key contributor to their ability to cross the BBB. Target specificity of phytochemical is another noticeable factor. For example some polyphenols acts on the hippocampal region of the brain while others target the striatum. The bioavailability of phytochemical depends on the physicochemical properties and the sugar moiety associated with each phytochemical (Kim et al. 2010; Essa et al. 2012; Venkatesan et al. 2015).

Phytochemicals are less expensive and less toxic than synthetic drugs, which produce many side effects. Unlike synthetic drugs that undergo extensive clinical trials in animals and humans prior to FDA approval, phytochemicals are not tested for their efficacy and safety. Plants contain complicated mixtures of phytochemicals (organic compound), the levels at which they may vary substantially depending upon many factors related to the growth of plants, harvesting, production, and storage conditions. These factors may cause alterations in chemical compositions of dietary phytochemicals resulting in batch variation. In recent years, research activity on phytochemicals has increased all over the world. New terms such as "nutraceutical and functional food", which include a broad spectrum of commercially available products as a part of diet have been introduced. Nutraceutical provide medical or health benefits and promote the prevention and treatment of disease. Nutraceuticals have no formal regulatory definition but they can be broadly defined to include functional foods, dietary supplements, and medical foods (Gupta and Prakash 2014).

9.2 Phytochemicals and Hormesis

Antioxidant and antiinflammatory effects of phytochemicals on visceral tissues and the in brain are due to hormesis, a process in which beneficial effects of phytochemical are observed at low doses, whereas at high doses the same phytochemical produces toxic effects (Mattson 2008). At the molecular level, the consumption of phytochemicals produce hormetic response involving the expression of adaptive stress-resistance genes responsible for encoding antioxidant enzymes, protein chaperones (heat-shock protein 70 and glucose-regulated protein 78), and neurotrophic factor (BDNF) (Fig. 9.2) (Mattson and Cheng 2006; Mattson 2008). Two mechanisms are associated with beneficial effects of phytochemicals. One mechanism involves Nrf2-mediated induction of 'phase 2' detoxifying enzymes (NQO1, SOD, HO-1, and glutathione peroxidase) (Wu et al. 2013) and the second mechanism involves activation of hypoxia inducible factor 1α (HIF-1α) and nuclear factor κB (NF-κB). In addition, phytochemicals not only stimulate the synthesis of adaptive enzymes along with modulation of angiogenesis, ionic homeostasis and redox imbalance through cross-talk signaling networks involved in controlling energy metabolism (Farooqui 2012). Side effects of phytochemicals at high doses include alterations in immune function and changes in metabolism. At low doses, phytochemicals not only improve neuronal and cognitive functions, ocular health, and memory formation but also protect genomic DNA integrity, inducing neuronal regeneration, and help strengthen the immune system. Long term consumption of

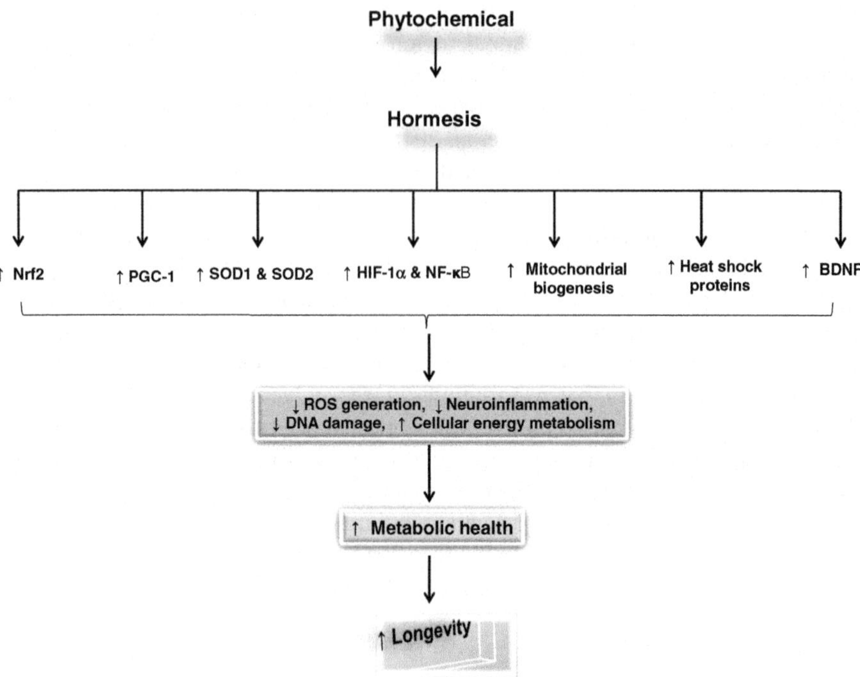

Fig. 9.2 Modulation of neuroprotective mechanisms by phytochemical-induced hormesis. Nuclear factor erythroid 2-related factor 2 (*Nrf2*); Peroxisome proliferator-activated receptor γ (*PPARγ*) coactivator 1α (*PGC-1α*); copper zinc-superoxide dismutase (*SOD1*); Mn-superoxide dismutase (*SOD2*); brain-derived neurotrophic factor (*BDNF*); reactive oxygen species (*ROS*); and deoxyribonucleic acid (*DNA*)

low doses of phytochemicals also decreases levels of circulating leptin and increases levels of adiponectin, improving energy metabolism and inducing resistance to disease processes. Hormesis not only provides a framework for explaining the dose response relationships between concentration of phytochemicals and beneficial effects, but also explains a key insight for improving the accuracy of the therapeutic dose of phytochemicals within the highly heterogeneous human population (Calabrese et al. 2010). Based on the stimulation of signal transduction network and adaptive stress-resistance genes, it is proposed that the use of phytochemicals from childhood to old age along with regular exercise 3 to 4 times per week is an inexpensive strategy for maintaining normal healthy aging and delaying onset of age-related neurological disorders including stroke, Alzheimer disease, AD, and Parkinson disease, PD (Farooqui 2010, 2012; Farooqui and Farooqui 2011). Risk factors for neurological disorders (stroke, AD, and PD) include old age, positive family history, unhealthy life style, and exposure to neurotoxins. These risk factors contribute not only to abnormal protein processing (oligomerization of unfolded proteins), generation of ROS, induction of neuroinflammation, and apoptotic cell death in neurological disorders (Farooqui 2010, 2012; Farooqui and Farooqui 2011). Diet is a major source of phytochemical intake and diets rich in fruits and vegetables, containing an abundance of various classes of phytochemicals, high levels of potassium, antioxidants, and dietary fibers. Long-term use of diet containing phytochemicals (enriched in fruits and vegetables) may retard the effect of unhealthy lifestyle by delaying or slowing the onset of stroke, AD and PD (Essa et al. 2012; Venkatesan et al. 2015). Brain responds to the effect of phytochemicals through receptors located on neuronal membranes by modulating neuronal and cognitive functions, ocular health, and memory formation.

9.3 Phytochemical and Alzheimer Disease

It is believed that phytochemicals are less toxic than novel synthetic drugs. However, since phytochemicals are commonly obtained from crude materials from roots, shoots, leaves, and fruits, there are many questions concerning their specific medicinal effects and reproducibility, mechanism of action, and the identity of the active ingredients (Kim et al. 2010). Therefore, recent studies have been focused on the identity and chemical structure of phytochemical, which produces beneficial effects in neurological disorder rather than on phytochemicals present in the crude mixture. Among phytochemicals, polyphenols, flavonoids, ginsenosides have been reported to reach the brain in sufficient amounts to modulate learning and memory systems not only by their pharmacological activity at receptors, but also by modulating lipid and protein kinases and transcription factors. Although the precise site of action of phytochemicals in the signaling pathways remains elusive, evidence indicates that phytochemicals act on in neural cells in a number of ways: (a) by binding to ATP sites on enzymes and receptors, (b) by modulating the activity of kinases (MAPKKK, MAPKK or MAPK) directly or indirectly, (3) by affecting the function of important phosphatases which act in opposition to kinases, (4) by preserving Ca^{2+} homeostasis, thereby preventing Ca^{2+}-dependent activation of kinases in neurons, and (5) by modulating signaling cascades lying downstream of kinases, i.e. transcription factor activation and binding to promoter sequences. In recent years, phytochemicals other than curcumin have been intensively studied to elucidate the mechanisms by which they mediate neuroprotection against neurological diseases (Farooqui 2012). Thus, resveratrol (from red grapes and certain types of nuts), epicatechins (from green tea and dark chocolate), ginkgo (from leaves of gingko plant), ginseng (from root of Panax species), and organosulfur compounds (from garlic bulb) are prominent examples of phytochemicals, which produce neuroprotective effects in neurological disorders (Fremont 2000; Wang et al. 2002; Scapagnini et al. 2006; Calabrese et al. 2010; Kensler et al. 2013; Nehlig 2013).

9.3.1 Resveratrol and Alzheimer Disease

Resveratrol (3,5,4′-trihydroxy-trans-stilbene) is a major nonflavonoid polyphenol (Fig. 9.3), which is not only present in the skin of red grapes and in wine, but also in blueberries, mulberries, cranberries, and peanuts. Resveratrol can cross the plasma membrane and is absorbed when given orally (de Santi et al. 2000a, b). It is metabolized in the body and can interact with and modulate phase I P450 enzymes CYP1A2, CYP3A4, and CYP2D6 and phase II enzymes glutathione S-transferase (GST) and catechol-O-methyltransferase (COMT) (de Santi et al. 2000a, b). The pharmacokinetic studies in humans have indicated that 25 mg of oral resveratrol is absorbed significantly via trans-epithelial diffusion. Resveratrol has a half-life of approximately 9 h and peak active metabolite plasma concentration of approximately 2 μM (Wen and Walle 2006; Walle et al. 2004; Wenzel and Somoza 2005; Wang et al. 2002; Ingram et al. 2006; Walle 2011). Resveratrol metabolites and polymers remain in the plasma much longer than unconverted resveratrol, whereas methylated resveratrol remains in the bloodstream for an even a longer period, a property that has been exploited in the drug development of resveratrol analogs (Pervaiz and Holme 2009). Resveratrol has attracted considerable scientific attention due to its potential health benefits related to its cardiovascular (French paradox), chemopreventive, antiobesity, antidiabetic, and neuroprotective properties.

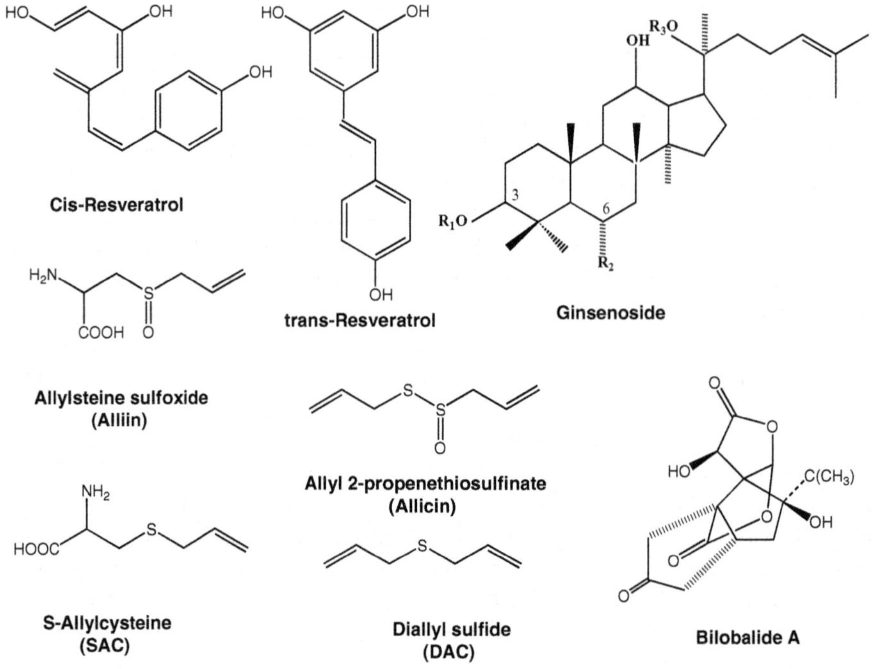

Fig. 9.3 Chemical structures of resveratrol, organosulfur compounds of garlic, ginsenoside, and ginkgo biloba (bilobalide)

However, in recent years resveratrol has been used for the treatment of AD in animal models. The antioxidant and free radical scavenging properties of resveratrol are related to its ability to transfer hydrogen atoms or electrons to the free radicals (Hussein 2011; Iuga et al. 2012; Petralia et al. 2014). To this end, the characteristic position of hydroxyl groups in resveratrol plays a major role, among which the 4′-hydroxyl group is the most reactive one (Stivala et al. 2001). Thus, antioxidant properties of resveratrol result from its chemical structure. In fact the molecule contains two phenol groups in which the presence of conjugated double bond makes the electrons more delocalized. The hydroxyl radical reacts with trans-resveratrol according to the following reaction.

$$\text{Resveratrol} + {}^\bullet\text{OH} \rightarrow \text{Resveratrol}^+ + \text{OH}^- \leftrightarrow \text{RV}(-\text{H})^\bullet + \text{H}_2\text{O}$$

Thus, antioxidant properties of resveratrol have been well demonstrated (Fremont 2000), with a wide range of biological and pharmacological effects including anti-cancer, anti-coagulant, anti-inflammatory, anti-aging, hypoglycemic, and hypolipidemic (Karuppagounder et al. 2009; Nosál' et al. 2014). Oxidative stress and neuroinflammation are closely associated with cognitive impairment during aging. Antiaging effect of resveratrol may be related to antioxidant effects involving NADPH oxidase and activation of heme oxygenase, where as antiinflammatory properties of resveratrol may be due to the inhibition of microglial cell activation and inhibition of NF-κB activation leading to decrease in the expression of TNF-α and IL-1β. Besides its antioxidant and anti-inflammatory properties, resveratrol can activate sirtuin 1 (SIRT1), which is a class of deacetylase (Albani et al. 2009). SIRT1 has recently emerged as a therapeutic target for the treatment of age-related degenerative diseases. Studies with neuronal cells showed that resveratrol acting as a SIRT1 activator protected SK-N-BE cells from oxidative stress and cytotoxicity by amyloid beta (Aβ) peptide (Albani et al. 2009).

The molecular mechanisms associated with neuroprotective effects of resveratrol remain elusive. However, recent studies have indicated that resveratrol has anti-inflammatory, blood-sugar-lowering potential, and Aβ anti-aggregative properties (Fig. 9.4) (Savouret and Quesne 2002; Ono et al. 2003; Ladiwala et al. 2010). In AD, it acts not only by decreasing levels of secreted and intracellular Aβ peptides, but also by increasing the blood flow (Pasinetti 2012; Kennedy et al. 2010). Resveratrol promotes intracellular degradation of Aβ, through a mechanism involving the proteasome activity (Marambaud et al. 2005). Moreover, studies in neuronal cell models of Aβ toxicity have indicated that the neuroprotective action of resveratrol against Aβ-induced toxicity may involve protein kinase C (Han et al. 2004). In the Tg2576 mouse model of AD, moderate consumption of Cabernet Sauvignon, a red wine enriched in resveratrol promotes non-amyloidogenic α-secretase-mediated APP processing leading to neuroprotection and preventing the generation of amyloidogenic Aβ1-42-mediated cognitive deterioration (Wang et al. 2006a). Similarly, in primary neuron cultures derived from Tg2576 embryos, Cabernet Sauvignon polyphenols exhibit Aβ-lowering activity through promotion of non-amyloidogenic processing of APP. Resveratrol treatment markedly inhibits polymerization of the

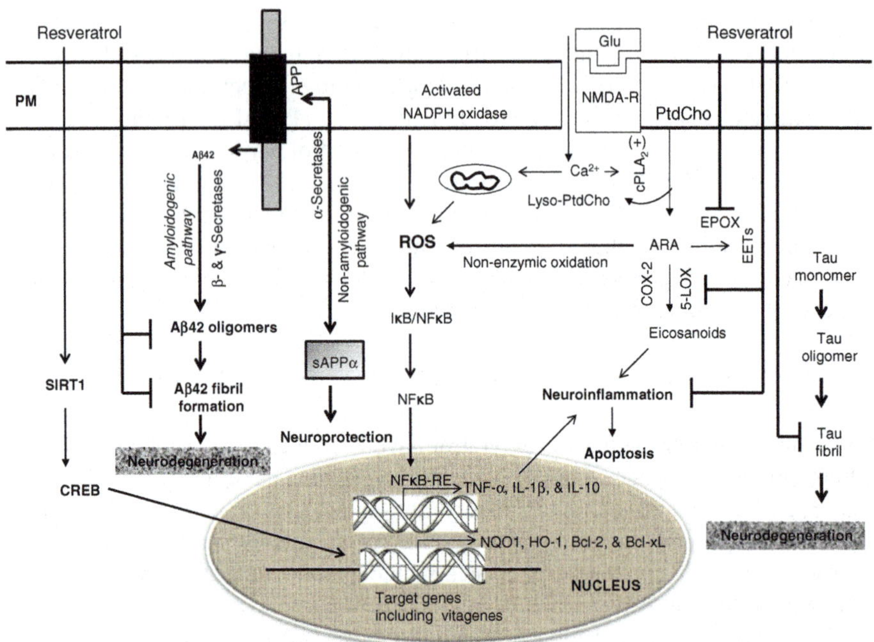

Fig. 9.4 Hypothetical diagram showing signaling pathways associated with neuroprotective effects of resveratrol. *N*-Methyl-D-aspartate receptor (*NMDA-R*); Glutamate (*Glu*); Phosphatidylcholine (*PtdCho*); cytosolic phospholipase A₂ (*cPLA₂*); lysophosphatidylcholine (*lyso-PtdCho*); arachidonic acid (*ARA*); CREB binding protein (*CBP/300*); sirtuin 1 (*SIRT1*); amyloid precursor protein (*APP*); soluble APP (*sAPP*); reactive oxygen species (*ROS*); nuclear factor-κB (*NF-κB*); nuclear factor-κB-response element (*NF-κB-RE*); inhibitory subunit of NF-κB (*I-κB*); tumor necrosis factor-α (*TNF-α*); interleukin-1β (*IL-1β*); and interleukin-10 (*IL-10*); NAD(P)H dehydrogenase [quinone] 1 (*NQO1*); heme oxygenase 1 (*HO-1*); B-cell lymphoma 2 (*Bcl-2*); and B-cell lymphoma-extra large (*Bcl-xl*)

β-amyloid peptide (Riviere et al. 2007) by a mechanism that does not involve β-amyloid production because resveratrol has no effect on activity of β- and γ-secretases but stimulates indirectly the proteosomal degradation of β-amyloid peptides (Marambaud et al. 2005). Moreover, in vitro and in vivo studies have also indicated that resveratrol reduces amyloid toxicity by decreasing Aβ production through sirtuin-dependent activation of a disintegrin and metalloproteinase domain-containing protein 10 (Donmez et al. 2010). Resveratrol does not inhibit Aβ production, because it has no effect on the Aβ-producing enzymes β- and γ-secretases, but promotes instead intracellular degradation of Aβ not only via a mechanism that involving the proteasome, but also promoting clearance and metabolism via an AMP-activated protein kinase-pathway and can induce autophagic and lysosomal Aβ degradation (Marambaud et al. 2005; Vingtdeux et al. 2010). Furthermore, the resveratrol-induced decrease of Aβ can be prevented by several selective proteasome inhibitors and by siRNA-directed silencing of the proteasome subunit β5. These findings support the view that resveratrol acts by a proteasome-dependent anti-amyloidogenic activity and suggest that this natural compound has a therapeutic potential in AD. Resveratrol may also act as an anti-aggregative agent and may prevent the formation of toxic $A\beta$ oligomers and protofibrillar intermediates (Fig. 9.4) (Jang and Surh 2003; Savaskan et al. 2003; Han et al. 2004). It is interesting to note that many resveratrol derivatives have been synthesized to treat AD. Amongst the synthesized compounds, 5d (E)-2-((4-(3,5-Dimethoxystyryl) phenylamino) methyl)-4-(dimethylamino) phenol and 10d (E)-5-(4-(5-(Dimethylamino)-2-hydroxybenzylamino) styryl)-benzene-1,3-diol not only significantly inhibit $A\beta$ aggregation, have ability to chelate metals, and disintegrate highly structured $A\beta$ fibrils, but also prevent Cu^{2+}-induced $A\beta$ aggregation and have antioxidant activity along with low neurotoxicity (Lu et al. 2013). These investigators have previously synthesized a series of stilbene derivatives based on the structure of resveratrol in which compound 7 1 (E)-5-(4-(isopropylamino)styryl) benzene-1-3-diol exerted potent β-amyloid aggregation inhibition activity (Lu et al. 2012). Novel synthetic compounds such as STACs confer remarkable health benefits in various animal models. SRT3025 is one such STAC, which penetrates the BBB; mimics the effects of calorie restriction on the brain and further reduces neurodegeneration (Hubbard and Sinclair 2014). Collective evidence suggests that resveratrol derivatives with improved bioavailability and neuroprotective effects can be used as novel multifunctional drugs in the treatment of AD.

In addition, hydroxylated but not methoxylated derivatives of resveratrol markedly inhibit cyclooxygenases. The most potent resveratrol derivatives are 3,3′,4′,5-tetra-trans-hydroxystilbene (COX-1: $IC_{50}=4.713$, COX-2: $IC_{50}=0.0113$ μM with selectivity index$=417.08$ and 3,3′,4,4′,5,5′-hexa-hydroxy-$trans$-stilbene, COX-1: $IC_{50}=0.748$, COX-2: $IC_{50}=0.00104$ μM, selectivity index$=719.23$) (Murias et al. 2004). The selectivity index values of hydroxylated derivatives of resveratrol is higher than celecoxib, a selective COX-2 inhibitor, which is available in the market (COX-1: $IC_{50}=19.026$, COX-2: $IC_{50}=0.03482$ μM with selectivity index$=546.41$) (Murias et al. 2004). Activation of cyclooxygenases, lipoxygenases, and epoxygenases during neurodegenerative process contribute to oxidative stress and neuroinflammation (Phillis et al. 2006). Availability of hydroxylated derivatives of resveratrol can be used to block the formation of $A\beta$ oligomer and fibril in animal models. Resveratrol can also produce beneficial effects on health by modulating also sirtuin-1 (SIRT1), a group of deacetylases (or deacylases) whose activities are dependent on and regulated by nicotinamide adenine dinucleotide (NAD^+). In vitro studies have indicated that resveratrol-mediated overexpression of SIRT1 not only leads to a reduction in oligomerization of $A\beta$ peptides in a concentration-dependent manner, but also ameliorate oxidative stress (Albani et al. 2009). Other studies have indicated that overexpression of SIRT1 prevents the activation of microglial cells by fibrillar $A\beta$ and the consequent production of neurotoxic chemokines, cytokines, and nitric oxide from the activated microglia through inhibition of NFκB signaling (Chen et al. 2005), which involves the SIRT1-mediated deacetylation of the lysine 310 residue of the RelA/p65 subunit of NFκB and blocks its transcriptional activity (Yeung et al. 2004). These findings support the view that epigenetic mechanisms are involved in SIRT1-mediated regulation of inflammatory responses. In embryonic mouse neuronal cultures activation of the

SIRT1/PGC-1 pathway by resveratrol protects against axonal degeneration and decreases the accumulation of Aβ. Resveratrol not only protects against the dysregulation of energy homeostasis observed in mouse models for metabolic syndromes (an important risk factor for AD) by a mechanism implicating the activation of SIRT1, PGC-1α, and the energy sensor protein kinase AMPK (AMP-activated protein kinase) (Dasgupta and Milbrandt 2007; Baur et al. 2006), but also enhances deacetylation of PGC-α. This process improves mitochondrial function and energy balance. Resveratrol-mediated stimulation of SIRT1 also leads to the direct deacetylation of acetylated tau, thereby promoting its proteasomal degradation (Min et al. 2010). In addition, resveratrol reduces phospho-tau toxicity (induced by cyclin-dependent kinase 5-p25 dependent tau phosphorylation) by favoring the deacetylation of peroxisome proliferator-activated receptor gamma, coactivator 1 alpha (PGC-1α) and p53 (Kim et al. 2007). Resveratrol produces neuroprotective effects not only against glutamate-mediated toxicity in neuronal cultures (Vieira de Almeida et al. 2007), but also downregulates the expression of glycogen synthase kinase 3 (GSK-3β) through the modulation of PtdIns 3 K/Akt pathway (Simão et al. 2012). This enzyme is involved in multiple signaling pathways and has several phosphorylation targets, which may protect from AD. A large number of studies have indicated that resveratrol modulates brain functions by improving glucose metabolism (Witte et al. 2014) and vasorelaxation by promoting eNOS and/or NO synthesis along with cerebral blood flow (Wightman et al. 2014; Kennedy et al. 2010).

It is well known that brain-derived neurotrophic factor (BDNF) plays a key role in neural cell development, growth. This growth factor also promotes synaptic plasticity in the hippocampus. Treatment of rats for 3, 10, and 30 days with resveratrol significantly and dose-dependently increases the levels of BDNF mRNA expression in hippocampal tissue suggesting that the resveratrol mediated neuroprotective impact may be related with activation of the BDNF pathway (Rahvar et al. 2011). Recently, grape powder extract was found to prevent oxidative stress-induced anxiety, memory impairment, and hypertension in rats by regulating also brain CREB and BDNF levels (Allam et al. 2013).

9.3.2 Green Tea and Alzheimer Disease

Green tea (*Camellia sinensis*) contains polyphenols (catechins), alkaloids (caffeine, theophylline, and theobromine), flavonols (quercetin, kaempferol and rutin), amino acids, carbohydrates, proteins, chlorophyll, volatile organic compounds that contribute to tea flavonoid, fluoride, aluminium, minerals and trace elements. In addition, green tea contains gallic acid, chlorogenic acid and caffeic acid (Wang and Ho 2009). The major catechins of green tea are (-)-epicatechin (EC), (-)-epicatechin-3-gallate (ECG), (-)-epigallocatechin (EGC) and (-)-epigallocatechin-3-gallate (EGCG) (Mukhtar and Ahmad 2000; Higdon and Frei 2003; Velayutham et al. 2008) (Fig. 9.5). The bioavailability of green tea catechins depends upon their structures. Catechin monomers can be easily absorbed through the gut barrier. In contrast

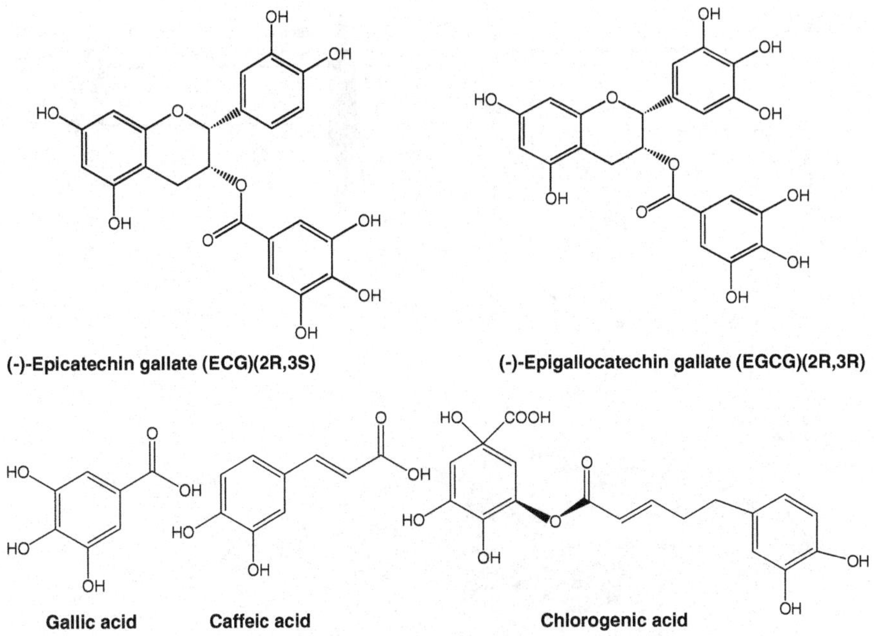

Fig. 9.5 Chemical structures of ECG, ECGC, gallic, chlorogenic, and caffeic acids

the large molecular weight catechins, such as EGCG are poorly absorbed. Efflux transporters Pgp, MRP1 and MRP2 are closely involved in the absorption and excretion of green tea catechins.

In AD, EGCG produces beneficial effects by regulating the proteolytic processing of APP both under in vitro and in vivo conditions (Levites et al. 2003) (Fig. 9.6). In neuronal cell cultures and mouse model of AD, EGCG enhances the nonamyloidogenic α-secretase pathway via PKC dependent activation of α-secretase (ADAM10) (Levites et al. 2003; Singh et al. 2008; Mandel et al. 2008; Smith et al. 2010) while EC reduces the formation of amyloid β-fibrils. EGCG modulates Aβ levels, either via translational inhibition of APP or by stimulating sAPPα generation and secretion. This process results in a significant reduction in cerebral Aβ levels and β-amyloid plaques. EGCG also inhibits the activation of extracellular signal-regulated kinase and nuclear transcription factor-kappaB in the Aβ-injected mouse brains and blocks Aβ-mediated apoptotic neuronal cell death in the brain. These studies strongly support the view that EGCG may contribute to the prevention of development or progression of AD in cell culture and animal models. In APP/PS1 mice, EGCG (2 or 6 mg/kg/day) ameliorates the impaired learning and memory in 4 weeks not only by inhibiting TNF-α/JNK signaling and increasing the phosphorylation of Akt and glycogen synthase kinase-3β in the hippocampus of APP/PS1 mice. In addition ECGC consumption may alleviate AD-related cognitive deficits by effectively attenuating central insulin resistance (Jia et al. 2013).

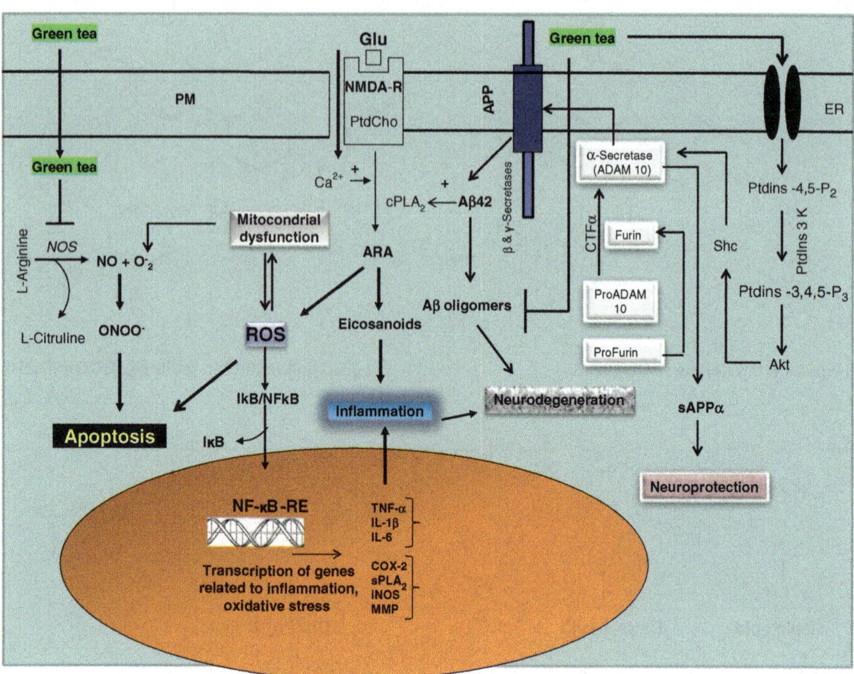

Fig. 9.6 Hypothetical diagram showing signaling pathways associated with neuroprotective effects of green tea. Plasma membrane (*PM*); *N*-Methyl-D-aspartate receptor (*NMDA-R*); estrogen receptor (*ER*); glutamate (*Glu*); phosphatidylcholine (*PtdCho*); cytosolic phospholipase A₂ (*cPLA₂*); secretory phospholipase A₂ (*sPLA₂*); cyclooxygenase (*COX-2*); matrix metalloproteinases (*MMPs*); inducible nitric oxide synthase (*iNOS*); arachidonic acid (*ARA*); reactive oxygen species (*ROS*); nuclear factor kappaB (*NF-κB*); nuclear factor kappaB response element (*NF-κB-RE*); inhibitory subunit of NFκB (*IκB*); tumor necrosis factor-α (*TNF-α*); interleukin-1β (*IL-1β*); interleukin-6 (*IL-6*); peroxynitrite (*ONOO⁻*); Superoxide (*O₂*); phosphatidylinositol-4,5-bisphosphate 3-kinase (*PtdIns 3K*); phosphatidylinositol 4,5-bisphosphate (*PtdIns-4,5-P₂*); phosphatidyl-inositoltrisphosphate (*PtdIns-3,4,5-P₃*); protein kinase B (*Akt*); amyloid precursor protein (*APP*); and soluble APP (*sAPP*); furin (paired basic amino acid cleaving enzyme). *Positive sign* (+) represents upregulation

As stated in Chap. 1, transition metals (Cu^{2+} and Zn^{2+}) may contribute to the pathogenesis od AD due to misregulation of their levels leading to aberrant neuronal function and neurodegeneration (Kepp 2012; Pithadia and Lim 2012). Treatment of Cu^{2+} or Zn^{2+}–Aβ species with EGCG results in formation of small amorphous Aβ aggregates, whereas EGCG-untreated samples produce mainly structured Aβ aggregates (Hyung et al. 2013). This reactivity with Aβ is more noticeable in preparations containing metal ions supporting the view that EGCG is capable of disrupting metal-mediated Aβ aggregation pathways compared with metal-free preparations. Furthermore under in vivo conditions, incubation of EGCG with metal-free or metal–Aβ bound complex enhances cell survival, which may represent EGCG inhibitory effect on aggregation in complex settings. IM-MS and NMR studies indicate that the overall mechanism of reactivity of EGCG with metal-free and metal-associated Aβ species may be driven by its ability to modulate the structure of Aβ monomer, dimer, and oligomer in the presence or absence of metal ions (Hyung et al. 2013). It is also reported that EGCG efficiently prevents the fibrillogenesis of Aβ by directly binding to the natively unfolded polypeptides and preventing their conversion into toxic aggregated intermediates (Wanker 2008). These observations support the view that green tea catechins produce a generic effect on aggregation and fibrillogenesis pathways in AD (Wanker 2008). Collectively these results indicate that EGCG is not only capable of modulating the aggregation and toxicity of metal–Aβ species forming small unstructured aggregates, but also possess multifunctional activities, such as antioxidant, antiinflammatory, mitochondrial membrane stabilizing and neuroprotective properties in a wide array of cellular and animal models of neurological disorders (Mandel et al. 2008). Attempts have been made to increase the bioavailability of EGCG through the synthesis of synthetic analogs of EGCG (Zaveri 2001). These analogs contain a trimethoxybenzoyl ester (D-ring) and are equally as potent as natural EGCG for their efficacy as antioxidants and anti-carcinogenic agents (Waleh et al. 2005). In addition, it is also reported that bioavailability of EGCG can also be increased by delivering EGCG using lipid nanocapsules and liposome encapsulation (Barras et al. 2009). These studies set the stage for large double blind clinical trials in AD patients.

Gallic acid is another component of green tea (Fig. 9.5). Like EGCG, octyl gallate drastically reduces Aβ generation, in concert with increased APP α-proteolysis in murine neuron-like cells transfected with human wild-type APP or "Swedish" mutant APP (Zhang et al. 2013). Octyl gallate markedly increases the formation of the neuroprotective amino-terminal APP cleavage product leading to the generation of soluble APP-α (sAPPα). It is proposed that these cleavage events are associated with increase in ADAM10 maturation and reduction by blockade of estrogen receptor-α (Erα)/PtdIns 3 K/Akt signaling (Zhang et al. 2013). To support these observations under in vivo conditions, investigators treated Aβ-overproducing Tg2576 mice with octyl gallate daily for one week by intracerebroventricular injection and noticed a decrease in Aβ levels associated with increased sAPPα (Zhang et al. 2013). These results support the view that octyl gallate increases anti-amyloidogenic APP α-secretase processing through the activation of ERα/PtdIns 3 K/Akt signaling and ADAM10 (Fernandez et al. 2010; Zhang et al. 2013).

9.3.3 Ginkgo Biloba and Alzheimer Disease

Ginkgo biloba (G. biloba) is an important herb, which is used in the Chinese traditional medicine system for several hundred years (Huh and Staba 1992). The herb shows memory enhancing action by increase the supply of oxygen, and helps the body to eliminate free radicals thereby improving memory. Phytoconstituents of ginkgo biloba include terpenoids bilobolide, ginkgolides, flavanoids, steroids (sitosterol and stigmasterol) and organic acids (ascorbic, benzoic shikimic and vanillic acid). Leaf extract contains 24 % of flavonoids and 6 % of terpenic lactones, giving this extract its unique polyvalent pharmacological action. The flavonoid fraction is mainly composed of three flavonols, quercetin, keampferol and isorhamnetin, whereas terpenic derivatives are represented by diterpenic lactones, the ginkgolides A, B, C, J and M, and a sesquiterpenic trilactone, the bilobalide (Fig. 9.3) (DeFeudis and Drieu 2000; Chandrasekaran et al. 2001). It has been developed and patented by Beaufour-Ipsen Pharma (Paris, France) and Willmar Schwabe Pharmaceuticals (Karlsruhe, Germany) (DeFeudis and Drieu 2000). The bioavailability of Egb 761 is relatively low due to limited absorption and rapid elimination (Goh and Barlow 2004). Unabsorbed Egb 761 is metabolized by colon bacterial enzymes, and then absorbed (DeFeudis and Drieu 2000). Once absorbed, Egb 76 reaches the liver where they are metabolized to conjugated derivatives (DeFeudis and Drieu 2000). EGb 761 not only acts as free radical-scavenger and inhibits lipid peroxidation, maintains ATP content by protecting the mitochondrial respiration and preserving oxidative phosphorylation, but also exerts arterial and venous vaso-regulating effects through the release of endothelial factors and controlling catecholaminergic system (Clostre 1999). At the molecular levels in brain, EGb 761 not only inhibits cPLA$_2$ (Zhao et al. 2011), but also induces antioxidant and anti-inflammatory effects (Chu et al. 2011; Lu et al. 2011), modulates gene expression (Smith and Luo 2004; Mahdy et al. 2011), and promotes memory (Mahdy et al. 2011), Moreover, it also modulates mitochondrial function and apoptotic cell death pathway (Fig. 9.7) (Rhein et al. 2010) by up-regulating anti-apoptotic Bcl-2 protein and down-regulating pro-apoptotic Bax protein, inhibiting cytochrome c release, reducing caspase 9 and caspase 3 activity (Rhein et al. 2010). Thus, EGb 761 has direct effects against apoptotic cell death of neurons and improves neural plasticity as evidenced in vestibular compensation.

Bilobalide and ginkgolides present in *Ginkgo biloba* have been classified as nootropic agents (Kumar 2006). Effect of EGb 761 has been studied in animal models of AD. Some human clinical trials of EGb 761 have also been performed in AD patients. EGb 761 reduces cognitive dysfunction, age-associated memory impairment, and dementia. The molecular mechanisms associated with the beneficial effects of EGb 761 are not fully understood. However, it is reported that EGb 761 stimulates α-secretase activity, an enzyme, which is involved in the generation of soluble N-terminal domain of APP (sAPPα). sAPPα contains a series of domains that include a growth factor domain (D1), a copper binding domain (D2), an acidic region (D3), and a carbohydrate domain (D5) (Storey and Cappai 1999; Reinhard et al. 2005). Importantly, this isoform does not contain either the KPI domain (D4) or the OX-2 domain (D5) (Sandbrink et al. 1996). Studies on crystal structure of sAPPα suggest that the growth factor like domain of sAPPα is similar to cysteine-rich growth factors (Rossjohn et al. 1999). Due to the presence of a growth factor like domain sAPPα produces neurotrophic and neuroprotective effects; therefore it prevents the formation of neurotoxic Aβ (Colciaghi et al. 2004; Postina 2008). In addition, EGb 761 has been reported to decrease Aβ levels as a downstream target

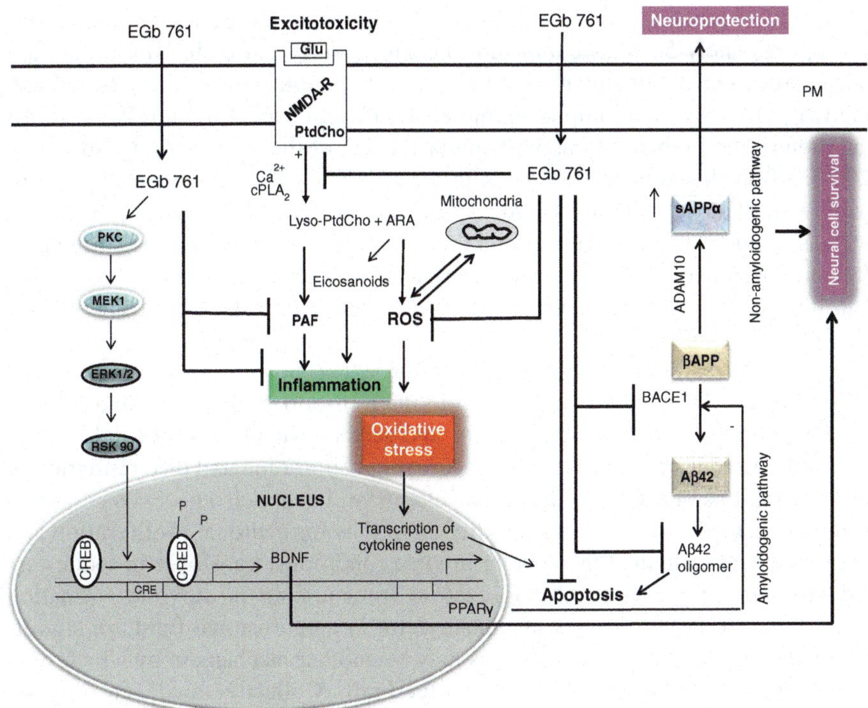

Fig. 9.7 Hypothetical diagram showing signaling pathways associated with neuroprotective effects of EGb 761. Plasma membrane (*PM*); *N*-Methyl-D-aspartate receptor (*NMDA-R*); estrogen receptor (*ER*); glutamate (*Glu*); phosphatidylcholine (*PtdCho*); Lyso-phosphatidylcholine (*Lyso-PtdCho*); platelet activating factor (*PAF*); cytosolic phospholipase A₂ (*cPLA₂*); arachidonic acid (*ARA*); protein kinase C (*PKC*); protein kinase of the STE7 kinase family (*MEK1*); extracellular signal-regulated protein kinases 1 and 2 (*ERK1/2*); p90 ribosomal S6 kinase (*RSK*); cAMP response element-binding protein; β amyloid precursor protein (*βAPP*); and soluble APP (*sAPP*); furin (paired basic amino acid cleaving enzyme). *Positive sign* (+) represents upregulation

of the activated PtdIns 3 K pathway (Shi et al. 2009, 2011). EGb 761 has no effect on BACE-1 or γ-secretase but inhibits the β-secretase activity of cathepsin B, suggesting that EGb 761-induced Aβ reduction is probably caused by the modulation of cathepsin B rather than BACE-1 activity. Similarly, inhibition of GSK3β has no effect on BACE-1 activity but decreases cathepsin B activity, supporting the view that the PtdIns 3 K-GSK3β pathway is probably associated with EGb 761-induced Aβ reduction (Shi et al. 2009, 2010, 2011). In addition, stimulation of α-secretase may be supported by phospholipase C, phosphatidylinositol 3-kinase and serine/threonine-specific kinases such as protein kinases C, and mitogen activated protein kinase signaling pathways. Thus, direct activation of protein kinase C and stimulation of distinct G protein-coupled receptors may contribute to an increase in α-secretase processing of APP (Postina 2008; Shi et al. 2011). Thus, long-term treatment (16 months) with EGb761 significantly lowers human APP protein levels by approximately 50 % as compared to controls in the cortex but not in the hippocampus. EGb 761 treatment in young mice has no effect on APP levels, indicating that APP may be an important molecular target of EGb761 in relation to the duration of the EGb 761 treatment and/or the age of the animals (Augustin et al. 2009). Aβ-mediated neurotoxicity in PC12 cells produces free radicals and treatment with EGb 761 not only retards Aβ-peptide-mediated free radical production, increases glucose uptake, but also inhibits apoptosis in a dose-dependent manner. It is proposed that EGb acts as a neuroprotectant against amyloid fibril formation and this process may be supported by MAP-kinase cascade, SIRT1 and NF-κB (Longpre et al. 2006).

EGb 761 treatment also up-regulates the expression of transthyretin in the hippocampus. This protein is involved in the transport of thyroxine and retinol-binding protein in cerebrospinal fluid and serum (Kuchler-Bopp et al. 2000). Thyroid hormones not only regulate neuronal proliferation and differentiation in discrete regions of the brain during development, but are also necessary for normal cytoskeletal assembly and stability as well as for neuronal proliferation and outgrowth (Porterfield 2000). Under in vitro conditions, transthyretin sequesters Aβ protein and prevents Aβ aggregation from arising in amyloid formation (Tsuzuki et al. 2000). Levels of transthyretin in cerebrospinal fluid are significantly decreased in AD patients. This may be another mechanism by which EGb 761 may exert its beneficial effects on the brain. Collective evidence suggests that EGb761 retards the generation of Aβ from APP and Aβ aggregation, which are crucial processes related with pathophysiology of AD pathogenesis (Bastianetto et al. 2000).

It is well known that energy deficiency and mitochondrial dysfunction are also early events associated with the pathogenesis of AD. Chronic exposure of human neuroblastoma cells over-expressing human wild-type APP to Aβ results not only in activity changes of complexes III and IV of the oxidative phosphorylation system, but also in a reduction in ATP levels, which may lead to the loss of synapses and neuronal cell death in AD. Treatment of neuroblastoma cells with standardized G. biloba extract LI 1370 results in optimal activity of complexes III and IV of the oxidative phosphorylation system as well as restoration of Aβ-induced mitochondrial dysfunction (Rhein et al. 2010). It is also reported that EGb 761 not only promotes basal proteasome activity (PA), but also enhances protein degradation presumably due to EGb 761-mediated enhanced expression of catalytic proteasome genes. This suggestion is supported by the activation of the Nrf2-KEAP1 pathway, which is not only required for the expression of proteasomal genes, but also contribute to the transcription of phase 2 genes (Kwak et al. 2003; Arlt et al. 2009; Liu et al. 2009a). Converging evidence suggests that in animal models of AD and AD patients, G. biloba may produce beneficial effects not only through the generation of sAPPα and mediating activation of Nrf2, but also through direct effects on the generation of Aβ.

9.3.4   Ginseng and Alzheimer Disease

Ginseng (Panax ginseng) belongs to the family Araliaceae. Ginseng roots, shoots, and leaves have been used as traditional herbal medicine in China, Korea, and Japan for more than 2000. The bioactive constituents of ginseng root include more than 60 ginsenosides (Rg), such as Rb1, Rb2, Rb3, Rc, Rd, Re, Rg1, Rg2, and Rg3 (Fig. 9.3), as well as polysaccharides, oligopeptides, polyacetylenic alcohols, and fatty acids (Qi et al. 2010). Ginsenosides are triterpene saponins that have a common 4 ring hydrophobic steroid-like structure with sugar moieties (monomeric, dimeric, or trimeric) attached mostly at the C-3, C-6, or C-20 position (Zhu et al. 2004; Nah et al. 2007). The bioavailability of ginsenosides is very low. The poor bioavailability of ginsenosides is not only because of low membrane permeability and active biliary excretion, but also due to biotransformation (Liu et al. 2009b). Intact ginsenosides are metabolized in the stomach (acid hydrolysis) and in the gastrointestinal tract (bacterial hydrolysis) or transformed to other ginsenosides, which are absorbed only from the intestines at a very low absorption rate and transferred to the blood. Ginsenosides can cross BBB and produce many neurochemical effects including modulation of ion channels and neurotransmitter receptors, preventing of oxidative stress and neuroinflammation, and retardation of memory deficit. Detailed investigations have indicated that ginsenosides regulate various types of ion channels by interacting with ligand-binding sites or channel pore sites in neuronal and heterologously expressed cells (Nah et al. 2007). Thus, ginsenosides inhibit voltage-dependent Ca^{2+}, K^+, and Na^+ channel activities in a stereospecific manner (Fig. 9.8) (Liu et al. 2010). Ginsenosides not only inhibit stimulation of N-methyl-D-aspartate receptor (NMDA-R), but also block some subtypes of nicotinic acetylcholine, and 5-hydroxytryptamine type 3 receptors (Chen et al. 2010; Kim et al. 2002). These effects may be responsible for retarding glutamate-mediated neurodegeneration by ginsenosides (Fig. 9.8). In addition, ginsenosides are used for the treatment of memory deficit in humans (Rudakewich et al. 2001). Furthermore, ginsenosides increase levels of dopamine and norepinephrine in the cerebral cortex (Itoh et al. 1989), which may explain the favorable effects of ginseng extract upon attention, cognitive processing, integrated sensory motor function, and auditory reaction time in healthy subjects (D'Angelo et al. 1986).

In cultured neurons, treatment with Radix notoginseng flavonol glycoside not only inhibits the aggregation of Aβ in a dose-dependent manner, but also abolishes the increase of Ca^{2+} and cell death triggered by Aβ in cultured neurons. The flavonol glycoside also reduces memory impairment in the passive avoidance task, in scopolamine-treated rats. It is suggested that the use of Radix notoginseng and its flavonol glycoside may be very useful in developing food supplements for the prevention, or potential treatment, of AD (Choi et al. 2010). Siberian ginseng also induces protective effects against Aβ-induced neurite atrophy. Twelve compounds have been isolated from the active fractions and identified. A novel lysophosphatidic acid receptor-activating ligand from ginseng, gintonin decreases Aβ1-42 release and attenuats Aβ1-42-induced cytotoxicity in SH-SY5Y cells. Gintonin also

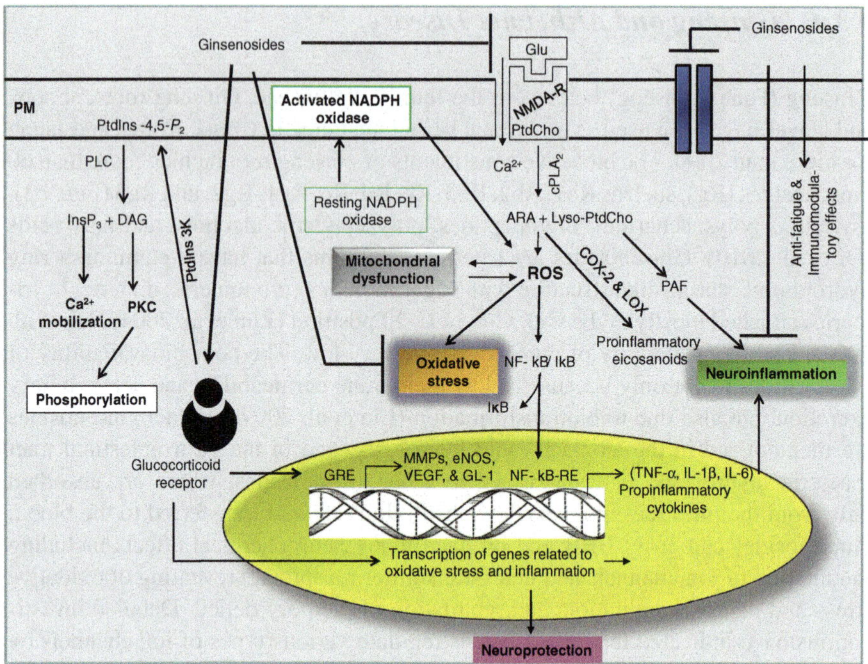

Fig. 9.8 Hypothetical diagram showing signaling pathways associated with neuroprotective effects of ginsenoside. Glutamate (*Glu*); *N*-methyl-D-aspartate receptors receptor (*NMDA-R*); plasma membrane (*PM*); phosphatidylcholine (*PtdCho*); arachidonic acid (*ARA*); lysophosphatidylcholine (*lyso-PtdCho*); platelet activating factor (*PAF*); phosphatidylinositol 4, 5-bisphosphate (*PtdIns-4,5-P₂*); cytosolic phospholipase A₂ (*cPLA₂*); cyclooxygenase-2 (*COX-2*); lipoxygenase (*LOX*); phospholipase C (*PLC*); diacylglycerol (*DAG*); inositol 1,4,5-trisphosphate (*InsP₃*); reactive oxygen species (*ROS*); nuclear factor-kappa B (*NF-κB*); nuclear factor-kappa B response element (*NF-κB-RE*); tumor necrosis factor-alpha (*TNF-α*); interleukin-1beta (*IL-1β*); interleukin-6 (*IL-6*); endothelial nitric oxide synthase (*eNOS*); nitric oxide (*NO*); peroxynitrite (*ONOO⁻*); Glucocorticoids (*GCs*); glucocorticoid receptor (*GR*); matrix metalloproteinases (*MMPs*), and endothelial growth factor (*VEGF*)

rescues Aβ1-42-induced cognitive dysfunction in mice. Moreover, in a transgenic mouse AD model, long-term oral administration of gintonin attenuates amyloid plaque deposition as well as short- and long-term memory impairment (Hwang et al. 2012). Fermented ginseng also reduces the level of soluble Aβ42 in HeLa cells, which stably express the Swedish mutant form of amyloid precursor protein, and decreases memory impairment in animal models of AD (Kim et al. 2013).

Similarly in aged transgenic AD mice (Tg mAPP) over expressing APP/Aβ treatment with Rg1 results in a significant reduction of cerebral Aβ levels, reversal of neuropathological changes, and preservation of spatial learning and memory, as compared to vehicle-treated mice. Rg1 treatment inhibits the activity of γ-secretase in both TgmAPP mice and B103-APP cells, indicating the involvement of Rg1 in APP regulation pathway. Furthermore, administration of Rg1 enhances PKA/CREB pathway activation in mAPP mice and in cultured cortical neurons exposed to Aβ or glutamate-mediated synaptic stress (Fang et al. 2012). It is also reported that Ginsenosides (CK, F1, Rh1 and Rh2) interact with BACE1 to produce their neuro-protective effects (Karpagam et al. 2013). Furthermore, Panax notoginseng saponins increases α-secretase activity perhaps by enhancing the level of ADAM9 expression, and decreases BACE1 protein level by downregulating the level of BACE1 gene expression and consequently precluding the activity of β secretase without modifying γ-secretase activity (Huang et al. 2014a). Similarly in tg2576 AD mice model, treatment with insenoside Rh2 improves learning and memory performance at 14 months of age and significantly reduces brain senile plaques. These results have been confirmed by In vitro studies, which indicate that Rh2 treatment not only increases soluble APPα (sAPPα) levels, but reduces Aβ 40 and 42 levels along with reduction in APP endocytosis (Qiu et al. 2014).

Panax ginseng also modulates tau phosphorylation. Total ginsenoside extracts from stems and leaves of Panax ginseng enhance the phosphatase activity of purified calcineurin. This may be useful in AD, since inhibition of calcineurin induces hyperphosphorylation of tau at multiple sites (Tu et al. 2009). It is reported that in SD rats, ginsenoside Rd reduces okadaic acid-induced neurotoxicity and tau hyperphosphorylation by enhancing the activities of PP-2A (Li et al. 2011). Rb1 also reverses aluminium-mediated increase in p-GSK and decrease in PP2A level and tau phosphorylation along with decrease in memory deficits (Zhao et al. 2013).

9.3.5 Garlic and Alzheimer Disease

Garlic (*Allium sativum*) belongs to the family Liliaceae. Garlic bulb has complex composition. The garlic bulb is enriched in organosulfur compounds. Two classes of organosulfur compounds, such as, (a) γ-glutamylcysteines, and (b) cysteine sulf-oxides have been reported to occur in whole garlic cloves. The γ-glutamylcysteine is hydrolyzed and oxidized to alliin (+S-allyl-L-cysteine sulfoxide). This compound is then converted to allicin (thio-2-propene-1-sulfinic acid S-allyl ester) by alliinase, which is released upon cutting, crushing, or chewing the garlic. Other important compounds present in garlic homogenate are 1-propenyl allyl thiosulfonate, allyl methyl thiosulfonate, (E,Z)-4,5,9-trithiadodeca-1,6,11-triene 9-oxide (ajoene), and y-L-glutamyl-S-alkyl-L-cysteine. The adenosine concentration increases several-fold as the homogenate is incubated at room temperature for several hours. Allicin not only interacts with thiol containing proteins, but also decomposes into 2-propenesulfenic acid, which has ability to bind the free-radicals and may contribute to antioxidant effects of garlic (Vaidya et al. 2009; Rabinkov et al. 1998). In addition to organosulfur compounds, garlic contains carbohydrates and proteins along with vitamins A, C, and E as well as selenium, a key element for the synthesis of the antioxidant enzyme glutathione peroxidase (Gorinstein Leontowicz et al. 2006). Constituents of garlic mediate anti-carcinogenic, antiatherosclerotic, anti-thrombotic, antimicrobial, antiinflammatory and antioxidant effects in rodents and humans. Many garlic constituents are easily absorbed in the gastrointestinal tract and others are not. Studies on bioavailability of water soluble constituent of garlic, S-allylcysteine (SAC) (Fig. 9.3), indicate that this constituent is rapidly absorbed and distributed in plasma, liver and kidney of rats, mice and dogs (Nagae et al. 1994). The bioavailability of SAC is about 103.0 % in mice, 98.2 % in rats and 87.2 % in dogs (Nagae et al. 1994). *N*-acetyltransferases converts SAC into *N*-Acetyl-SAC, which can be detected in the urine of dogs and humans (Steiner and Li 2001). Fresh garlic extracted over a prolonged period (up to 20 months) produces odourless aged garlic extract (AGE) containing stable and water soluble organosulphur compounds (s-allycisteine, s-allymercaptocysteine, allicin, and diallosulfides) that prevent oxidative damage by scavenging free radicals.

The occurrence and identification of two new compounds, tetrahydro-beta-carbolines (1-methyl-1,2,3,4-tetrahydro-beta-carboline-3-carboxylic acid and 1-methyl-1,2,3,4-tetrahydro-beta-carboline-1,3-dicarboxylic acid) as well as Nα-(1-deoxy-D-fructos-1-yl)-L-arginine, have also been reported in AGE. These compounds are increased during the garlic aging process and play an important role as antioxidants (Ryu et al. 2001; Ichikawa et al. 2002, 2006). Indeed, tetrahydro-β-carbolines are biologically active alkaloids and are structurally similar to flavonoids. Nα-(1-deoxy-D-fructos-1-yl)-L-arginine is only found in AGE and no other garlic products.

Little information is available on the bioavailability and half lives of organosulfur compounds of garlic in the brain. It is not known how many garlic constituents can cross BBB. In vitro studies have indicated that AGE and S-allyl-L-cysteine (SAC) mediate antioxidant effects by blocking the production of superoxide anion ($O_2{}^{\bullet-}$), hydrogen peroxide (H_2O_2), hydroxyl radical ($OH^{\bullet}$), peroxynitrite radical ($ONOO^-$), and peroxyl radical ($LOO^{\bullet}$) as well as hypochlorous acid ($HOCl$) and singlet oxygen ($^1O_2$) in a variety of neural and nonneural cells (Fig. 9.9). Thus, AGE and SAC lower Aβ levels and neurotoxicity in neural cell cultures (Ray et al. 2011a, b). Treatment of neuronal cultures with AGE and SAC provides protection against H_2O_2-mediated oxidative stress. Pretreatment with AGE alone also induces neuro-preservation in 80 % neurons in cultures from ROS-mediated oxidative damage. In addition, AGE also preserves synaptosome associated protein of 25 kDa (SNAP25) from ROS-mediated damage. Thus, treatment with AGE and SAC independently enhances SNAP25 levels ($\sim$70 %) and synaptophysin in AD amyloid precursor protein-transgenic mice, which are significantly decreased in AD (Ray et al. 2011a, b). In vivo studies have indicated that Aβ-mediated increase in lipid peroxidation can be prevented by pretreatment with S-allylcysteine (Pérez-Severiano et al. 2004). In addition, Aβ-induced abnormalities in learning and memory can also be partially restored by SAC treatment. Furthermore, SAC treatment not only retards Aβ aggregation, but also blocks Aβ fibrillation in a dose-dependent (Gupta and Rao 2007). SAC has been reported to protect neurons against the caspase-12 dependent neurotoxicity induced by Aβ (Kosuge et al. 2003). It is well known that hypercholesterolemia is a risk factor for AD. In Alzheimer's transgenic model Tg2576, it is suggested that SAC may act as HMG CoA reductase inhibitor and mediate its beneficial effect by lowering cholesterol levels (Chauhan 2006). Garlic compounds are known to reduce Aβ induced neuronal apoptosis, possibly by enhancing the endogenous

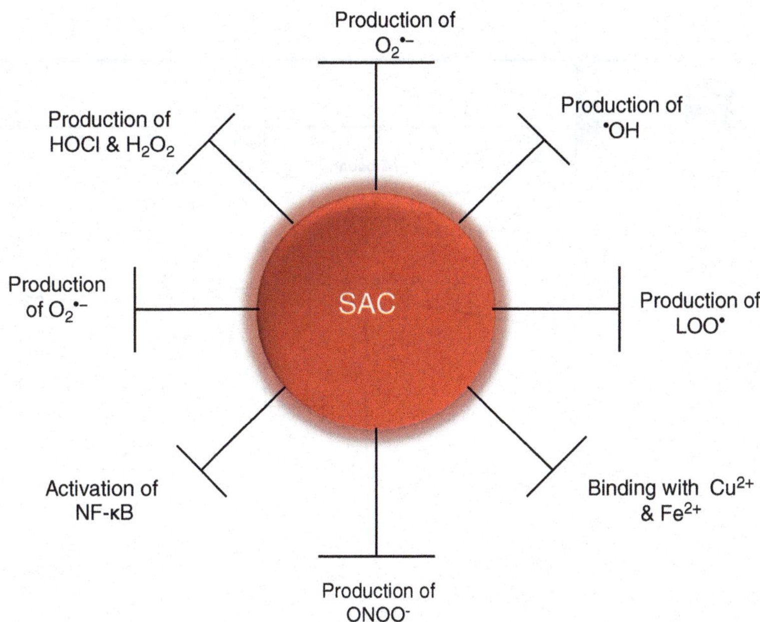

Fig. 9.9 Inhibition of reactive oxygen and nitrogen species by S-allylcysteine (*SAC*)

antioxidant defenses (Fig. 9.10) (Peng et al. 2002). SAC has been shown to exhibit neurotrophic activity in cultured neurons (Moriguchi et al. 1997a, b). There are no data on the effect of SAC on Nrf2 in the brain. However, it is reported that treatment of Balb/cA mice with SAC (1 g/L in drinking water for 4 weeks) increases glutathione levels in kidney and liver when compared with controls. Moreover, SAC enhanced catalase and glutathione peroxidase activities in kidney and liver (Hsu et al. 2004) supporting the view that SAC may act through different mechanisms, including radical scavenging, induction of antioxidant enzymes, binding with Cu^{2+} and Fe^{2+}, and inhibiting the activation of NFκB. Thus, accumulating evidence suggests that SAC may prevent the progression of AD by multiple mechanisms in vivo.

Cerebroprotective effects of dietary garlic may also be mediated through the generation of H_2S, an indigenous gaseous molecule, which supports vascular function and retards the progression of heart disease through the relaxation of vascular smooth muscle and induction of vasodilation of blood vessels (Benavides et al. 2007). In brain, H_2S, which is generated from cysteine by cystathionine-synthase, modulates several signal transduction pathways. Thus, H2S regulates (a) protein kinase A, receptor tyrosine kinases, mitogen kinases and oxidative stress signaling, (b) ion channels such as calcium (L-type, T-type and intracellular stores), potassium (K(ATP) and small conductance channels) and cystic fibrosis transmembrane conductance regulator chloride channels, (c) the release and function of neurotransmitters such as γ-aminobutyric acid, *N*-methyl-D-aspartate, and catecholamines (Tan et al. 2010). Low levels of H2S are risk factor for the development of AD (Seshadri

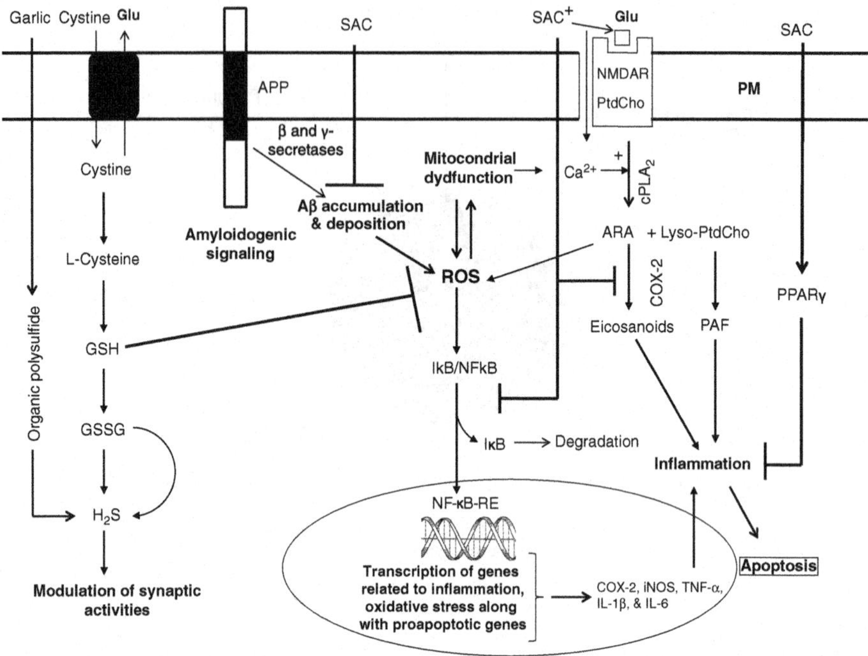

Fig. 9.10 Hypothetical diagram showing signaling pathways associated with neuroprotective effects of S-allycysteine. Plasma membrane (*PM*); *N*-Methyl-D-aspartate receptor (*NMDA-R*); estrogen receptor (*ER*); peroxisome proliferator-activated receptor γ (*PPARγ*); glutamate (*Glu*); phosphatidylcholine (*PtdCho*); cytosolic phospholipase A$_2$ (*cPLA$_2$*); secretory phospholipase A$_2$ (*sPLA$_2$*); cyclooxygenase (*COX-2*); inducible nitric oxide synthase (*iNOS*); arachidonic acid (*ARA*); reactive oxygen species (*ROS*); nuclear factor kappaB (*NF-κB*); nuclear factor kappaB response element (*NF-κB-RE*); inhibitory subunit of NFκB (*IκB*); tumor necrosis factor-α (*TNF-α*); interleukin-1β (*IL-1β*); interleukin-6 (*IL-6*); amyloid precursor protein (*APP*); reduced glutathione (*GSH*); oxidized glutathione (*GSSG*); hydrogen sulfide (*H$_2$S*)

et al. 2002; Tang et al. 2010). Localized increases in H$_2$S may delay the aggravation and exacerbation of symptoms in patients with AD (Tan et al. 2010; Whiteman and Winyard 2011). H$_2$S also has protective effects against Aβ-induced cell injury by inhibiting inflammation, promoting cell growth, and preserving mitochondrial function (Liu and Bian 2010). Moreover, H$_2$S protects neurons from oxidative stress by enhancing the activities of γ-GCS and cystine transport, which results in increasing glutathione levels (Kimura and Kimura 2004). Garlic-derived organic polysulfides are transformed by erythrocytes into H$_2$S that decreases the risk of AD by lowering cholesterol levels, inhibiting neuroinflammation, reducing homocysteine, preventing oxidative brain injury, and protecting neurons against apoptosis triggered by oxidative stress along with increase in cerebral blood flow. Garlic also stimulates production of NO, another gaseous molecule that is widely accepted for its role in promoting cerebral blood flow in the brain and heart. Low levels of NO and H$_2$S has been reported to improve neural cell survival after ischemic injury (Farooqui 2014).

AGE retards the onset of the frontal brain atrophy in early senescence mice models leading to improvement in learning and memory retention and increase in longevity (Moriguchi et al. 1997a). It is also shown that allixin, a component of garlic and AGE enhances the survival of neurons and increase the branching points in axons of hippocampus neurons (Moriguchi et al. 1997b).

9.3.6 Huperzine and Alzheimer Disease

AD is also accompanied by the loss of basal forebrain cholinergic neurons, which project from basal forebrain to the cerebral cortex and hippocampus (Zaborszky and Duque 2000). Impairment in cortical cholinergic neurotransmission not only promotes $A\beta$ pathology, but also increases phosphorylation of tau protein (Yan and Feng 2004). Huperzine A (HupA) is a novel *Lycopodium* alkaloid isolated from Chinese herb Huperzia serrata. It is a potent, highly specific and reversible inhibitor of acetylcholinesterase, a serine hydrolase, which is responsible for the termination of impulse signaling at cholinergic brain synapses (Rosenberry 2006). The pharmacokinetics studies have indicated that HupA is absorbed rapidly and distributed widely in various body tissues (Wang et al. 2006b; Rafii et al. 2011; Yang et al. 2013). It is eliminated at a moderate rate from the body. The oral bioavailability is 96.9 % in mice, with the highest radioactivities in the kidney and liver. The majority of the radioactivity is excreted in the urine 24 h after iv administration of [^{3}H] HupA. Only 2.4 % can be recovered from the feces. HupA can cross BBB. Autoradiographic studies in mice have shown that HupA has good brain penetration and is relatively free of cholinergic toxicity (Rafii et al. 2011). HupA is present in all regions of the brain, but is particularly concentrated in the frontoparietal cortex, striatal cortex, hippocampus, and nucleus accumbens after iv injection (Tang et al. 1994; Rafii et al. 2011).

In addition to acting as an acetylcholinesterase inhibitor, HupA possesses neuroprotective properties. Neuroprotective properties of HupA are due to its effects variety of parameters. Thus, HupA produces its effects by inhibiting oxidative stress and neuroinflammation, decreasing apoptotic cell death, decreasing $A\beta$ production, preventing NMDA receptor activation, and binding with Fe^{2+} (Fig. 9.11) (Yang et al. 2012; Huang et al. 2014b). HupA has the ability to effectively ameliorate the mitochondrial malfunction. Treatment with $A\beta$ leads to a rapid decline in ATP level, an obvious disruption of mitochondrial membrane homeostasis and integrity, a reduction in key enzyme activities in the electron transport chain and the TCA cycle, and an increase in ROS in PC12 cells. HupA not only attenuates $A\beta$-mediated cellular stress but also enhances ATP concentration and reduces ROS accumulation (Gao and Tang 2006; Gao et al. 2009). In isolated rat brain mitochondria, HupA effectively retards $A\beta$-mediated mitochondrial swelling, increase in ROS, and cytochrome c release. There is no evidence for the existence of cholinergic system in isolated brain mitochondria. Therefore, the mitochondria-targeted effects of HupA are clearly independent of cholinergic system. In addition, HupA also inhibits the

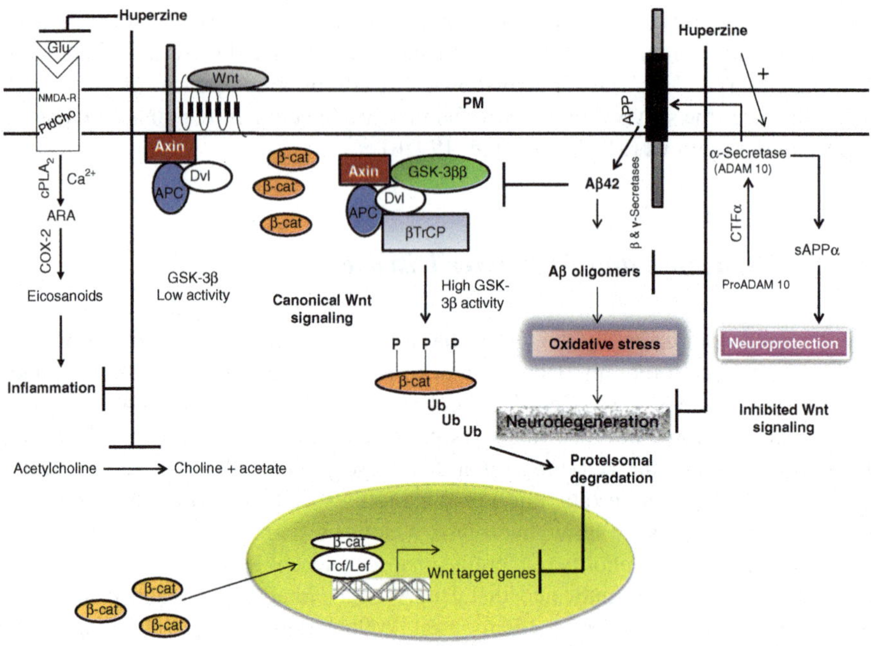

Fig. 9.11 Hypothetical diagram showing signaling pathways associated with neuroprotective effects of Huperzine. Plasma)membrane (*PM*); *N*-Methyl-D-aspartate receptor (*NMDA-R*); glutamate (*Glu*); phosphatidylcholine (*PtdCho*); cytosolic phospholipase A₂ (*cPLA₂*); cyclooxygenase (*COX-2*); Frizzled class of 7-transmembrane receptor (*Fz*); glycogen synthase kinase 3β (*GSK-3β*); β-catenin (*β-cat*)

penetration of Aβ into mitochondria and ameliorates Aβ-mediated dysfunction of TCA cycle in isolated brain cortical mitochondria (Yang et al. 2012). Furthermore, Hup A mediates its effects not only by activating Wnt/β-catenin signaling and enhancing the nonamyloidogenic pathway in an Alzheimer transgenic mouse model, but also by increasing the expression of antiapoptotic proteins Bcl-2, Bax, P53, decreasing caspase-3 activity, and normalization of mitochondrial function (Fig. 9.11) (Wang et al. 2011).

The beneficial effects of HupA can be abolished by feeding the animals with a high iron diet. HupA decreases iron content in the brain. HupA also down-regulates the expression of transferrin-receptor 1 as well as the transferrin-bound iron uptake in cultured neurons. These findings support the view that HupA may produce neuroprotective effects by reducing iron levels in the AD brain (Huang et al. 2014a, b). Based on these neurochemical effects it is stated that HupA may reverse or attenuate cognitive deficits in a broad range of animal models of AD. The phase IV clinical trials conducted in China have demonstrated that HupA induces significant improvement in the memory of elderly people and patients with AD and vascular dementia (VD) without any notable side effects (Zhang et al. 2002; Qian and Ke 2014). In USA, HupA is sold as nutraceutical.

9.4 Conclusion

Oxidative stress and neuroinflammation are closely associated with the pathogenesis of AD. It is becoming increasingly evident that the use of phytochemicals other than curcumin also produces beneficial effects in animal and cell culture models of AD. Like curcumin, many of these phytochemicals contain polyphenolic hydroxyl groups, which not only have antioxidant, anti-inflammatory properties, but also promote downregulation of innate immune cell functions. It is becoming increasingly evident that the ROS scavenge capacity of phytochemical depends on the number and position of the hydroxyl group and substituent, as well as glycosylation of phytochemical molecules. In general, phytochemicals with more hydroxyl groups have a stronger antioxidant and anti-inflammatory capacity. The molecular mechanisms associated with beneficial effects of phytochemicals are not fully understood. However, it is proposed that phytochemicals have ability to delay the initiation of and/or slow the progression of AD-like pathology and related neurodegenerative disorders, including a potential to inhibit neuronal apoptosis triggered by neurotoxic species (mediators of oxidative stress and neuroinflammation) or disrupt amyloid β aggregation and effects on amyloid precursor protein processing through the inhibition of β-secretase (BACE-1) and/or activation of α-secretase (ADAM10). Phytochemicals may also act by triggering the activation of Nrf2 and increasing the expression of antioxidant enzymes and reducing the production of inflammatory cytokines and chemokines. In addition, some phytochemicals may contribute specific biochemical effects that are beyond their antioxidant and radical-scavenging properties, such as induction of "vitagene genes". These effects may have an impact on the onset and progression of neurodegenerative diseases and aging. Some polyphenolic hydroxyl groups containing phytochemicals may also modulate adaptive T-cell mediated immune function by downregulating innate immune stimulating cytokines that promote Th1 immunity (e.g., TNF-α) and by promoting Th2 cytokines. These effects are thought to be mediated in part via downregulation of NFκB signaling. Together, these processes act to maintain the number and quality of synaptic connections in key brain regions and thus phytochemicals other than curcumin have the potential to prevent the progression of AD pathologies and to promote cognitive performance.

References

Albani D, Polito L, Batelli S, De Mauro S, Fracasso C, Martelli G, Colombo L, Manzoni C, Salmona M, Caccia S, Negro A, Forloni G (2009) The SIRT1 activator resveratrol protects SK-N-BE cells from oxidative stress and against toxicity caused by alpha-synuclein or amyloid-beta (1-42) peptide. J Neurochem 110:1445–1456

Allam F, Dao AT, Chugh G, Bohat R, Jafri F, Patki G, Mowrey C, Asghar M, Alkadhi KA, Salim S (2013) Grape powder supplementation prevents oxidative stress-induced anxiety-like behavior, memory impairment, and high blood pressure in rats. J Nutr 143:835–842

Arlt A, Bauer I, Schafmayer C, Tepel J, Müerköster SS, Brosch M, Röder C, Kalthoff H, Hampe J, Moyer MP, Fölsch UR, Schäfer H (2009) Increased proteasome subunit protein expression and proteasome activity in colon cancer relate to an enhanced activation of nuclear factor E2-related factor 2 (Nrf2). Oncogene 28:3983–3996

Augustin S, Rimbach G, Augustin K, Schliebs R, Wolffram S, Cermak R (2009) Effect of a short- and long-term treatment with Ginkgo biloba extract on amyloid precursor protein levels in a transgenic mouse model relevant to Alzheimer's disease. Arch Biochem Biophys 481:177–182

Barras A, Mezzetti A, Richard A, Lazzaroni S, Roux S, Melnyk P, Betbeder D, Monfilliette-Dupont N (2009) Formulation and characterization of polyphenol loaded lipid nanocapsules. Int J Pharm 379:270–277

Bastianetto S, Ramassamy C, Dore S, Christen Y, Poirier J, Quirion R (2000) The Ginkgo biloba extract (EGb 761) protects hippocampal neurons against cell death induced by betaamyloid. Eur J Neurosci 12:1882–1890

Baur JA, Pearson KJ, Price NL, Jamieson HA, Lerin C, Kalra A, Prabhu VV, Allard JS, Lopez-Lluch G, Lewis K, Pistell PJ, Poosala S, Becker KG, Boss O, Gwinn D, Wang M, Ramaswamy S, Fishbein KW, Spencer RG, Lakatta EG, Le Couteur D, Shaw RJ, Navas P, Puigserver P, Ingram DK, de Cabo R, Sinclair DA (2006) Resveratrol improves health and survival of mice on a high-calorie diet. Nature 444:337–342

Benavides GA, Squadrito GL, Mills RW, Patel HD, Isbell TS, Patel RP, Darley-Usmar VM, Doeller JE, Kraus DW (2007) Hydrogen sulfide mediates the vasoactivity of garlic. Proc Natl Acad Sci U S A 104:17977–17982

Calabrese EJ, Mattson MP, Calabrese V (2010) Resveratrol commonly displays hormesis: occurrence and biomedical significance. Hum Exp Toxicol 29:980–1015

Chandra V, Pandav R, Dodge HH, Johnston JM, Belle SH, DeKosky ST, Ganguli M (2001) Incidence of Alzheimer's disease in a rural community in India: the Indo-US study. Neurology 57:985–989

Chandrasekaran K, Mehrabian Z, Spinnewyan B, Drieu K, Kiskum G (2001) Neuroprotective effects of bilobalide, a component of Ginkgo biloba extract (EGb 761), in gerbil global brain ischemia. Brain Res 922:282–292

Chauhan NB (2006) Effect of aged garlic extract on APP processing and tau phosphorylation in Alzheimer's transgenic model Tg2576. J Ethnopharmacol 108:385–394

Chen JC, Ho FM, Pei-Dawn Lee Chao, Chen CP, Jeng KC, Hsu HB, Lee ST, Wen Tung Wu, Lin WW (2005) Inhibition of iNOS gene expression by quercetin is mediated by the inhibition of IkappaB kinase, nuclear factor-kappa B and STAT1, and depends on heme oxygenase-1 induction in mouse BV-2 microglia. Eur J Pharmacol 521:9–20

Chen Z, Lu T, Yue X, Wei N, Jiang Y, Chen M, Ni G, Liu X, Xu G (2010) Neuroprotective effect of ginsenoside Rb1 on glutamate-induced neurotoxicity: with emphasis on autophagy. Neurosci Lett 482:264–268

Chen R, Hu Z, Wei L, Ma Y, Liu Z, Copeland JR (2011) Incident dementia in a defined older Chinese population. PLoS One 6, e24817

Choi RC, Zhu JT, Leung KW, Chu G, Xie HQ, Chen VP, Zheng KY, Lau DT, Dong TT, Chow PC, Han YF, Wang ZT, Tsim KW (2010) A flavonol glycoside, isolated from roots of Panax notoginseng, reduces amyloid-beta-induced neurotoxicity in cultured neurons: signaling transduction and drug development for Alzheimer's disease. J Alzheimers Dis 19:795–811

Chu X, Ci X, He J, Wei M, Yang X, Cao Q, Li H, Guan S, Deng Y, Pang D, Deng X (2011) A novel anti-inflammatory role for ginkgolide B in asthma via inhibition of the ERK/MAPK signaling pathway. Molecules 16:7634–7648

Clostre F (1999) Ginkgo biloba extract (EGb 761): state of knowledge in the dawn of the year 2000. Ann Pharm Fr 57(Suppl 1):S8–S88

Colciaghi F, Borroni B, Zimmermann M, Bellone C, Longhi A, Padovani A, Cattabene F, Christen Y, Di Luca M (2004) Amyloid precursor protein metabolism is regulated toward alpha-secretase pathway by Ginkgo biloba extracts. Neurobiol Dis 16:454–466

D'Angelo L, Grimaldi R, Caravaggi M, Marcoli M, Perucca E, Lecchini S, Frigo GM, Crema A (1986) A double-blind, placebo-controlled clinical study on the effect of a standardized ginseng extract on psychomotor performance in healthy volunteers. J Ethnopharmacol 16:15–22

Dasgupta B, Milbrandt J (2007) Resveratrol stimulates AMP kinase activity in neurons. Proc Natl Acad Sci U S A 104:7217–7222

De Santi C, Pietrabissa A, Spisni R, Mosca F, Pacifici GM (2000a) Sulphation of resveratrol, a natural compound present in wine, and its inhibition by natural flavonoids. Xenobiotica 30:857–866

De Santi C, Pietrabissa A, Mosca F, Pacifici GM (2000b) Glucuronidation of resveratrol, a natural product present in grape and wine, in the human liver. Xenobiotica 30:1047–1054

DeFeudis FV, Drieu K (2000) *Ginkgo biloba* extract (EGb761) and CNS functions: basic studies and clinical applications. Curr Drug Targets 1:25–58

Donmez G, Wang D, Cohen DE, Guarente L (2010) SIRT1 suppresses beta-amyloid production by activating the alpha-secretase gene ADAM10. Cell 142:320–332

Eggler AL, Gay KA, Mesecar AD (2008) Molecular mechanisms of natural products in chemoprevention: induction of cytoprotective enzymes by Nrf2. Mol Nutr Food Res 52(Suppl. 1):S84–S94

Essa MM, Vijayan RK, Castellano-Gonzalez G, Memon MA, Braidy N, Guillemin GJ (2012) Neuroprotective effect of natural products against Alzheimer's disease. Neurochem Res 37:1829–1842

Fang F, Chen X, Huang T, Lue LF, Luddy JS, Yan SS (2012) Multi-faced neuroprotective effects of Ginsenoside Rg1 in an Alzheimer mouse model. Biochim Biophys Acta 1822:286–292

Farooqui AA (2010) Neurochemical aspects of neurotraumatic and neurodegenerative diseases. Springer, New York

Farooqui AA (2012) Phytochemicals, signal transduction, and neurological disorders. Springer, New York

Farooqui AA (2014) Inflammation and oxidative stress in neurological disorders. Springer International Publishing, Switzerland

Farooqui T, Farooqui AA (2011) Lipid-mediated oxidative stress and inflammation in the pathogenesis of Parkinson's disease. Parkinsons Dis 2011:247467

Fernandez JW, Rezai-Zadeh K, Obregon D, Tan J (2010) EGCG functions through estrogen receptor-mediated activation of ADAM10 in the promotion of non-amyloidogenic processing of APP. FEBS Lett 584:4259–4267

Fremont L (2000) Biological effects of resveratrol. Life Sci 66:663–673

Gao X, Tang XC (2006) Huperzine A attenuates mitochondrial dysfunction in beta-amyloid-treated PC12 cells by reducing oxygen free radicals accumulation and improving mitochondrial energy metabolism. J Neurosci Res 83:1048–1057

Gao X, Zheng CY, Yang L, Tang XC, Zhang HY (2009) Huperzine A protects isolated rat brain mitochondria against beta-amyloid peptide. Free Radic Biol Med 46:1454–1462

Goh LML, Barlow PJ (2004) Flavonoid recovery and stability from *Ginkgo biloba* subjected to a simulated digestion process. Food Chem 86:195–202

Gorinstein Leontowicz H, Leontowicz M, Drzewiecki J, Najman K, Katrich E, Barasch D, Yamamoto K, Trakhtenberg SS (2006) Raw and boiled garlic enhances plasma antioxidant activity and improves plasma lipid metabolism in cholesterol-fed rats. Life Sci 78:655–663

Gupta C, Prakash D (2014) Nutraceuticals for geriatrics. J Tradit Complement Med 5:5–14

Gupta VB, Rao KS (2007) Anti-amyloidogenic activity of S-allyl-L-cysteine and its activity to destabilize Alzheimer's beta-amyloid fibrils in vitro. Neurosci Lett 429:75–80

Han YS, Zheng WH, Bastianetto S, Chabot JG, Quirion R (2004) Neuroprotective effects of resveratrol against beta-amyloid-induced neurotoxicity in rat hippocampal neurons: involvement of protein kinase C. Br J Pharmacol 141:997–1005

Higdon JV, Frei B (2003) Tea catechins and polyphenols: health effects, metabolism, and antioxidant functions. Crit Rev Food Sci Nutr 43:89–143

Hsu CC, Huang CN, Hung YC, Yin MC (2004) Five cysteine-containing compounds have antioxidative activity in Balb/cA mice. J Nutr 134:149–152

Huang J, Wu D, Wang J, Li F, Lu L, Gao Y, Zhong Z (2014a) Effects of Panax notoginseng saponin on alpha, beta, and gamma secretase involved in Abeta deposition in SAMP8 mice. Neuroreport 25:89–93

Huang XT, Qian ZM, He X, Gong Q, Wu KC, Jiang LR, Lu LN, Zhu ZJ, Zhang HY, Yung WH, Ke Y (2014b) Reducing iron in the brain: a novel pharmacologic mechanism of huperzine A in the treatment of Alzheimer's disease. Neurobiol Aging 35:1045–1054

Hubbard BP, Sinclair DA (2014) Small molecule SIRT1 activators for the treatment of aging and age-related diseases. Trends Pharmacol Sci 35:146–154

Huh H, Staba EJ (1992) Botany and chemistry of Ginkgo biloba L. J Herbs Spices Med Plants 1:92–124

Hussein MA (2011) A convenient mechanism for the free radical scavenging activity of resveratrol. Int J Phytomed 3:459–469

Hwang SH, Shin EJ, Shin TJ, Lee BH, Choi SH, Kang J, Kim HJ, Kwon SH, Jang CG, Lee JH, Kim HC, Nah SY (2012) Gintonin, a ginseng-derived lysophosphatidic acid receptor ligand, attenuates Alzheimer's disease-related neuropathies: involvement of non-amyloidogenic processing. J Alzheimers Dis 31:207–223

Hyung SJ, DeToma AS, Brender JR, Lee S, Vivekanandan S, Kochi A, Choi JS, Ramamoorthy A, Ruotolo BT, Lim MH (2013) Insights into antiamyloidogenic properties of the green tea extract (-)-epigallocatechin-3-gallate toward metal-associated amyloid-β species. Proc Natl Acad Sci U S A 110:3743–3748

Ichikawa M, Ryu K, Yoshida J et al (2002) Antioxidant effects of tetrahydro-β-carboline derivatives identified in aged garlic extract. Biofactors 16:57–72

Ichikawa M, Yoshida J, Ide N, Sasaoka T, Yamaguchi H, Ono K (2006) Tetrahydro-β-carboline derivatives in aged garlic extract show antioxidant properties. J Nutr 136(Suppl 3):726S–731S

Ingram DK, Roth GS, Lane MA, Ottinger MA, Zou S, de Cabo R, Mattison JA (2006) The potential for dietary restriction to increase longevity in humans: extrapolation from monkey studies. Biogerontology 7:143–148

Itoh T, Zang YF, Murai S, Saito H (1989) Effects of Panax ginseng root on the vertical and horizontal motor activities and on brain monoamine-related substances in mice. Planta Med 55:429–433

Iuga C, Alvarez-Idaboy JR, Russo N (2012) Antioxidant activity of trans-resveratrol toward hydroxyl and hydroperoxyl radicals: a quantum chemical and computational kinetics study. J Org Chem 77:3868–3877

Jang JH, Surh YJ (2003) Protective effect of resveratrol on beta-amyloid-induced oxidative PC12 cell death. Free Radic Biol Med 34:1100–1110

Jia N, Han K, Kong JJ, Zhang XM, Sha S, Ren GR, Cao YP (2013) (-)-Epigallocatechin-3-gallate alleviates spatial memory impairment in APP/PS1 mice by restoring IRS-1 signaling defects in the hippocampus. Mol Cell Biochem 380:211–218

Kannappan R, Gupta SC, Kim JH, Reuter S, Aggarwal BB (2011) Neuroprotection by spice-derived nutraceuticals: you are what you eat! Mol Neurobiol 44:142–159

Karpagam V, Sathishkumar N, Sathiyamoorthy S, Rasappan P, Shila S, Kim YJ, Yang DC (2013) Identification of BACE1 inhibitors from Panax ginseng saponins-An Insilco approach. Comput Biol Med 43:1037–1044

Karuppagounder SS, Pinto JT, Xu H, Chen HL, Beal MF, Gibson GE (2009) Dietary supplementation with resveratrol reduces plaque pathology in a transgenic model of Alzheimer's disease. Neurochem Int 54:111–118

Kennedy DO, Wightman EL, Reay JL, Lietz G, Okello EJ, Wilde A, Haskell CF (2010) Effects of resveratrol on cerebral blood flow variables and cognitive performance in humans: a double-blind, placebo-controlled, crossover investigation. Am J Clin Nutr 91:1590–1597

Kensler TW, Egner PA, Agyeman AS, Visvanathan K, Groopman JD, Chen JG, Chen TY, Fahey JW, Talalay P (2013) Keap1-nrf2 signaling: a target for cancer prevention by sulforaphane. Top Curr Chem 329:163–177

Kepp KP (2012) Bioinorganic chemistry of Alzheimer's disease. Chem Rev 112:5193–5239

Kim S, Ahn K, Oh TH, Nah SY, Rhim H (2002) Inhibitory effect of ginsenosides on NMDA receptor-mediated signals in rat hippocampal neurons. Biochem Biophys Res Commun 296:247–254

Kim D, Nguyen MD, Dobbin MM, Fischer A, Sananbenesi F, Rodgers JT, Delalle I, Baur JA, Sui G, Armour SM, Puigserver P, Sinclair DA, Tsai LH (2007) SIRT1 deacetylase protects against neurodegeneration in models for Alzheimer's disease and amyotrophic lateral sclerosis. EMBO J 26:3169–3179

Kim J, Lee HJ, Lee KW (2010) Naturally occurring phytochemicals for the prevention of Alzheimer's disease. J Neurochem 112:1415–1430

Kim J, Kim SH, Lee DS, Lee DJ, Kim SH, Chung S, Yang HO (2013) Effects of fermented ginseng on memory impairment and beta-amyloid reduction in Alzheimer's disease experimental models. J Ginseng Res 37:100–107

Kimura Y, Kimura H (2004) Hydrogen sulfide protects neurons from oxidative stress. FASEB J 18:1165–1167

Kosuge Y, Koen Y, Ishige K, Minami K, Urasawa H, Saito H, Ito Y (2003) S-allyl-L-cysteine selectively protects cultured rat hippocampal neurons from amyloid beta-protein- and tunicamycin-induced neuronal death. Neuroscience 122:885–895

Kuchler-Bopp S, Dietrich J-B, Zaepfel M, Delaunoy J-P (2000) Receptor-mediated endocytosis of transthyretin by ependymoma cells. Brain Res 7:185–194

Kumar V (2006) Potential medicinal plants for CNS disorders: an overview. Phytother Res 20:1023–1035

Kwak M-K, Wakabayashi N, Greenlaw JL, Yamamoto M, Kensler TW (2003) Antioxidants enhance mammalian proteasome expression through the Keap1-Nrf2 signaling pathway. Mol Cell Biol 23:8786–8794

Ladiwala AR, Lin JC, Bale SS, Marcelino-Cruz AM, Bhattacharya M, Dordick JS, Tessier PM (2010) Resveratrol selectively remodels soluble oligomers and fibrils of amyloid Abeta into off-pathway conformers. J Biol Chem 285:24228–24237

Levites Y, Amit T, Mandel S, Youdim MB (2003) Neuroprotection and neurorescue against Abeta toxicity and PKC-dependent release of nonamyloidogenic soluble precursor protein by green tea polyphenol (-)-epigallocatechin-3-gallate. FASEB J 17:952–954

Li L, Liu J, Yan X et al (2011) Protective effects of ginsenoside Rd against okadaic acid-induced neurotoxicity in vivo and in vitro. J Ethnopharmacol 138:135–141

Liu Y-Y, Bian J-S (2010) Hydrogen sulfide protects amyloid-β induced cell toxicity in microglia. J Alzheimers Dis 22:1189–1200

Liu X-P, Goldring CEP, Wang H-Y, Copple IM, Kitteringham NR, Park BK (2009a) Extract of *Ginkgo biloba* induces glutathione-S-transferase subunit-P1 in vitro. Phytomedicine 16:451–455

Liu H, Yang J, Du F, Gao X, Ma X, Huang Y, Xu F, Niu W, Wang F, Mao Y (2009b) Absorption and disposition of ginsenosides after oral administration of Panax notoginseng extract to rats. Drug Metab Dispos 37:2290–2298

Liu ZJ, Zhao M, Zhang Y, Xue JF, Chen NH (2010) Ginsenoside Rg1 promotes glutamate release via a calcium/calmodulin-dependent protein kinase II-dependent signaling pathway. Brain Res 1333:1–8

Longpre F, Garneau P, Christen Y, Ramassamy C (2006) Protection by EGb 761 against beta-amyloid-induced neurotoxicity: involvement of NF-kappaB, SIRT1, and MAPKs pathways and inhibition of amyloid fibril formation. Free Radic Biol Med 41:1781–1794

Lu S, Guo X, Zhao P (2011) Effect of ginkgo biloba extract 50 on immunity and antioxidant enzyme activities in ischemia reperfusion rats. Molecules 16:9194–9206

Lu C, Guo Y, Li J, Yao M, Liao Q, Xie Z, Li X (2012) Design, synthesis, and evaluation of resveratrol derivatives as Aβ1-42 aggregation inhibitors, antioxidants, and neuroprotective agents. Bioorg Med Chem Lett 22:7683–7687

Lu C, Guo Y, Yan J, Luo Z, Luo H-B, Yan M, Huang L, Li X (2013) Design, synthesis, and evaluation of multitarget-directed resveratrol derivatives for the treatment of Alzheimer's disease. J Med Chem 56:5843–5859

Mahdy HM, Tadros MG, Mohamed MR, Karim AM, Khalifa AE (2011) The effect of Ginkgo biloba extract on 3-nitropropionic acid-induced neurotoxicity in rats. Neurochem Int 59:770–778

Mandel SA, Amit T, Kalfon L, Reznichenko L, Youdim MB (2008) Targeting multiple neurodegenerative diseases etiologies with multimodal-acting green tea catechins. J Nutr 138:1578S–1583S

Marambaud P, Zhao H, Davies P (2005) Resveratrol promotes clearance of Alzheimer's disease amyloid-beta peptides. J Biol Chem 280:37377–37382

Mattson MP (2008) Awareness of hormesis will enhance future research in basic and applied neuroscience. Crit Rev Toxicol 38:633–639

Mattson MP, Cheng A (2006) Neurohormetic phytochemicals: Low-dose toxins that induce adaptive neuronal stress responses. Trends Neurosci 29:632–639

Min SW, Cho SH, Zhou Y, Schroeder S, Haroutunian V, Seeley WW, Huang EJ, Shen Y, Masliah E, Mukherjee C, Meyers D, Cole PA, Ott M, Gan L (2010) Acetylation of tau inhibits its degradation and contributes to tauopathy. Neuron 67:953–966

Moriguchi T, Saito H, Nishyama N (1997a) Anti-aging effect of aged garlic extract in the inbred brain atrophy mouse model. Clin Exp Pharmacol Physiol 24:235–242

Moriguchi T, Matsuura H, Itakura Y, Katsuki H, Saito H, Nishiyama N (1997b) Allixin, a phytoalexin produced by garlic, and its analogues as novel exogenous substances with neurotrophic activity. Life Sci 61:1413–1420

Mukhtar H, Ahmad N (2000) Tea polyphenols: prevention of cancer and optimizing health. Am J Clin Nutr 71(6 Suppl):1698S–1702S

Murias M, Handler N, Erker T, Pleban K, Ecker G, Saiko P, Szekeres T, Jäger W (2004) Resveratrol analogues as selective cyclooxygenase-2 inhibitors: synthesis and structure-activity relationship. Bioorg Med Chem 12:5571–5578

Nagae S, Ushijima M, Hatono S, Imai J, Kasuga S, Matsuura H, Itakura Y, Higashi Y (1994) Pharmacokinetics of the garlic compound S-allylcysteine. Planta Med 60:214–217

Nah SY, Kim DH, Rhim H (2007) Ginsenosides: are any of them candidates for drugs acting on the central nervous system? CNS Drug Rev 13:381–404

Nehlig A (2013) The neuroprotective effects of cocoa flavanol and its influence on cognitive performance. Br J Clin Pharmacol 75:716–727

Nosál' R, Drábiková K, Jančinová V, Perečko T, Ambrožová G, Číž M, Lojek A, Pekarová M, Šmidrkal J, Harmatha J (2014) On the molecular pharmacology of resveratrol on oxidative burst inhibition in professional phagocytes. Oxid Med Cell Longev 2014:706269

Ono K, Yoshiike Y, Takashima A, Hasegawa K, Naiki H, Yamada M (2003) Potent anti-amyloidogenic and fibril-destabilizing effects of polyphenols in vitro: implications for the prevention and therapeutics of Alzheimer's disease. J Neurochem 87:172–181

Pasinetti GM (2012) Novel role of red wine-derived polyphenols in the prevention of Alzheimer's disease dementia and brain pathology: experimental approaches and clinical implications. Planta Med 78:1614–1619

Peng Q, Buz'Zard AR, Lau BH (2002) Neuroprotective effect of garlic compounds in amyloid-beta peptide-induced apoptosis in vitro. Med Sci Monit 8:28–37

Pérez-Severiano F, Rodríguez-Pérez M, Pedraza-Chaverrí J, Maldonado PD, Medina-Campos ON, Ortíz-Plata A, Sánchez-García A, Villeda-Hernández J, Galván-Arzate S, Aguilera P, Santamaría A (2004) S-Allylcysteine prevents amyloid-beta peptide-induced oxidative stress in rat hippocampus and ameliorates learning deficits. Eur J Pharmacol 489:197–202

Pervaiz S, Holme AL (2009) Resveratrol: its biologic targets and functional activity. Antioxid Redox Signal 11:2851–2897

Petralia S, Spatafora C, Tringali C, Foti MC, Sortino S (2004) Hydrogen atom abstraction from resveratrol and two lipophilic derivatives by tert-butoxyl radicals. A laser flash photolysis study. New J Chem 28:1484–1487

Phillis JW, Horrocks LA, Farooqui AA (2006) Cyclooxygenases, lipoxygenases, and epoxygenases in CNS: their role and involvement in neurological disorders. Brain Res Rev 52:201–243

Pithadia AS, Lim MH (2012) Metal-associated amyloid-β species in Alzheimer's disease. Curr Opin Chem Biol 16:67–73

Porterfield SP (2000) Thyroidal dysfunction and environmental chemicals—potential impact on brain development. Environ Health Perspect 108:433–438

Postina R (2008) A closer look at alpha-secretase. Curr Alzheimer Res 5:179–186

Qi X, Ignatova S, Luo G, Liang Q, Jun FW, Wang Y, Sutherland I (2010) Preparative isolation and purification of ginsenosides Rf, Re, Rd and Rb1 from the roots of Panax ginseng with a salt/containing solvent system and flow step-gradient by high performance counter-current chromatography coupled with an evaporative light scattering detector. J Chromatogr A 1217:1995–2001

Qian ZM, Ke Y (2014) Huperzine A: is it an effective disease-modifying drug for Alzheimer's disease? Front Aging Neurosci 6:216

Qiu J, Li W, Feng SH, Wang M, He ZY (2014) Ginsenoside Rh2 promotes nonamyloidgenic cleavage of amyloid precursor protein via a cholesterol-dependent pathway. Genet Mol Res 13:3586–3598

Rabinkov A, Miron T, Konstantinovski L, Wilchek M, Mirelman D, Weiner L (1998) The mode of action of allicin: trapping of radicals and interaction with thiol containing proteins. Biochim Biophys Acta 1379:233–244

Rafii MS, Walsh S, Little JT, Behan K, Reynolds B, Ward C, Jin S, Thomas R, Aisen PS; Alzheimer's Disease Cooperative Study (2011) A phase II trial of huperzine A in mild to moderate Alzheimer disease. Neurology 76: 1389–1394

Rahvar M, Nikseresht M, Shafiee SM, Naghibalhossaini F, Rasti M, Panjehshahin MR, Owji AA (2011) Effect of oral resveratrol on the BDNF gene expression in the hippocampus of the rat brain. Neurochem Res 36:761–765

Ray B, Chauhan NB, Lahiri DK (2011a) The "aged garlic extract:" (AGE) and one of its active ingredients S-allyl-L-cysteine (SAC) as potential preventive and therapeutic agents for Alzheimer's disease (AD). Curr Med Chem 18:3306–3313

Ray B, Chauhan NB, Lahiri DK (2011b) Oxidative insults to neurons and synapse are prevented by aged garlic extract and S-allyl-L-cysteine treatment in the neuronal culture and APP-Tg mouse model. J Neurochem 117:388–402

Reinhard C, Hebert SS, De Strooper B (2005) The amyloid-beta precursor protein: integrating structure with biological function. EMBO J 24:3996–4006

Rhein V, Giese M, Baysang G, Meier F, Rao S, Schulz KL, Hamburger M, Eckert A (2010) Ginkgo biloba extract ameliorates oxidative phosphorylation performance and rescues abeta-induced failure. PLoS One 5, e12359

Riviere C, Richard T, Quentin L, Krisa S, Mérillon JM, Monti JP (2007) Inhibitory activity of stilbenes on Alzheimer's beta-amyloid fibrils in vitro. Bioorg Med Chem 15:1160–1167

Rosenberry TL (2006) Acetylcholinesterase. Adv Enzymol Relat Areas Mol Biol 43:103–218

Rossjohn J, Cappai R, Feil SC, Henry A, Mckinstry WJ, Galatis D, Hesse L, Multhaup G, Beyreuther K, Masters CL, Parker MW (1999) Crystal structure of the N-terminal, growth factor-like domain of Alzheimer amyloid precursor protein. Nat Struct Biol 6:327–331

Rudakewich M, Ba F, Benishin CG (2001) Neurotrophic and neuroprotective actions of ginsenosides Rb1 and Rg1. Planta Med 67:533–537

Ryu K, Ide N, Matsuura H, Itakura Y (2001) Nα-(1-deoxy-D-fructos-1-yl)-L-arginine, an antioxidant compound identified in aged garlic extract. J Nutr 131(Suppl 3):972S–976S

Sandbrink R, Masters CL, Beyreuther K (1996) APP gene family: alternative splicing generates functionally related isoforms. Ann N Y Acad Sci 777:281–287

Savaskan E, Olivieri G, Meier F, Seifritz E, Wirz-Justice A, Muller-Spahn F (2003) Red wine ingredient resveratrol protects from beta-amyloid neurotoxicity. Gerontology 49:380–383

Savouret JF, Quesne M (2002) Resveratrol and cancer: a review. Biomed Pharmacother 56:84–87

Scapagnini G, Colombrita C, Amadio M, D'Agata V, Arcelli E, Sapienza M, Quattrone A, Calabrese V (2006) Curcumin activates defensive genes and protects neurons against oxidative stress. Antioxid Redox Signal 8:395–403

Seshadri S, Beiser A, Selhub J, Jacques PF, Rosenberg IH, D'Agostino RB, Wilson PW, Wolf PA (2002) Plasma homocysteine as a risk factor for dementia and Alzheimer's disease. N Engl J Med 346:476–483

Shi C, Zhao L, Zhu B, Li Q, Yew DT, Yao Z, Xu J (2009) Protective effects of *Ginkgo biloba* extract (EGb761) and its constituents quercetin and ginkgolide B against beta-amyloid peptideinduced toxicity in SH-SY5Y cells. Chem Biol Interact 181:115–123

Shi C, Liu J, Wu F, Yew DT (2010) *Ginkgo biloba* extract in Alzheimer's disease: from action mechanisms to medical practice. Int J Mol Sci 11:107–123

Shi C, Zheng DD, Wu FM, Liu J, Xu J (2011) The phosphatidyl inositol 3 kinase-glycogen synthase kinase 3β pathway mediates bilobalide-induced reduction in amyloid β-peptide. Neurochem Res 37:298–306

Simão F, Matté A, Pagnussat AS, Netto CA, Salbego CG (2012) Resveratrol prevents CA1 neurons against ischemic injury by parallel modulation of both GSK-3β and CREB through PI3-K/Akt pathways. Eur J Neurosci 36:2899–2905

Singh M, Arseneault M, Sanderson T, Murthy V, Ramassamy C (2008) Challenges for research on polyphenols from foods in Alzheimer's disease: bioavailability, metabolism, and cellular and molecular mechanisms. J Agric Food Chem 56:4855–4873

Smith JV, Luo Y (2004) Studies on molecular mechanisms of Ginkgo biloba extract. Appl Microbiol Biotechnol 64:465–472

Smith A, Giunta B, Bickford PC, Fountain M, Tan J, Shytle RD (2010) Nanolipidic particles improve the bioavailability and alpha-secretase inducing ability of epigallocatechin-3-gallate (EGCG) for the treatment of Alzheimer's disease. Int J Pharm 389:207–212

Spencer JP (2009) The impact of flavonoids on memory: physiological and molecular considerations. Chem Soc Rev 38:1152–1161

Spencer JP (2010) Beyond antioxidants: the cellular and molecular interactions of flavonoids and how these underpin their actions on the brain. Proc Nutr Soc 69:244–260

Steiner M, Li W (2001) Aged garlic extract, a modulator of cardiovascular risk factors: a dose-finding study on the effects of AGE on platelet functions. J Nutr 131(3s):980S–984S

Stivala LA, Savio M, Carafoli F, Perucca P, Bianchi L, Maga G, Forti L, Pagnoni UM, Albini A, Prosperi E, Vannini V (2001) Specific structural determinants are responsible for the antioxidant activity and the cell cycle effects of resveratrol. J Biol Chem 276:22586–22594

Storey E, Cappai R (1999) The amyloid precursor protein of Alzheimer's disease and the Abeta peptide. Neuropathol Appl Neurobiol 25:81–97

Tan BH, Wong PT-H, Bian J-S (2010) Hydrogen sulfide: a novel signaling molecule in the central nervous system. Neurochem Int 56:3–10

Tang XC, Kindel GH, Kozikowski AP, Hanin I (1994) Comparison of the effects of natural and synthetic huperzine A on rat brain cholinergic function in vitro and in vivo. J Ethnopharmacol 44:147–155

Tang X-Q, Shen X-T, Huang Y-E et al (2010) Hydrogen sulfide antagonizes homocysteine-induced neurotoxicity in PC12 cells. Neurosci Res 68:241–249

Tsuzuki K, Fukatsu R, Yamaguchi H, Tateno M, Imai K, Fujii N, Yamauchi T (2000) Transthyretin binds amyloid beta peptides, Abeta1-42 and Abeta1-40 to form complex in the autopsied human kidney – possible role of transthyretin for abeta sequestration. Neurosci Lett 281:171–174

Tu LH, Ma J, Liu HP, Wang RR, Luo J (2009) The neuroprotective effects of ginsenosides on calcineurin activity and tau phosphorylation in SY5Y cells. Cell Mol Neurobiol 29:1257–1264

Vaidya V, Ingold KU, Pratt DA (2009) Garlic: Source of the ultimate antioxidants – sulfenic acids. Angew Chem Int Ed Engl 48:157–160

Velayutham P, Babu A, Liu D (2008) Green tea catechins and cardiovascular health: an update. Curr Med Chem 15:1840–1850

Venkatesan R, Ji E, Kim SY (2015) Phytochemicals that regulate neurodegenerative disease by targeting neurotrophins: a comprehensive review. Biomed Res Int 2015:814068

Vieira de Almeida LM, Piñeiro CC, Leite MC, Brolese G, Tramontina F, Feoli AM, Gottfried C, Gonçalves CA (2007) Resveratrol increases glutamate uptake, glutathione content, and S100B secretion in cortical astrocyte cultures. Cell Mol Neurobiol 27:661–668

Vingtdeux V, Giliberto L, Zhao H, Chandakkar P, Wu Q, Simon JE, Janle EM, Lobo J, Ferruzzi MG, Davies P, Marambaud P (2010) AMP-activated protein kinase signaling activation by resveratrol modulates amyloid-beta peptide metabolism. J Biol Chem 285:9100–9113

Waleh NS, Chao WR, Bensari A, Zaveri NT (2005) Novel D-ring analog of epigallocatechin-3-gallate inhibits tumor growth and VEGF expression in breast carcinoma cells. Anticancer Res 25:397–402

Walle T (2011) Bioavailability of resveratrol. Ann N Y Acad Sci 1215:9–15

Walle T, Hsieh F, DeLegge MH, Oatis JE Jr, Walle UK (2004) High absorption but very low bioavailability of oral resveratrol in humans. Drug Metab Dispos 32:1377–1382

Wang Y, Ho CT (2009) Polyphenolic chemistry of tea and coffee: a century of progress. J Agric Food Chem 57:8109–8114

Wang Q, Xu J, Rottinghaus GE (2002) Resveratrol protects against global cerebral ischemic injury in gerbils. Brain Res 958:439–447

Wang J, Ho L, Zhao Z, Seror I, Humala N, Dickstein DL, Thiyagarajan M, Percival SS, Talcott ST, Pasinetti GM (2006a) Moderate consumption of Cabernet Sauvignon attenuates Abeta neuropathology in a mouse model of Alzheimer's disease. FASEB J 20:2313–2320

Wang R, Yan H, Tang XC (2006b) Progress in studies of huperzine A, a natural cholinesterase inhibitor from Chinese herbal medicine. Acta Pharmacol Sin 27:1–26

Wang J, Ho L, Zhao W, Ono K, Rosensweig C, Chen L, Humala N, Teplow DB, Pasinetti GM (2008) Grape-derived polyphenolics prevent Abeta oligomerization and attenuate cognitive deterioration in a mouse model of Alzheimer's disease. J Neurosci 28:6388–6392

Wang CY, Zheng W, Wang T, Xie JW, Wang SL, Zhao BL, Teng WP, Wang ZY (2011) Huperzine A activates Wnt/β-catenin signaling and enhances the nonamyloidogenic pathway in an Alzheimer transgenic mouse model. Neuropsychopharmacology 36:1073–1089

Wanker EE (2008) EGCG redirects amyloidogenic polypeptides into unstructured, off-pathway oligomers. Nat Struct Mol Biol 15:558–566

Wen X, Walle T (2006) Methylated flavonoids have greatly improved intestinal absorption and metabolic stability. Drug Metab Dispos 34:1786–1792

Wenzel E, Somoza V (2005) Metabolism and bioavailability of trans-resveratrol. Mol Nutr Food Res 49:472–481

Whiteman M, Winyard PG (2011) Hydrogen sulfide and inflammation: the good, the bad, the ugly and the promising. Expert Rev Clin Pharmacol 4:13–32

Wightman EL, Reay JL, Haskell CF, Williamson G, Dew TP, Kennedy DO (2014) Effects of resveratrol alone or in combination with piperine on cerebral blood flow parameters and cognitive performance in human subjects: a randomised, double-blind, placebo-controlled, cross-over investigation. Br J Nutr 112:203–213

Williams RJ, Spencer JP (2012) Flavonoids, cognition, and dementia: actions, mechanisms, and potential therapeutic utility for Alzheimer disease. Free Radic Biol Med 52:35–45

Witte AV, Kerti L, Margulies DS, Flöel A (2014) Effects of resveratrol on memory performance, hippocampal functional connectivity, and glucose metabolism in healthy older adults. J Neurosci 34:7862–7870

Wu W, Qiu Q, Wang H, Whitman SA, Fang D, Lian F, Zhang DD (2013) Nrf2 is crucial to graft survival in a rodent model of heart transplantation. Oxid Med Cell Longev 2013:919313

Yan Z, Feng J (2004) Alzheimer's disease: interactions between cholinergic functions and β-amyloid. Curr Alzheimer Res 1:241–248

Yang L, Ye CY, Huang XT, Tang XC, Zhang HY (2012) Decreased accumulation of subcellular amyloid-β with improved mitochondrial function mediates the neuroprotective effect of huperzine A. J Alzheimers Dis 31:131–142

Yang G, Wang Y, Tian J, Liu J-P (2013) Huperzine A for Alzheimer's disease: A systematic review and meta-analysis of randomized clinical trials. PLoS One 8, e74916

Yeung F, Hoberg JE, Ramsey CS, Keller MD, Jones DR, Frye RA, Mayo MW (2004) Modulation of NF-kB-dependent transcription and cell survival by the SIRT1 deacetylase. EMBO J 23:2369–2380

Zaborszky L, Duque A (2000) Local synaptic connections of basal forebrain neurons. Behav Brain Res 115:143–158

Zaveri NT (2001) Synthesis of a 3,4,5-trimethoxybenzoyl ester analogue of epigallocatechin-3-gallate (EGCG): a potential route to the natural product green tea catechin, EGCG. Org Lett 3:843–846

Zhang ZX, Wang XD, Chen QT, Shu L, Wang JZ, Shan GL (2002) Clinical efficacy and safety of huperzine alpha in treatment of mild to moderate Alzheimer disease, a placebo-controlled, doubleblind, randomized trial. Natl Med J China 82:941–944

Zhang SQ, Sawmiller D, Li S, Rezai-Zadeh K, Hou H, Zhou S, Shytle D, Giunta B, Fernandez F, Mori T, Tan J (2013) Octyl gallate markedly promotes anti-amyloidogenic processing of APP through estrogen receptor-mediated ADAM10 activation. PLoS One 8, e71913

Zhao Z, Liu N, Huang J, Lu RH, Xu XM (2011) Inhibition of cPLA$_2$ activation by Ginkgo biloba extract protects spinal cord neurons from glutamate excitotoxicity and oxidative stress-induced cell death. J Neurochem 116:1057–1065

Zhao HH, Di J, Liu WS, Liu HL, Lai H, Lu YL (2013) Involvement of GSK3 and PP2A in ginsenoside Rb1's attenuation of aluminum-induced tau hyperphosphorylation. Behav Brain Res 241:228–234

Zhu S, Zou K, Cai S, Meselhy MR, Komatsu K (2004) Simultaneous determination of triterpene saponins in ginseng drugs by high-performance liquid chromatography. Chem Pharm Bull (Tokyo) 52:995–998

Chapter 10
Summary, Perspective and Direction for Future Research

10.1 Introduction

Alzheimer disease (AD) is a devastating neurodegenerative disorder characterized by the deposition of senile plaques, neurofibrillary tangles, the loss of neurons, and synapses, along with impairment of neuronal functions in AD patients. Plaques and tangles appear in the brain before the onset of the AD symptoms indicating that AD can remain asymptomatic for several years (Savva et al. 2009; Ballard et al. 2011). The causes and pathogenesis of AD remains unknown. However, it is becoming increasingly evident that AD is accompanied by increase in age along with amyloidosis, metal toxicity, oxidative stress and neuroinflammation (Farooqui 2010; Petrou et al. 2012). Neurochemical changes mentioned above are interrelated. They cause disruption of normal neuronal function eventually leading to neural cell death. In addition, neurodegenerative process is complicated by the fact that degenerating neurons also mount prosurvival responses to protect themselves from above neurochemical and neuropathological disease-related changes. This results in a very complex situation in which degenerating neurons go through a struggle between prodeath factors and prosurvival responses. Although, oxidative stress and neuroinflammation are instigators of AD, in reality oxidative stress and neuroinflammation are key features of redox balance (status) of neural cells and immune system functioning to maintain cellular homeostasis, which defends and preserves the integrity of brain tissue and maintains normal cellular function.

Oxidative stress and neuroinflammation are induced and maintained by distinct biochemical cascades, which are closely intertwined and generally function in parallel particularly in the brain, an organ, which plays a critical role in regulating energy balance and due to the presence of high levels of polyunsaturated fatty acids is prone to oxidative stress. The complex interplay between markers of oxidative stress (MDA, acroline, 4-HNE, isoprostanes) and inflammatory mediators (PGs, TNF-α, IL-1β, and IL-6) caused by the long term consumption of high calorie western style diet has been proposed to regulate the progression of chronic neurodegeneration not only in healthy subjects (Francis and Stevenson 2011), but also in many neurodegenerative diseases

© Springer International Publishing Switzerland 2016

A.A. Farooqui, *Therapeutic Potentials of Curcumin for Alzheimer Disease,*
DOI 10.1007/978-3-319-15889-1_10

including AD (Farooqui 2010, 2014). Long term consumption of high calorie western diet leads to changes in neurotransmitters involved in the hedonic appraisal of foods, indicative of an addiction-like capacity of diet high in fat and/or sugar. Importantly, withdrawal of this type of diet leads to a stress-like response. Furthermore, long term consumption of high calorie diet attenuates the physiological effects of acute stress (restraint). All these processes may be associated with risk factors that promote the pathogenesis AD and depression (Farooqui 2014).

In AD, aggregation and accumulation of the amyloid β (Aβ) protein and microtubule associated protein Tau act as 'triggers' for the neurodegenerative process. Aβ accumulates in senile plaques, cerebral vessels, and, to a more limited extent, within neurons (Golde 2006) where as hyperphosphorylated Tau accumulates inside the cells as neurofibrillary tangles and Tau neurites (Lee et al. 2001). Misfolded Aβ and Tau aggregates are known to activate the adaptive immune system. In AD patients and animal models of AD, neuroinflammation appears in the earliest stages of the disease process. This may be related to microglial activation and increase in levels of proinflammatory mediators including cytokines, chemokines, complement components, increase in proinflammatory eicosanoids, and free radicals (Yoshiyama et al. 2007). Thus, converging hypotheses to explain the pathogenesis of AD include interactions among Aβ-mediated oxidative stress, cytokine and chemokine-mediated neuroinflammation, protein aggregation-mediated mitochondrial dysfunction, glutamate receptor-mediated elevation in calcium, proteasomal dysfunction, decrease in blood flow along with alterations in blood brain barrier, and onset of cognitive decline (Golde 2009; Farooqui 2010; Zhu et al. 2013). However, placing these pathways in the proper order and relationship to the onset, time course, and progress of neurodegeneration and its relationship to cytoskeletal pathology in the brain are challenging issues on which progress is urgently needed (Golde 2009). It must be recognized that neurochemical changes mentioned above are interrelated supporting the view that neurodegenerative process in AD is multifactorial, which is complicated by the fact that degenerating neurons also mount prosurvival responses to protect themselves from above neurochemical and neuropathological disease-related changes. This results in a very complex situation in which degenerating neurons go through a struggle between prodeath factors and prosurvival responses causing the disruption of normal neuronal function, but eventually leading to neural cell death.

10.2 Curcumin as Therapeutic Agent for AD

The number of AD patients in the developed countries and United States is expected to double and triple in both India and China over the next three decades (Ferri et al. 2005). Currently there are no effective means to treat AD or even to slow its progression. Studies using MRI and PET brain scans have revealed early signs of AD pathology in patients appear ~4–17 years before the onset of dementia (Villemagne et al. 2013). Some drugs are available to treat symptoms of AD. These drugs include acetylcholinesterase (AChE) inhibitors (tacrine, donepezil, rivastigmine, metrifonate, and

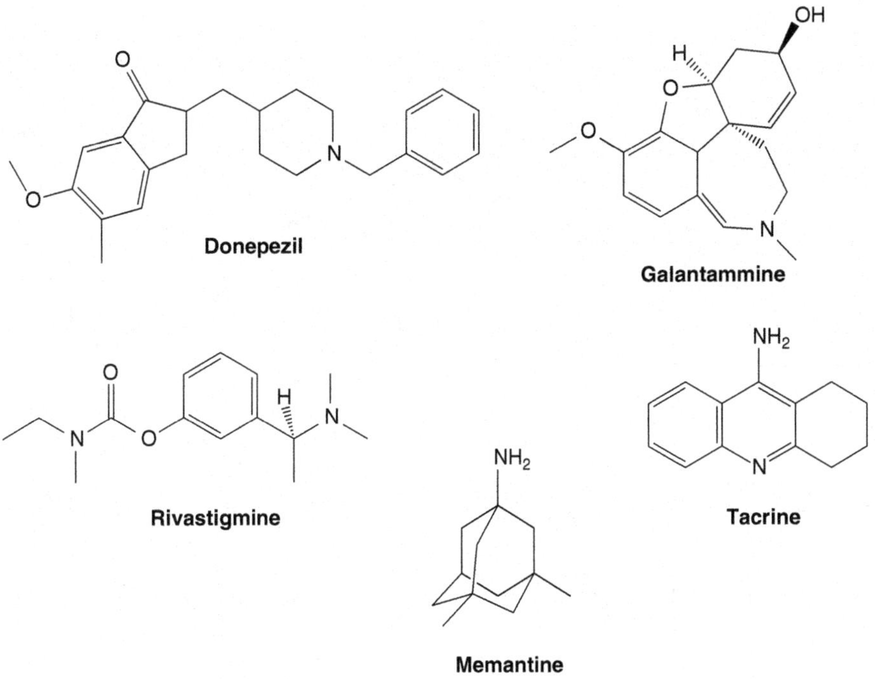

Donepezil

Galantammine

Rivastigmine

Memantine

Tacrine

Fig. 10.1 Chemical structures of tacrine, donepezil, rivastigmine, metrifonate, galantamine and memantine

galantamine) and the NMDA receptor antagonists (memantine) (Fig. 10.1). AChE inhibitors and memantine have only modest effects in delaying the progression of AD (Hansen et al. 2007; Raina et al. 2008). Furthermore, the tolerability of these drugs is compromised by their side effects, for instance dizziness, anorexia, vomiting, and diarrhea (Alva and Cummings 2008). Over the past two decades, considerable efforts have been made to develop safe and effective pharmacological treatment of AD (Yao and Xue 2014). These efforts have failed and therefore the treatment of AD still remains a great challenge. The development of new strategies is an active area of research on this topic. Epidemio-logical studies have revealed that in India, where dietary curcumin is consumed daily in the form of curry than in the United States, the morbidity rate attributed to AD for Indian elders (70–79 years old) is 4.4 times lower compared to the same age group of Americans (Jorm and Jolley 1998; Ganguli et al. 2000). The consumption of curry containing food by healthy elderly individuals results in a better cognitive performance (Ng et al. 2006) than seniors who did not consume curry (Ng et al. 2006). For these reasons, curcumin and its derivatives have been extensively investigated for their potential for AD prevention over decades in animal model and cell culture systems (Gupta et al. 2012; Valera et al. 2013; Jayaraj et al. 2014).

As mentioned in earlier chapters, curcumin is a highly promising natural polyphenol with antioxidant and anti-inflammatory properties (Menon and Sudheer 2007; Gupta et al. 2012). It can not only cross blood brain barrier (BBB), but can pass

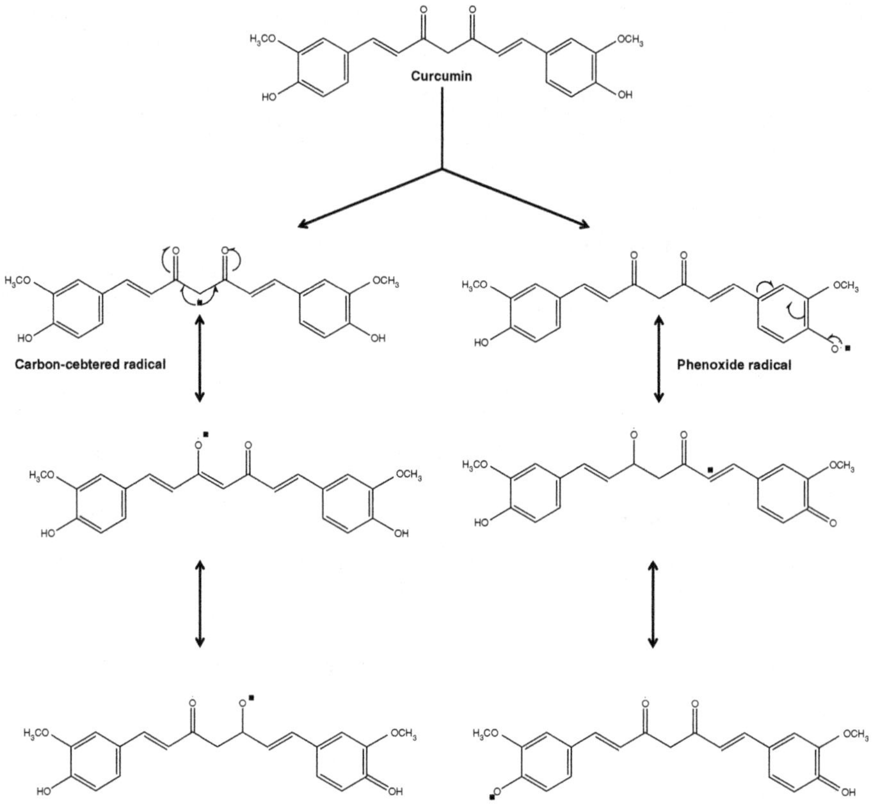

Fig. 10.2 Sites contributing to radical scavenging activity and antioxidant effects of curcumin

through all cell membranes inducing its intracellular effects. The antioxidant activity of curcumin is due to its chemical structure. The phenolic and the methoxy group on the phenyl ring and the 1,3-diketone system contribute to the antioxidant activity of curcumin (Fig. 10.2). As stated above, the antioxidant activity of curcumin increases when the phenolic group with a methoxy group is at the ortho position (Itokawa et al. 2008; Motterlini et al. 2000). The orthomethoxy group can form an intramolecular hydrogen bond with the phenolic hydrogen, making the H-atom abstraction from the orthomethoxyphenols surprisingly easy (Chen et al. 2011). The H abstraction from these groups is responsible for the remarkable antioxidant activity of curcumin. Moreover, the reactions of curcumin with free radicals produce a phenoxyl radicals and a carbon-centered radical at the methylene CH_2 group (Barzegar and Moosavi-Movahedi 2011). The degradation of curcumin in vivo produces smaller phenols like ferulic acid (*trans*-4-hydroxy-3-methoxycinnamic acid) (Priyadarsini et al. 2003; Ghosh et al. 2015), which is capable of mediating neuroprotective effects. Recently, investigators are focusing their attention on modifications of the basic structure of curcumin for examining the effect of these changes on Aβ aggregation, neuroinflammation, and Aβ-induced neurotoxicity (Begum et al. 2008; Narlawar et al. 2008; Park and Kim 2002; Priyadarsini 2013). Results have indicated that replacement of the 1,3-dicarbonyl moiety in curcumin with isosteric isoxazoles and pyrazoles generated compounds that inhibit γ-secretase activity (Narlawar et al. 2007) and prevented both Aβ and Tau aggregation (Narlawar et al. 2008). More modest alterations in the curcumin structure have no effects on neuroprotective activity toward Aβ-induced neurotoxicity (Park and Kim 2002; Lee et al. 2011); however, some changes, such as saturation of the 7-carbon linker to generate tetrahydrocurcumin, abolish Aβ aggregation inhibitory activity, but retained anti-neuroinflammation activity (Begum et al. 2008). Although these observations clearly indicate that the base structure of curcumin can be modified without compromising certain properties of its bioactivity, none of the compounds tested display significant improvement as Aβ aggregation inhibitors when compared to native curcumin. To further explore if modifications to the native structure of curcumin can result in the identification of improved inhibitors of Aβ aggregation, more analogs of curcumin have been have been synthesized with various modifications and substitutions on the phenolic rings, varying degrees of unsaturation of the spacer between aromatic rings, as well as compounds that contain either 5- or 7-carbon spacers to determine if spatial variations between phenols affects anti-Aβ aggregation activity (Weber et al. 2005; Lee et al. 2011). Studies on new curcumin analogs have indicated that anti-oligomerization activity of curcumin and its analogs is not only modulated by (a) at least one enone group in the spacer between aryl rings is necessary for measureable anti-Aβ aggregation activity, (b) an unsaturated carbon spacer between aryl rings is essential for inhibitory activity, but also by methoxyl and hydroxyl substitutions in the meta- and para-positions on the aryl rings in curcumin. The best anti-oligomerization activity is mediated by analogs that have either their meta- and para-substituted methoxyl and hydroxyl groups or methoxyl or hydroxyl groups placed in both positions. The simple substitution of the para-hydroxy group on curcumin with a methoxy substitution improves inhibitor function by sixfold to sevenfold over that measured for curcumin (Ray et al. 2011; Orlando et al. 2012).

As stated in Chap. 1, elevated levels of redox active Fe^{3+} and Cu^{2+} in AD generate ROS causing DNA damage in neural cell, by producing hydroxyl and superoxide radicals via Fenton reaction. Curcumin is a powerful scavenger of the superoxide anion and the hydroxyl radical. Curcumin has ability to bind Cu^{2+} and Fe^{3+} and form tight and inactive complexes (Fig. 10.3), and can protect neural cell DNA against ROS and singlet oxygen-induced strand breaks (Kim et al. 2005a). In the scavenging ability the position and the number of phenolic –OH groups play a role through donation of a hydrogen atom from their hydroxyl groups to radicals, resulting in radical moiety elimination and in formation of inactive complexes with divalent metal ions. During this reaction, phenoxyl radical is formed that can form stable compound and terminates radical reaction via reaction with another radical. Both ROS and metals produce a multitude of oxidative modifications in DNA bases and sugar moieties, including DNA strand breaks. Persistent accumulation of this damage may lead to secondary double-strand breaks, which are the most toxic genomic damage (Tung et al. 2012). Redox-active Fe^{3+} and Cu^{2+} also specifically inhibit key

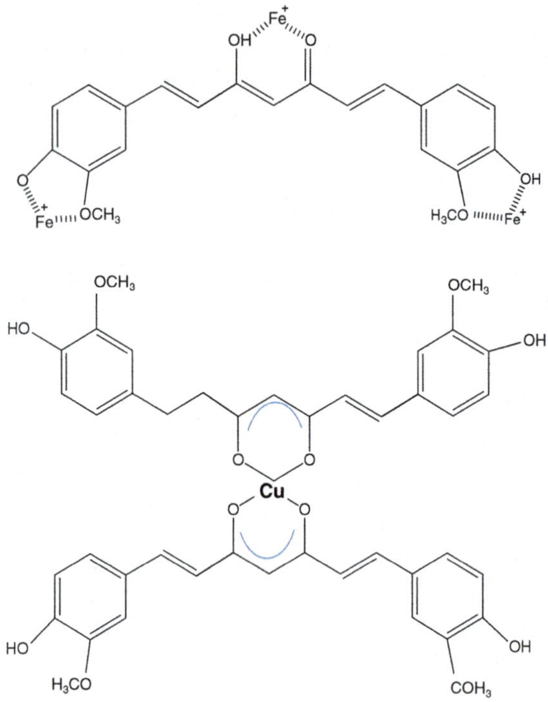

Fig. 10.3 Iron and copper chelation by curcumin

enzymes in the oxidative genome damage machinery by both altering their structure and reversibly oxidizing cysteine residues in these proteins. Curcumin also decreases levels of mediators associated with the initiation and maintenance of oxidative stress. In a neuronal cell culture, curcumin suppresses Aβ formation by downregulating BACE-1 (Shimmyo et al. 2008; Lin et al. 2008). Thus, at 30 μM curcumin attenuates the production of Aβ-mediated ROS formation and 20 μM curcumin blocks structural changes in Aβ toward β-sheet-rich secondary structures (Shimmyo et al. 2008; Lin et al. 2008).

When fed to aged mice with advanced plaque deposits similar to those of AD, curcumin not only reduces the amount of plaque deposition in mice cortex and hippocampus, but also decreases levels of proinflammatory cytokines (Begum et al. 2008; Yang et al. 2005) and increases levels of anti-inflammatory cytokine (IL-4) in microglial cell cultures from curcumin treated mice (Shytle et al. 2012). This cytokine not only reduces the production of TNF-α and MCP-1 (Chao et al. 1993), but also inhibits microglial cell activation and subsequently inflammation caused by Aβ (Lyons et al. 2007). It is also reported that sustained expression of IL-4 reduces Aβ oligomerization and deposition and improves neurogenesis by increasing the expression of BDNF (Fig. 10.4) (Kiyota et al. 2010), supporting the view that curcumin may retard neuroinflammation and neuronal cell death via IL-4 production. This increase in IL-4 may also promote the clearance of Aβ in vivo leading to a decrease

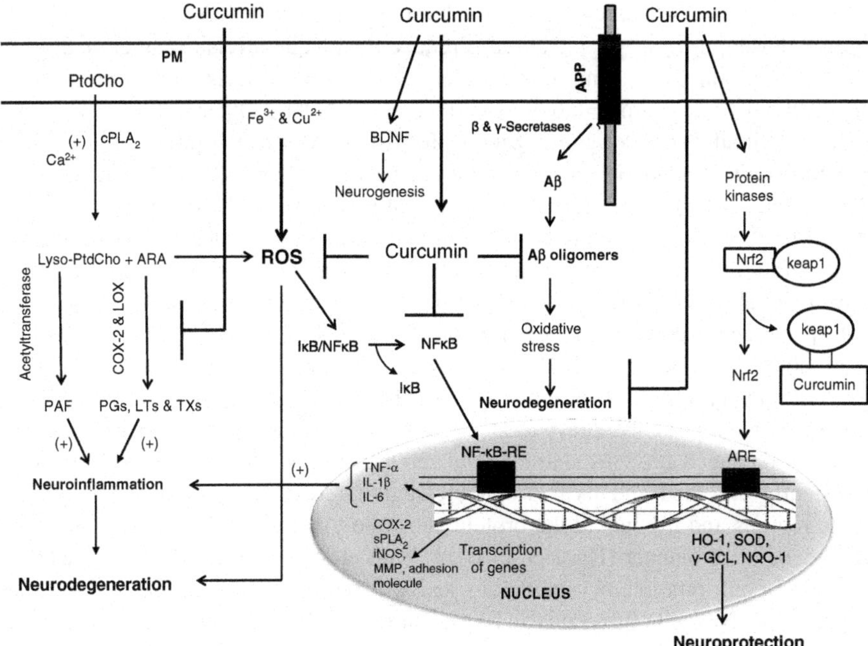

Fig. 10.4 Hypothetical diagram showing molecular mechanism involved in antioxidant, antiamyloidogenic, and anti-inflammatory effects of curcumin. Plasma membrane (*PM*); phosphatidylcholine (*PtdCho*); lyso-phosphatidylcholine (*Lyso-PtdCho*); arachidonic acid (*ARA*); cytosolic phospholipase A_2 (*cPLA$_2$*); cyclooxygenase (*COX-2*); lipoxygenase (*LOX*); platelet activating factor (*PAF*); prostaglandinds (*PGs*); leukotrienes (*LTs*); thromboxanes (*TXs*); reactive oxygen species (*ROS*); nuclear factor kappaB (*NF-κB*); nuclear factor kappaB response element (*NF-κB-RE*); inhibitory subunit of NFκB (*IκB*); tumor necrosis factor-α (*TNF-α*); interleukin-1β (*IL-1β*); interleukin-6 (*IL-6*); secretory phospholipase A2 (*sPLA2*); inducible nitric oxide synthase (*iNOS*); matrix metalloproteinase (*MMP*); amyloid precursor protein (*APP*); amyloid β (*Aβ*); NF-E2 related factor 2 (*Nrf2*); kelch-like erythroid Cap'n'Collar homologue-associated protein 1 (*Keap1*); antioxidant response-element (*ARE*); heme oxygenase (*HO-1*); superoxide Dismutase (*SOD*); NADPH quinine oxidoreductase (*NQO-1*); and γ-glutamate cysteine ligase (*γ-GCL*)

in neurotoxicity (Shytle et al. 2012). Antioxidant properties of curcumin are promoted by the upregulation of PtdIns 3K/AKT/Nrf2 pathway (Fig. 10.4). This pathway can be blocked by 2-(4-morpholinyl)-8-phenyl-4H-1-benzopyran-4-one (LY294002), a selective PtdIns 3K inhibitor, supporting the view that PtdIns 3K/AKT/Nrf2 pathway plays an important role in curcumin mediated cytoprotection (Li et al. 2007). Furthermore, curcumin-mediated increase in PtdIns 3K/AKT/Nrf2 pathway can also be reversed by the Nrf2 siRNA strongly indicating that cytoprotection after curcumin treatment can also be mediated by the PtdIns 3K/AKT/Nrf2 signaling pathway (Yin et al. 2012). Collective evidence suggests that curcumin not only reduces oxidative damage and neuroinflammation, but also reverses the amyloid pathology in an AD transgenic mouse (Lim et al. 2001; Yang et al. 2005; Ye and Zhang 2012; Potter 2013).

Direct injections of curcumin into the brains of the mice with AD not only hamper further development of plaque but also reduce the plaque levels (Yang et al. 2005). Curcumin produces a strong inhibitory effect on Aβ fibril formation in an in vitro assay (IC50=0.25 µg/mL=0.679 µM) (Kim et al. 2005b). Other studies report that curcumin inhibits Aβ oligomerization (IC50=361.11±38.91 µM) but does not inhibit fibrillization in vitro at concentrations between 30 and 300 µM (Necula et al. 2007). The reason for this discrepancy is not fully understood. However, it is possible that differences in concentration of curcumin used in these studies may contribute to concentration-dependent multiphasic behavior on modulation of Aβ aggregation (Necula et al. 2007). Another major curcumin-mediated defense against intraneuronal Aβ aggregate formation is the induction of heat shock proteins (HSPs) (Scapagnini et al. 2006; Yuan et al. 2008; Hu et al. 2015). HSPs are associated with folding of nascent polypeptides to their appropriate conformation and refolding of mild denatured/damaged proteins, prevention of protein aggregation, and degradation of severely damaged proteins (Parodi et al. 2006). As a protein is synthesized, it is transiently unfolded and its hydrophobic regions are exposed. Hsp70 recognizes these regions and it binds to the protein substrate via its peptide-binding site in an ATP-dependent manner (Hartl et al. 2011; Kim et al. 2013). Hsp70 promotes this extended conformation by stabilizing and preventing premature misfolding and aggregation. Next, the substrate can be transferred to another chaperone system, such as the chaperonins (a large, cylindrical complexes that provide a central compartment for a single protein chain to fold unimpaired by aggregation), where folding takes place and a three-dimensional structure is acquired (Hartl et al. 2011; Kim et al. 2013). HSPs also help to degrade proteins by delivering them to the ubiquitin proteasome system (Fang et al. 2005). Induction of HSPs by curcumin may limit β-amyloid accumulation and protect against amyloid peptide-mediated neurotoxicity (Maiti et al. 2014). Regular supplementation of curcumin not only inhibits neuroinflammation, but also improves cognitive function (Dong et al. 2012). Additionally, curcumin administration in various animal models of memory impairment reverses memory deficits (Dong et al. 2012). These outcomes may be due to curcumin's effects on oxidative stress, expression of BDNF, extracellular signal-regulated kinase (ERK)/P38 signaling pathways and degradation of PKCδ (Dong et al. 2012).

Curcumin has been potentially used for retarding or delaying the onset of AD not only in animal models of AD, but also in animal models of other neurological disorders including stroke, spinal cord injury, Parkinson disease, and depression (Valera et al. 2013; Jayaraj et al. 2014). In animal models of above mentioned neurological disorders, curcumin acts by downregulating multiple molecular signaling pathways associated with the pathogenesis of neurotraumatic and neurodegenerative diseases. It exerts antioxidant, antiinflammatory, antiamyloidogenic, antiangigenic, antiproliferative, and antiapoptotic effects in the brain. As stated above, curcumin can pass through BBB and neural cell membranes and induce its intracellular effects. It inhibits cyclooxygenase-2 (COX-2), 5-lipoxygenase (5-LOX), phospholipases, transcription factors (AP-1 and NF-κB) and other enzymes involved in metabolizing the neural membrane phospholipids into prostaglandins, a group of lipid mediators, which contribute to neuroinflammation (Farooqui et al. 2006; Ong et al. 2015) (Fig. 10.5).

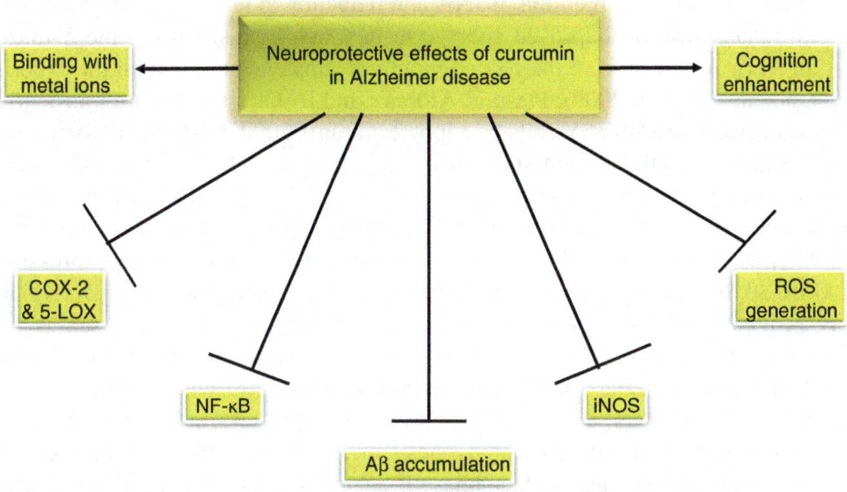

Fig. 10.5 Antioxidant, antiamyloidogenic, and anti-inflammatory effects of curcumin

Furthermore, curcumin suppresses the activation of NF-κB leading to downregulation of proinflammatory cytokines expression (TNF-α and IL-1β) (Kim et al. 2005a). In PBM and THP-1 cells, curcumin (12.5–25 μM) retards early growth response-1 (Egr-1) activation, which increases the expression of cytokines (TNF-α and IL-1β) and chemokines (MIP-1β, MCP-1, and IL-8) in monocytes through the interactions with Aβ or fibrillar Aβ (Giri et al. 2004), and reduces the expression of proinflammatory cytokines and chemokines (Giri et al. 2004) supporting the view that curcumin is a natural anti-inflammatory agent.

The potential effect curcumin on Tau metabolism has not been studied as extensively as the anti-amyloidogenic effects (Yin et al. 2012). However, it is known that coincubation of curcumin with Aβ in PC12 cells attenuates the effects of Aβ on the phosphorylation of Tau at serine 202. Curcumin also modulates the phosphorylation of Tau independent from Aβ. It is also reported that curcumin is a potent inhibitor of GSK3β, with an IC$_{50}$ value of 66.3 nM compared to thiadiazolindione 8, which is a standard inhibitor of GSK3β (IC$_{50}$ of 1.5 μM). Furthermore, based on stimulated molecular docking studies, it is suggested that curcumin fits within the binding pocket of GSK3β via interactions with key amino acids (Bustanji et al. 2009). As GSK3β is one of the kinase associated with the phosphorylation of Tau, the inhibition of GSK3β by curcumin may be one possible mechanism by which curcumin reduces Tau phosphorylation. Dietary supplementation of 0.05 % curcumin for 4 months not only suppresses the c-jun N-terminal kinase (JNK) activity, but also significantly reduces levels of phosphorylated Tau in 5 month old 3xTg-AD mice (Ma et al. 2009). JNK is a MAPK that phosphorylates tau at serine 422 and is implicated in the progression of neurodegeneration (Ma et al. 2009). Curcumin inhibits JNK (Suh et al. 2007; Zhang et al. 2012), indicating an alternate pathway through which curcumin can induce inhibition of Tau phosphorylation. It is reported that the curcumin dependent suppression of JNK and reduction in levels of phosphorylated Tau correlates with an improved short-term memory, assessed using the Y-maze test, in 3xTg-AD mice from the second month of supplementation until the experiment ended (Ma et al. 2009). Plasmid APPswe and BACE1-mychis are transiently co-transfected into SHSY5Y cells by Liposfectamin™2000. RT-PCR, Western blot assay, and immunofluorescent staining of curcumin treated cells shows that curcumin treatment results in decreased expression of GSK-3β mRNA and protein in the transfected cells (Zhang et al. 2011a). It is proposed that curcumin activates the Wnt/β-catenin signaling pathway not only through downregulating the expression of GSK-3β, but also by inducing the expression of β-catenin and CyclinD$_1$, which may provide a new theory for the treatment of neurodegenerative diseases by Curcumin (Zhang et al. 2011b). Collectively, these studies indicate that curcumin fulfills the characteristics for an ideal neuroprotective agent with its low toxicity, affordability, and easy accessibility. Curcumin decreases most neurochemical parameters associated with the pathogenesis of AD (Table 10.1) Neuroprotective effects of curcumin can also induce phase II enzymes in astrocytes and heme oxygenase-1 in neurons in vitro (Jiao et al. 2006). Finally, as stated above, one of the major defenses against intraneuronal protein aggregate formation, which occurs in AD is the induction of heat shock proteins (HSPs) that functions as molecular chaperones to block protein aggregate formation (Scapagnini et al. 2006; Hu et al. 2015). Increased HSP expression from transgenes clearly protects from neurotoxicity arising from intraneuronal protein aggregates (Ohtsukaand and Suzuki 2000).

10.3 Pharmacokinetics of Curcumin Metabolism

Pharmacokinetic studies have indicated that curcumin undergoes rapid metabolism through glucuronidation and sulfation in the liver and then excretion in the feces, which accounts for its poor systemic and brain bioavailability. It does not produce any adverse effects, even up to doses as high as 8 g/day in humans. The clinical application of curcumin for the treatment of AD and other neurological disorders in humans is compromised not only due to its poor absorption, but also by its bioavailability in vivo. This suggests that in order to make curcumin as a viable therapeutic agent for AD one needs to address two shortcomings of curcumin, one being its low bioavailability and the other concerning its rapid metabolism. Attempts have been made to solve these problems by adopting two strategies: (a) employing novel drug delivery systems and (b) synthesizing curcumin analogs through modification of its structural motif. The bioavailability of curcumin through delivery systems can be increased using liposomes, micelles, poly(lactic-*co*-glycolic acid) (PLGA) nanoparticles, polysaccharides, phospholipid complexes, inhalant, and nanoparticles that can not only result in a physiological distribution in visceral tissues, but can also overcome BBB permeability (Fig. 10.6) (Wang et al. 2014a; McClure et al. 2015). These delivery systems have successfully improved the aqueous solubility and chemical stability of curcumin (Xie et al. 2011; Nair et al. 2012). In addition, these delivery systems exhibit a sustained release of curcumin in various tissues (Xie et al. 2011; Nair et al. 2012). After oral administration of 1 g/kg

Table 10.1 Effects of curcumin on neurochemical parameters in Alzheimer disease

Neurochemical parameters in Alzheimer disease	Effect of curcumin treatment	Reference
Increase in amyloidgenesis	Decrease in amyloidgenesis	Liu et al. (2010)
Induction of β-sheet formation	Decrease in β-sheet formation	Orlando et al. (2012)
Increase in neurodegeneration	Decrease in neurodegeneration	Kim et al. (2001)
Increase in NF-κB activation	Decrease in NF-κB activation	Kuner et al. (1998)
Activation in ERK1/2 activity	Decrease in ERK1/2 activity	Giri et al. (2004)
Enhancement in γ-secretase activity	Decrease in γ-secretase activity	Zhang et al. (2011a, b)
Enhancement in TNF-α, IL-1β & IL-6 levels	Decrease in TNF-α, IL-1β & IL-6 levels	Lim et al. (2001)
Enhancement in GSK-3β activity	Decrease in GSK-3β activity	Zhang et al. (2011a, b)
Increase in caspase-3 activity	Decrease in caspase-3 activity	Qin et al. (2010)
Increase in apoptotic cell death	Decrease in apoptotic cell death	Qin et al. (2010)
Basal levels of HSPs	Induction of HSPs	Maiti et al. (2014)
Increase in levels of Fe^{3+} and Cu^{2+}	Decrease in levels of Fe^{3+} and Cu^{2+}	Jiao et al. (2006, 2009)
Phase II detoxification enzymes	Induction of Phase II detoxification enzyme	Jiao et al. (2006, 2009)
Elevation in cholesterol	Decrease in cholesterol	Alwi et al. (2008)

Extracellular signal-regulated kinase-1/2 (*ERK1/2*); Glycogen synthase kinase-3β (*GSK-3β*), tumor necrosis factor-α (*TNF-α*); interleukin-1β (*IL-1β*), nuclear factor-kappaB (*NF-κB*); heat shock proteins (*HSPs*); and phase II detoxification enzymes (superoxide dismutase, glutathione peroxidases, heme oxygenase-1, and NAD(P)H:quinone oxidoreductase-1)

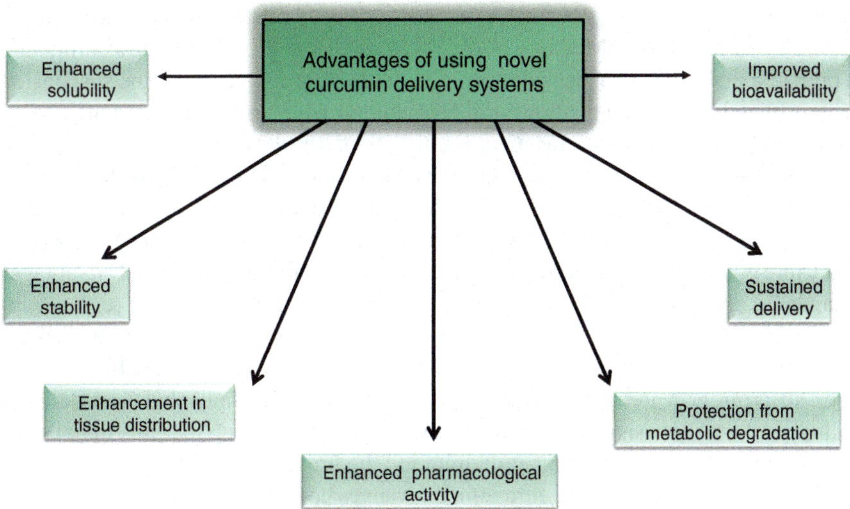

Fig. 10.6 Advantages of curcumin delivery by various formulations used for curcumin delivery to the brain and other visceral tissues

body weight and 2 g/kg body weight of curcumin to rats, the peak blood concentrations detected in the blood were 1.4 µM and 3.7 µM, respectively (Wahlstrom and Blennow 1978; Anand et al. 2007). In humans, the consumption of 4–12 g of curcumin gives the peak blood curcumin concentrations of about 4.0 µM (Anand et al. 2007; Lao et al. 2006). After encapsulating curcumin in various delivery systems, the oral bioavailability can be increased by more than twofold (Gao et al. 2012; Yu and Huang 2012; Kakkar et al. 2011; Shaikh et al. 2009). After intravenous administration of nano-curcumin formulations, blood curcumin concentrations and its circulation time are also significantly increased (Mohanty and Sahoo 2010; Anand et al. 2010; Tsai et al. 2011; Gota et al. 2010; Dadhaniya et al. 2011; Ghalandarlaki et al. 2014).

10.4   Curcumin Analogs as Imaging Probes

Recently, inhalable form of curcumin has been developed not only as therapeutic agent, but also as an intrinsic fluorescence signal that can be used to image its binding optically with Aβ plaques in the brain (McClure et al. 2015). For example, the binding of curcumin with Aβ and clearance of the existing plaques in a transgenic mouse model of AD can be detected by two-photon microscopy (Garcia-Alloza et al. 2007; Yang et al. 2005). Other investigators have used radiolabeled curcumin derivatives for Aβ imaging (Ryu et al. 2006). Recently, perfluoro curcumin analog (FMeC1) has been synthesized and used for ^{19}F NMR imaging (Fig. 10.7) to facilitate in vivo visualization of inhalable curcumin binding with Aβ (Yanagisawa et al. 2011). As stated above, the clinical use of curcumin is severely limited due to its poor aqueous solubility and relatively low penetration across the BBB (Anand et al. 2007; Prasad et al. 2014). Physicochemical properties of curcumin analogs are not very different from native curcumin. This makes curcumin analogs unsuitable for human trials (Yanagisawa et al. 2014). Recently, attempts have been made to deliver curcumin to the brain via inhalation (McClure et al. 2015). In this approach, FMeC1 compound is prepared in the same suspension formulation used for iv injection. The suspension is aerosolized using a center-flow atomizer, diluted with air and subsequently delivered by nose-only inhalation (McClure et al. 2015). Results indicate that FMeC1 can be delivered across BBB. Its concentration can be detected by ^{19}F NMR. In the brain inhaled FMeC1 analog is colocalized with immuno-stained Aβ plaques in the cortex and hippocampal regions of the 5XFAD mouse brain under fluorescence microscopy (McClure et al. 2015). The delivery of curcumin by inhalation has several advantages such as (a) effective delivery of curcumin to the pathologically relevant region (hippocampus); (b) protection from the glucuronidation and biotransformation (Mistry et al. 2009; Ringman et al. 2012); (c) protection from the low bioavailability due to poor absorption in the gut and (d) negligible toxicity. During inhalation delivery animals do not show signs of toxicity or discomfort, suggesting that this may be the most efficient way of delivering curcumin into the brain (Subhashini et al. 2013).

In addition to FMeCl, other curcumin derivatives have also been synthesized, characterized, and used for the treatment and in vivo imaging of Aβ in AD. These curcumin derivatives (CNB001, curcumin-glucoside, curcumin isozole, and curcumin

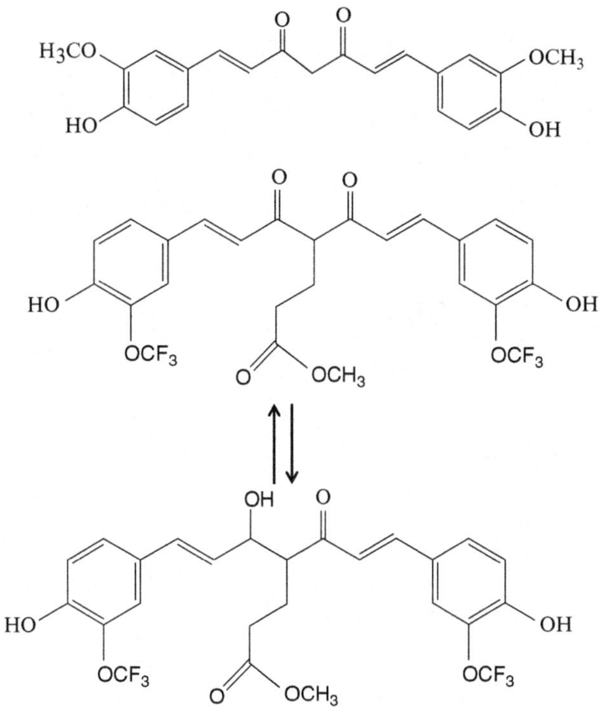

Curcumin 1 (keto form)

Perfluoro curcumin analog

Perfluoro curcumin analog

Fig. 10.7 Chemical structures of curcumin and perfluoro curcumin analogs

pyrazole) not only show improved stability, better BBB permeability, and potent anti-amyloidogenic properties than native curcumin (Fig. 10.8) (Liu et al. 2008; Ran et al. 2009; Zhang et al. 2013). These derivatives have been used as neuroprotective agents not only in animal models of AD, but also in animal model of stroke and PD (Liu et al. 2008; Valera et al. 2013; Jayaraj et al. 2014; Ahsan et al. 2015).

Synthesis of curcumin analogs (CRANAD-58, CRANAD-17, and CRANAD-2) for in vivo imaging of insoluble Aβ species by fluorescent and near-infrared fluorescence (NIRF) imaging indicates that these probes not only attenuate the cross-linking of Aβ induced by copper, but also induce significant fluorescence property that changes upon mixing with both soluble and insoluble Aβ species in vitro (Ran et al. 2009; Zhang et al. 2013; Cui et al. 2014). In vivo NIRF imaging have indicated that CRANAD-58 is capable of differentiating transgenic with wild type mice as young as 4-months old, the age that lacks apparently visible Aβ plaques and Aβ is likely in its soluble forms (Ran et al. 2009; Zhang et al. 2013). However, its sensitivity for in vivo imaging of Aβ is lower than that of newly synthesized NIRF analog called CRANAD-3. This new analog is capable of detecting soluble and insoluble Aβ species in vivo with greater sensitivity (Zhang et al. 2015). In vivo imaging studies have indicated that the NIRF signal of CRANAD-3 from 4-month-old transgenic AD (APP/PS1) mice is 2.29-fold higher than that from age-matched wild-type mice, suggesting that CRANAD-3 is capable of detecting early molecular pathology (Zhang et al. 2015). To verify the feasibility of CRANAD-3 for monitor-

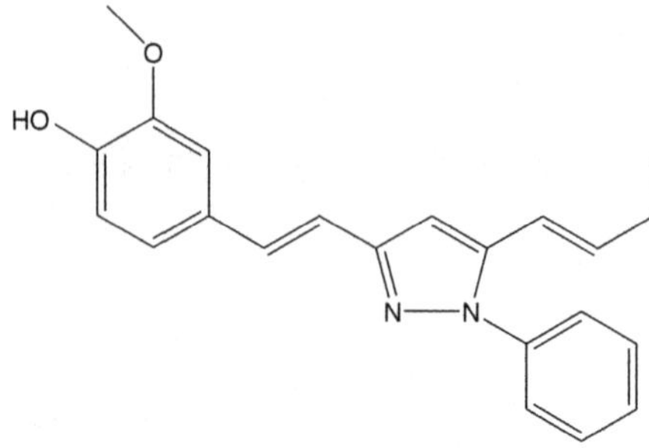

CNB001

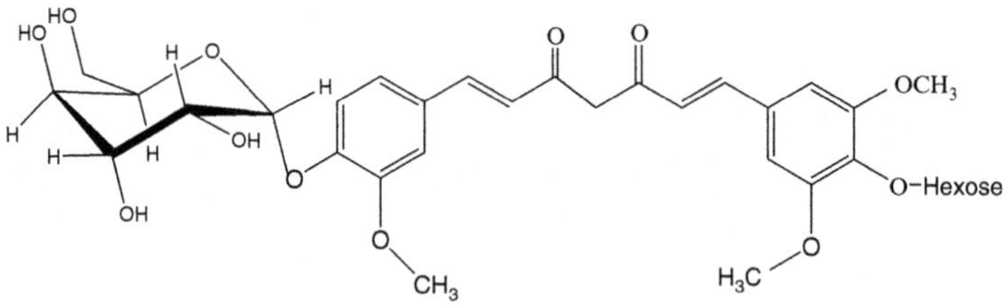

Curcumin-bis-αD-glucoside

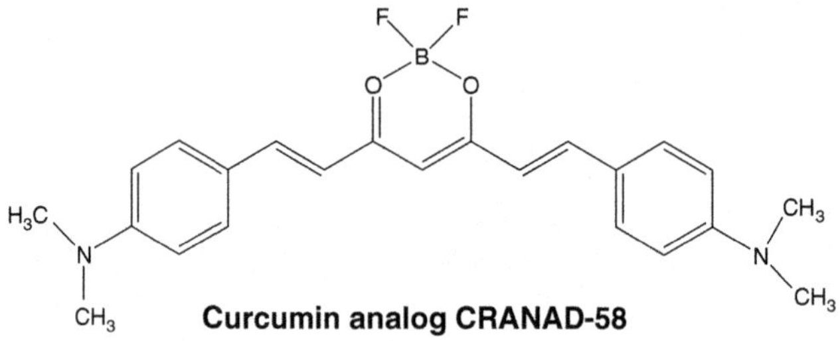

Curcumin analog CRANAD-58

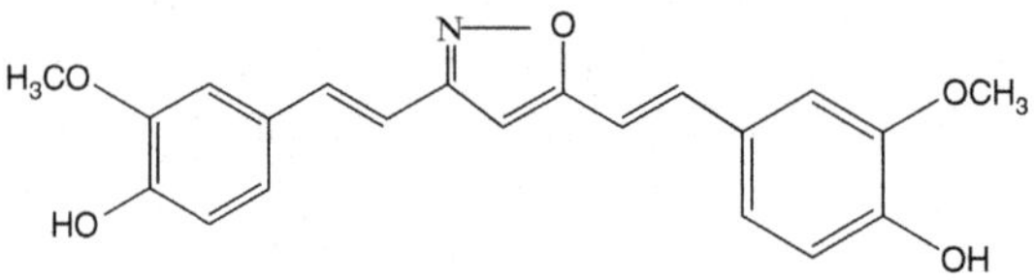

Curcumin isozole

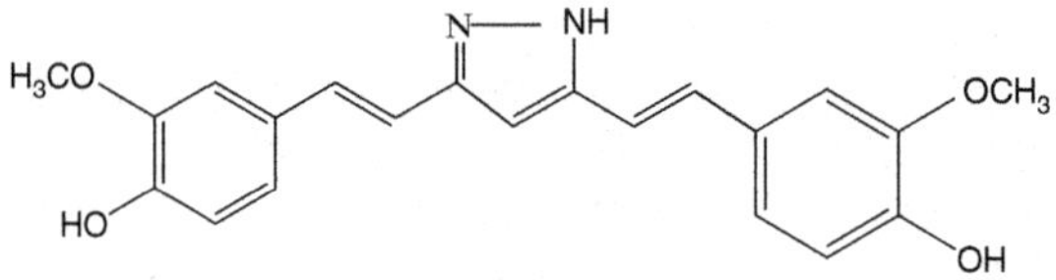

Curcumin pyrazole

Fig. 10.8 Chemical structures of recently synthesized curcumin analogs ing therapy, LY2811376 has been used to lower Aβ in APP/PS1 mice and it is shown that CRANAD-3 can be used to monitor the decrease in Aβ after LY2811376 treatment. To confirm the imaging capacity of CRANAD-3, therapeutic effect of CRANAD-17 was studied on Aβ cross-linking. The imaging data indicated that the fluorescence signal in the CRANAD-17–treated group is significantly lower than that in the control group, and the result correlate with ELISA analysis of brain extraction and Aβ plaque counting. These observations support the view that NIRF can be used to monitor AD therapy (Zhang et al. 2015).

After crossing BBB in brains of animal models of AD, curcumin accumulates in the local vicinity of senile plaques and reduces plaque burden by preventing the formation of new deposits, clearing existing deposits, and reducing the size of remaining deposits (Garcia-Alloza et al. 2007) through the reduction in the expression of the γ-secretase component presenilin-2; and increase in the expression of β-amyloid-degrading enzymes, including insulin-degrading enzymes and neprilysin (Wang et al. 2014b). Curcumin not only binds and inhibits the aggregation of cerebral Aβ peptide, but also blocks ROS-mediated activation of NF-κB (Mukhopadhyay et al. 2002; Wyss-Coray et al. 2001; Yang et al. 2005). Curcumin treatment also reverses the change of synaptophysin and postsynaptic density 95 (PSD-95) as well as impairs the performance in water maze test (Ringman et al. 2005). Curcumin binds to senile plaques in the brain tissues when fed or injected in the carotid artery in Tg2576 mice (Ringman et al. 2005). Three months curcumin administration in animal models of AD not only results in reduction in Aβ levels and decrease in Aβ aggregation of Aβ in the mouse hippocampal CA1 (Fig. 10.9), but also produces reduction in the expression of the γ-secretase component presenilin-2; and increase in the expression of β-amyloid-degrading enzymes, including insulin-degrading enzymes and neprilysin (Wang et al. 2014b). Recent studies have indicated that standardize turmeric extract, HSS-888, shows strong inhibition of Aβ aggregation and secretion in vitro, suggesting that HSS-888 can be used for the treatment of AD. Treatment of Tg2576 mice, which over-expressing Aβ protein for 6 months with HSS-888 (5mg/mouse/day) results in significant reduction in brain levels of soluble (~40 %) and insoluble (~20 %) Aβ as well as phosphorylated Tau protein (~80 %) compared to a control mice, which consumed customized animal feed pellets (0.1 % w/w treatment). In addition, treatment of primary cultures of microglia from these mice shows increase in expression of the cytokines IL-4 and IL-2. In contrast, tetrahydrocurcumin treatment only weakly reduces phosphorylated Tau protein and failed to significantly alter plaque burden and cytokine expression. These findings support the view that the development of optimized turmeric extract HSS-888 represents an important step in phytochemical based therapy for AD (Shytle et al. 2012). Finally, curcumin regulates lipid metabolism, which can affect the activity of lipoprotein related metabolic enzymes, increases the content of lipoprotein, mobilizes the reverse transportation of cholesterol and accelerates the cholesterol metabolism pathway to decrease total cholesterol, triglyceride, low density lipoprotein cholesterol and increase high density lipoprotein cholesterol (Alwi et al. 2008). These results along with lack of curcumin toxicity support the view that curcumin qualifies for the treatment of AD. To properly evaluate clinical relevance of curcumin and its analogs in humans, large, double blind trials with longer duration have to be performed. Importantly, a better understanding of both effective

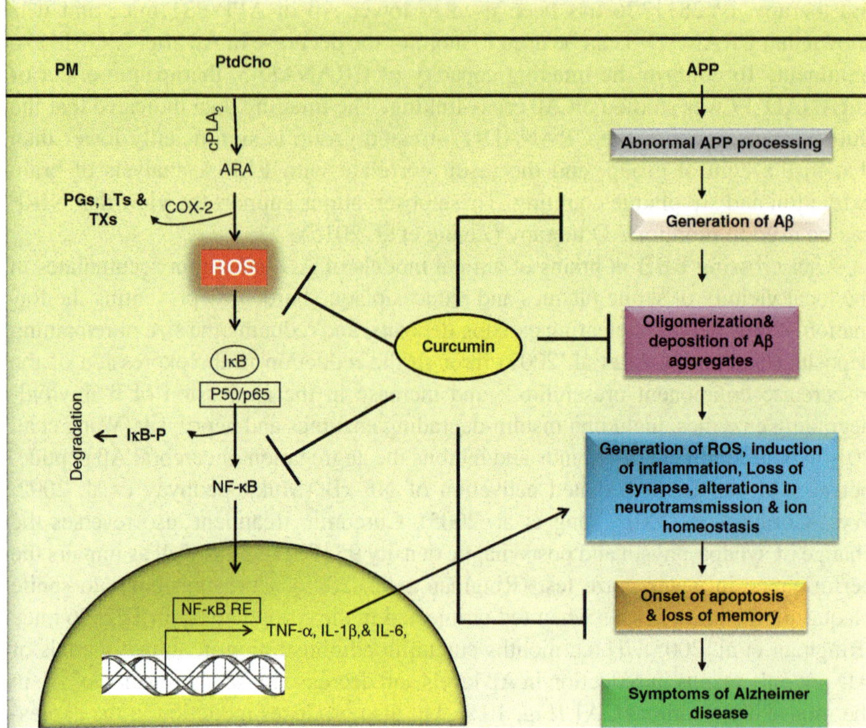

Fig. 10.9 Sites of action of curcumin in Alzheimer disease. Plasma membrane (*PM*); phosphatidylcholine (*PtdCho*); arachidonic acid (*ARA*); cytosolic phospholipase A$_2$ (*cPLA$_2$*); cyclooxygenase (*COX-2*); prostaglandinds (*PGs*); leukotrienes (*LTs*); thromboxanes (*TXs*); reactive oxygen species (*ROS*); nuclear factor kappaB (*NF-κB*); nuclear factor kappaB response element (*NF-κB-RE*); inhibitory subunit of NFκB (*IκB*); tumor necrosis factor-α (*TNF-α*); interleukin-1β (*IL-1β*); amyloid precursor protein (*APP*); and beta- amyloid (*Aβ*)

dose and bioavailability will be fundamental importance for appropriate evaluation. A phase 1 clinical trial of curcumin has indicated that even with a dose of 8 g/day, the amount of curcumin found in plasma is very low (1.77 µM). This low concentration may be due to rapid degradation when pH>7 (Toennesen and Karlsen 1985) and the high propensity of curcumin to be metabolized (Anand et al. 2007). Only 30 % of subjects experienced a minor toxicity (headache, diarrhea, and rash) which is not dose-related (Lao et al. 2006). Several phase II, double-blind, placebo-controlled studies of curcumin safety and tolerability in human AD patients are undergoing nowadays. Results of these trials in humans have not been published.

10.5   Biomarkers to Follow the Effect of Curcumin Treatment

The biomarkers are metabolites whose concentration, presence, and activity are closely associated with the pathogenesis of the disease processes. Biomarkers can be detected in patients with the disease for not only for early detection of the disease (preclinical stage) and monitoring the disease progression, but also for following the treatment response more sensitively and objectively. At present information on an ideal and specific biomarker for AD is not available. The discovery of an ideal and specific biomarker will not only improve the differential diagnosis of AD, but also track progression of AD and measure efficacy of curcumin treatment. This means that there is an urgent need to develop biomarkers that are sensitive and specific to AD pathology with positive and negative predictive value for the disorder (Grimes and Schulz 2002). The major imaging biomarkers for the diagnosis and prognosis of AD include positron emission tomography (PET) neuroimaging of β-amyloid (Aβ) protein deposition, magnetic resonance imaging (MRI) of volume hippocampus and other brain structures, and in vivo imaging of insoluble Aβ species by fluorescent and near-infrared fluorescence (NIRF) imaging.

Neurofunctional imaging modalities, such as FDG-PET and regional cerebral blood flow imaging with single photon emission computed tomography (SPECT) have been used to provide information about regional glucose metabolism and brain perfusion. Converging evidence suggests that neuroimaging methods are useful not only in the early diagnosis of AD, but also in differentiating AD from other neurodegenerative diseases (Westman et al. 2011). Neuroimaging techniques can also be used for the prediction of conversion of MCI to AD.

There are many neurochemical biomarkers (CSF Aβ, BACE1, total tau, p-tau, and isoprostanes) which are substantially altered in AD. However, these markers are not very specific for AD. These biomarkers are also altered in dementia with Lewy bodies and in cognitively intact elderly subjects (Rowe et al. 2007; Edison et al. 2008). Their quantification by CSF requires quantitative proteomics strategies, and finally emerging approaches, such as trascriptome analysis. Detection and determination of these biomarkers require advanced technology.

10.6 Side Effects and Future Studies on Therapeutic Potential of Curcumin

Food and drug administration has declared that curcumin is a safe drug (supplement), which exhibits a wide variety of pharmacological activities in animals and humans. However, most studies on curcumin have been performed on prevention of carcinogenesis under in vitro conditions at the micromolar range, which is hard to achieve under in vivo conditions (Burgos-Morón et al. 2010). Although, many studies have been published on the beneficial effects of curcumin in animal models of AD, but studies on the efficacy of curcumin in human beings are lacking. There are no reports on randomized double blind clinical trials in large human populations. Cell and animal models offer an appealing and ethical alternative to human experimentation for initial efficacy and molecular mechanism (Hatcher et al. 2008). Human studies on pharmacokinetics and molecular mechanism of curcumin and its analogs action in humans are not available. Very little is known about half-life and ability of curcumin to reach and accumulate at the specific site (subcellular organelle) within the brain in humans. Furthermore, information on optimal levels of curcumin at the specific site, which can delay or retard oxidative stress and neuroinflammation, remains unknown. Detailed investigations on metabolic fate of absorbed curcumin and its levels in human tissues in general and brain in particular have not been performed. Answers to above questions are essential not only for understanding the beneficial effects of curcumin in human brain, but also for determining the ability of curcumin to delay the onset of oxidative stress and neuroinflammation in AD.

Curcumin has been reported to bind iron (Baum and Ng 2004). This may cause iron deficiency in people with low iron stores, cancer, or other chronic diseases (Jiao et al. 2009). Curcumin also has blood thinner properties, thus people who are undergoing surgery should avoid it. High (micromolar) concentrations of curcumin have been reported to induce oxidative stress that may mediate its ability to trigger apoptosis in cancer cells (Salvioli et al. 2007). Curcumin attenuates hepcidin biosynthesis, a regulatory protein involved in iron transport (Jiao et al. 2009). The chronic use of curcumin has been reported to produce liver toxicity. For this reason, turmeric products should be avoided by subjects with liver disease, heavy drinkers and those who take prescription medications that are metabolized by liver.

10.7   Conclusion

Curcumin is derived from the rhizome of the East Indian plant *Curcuma longa*. It has been consumed as a dietary supplement for centuries and is considered pharmacologically safe.

Many studies have indicated that curcumin mediates antioxidant, antiinflammatory, anti-amyloidogenic, and antiapoptotic properties. These properties are particularly relevant for the treatment of AD. Thus, by binding to senile plaques, reducing Aβ levels and plaque burden, blocking aggregation and fibril formation, lowering oxidative stress and reducing neuroinflammation, and reversing cognitive deficits, curcumin alleviates cell death and neuronal loss in cellular and animal models of AD. Curcumin reduces soluble tau protein and elevates heat shock proteins, which promote tau protein clearance even after the formation of tangles. These curcumin-mediated processes may reverse tau protein-dependent behavioral abnormalities and synaptic deficits. Other activities of curcumin include reduction of blood cholesterol level, prevention of low density lipoprotein (LDL) oxidation, inhibition of platelet aggregation, suppression of thrombosis and myocardial infarction.

References

Ahsan N, Mishra S, Jain MK, Surolia A, Gupta S (2015) Curcumin PYRAZOLE and its derivative (N-(3-Nitrophenylpyrazole) Curcumin inhibit aggregation, disrupt fibrils and modulate toxicity of wild type and mutant α-synuclein. Sci Rep 5:9862

Alva G, Cummings JL (2008) Relative tolerability of Alzheimer's disease treatments. Psychiatry (Edgmont) 5:27–36

Alwi I, Santoso T, Suyono S, Sutrisna B, Suyatna FD, Kresno SB, Ernie S (2008) The effect of curcumin on lipid level in patients with acute coronary syndrome. Acta Med Indones 40:201–210

Anand P, Kunnumakkara AB, Newman RA, Aggarwal BB (2007) Bioavailability of curcumin: problems and promises. Mol Pharm 4:807–818

Anand P, Nair HB, Sung B, Kunnumakkara AB, Yadav VR, Tekmal RR, Aggarwal BB (2010) Design of curcumin-loaded PLGA nanoparticles formulation with enhanced cellular uptake, and increased bioactivity in vitro and superior bioavailability in vivo. Biochem Pharmacol 79:330–338

Ballard C, Gauthier S, Corbett A, Brayne C, Aarsland D, Jones E (2011) Alzheimer's disease. Lancet 377:1019–1031

Barzegar A, Moosavi-Movahedi AA (2011) Intracellular ROS protection efficiency and free radical-scavenging activity of curcumin. PLoS One 6, e26012

Baum L, Ng A (2004) Curcumin interaction with copper and iron suggests one possible mechanism of action in Alzheimer's disease animal models. J Alzheimers Dis 6:367–377

Begum AN, Jones MR, Lim GP, Morihara T, Kim P, Heath DD, Rock CL, Pruitt MA, Yang F, Hudspeth B, Hu S, Faull KF, Teter B, Cole GM, Frautschy SA (2008) Curcumin structure-function, bioavailability, and efficacy in models of neuroinflammation and Alzheimer's disease. J Pharmacol Exp Ther 326:196–208

Burgos-Morón E, Calderón-Montaño JM, Salvador J, Robles A, López-Lázaro M (2010) The dark side of curcumin. Int J Cancer 126:1771–1775

Bustanji Y, Taha MO, Almasri IM, Al-Ghussein MA, Mohammad MK, Alkhatib HS (2009) Inhibition of glycogen synthase kinase by curcumin: investigation by simulated molecular docking and subsequent in vitro/in vivo evaluation. J Enzyme Inhib Med Chem 24:771–778

Chao CC, Molitor TW, Hu S (1993) Neuroprotective role of IL-4 against activated microglia. J Immunol 151:1473–1481

Chen WF, Deng SL, Zhou B, Yang L, Liu ZL (2011) Curcumin and its analogues as potent inhibitors of low density lipoprotein oxidation: H-atom abstraction from the phenolic groups and possible involvement of the 4-hydroxy-3-methoxyphenyl groups. Free Radic Biol Med 40:526–535

Cui M, Ono M, Watanabe H, Kimura H, Liu B, Saji H (2014) Smart near-infrared fluorescence probes with donor-acceptor structure for in vivo detection of β-amyloid deposits. J Am Chem Soc 136:3388–3394

Dadhaniya P, Patel C, Muchhara J, Bhadja N, Mathuria N, Vachhani K, Soni MG (2011) Safety assessment of a solid lipid curcumin particle preparation: acute and subchronic toxicity studies. Food Chem Toxicol 49:1834–1842

Dong S, Zeng Q, Mitchell ES, Xiu J, Duan Y, Li C, Tiwari JK, Hu Y, Cao X, Zhao Z (2012) Curcumin enhances neurogenesis and cognition in aged rats: implications for transcriptional interactions related to growth and synaptic plasticity. PLoS One 7, e31211

Edison P, Rowe CC, Rinne JO, Ng S, Ahmed I, Kemppainen N, Villemagne VL, O'Keefe G, Någren K, Chaudhury KR, Masters CL, Brooks DJ (2008) Amyloid load in Parkinson's disease dementia and Lewy body dementia measured with [11C]PIB positron emission tomography. J Neurol Neurosurg Psychiatry 79:1331–1338

Fang J, Lu J, Holmgren A (2005) Thioredoxin reductase is irreversibly modified by curcumin: a novel molecular mechanism for its anticancer activity. J Biol Chem 280:25284–25290

Farooqui AA (2010) Neurochemical aspects of neurotraumatic and neurodegenerative diseases. Springer, New York

Farooqui AA (2014) Inflammation and oxidative stress in neurological disorders. Springer, Switzerland

Farooqui AA, Ong WY, Horrocks LA (2006) Inhibitors of brain phospholipase A2 activity: their neuropharmacological effects and therapeutic importance for the treatment of neurologic disorders. Pharmacol Rev 58:591–620

Ferri CP, Prince M, Brayne C, Brodaty H, Fratiglioni L, Ganguli M, Hall K, Hasegawa K, Hendrie H, Huang Y, Jorm A, Mathers C, Menezes PR, Rimmer E, Scazufca M; Alzheimer's Disease International (2005) Global prevalence of dementia: a Delphi consensus study. Lancet 366:2112–2117

Francis HM, Stevenson RJ (2011) Higher reported saturated fat and refined sugar intake is associated with reduced hippocampal-dependent memory and sensitivity to interoceptive signals. Behav Neurosci 125:943–955

Ganguli M, Chandra V, Kamboh MI, Johnston JM, Dodge HH, Thelma BK, Juyal RC, Pandav R, Belle SH, DeKosky ST (2000) Apolipoprotein E polymorphism and Alzheimer disease: The Indo-US Cross-National Dementia Study. Arch Neurol 57:824–830

Gao Y, Wang C, Sun M, Wang X, Yu A, Li A, Zhai G (2012) In vivo evaluation of curcumin loaded nanosuspensions by oral administration. J Biomed Nanotechnol 8:659–668

Garcia-Alloza M, Borrelli LA, Rozkalne A, Hyman BT, Bacskai BJ (2007) Curcumin labels amyloid pathology in vivo, disrupts existing plaques, and partially restores distorted neurites in an Alzheimer mouse model. J Neurochem 102:1095–1104

Ghalandarlaki N, Alizadeh AM, Ashkani-Esfahani S (2014) Nanotechnology-applied curcumin for different diseases therapy. Biomed Res Int 2014:394264

Ghosh S, Banerjee S, Sil PC (2015) The beneficial role of curcumin on inflammation, diabetes and neurodegenerative disease: a recent update. Food Chem Toxicol 83:111–124

Giri RK, Rajagopal V, Kalra VK (2004) Curcumin, the active constituent of turmeric, inhibits amyloid peptide-induced cytochemokine gene expression and CCR5-mediated chemotaxis of THP-1 monocytes by modulating early growth response-1 transcription factor. J Neurochem 91:1199–1210

Golde TE (2006) Disease modifying therapy for AD? J Neurochem 99:689–707

Golde TE (2009) The therapeutic importance of understanding mechanisms of neuronal cell death in neurodegenerative disease. Mol Neurodegener 4:8

Gota VS, Maru GB, Soni TG, Gandhi TR, Kochar N, Agarwal MG (2010) Safety and pharmacokinetics of a solid lipid curcumin particle formulation in osteosarcoma patients and healthy volunteers. J Agric Food Chem 58:2095–2099

Grimes DA, Schulz KF (2002) Uses and abuses of screening tests. Lancet 359:881–884

Gupta SC, Patchva S, Koh W, Aggarwal BB (2012) Discovery of curcumin, a component of golden spice, and its miraculous biological activities. Clin Exp Pharmacol Physiol 39:283–299

Hansen RA, Gartlehner G, Lohr KN, Kaufer DI (2007) Functional outcomes of drug treatment in Alzheimer's disease: a systematic review and meta-analysis. Drugs Aging 24:155–167

Hartl FU, Bracher A, Hayer-Hartl M (2011) Molecular chaperones in protein folding and proteostasis. Nature 475:324–332

Hatcher H, Planalp R, Cho J, Torti FM, Torti SV (2008) Curcumin: from ancient medicine to current clinical trials. Cell Mol Life Sci 65:1631–1652

Hu S, Maiti P, Ma Q, Zuo X, Jones MR, Cole GM, Frautschy SA (2015) Clinical development of curcumin in neurodegenerative disease. Expert Rev Neurother 15:629–637

Itokawa H, Shi Q, Akiyama T, Morris-Natschke SL, Lee KH (2008) Recent advances in the investigation of curcuminoids. Chin Med 3:11

Jayaraj RL, Elangovan N, Manigandan K, Singh S, Shukla S (2014) CNB-001 a novel curcumin derivative, guards dopamine neurons in MPTP model of Parkinson's disease. Biomed Res Int 2014:236182

Jiao Y, Wilkinson JT, Christine Pietsch E, Buss JL, Wang W, Planalp R, Torti FM, Torti SV (2006) Iron chelation in the biological activity of curcumin. Free Radic Biol Med 40:1152–1160

Jiao Y, Wilkinson J, Di X, Wang W, Hatcher H, Kock ND, D'Agostino R Jr (2009) Curcumin, a cancer chemopreventive and chemotherapeutic agent, is a biologically active iron chelator. Blood 113:462–469

Jorm AF, Jolley D (1998) The incidence of dementia: a meta analysis. Neurology 51:728–733

Kakkar V, Singh S, Singla D, Kaur IP (2011) Exploring solid lipid nanoparticles to enhance the oral bioavailability of curcumin. Mol Nutr Food Res 55:495–503

Kim DS, Park SY, Kim JK (2001) Curcuminoids from Curcuma longa L. (Zingiberaceae) that protect PC12 rat pheochromocytoma and normal human umbilical vein endothelial cells from βA(1–42) insult. Neurosci Lett 303:57–61

Kim GY, Kim KH, Lee SH, Yoon MS, Lee HJ, Moon DO (2005a) Curcumin inhibits immunostimulatory function of dendritic cells: MAPKs and translocation of NF-B as potential targets. J Immunol 174:8116–8124

Kim H, Park BS, Lee KG, Choi CY, Jang SS, Kim YH, Lee SE (2005b) Effects of naturally occurring compounds on fibril formation and oxidative stress of beta-amyloid. J Agric Food Chem 53:8537–8541

Kim YE, Hipp MS, Bracher A, Hayer-Hartl M, Hartl FU (2013) Molecular chaperone functions in protein folding and proteostasis. Annu Rev Biochem 82:323–355

Kiyota T, Okuyama S, Swan RJ, Jacobsen MT, Gendelman HE, Ikezu T (2010) CNS expression of anti-inflammatory cytokine interleukin-4 attenuates Alzheimer's disease-like pathogenesis in APP+PS1 bigenic mice. FASEB J 24:3093–3102

Kuner P, Schubenel R, Hertel C (1998) β-Amyloid binds to p57NTR and activates NFκB in human neuroblastoma cells. J Neurosci Res 54:798–804

Lao CD, Ruffin MT, Normolle D, Heath DD, Murray SI, Bailey JM, Boggs ME, Crowell J, Rock CL, Brenner DE (2006) Dose escalation of a curcuminoid formulation. BMC Complement Altern Med 6:10

Lee VM, Goedert M, Trojanowski JQ (2001) Neurodegenerative tauopathies. Annu Rev Neurosci 24:1121–1159

Lee I, Yang J, Lee JH, Choe YS (2011) Synthesis and evaluation of 1-(4-[(1)(8)F]fluoroethyl)-7-(4′-methyl)curcumin with improved brain permeability for beta-amyloid plaque imaging. Bioorg Med Chem Lett 21:5765–5769

Li MH, Jang JH, Na HK, Cha YN, Surh YJ (2007) Carbon monoxide produced by heme oxygenase-1 in response to nitrosative stress induces expression of glutamatecysteine ligase in PC12 cells via activation of phosphatidylinositol 3-kinase and Nrf2 signaling. J Biol Chem 282:28577–28586

Lim GP, Chu T, Yang F, Beech W, Frautschy SA, Cole GM (2001) The curry spice curcumin reduces oxidative damage and amyloid pathology in an Alzheimer transgenic mouse. J Neurosci 21:8370–8377

Lin R, Chen X, Li W, Han Y, Liu P, Pi R (2008) Exposure to metal ions regulates mRNA levels of APP and BACE1 in PC12 cells: blockage by curcumin. Neurosci Lett 440:344–347

Liu Y, Dargusch R, Maher P, Schubert D (2008) A broadly neuroprotective derivative of curcumin. J Neurochem 105:1336–1345

Liu H, Li Z, Qiu D, Gu Q, Lei Q, Mao L (2010) The inhibitory effects of different curcuminoids on β-amyloid protein, β-amyloid precursor protein and β-site amyloid precursor protein cleaving enzyme 1 in swAPP HEK293 cells. Neurosci Lett 485:83–88

Lyons A, Griffin RJ, Costelloe CE, Clarke RM, Lynch MA (2007) IL-4 attenuates the neuroinflammation induced by amyloid-beta in vivo and in vitro. J Neurochem 101:771–781

Ma QL, Yang F, Rosario ER, Ubeda OJ, Beech W, Gant DJ, Chen PP, Hudspeth B, Chen C, Zhao Y, Vinters HV, Frautschy SA, Cole GM (2009) Beta-amyloid oligomers induce phosphorylation of tau and inactivation of insulin receptor substrate via c-Jun N-terminal kinase signaling: suppression by omega-3 fatty acids and curcumin. J Neurosci 29:9078–9089

Maiti P, Manna J, Veleri S, Frautschy S (2014) Molecular chaperone dysfunction in neurodegenerative diseases and effects of curcumin. Biomed Res Int 2014:495091

McClure R, Yanagisawa D, Stec D, Abdollahian D, Koktysh D, Xhillari D, Jaeger R, Stanwood G, Chekmenev E, Tooyama I, Gore JC, Pham W (2015) Inhalable curcumin: offering the potential for translation to imaging and treatment of Alzheimer's disease. J Alzheimers Dis 44:283–295

Menon VP, Sudheer AR (2007) Antioxidant and anti-inflammatory properties of curcumin. Adv Exp Med Biol 595:105–125

Mistry A, Glud SZ, Kjems J, Randel J, Howard KA, Stolnik S, Illum L (2009) Effect of physicochemical properties on intranasal nanoparticle transit into murine olfactory epithelium. J Drug Target 17:543–552

Mohanty C, Sahoo SK (2010) The in vitro stability and in vivo pharmacokinetics of curcumin prepared as an aqueous nanoparticulate formulation. Biomaterials 31:6597–6611

Motterlini R, Foresti R, Bassi R, Green CJ (2000) Curcumin, an antioxidant and anti-inflammatory agent, induces heme oxygenase-1 and protects endothelial cells against oxidative stress. Free Radic Biol Med 28:1303–1312

Mukhopadhyay AS, Banerjee LJ, Stafford C, Xia M, Liu BB, Aggarwal BB (2002) Curcumin induced suppression of cell proliferation correlates with downregulation of cyclin D1 expression and CDK4-mediated retinoblastoma protein phosphorylation. Oncogene 21:8852–8861

Nair KL, Thulasidasan AK, Deepa G, Anto RJ, Kumar GS (2012) Purely aqueous PLGA nanoparticulate formulations of curcumin exhibit enhanced anticancer activity with dependence on the combination of the carrier. Int J Pharm 425:44–52

Narlawar R, Baumann K, Czech C, Schmidt B (2007) Conversion of the LXR-agonist TO-901317– from inverse to normal modulation of gamma-secretase by addition of a carboxylic acid and a lipophilic anchor. Bioorg Med Chem Lett 17:5428–5431

Narlawar R, Pickhardt M, Leuchtenberger S, Baumann K, Krause S, Dyrks T, Weggen S, Mandelkow E, Schmidt B (2008) Curcumin-derived pyrazoles and isoxazoles: Swiss army knives or blunt tools for Alzheimer's disease? ChemMedChem 3:165–172

Necula M, Kayed R, Milton S, Glabe CG (2007) Small molecule inhibitors of aggregation indicate that amyloid beta oligomerization and fibrillization pathways are independent and distinct. J Biol Chem 282:10311–10324

Ng TP, Chiam PC, Lee T, Chua HC, Lim L, Kua EH (2006) Curry consumption and cognitive function in the elderly. Am J Epidemiol 164:898–906

Ohtsukaand K, Suzuki T (2000) Roles of molecular chaperones in the nervous system. Brain Res Bull 53:141–146

Ong WY, Farooqui T, Kokotos G, Farooqui AA (2015) Synthetic and natural inhibitors of phospholipases A_2: their importance for understanding and treatment of neurological disorders. ACS Chem Neurosci 6:814–831

Orlando RA, Gonzales AM, Royer RE, Deck LM, Vander Jagt DL (2012) A chemical analog of curcumin as an improved inhibitor of amyloid Aβ oligomerization. PLoS One 7, e31869

Park SY, Kim DS (2002) Discovery of natural products from Curcuma longa that protect cells from beta-amyloid insult: a drug discovery effort against Alzheimer's disease. J Nat Prod 65:1227–1231

Parodi FE, Mao D, Ennis TL, Pagano MB, Thompson RW (2006) Oral administration of diferuloylmethane (curcumin) suppresses proinflammatory cytokines and destructive connective tissue remodeling in experimental abdominal aortic aneurysms. Ann Vasc Surg 20:360–368

Petrou M, Bohnen NI, Müller ML, Koeppe RA, Albin RL, Frey KA (2012) Aβ-amyloid deposition in patients with Parkinson disease at risk for development of dementia. Neurology 79:1161–1167

Potter PE (2013) Curcumin: a natural substance with potential efficacy in Alzheimer's disease. J Exp Pharmacol 5:23–31

Prasad S, Tyagi AK, Aggarwal BB (2014) Recent developments in delivery, bioavailability, absorption and metabolism of curcumin: the golden pigment from golden spice. Cancer Res Treat 46:2–18

Priyadarsini KI (2013) Chemical and structural features influencing the biological activity of curcumin. Curr Pharm Des 19:2093–2100

Priyadarsini KI, Maity DK, Naik GH, Kumar MS, Unnikrishnan MK, Satav JG, Mohan H (2003) Role of phenolic O-H and methylene hydrogen on the free radical reactions and antioxidant activity of curcumin. Free Radic Biol Med 35:475e484

Qin XY, Cheng Y, Yu LC (2010) Potential protection of curcumin against intracellular amyloid β-induced toxicity in cultured rat prefrontal cortical neurons. Neurosci Lett 480:21–24

Raina P, Santaguida P, Ismaila A, Patterson C, Cowan D, Levine M, Booker L, Oremus M (2008) Effectiveness of cholinesterase inhibitors and memantine for treating dementia: evidence review for a clinical practice guideline. Ann Intern Med 148:379–397

Ran C, Xu X, Raymond SB, Ferrara BJ, Neal K, Bacskai BJ, Medarova Z, Moore A (2009) Design, synthesis, and testing of difluoroboron-derivatized curcumins as near-infrared probes for in vivo detection of amyloid-beta deposits. J Am Chem Soc 131:15257–15261

Ray B, Bisht S, Maitra A, Maitra A, Lahiri DK (2011) Neuroprotective and neurorescue effects of a novel polymeric nanoparticle formulation of curcumin (NanoCurc™) in the neuronal cell culture and animal model: implications for Alzheimer's disease. J Alzheimers Dis 23:61–77

Ringman JM, Frautschy SA, Cole GM, Masterman DL, Cummings JL (2005) A potential role of the curry spice curcumin in Alzheimer's disease. Curr Alzheimer Res 2:131–136

Ringman JM, Frautschy SA, Teng E, Begum AN, Bardens J, Beigi M, Gylys KH, Badmaev V, Heath DD, Apostolova LG, Porter V, Vanek Z, Marshall GA, Hellemann G, Sugar C, Masterman DL, Montine TJ, Cummings JL, Cole GM (2012) Oral curcumin for Alzheimer's disease: tolerability and efficacy in a 24-week randomized, double blind, placebo-controlled study. Alzheimers Res Ther 4:43

Rowe CC, Ng S, Ackermann U, Gong SJ, Pike K, Savage G, Cowie TF, Dickinson KL, Maruff P, Darby D, Smith C, Woodward M, Merory J, Tochon-Danguy H, O'Keefe G, Klunk WE, Mathis CA, Price JC, Masters CL, Villemagne VL (2007) Imaging beta-amyloid burden in aging and dementia. Neurology 68:1718–1725

Ryu EK, Choe YS, Lee KH, Choi Y, Kim BT (2006) Curcumin and dehydrozingerone derivatives: synthesis, radiolabeling, and evaluation for beta-amyloid plaque imaging. J Med Chem 49:6111–6119

Salvioli S, Sikora E, Cooper EL, Franceschi C (2007) Curcumin in cell death processes: a challenge for CAM of age-related pathologies. Evid Based Complement Alternat Med 4:181–190

Savva GM, Wharton SB, Ince PG, Forster G, Matthews FE, Brayne C (2009) Age, neuropathology, and dementia. N Engl J Med 360:2302–2309

Scapagnini G, Colombrita C, Amadio M, D'Agata V, Arcelli E, Sapienza M, Quattrone A, Calabrese V (2006) Curcumin activates defensive genes and protects neurons against oxidative stress. Antioxid Redox Signal 8:395–403

Shaikh J, Ankola DD, Beniwal V, Singh D, Kumar MN (2009) Nanoparticle encapsulation improves oral bioavailability of curcumin by at least 9-fold when compared to curcumin administered with piperine as absorption enhancer. Eur J Pharm Sci 37:223–230

Shimmyo Y, Kihara T, Akaike A, Niidome T, Sugimoto H (2008) Epigallocatechin-3-gallate and curcumin suppress amyloid beta-induced beta-site APP cleaving enzyme-1 upregulation. Neuroreport 19:1329–1333

Shytle RD, Tan J, Bickford PC, Rezai-Zadeh K, Hou L, Zeng J, Sanberg PR, Sanberg CD, Alberte RS, Fink RC, Roschek B Jr (2012) Optimized turmeric extract reduces β-Amyloid and phosphorylated Tau protein burden in Alzheimer's transgenic mice. Curr Alzheimer Res 9:500–506

Subhashini, Chauhan PS, Kumari S, Kumar JP, Chawla R, Dash D, Singh M, Singh R (2013) Intranasal curcumin and its evaluation in murine model of asthma. Int Immunopharmacol 17:733–743

Suh HW, Kang S, Kwon KS (2007) Curcumin attenuates glutamateinduced HT22 cell death by suppressing MAP kinase signaling. Mol Cell Biochem 298:187–194

Toennesen HH, Karlsen J (1985) Studies on curcumin and curcuminoids. VI. Kinetics of curcumin degradation in aqueous solution. Z Lebensm Unters Forsch 180:402–404

Tsai YM, Chien CF, Lin LC, Tsai TH (2011) Curcumin, its nano-formulation: the kinetics of tissue distribution and blood-brain barrier penetration. Int J Pharm 416:331–338

Tung EW, Philbrook NA, Macdonald KD, Winn LM (2012) DNA double-strand breaks and DNA recombination in benzene metabolite-induced genotoxicity. Toxicol Sci 126:569–577

Valera E, Dargusch R, Maher PA, Schubert D (2013) Modulation of 5-lipoxygenase in proteotoxicity and Alzheimer's disease. J Neurosci 33:10512–10525

Villemagne VL, Burnham S, Bourgeat P, Brown B, Ellis KA, Salvado O, Szoeke C, Macaulay SL, Martins R, Maruff P, Ames D, Rowe CC, Masters CL; Australian Imaging Biomarkers and Lifestyle (AIBL) Research Group (2013) Amyloid β deposition, neurodegeneration, and cognitive decline in sporadic Alzheimer's disease: a prospective cohort study. Lancet Neurol 12:357–367

Wahlstrom B, Blennow G (1978) A study on the fate of curcumin in the rat. Acta Pharmacol Toxicol (Copenh) 43:86–92

Wang S, Su R, Nie S, Sun M, Zhang J, Wu D, Moustaid-Moussa N (2014a) Application of nanotechnology in improving bioavailability and bioactivity of diet-derived phytochemicals. J Nutr Biochem 25:363–376

Wang P, Su C, Li R, Wang H, Ren Y, Sun H, Yang J, Sun J, Shi J, Tian J, Jiang S (2014b) Mechanisms and effects of curcumin on spatial learning and memory improvement in APPswe/PS1dE9 mice. J Neurosci Res 92:218–231

Weber WM, Hunsaker LA, Abcouwer SF, Deck LM, Vander Jagt DL (2005) Anti-oxidant activities of curcumin and related enones. Bioorg Med Chem 13:3811–3820

Westman E, Wahlund LO, Foy C, Poppe M, Cooper A, Murphy D, Spenger C, Lovestone S, Simmons A (2011) Magnetic resonance imaging and magnetic resonance spectroscopy for detection of early Alzheimer's disease. J Alzheimers Dis 26(Suppl 3):307–319

Wyss-Coray T, Lin C, Yan F, Yu GQ, Rohde M, McConlogue L, Masliah E, Mucke L (2001) TGF-beta1 promotes microglial amyloid-beta clearance and reduces plaque burden in transgenic mice. Nat Med 7:612–618

Xie X, Tao Q, Zou Y, Zhang F, Guo M, Wang Y, Wang H, Zhou Q, Yu S (2011) PLGA nanoparticles improve the oral bioavailability of curcumin in rats: characterizations and mechanisms. J Agric Food Chem 59:9280–9289

Yanagisawa D, Amatsubo T, Morikawa S, Taguchi H, Urushitani M, Shirai N, Hirao K, Shiino A, Inubushi T, Tooyama I (2011) In vivo detection of amyloid beta deposition using (1)(9)F magnetic resonance imaging with a (1)(9)F-containing curcumin derivative in a mouse model of Alzheimer's disease. Neuroscience 184:120–127

Yanagisawa D, Taguchi H, Ibrahim NF, Morikawa S, Shiino A, Inubushi T, Hirao K, Shirai N, Sogabe T, Tooyama I (2014) Preferred features of a fluorine-19 MRI probe for amyloid detection in the brain. J Alzheimers Dis 39:617–631

Yang F, Lim GP, Begum AN, Ubeda OJ, Simmons MR, Ambegaokar SS, Chen PP, Kayed R, Glabe CG, Frautschy SA, Cole GM (2005) Curcumin inhibits formation of amyloid beta oligomers and fibrils, binds plaques, and reduces amyloid in vivo. J Biol Chem 280:5892–5901

Yao EC, Xue L (2014) Therapeutic effects of curcumin on Alzheimer's disease. Adv Alzheimer Dis 3:145–159

Ye J, Zhang Y (2012) Curcumin protects against intracellular amyloid toxicity in rat primary neurons. Int J Clin Exp Med 5:44–49

Yin W, Zhang X, Li Y (2012) Protective effects of curcumin in *APPswe* transfected SH-SY5Y cells. Neural Regen Res 7:405–412

Yoshiyama Y, Higuchi M, Zhang B, Huang SM, Iwata N, Saido TC, Maeda J, Suhara T, Trojanowski JQ, Lee VM (2007) Synapse loss and microglial activation precede tangles in a P301S tauopathy mouse model. Neuron 53:337–351

Yu H, Huang Q (2012) Improving the oral bioavailability of curcumin using novel organogel-based nanoemulsions. J Agric Food Chem 60:5373–5379

Yuan HY, Kuang SY, Zheng X, Ling HY, Yang YB, Yan PK, Li K, Liao DF (2008) Curcumin inhibits cellular cholesterol accumulation by regulating SREBP-1/caveolin-1 signaling pathway in vascular smooth muscle cells. Acta Pharmacol Sin 29:555–563

Zhang X, Yin WK, Shi XD, Li Y (2011a) Curcumin activates Wnt/β-catenin signaling pathway through inhibiting the activity of GSK-3β in APPswe transfected SY5Y cells. Eur J Pharm Sci 42:540–546

Zhang X, Zhang HM, Si L, Li Y (2011b) Curcumin mediates presenilin-1 activity to reduce β-amyloid production in a model of Alzheimer's disease. Pharmacol Rep 63:1101–1108

Zhang ZJ, Zhao LX, Cao DL, Zhang X, Gao YJ, Xia C (2012) Curcumin inhibits LPS-induced CCL2 expression via JNK pathway in C6 rat astrocytoma cells. Cell Mol Neurobiol 32:1003–1010

Zhang X, Tian Y, Li Z, Tian X, Sun H, Liu H, Moore A, Ran C (2013) Design and synthesis of curcumin analogues for in vivo fluorescence imaging and inhibiting copper-induced cross-linking of amyloid beta species in Alzheimer's disease. J Am Chem Soc 135:16397–16409

Zhang X, Tian Y, Zhang C, Tian X, Ross AW, Moir RD, Sun H, Tanzi RE, Moore A, Ran C (2015) Near-infrared fluorescence molecular imaging of amyloid beta species and monitoring therapy in animal models of Alzheimer's disease. Proc Natl Acad Sci U S A. pii: 201505420

Zhu X, Perry G, Smith MA, Wang X (2013) Abnormal mitochondrial dynamics in the pathogenesis of Alzheimer's disease. J Alzheimers Dis 33(Suppl 1):S253–S262

Index

© Springer International Publishing Switzerland 2016

393

A.A. Farooqui, *Therapeutic Potentials of Curcumin for Alzheimer Disease*,
DOI 10.1007/978-3-319-15889-1

CPSIA information can be obtained
at www.ICGtesting.com
Printed in the USA
LVOW05*2127151216
517432LV00002B/39/P